ELASTICITY

CIVIL ENGINEERING AND ENGINEERING MECHANICS SERIES

N. M. Newmark and W. J. Hall, *editors*

ELASTICITY

Robert Wm. Little

Professor and Chairman
Department of Mechanical Engineering
Michigan State University

PRENTICE-HALL, INC., Englewood Cliffs, New Jersey

Library of Congress Cataloging in Publication Data

LITTLE, ROBERT WILLIAM,
Elasticity.

Includes bibliographical references.
1. Elasticity. I. Title.
QA931.L58 531'.3823 72-13136
ISBN 0-13-246884-0

10 9 8 7 6 5 4 3 2 1

Printed in the United States of America

PRENTICE-HALL INTERNATIONAL, INC., London
PRENTICE-HALL OF AUSTRALIA, PTY. LTD., Sydney
PRENTICE-HALL OF CANADA, LTD., Toronto
PRENTICE-HALL OF JAPAN, INC., Tokyo
PRENTICE-HALL OF INDIA PRIVATE LTD., New Delhi

To

Professor Gerald Pickett

CONTENTS

PART II
Two-Dimensional Elasticity

8
Problems in Polar Coordinates 154

9
Complex Variable Solutions 194

10
Finite Difference and Finite Element Methods 228

13
Navier Equation and the Galerkin Vector 315

14
Papkovich-Neuber Solution 376

PREFACE

The major formulation of the mathematical theory of elasticity was developed in the 19th century after the first concept was proposed by Robert Hooke in 1678. Although Hooke and, independently, Mariotte (1680) introduced the concept of force-deformation relations, Galileo had earlier considered the nature of the resistance of solids. During the interval between Hooke's discovery and Navier's development of the general differential equations, such famous mathematicians as James and Daniel Bernoulli, Euler and Lagrange applied their talents to develop the methods of analysis of slender members.

In 1821 Navier presented the results of his investigations of the general equations of equilibrium. The next year Cauchy presented the notation of stress at a point and, using methods different from Navier's, developed similar equations of elasticity. Cauchy's equations contained two constants to describe elastic behavior, rather than Navier's single constant. A long list of researchers developed and applied the theory during the remainder of the 19th century. This list includes Lord Kelvin, Poisson, Lamé, Clapeyron, Green, Saint-Venant, Betti, Cerruti, Airy, Kirchhoff, Boussinesq, and others. Details of their investigations may be found in the introduction to A.E.H. Love's *A Treatise on the Mathematical Theory of Elasticity*, Timoshenko's *A History of Strength of Materials* and Todhunter and Pearson's *History of the Theory of Elasticity*.

In the 20th century, the research of Lord Rayleigh, Kolosoff, Timoshenko, Galerkin, and Muskhelishvilli led to the development of the mathematical means of solving the equations presented earlier. Contemporary researchers in the field of elasticity are too numerous to document in a work of this type, and fear of oversight forbids any such attempt. The emphasis on the general field of continuum mechanics has led to a review of the foundations of elasticity and placed the theory within this general framework.

The student of elasticity may view the subject in many different ways. To some, it is a dusty classical subject serving as the mother or verifier of beam, plate, or shell theory rather than as an entity itself. To others, it is the paradise of mathematics, while to still others it is a wasteland of algebra, numerical methods, and approximations. In this book a survey of the methods and theories of linear elasticity will be conducted, starting from the approach currently used in continuum mechanics. It is not expected that this will be a definitive treatment, but it is hoped that the reader will become a friend of the subject. This friendship will be initiated by application.

A brief review of the mathematics and basic continuum theories is presented in Part I. This review is not meant to serve as a substitute for a comprehensive course in continuum mechanics; such a course should preceed one in elasticity that uses this book. Part II presents the theory of plane elasticity, whereas three-dimensional problems are considered in Part III. This book may be used as a first course in elasticity by employing selected sections of each part. It has also been used by the author in a two-course sequence.

This book is dedicated to the late Professor Gerald Pickett of the University of Wisconsin, who first showed the author the beauty of the subject and gave so much of his time to instill the necessary mathematical self-confidence in his students. The author would also like to thank Professors James Lubkin, James Klemm, and S. B. Childs; the many graduate students who offered suggestions; and his wife and daughter for their help.

East Lansing, Michigan ROBERT WM. LITTLE

PART I

REVIEW OF MATHEMATICAL NOTATION AND CONTINUUM MECHANICS

BASIC EQUATIONS OF THE LINEAR THEORY OF ELASTICITY

CHAPTER 1

VECTORS AND TENSORS

1 Notation

The discussion of notation and an introductory knowledge of methods of manipulation of tensorial quantities are essential to the study of any area of continuum mechanics. The tensorial characteristics of the physical quantities reflect the fact that all physical laws must be invariant in form under change of coordinate systems. This means that the quantity is independent of the coordinate system used, but its mathematical description may appear different. The ability to change coordinate systems facilitates the mathematical approach to the physical problem being considered. A particular coordinate system may be chosen to simplify the form of the differential equation or to place the boundary conditions in a more tractable form.

Four types of notation will be used in this discussion, depending upon which is the most convenient at the time. We shall not be bound to any single convention but shall use each as a tool. Many readers will be familiar with the conventions to one degree or another, and it is hoped that they will gain such proficiency in each that no confusion will result in shifting from one to another. The presentation here will be brief, and the reader is referred for further details to the many texts on each of the notations mentioned. The four notations to be used are symbolic notation, scalar notation or expanded notation, Cartesian tensor notation or indicial notation, and dyadic or extended Gibb's notation.

2 Vectors

The differences in these notations may be examined by reviewing vectors and then proceeding to higher-order tensors. A vector may be represented pictorially by an arrow and symbolically by a barred letter or the use of bold type, $\bar{A}$ or **A**. Vector operations require that the definition of each be clearly understood before a symbol is given designating this operation. For example, vector addition may be written

$$\bar{A} + \bar{B} = \bar{C}. \tag{1-2.1}$$

If the vector is referred to a coordinate system, then the components of the vector may be designated. In Cartesian coordinates, for example, they may be expressed as A_x, A_y, and A_z (Figure 1-1).* The magnitude of the vector

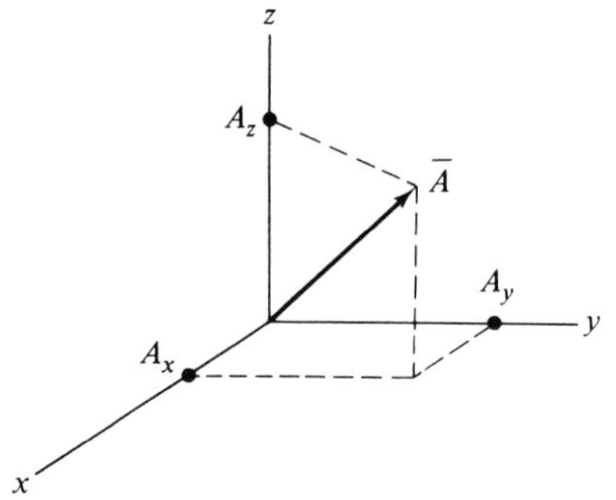

Figure 1-1

A may be written

$$|\mathbf{A}| = \sqrt{(A_x^2 + A_y^2 + A_z^2)}. \tag{1-2.2}$$

Vector addition now may be written in expanded notation as follows:

$$\mathbf{A} + \mathbf{B} = \mathbf{C}, \Longrightarrow \begin{aligned} A_x + B_x &= C_x, \\ A_y + B_y &= C_y, \\ A_z + B_z &= C_z. \end{aligned} \tag{1-2.3}$$

* For figure, section, and equation cross references, the first number refers to the chapter and the following numbers to the sequence within chapters.

This notation may be simplified by numbering the axes x_1, x_2, and x_3 instead of lettering them x, y and z. The component notation is then

$$A_i + B_i = C_i, \qquad i = 1, 2, 3. \tag{1-2.4}$$

The designation that the equation is valid for $i = 1$, 2, and 3 is usually omitted and the Cartesian tensor notation or indicial notation becomes

$$A_i + B_i = C_i. \tag{1-2.5}$$

This equation may be viewed as a single equation for each component, or as a collection of equations describing the entire vector.

The Gibb's notation in vector analysis utilizes unit vectors along the coordinate axes. If these unit vectors are designated by $\hat{i}_{(x)}$, $\hat{i}_{(y)}$, and $\hat{i}_{(z)}$ along the x, y, and z axes, respectively, then the entire vector $\bar{A}$ may be written as

$$\bar{A} = A_x\hat{i}_{(x)} + A_y\hat{i}_{(y)} + A_z\hat{i}_{(z)}. \tag{1-2.6}$$

Note that $\hat{i}$, $\hat{j}$, and $\hat{k}$ are usually used instead of $\hat{i}_{(x)}$, $\hat{i}_{(y)}$, and $\hat{i}_{(z)}$, but for the summation convention to be developed the subscripted unit vectors are more convenient. If the axes are numbered, the vector becomes

$$\bar{A} = A_1\hat{i}_{(1)} + A_2\hat{i}_{(2)} + A_3\hat{i}_{(3)} = \sum_{n=1}^{3} A_n\hat{i}_{(n)}. \tag{1-2.7}$$

Note that $\hat{i}_{(x)}$ or $\hat{i}_{(1)}$ does not designate the component of $\hat{i}_{(x)}$ in the x or 1 direction, but an entire unit vector in that direction. The components of this vector are $i_{(x)x}$ or $i_{(m)n}$, and one can now see the reason for setting off the first subscript in parentheses. It is also clear that

$$i_{(i)j} = \begin{cases} 1 & \text{if } i = j \\ 0 & \text{if } i \neq j \end{cases}. \tag{1-2.8}$$

Before proceeding further into a notational discussion, it is worthwhile to introduce two special symbols:

$$\delta_{ij} = \begin{cases} 1 & \text{if } i = j \\ 0 & \text{if } i \neq j \end{cases}. \tag{1-2.9}$$

This is called the Kronecker delta or the unit tensor. Symbolically, it may be designated by $\bar{\bar{\mathbf{I}}}$. The second symbol is the permutation constant

$$e_{ijk} = \begin{cases} +1 & \text{for 123 subscripts or even permutation thereof} \\ -1 & \text{for odd permutations of 123} \\ 0 & \text{if any two subscripts are equal} \end{cases} \tag{1-2.10}$$

This constant is also called the alternating tensor or the Levi-Civita density.

Vector multiplication may be of three types, the scalar product, the vector product, and the tensor product. The scalar or dot product is defined as

$$\bar{A} \cdot \bar{B} = \lambda \qquad \text{(scalar)}, \tag{1-2.11}$$

$$A_x B_x + A_y B_y + A_z B_z = \lambda. \tag{1-2.12}$$

This may be written in component notation as

$$\sum_{i=1}^{3} A_i B_i = \lambda. \tag{1-2.13}$$

If the summation convention is adopted, we can simplify this expression. This convention requires that unless otherwise stated any repeated indices within a term require a sum on those indices from 1 to 3. If Greek letters are used as indices, the sum is usually taken from 1 to 2, which furnishes a modification useful for two-dimensional work. Equation 1-2.13 may be written

$$A_i B_i = \lambda. \tag{1-2.14}$$

The scalar product may be viewed pictorially as the product of the magnitude of the two vectors times the cosine of the angle between them (Figure 1-2):

$$\bar{A} \cdot \bar{B} = |A||B| \cos\theta. \tag{1-2.15}$$

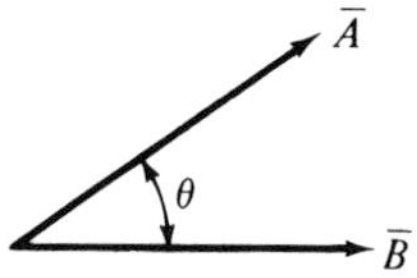

Figure 1-2

It may be verified that the scalar product between orthogonal unit vectors is one or zero:

$$\hat{i}_{(i)} \cdot \hat{i}_{(j)} = \delta_{ij}. \tag{1-2.16}$$

We may combine notations between the Gibb's notation and the component notation as follows:

$$A_i \hat{i}_{(i)} \cdot B_j \hat{i}_{(j)} = A_i B_j \hat{i}_{(i)} \cdot \hat{i}_{(j)} = A_i B_j \delta_{ij} = A_i B_i. \tag{1-2.17}$$

The vector product gives as the result a vector perpendicular to the original two vectors, and it is a little more difficult to define. Symbolically, this product is written with a $\times$ between vectors, explaining the name *cross product:*

$$\bar{C} = \bar{A} \times \bar{B} = \hat{n}|A||B| \sin\theta, \tag{1-2.18}$$

where $\hat{n}$ is a unit vector perpendicular to the plane of $\bar{A}$ and $\bar{B}$ thus forming a right-handed system ($\bar{A}$, $\bar{B}$, $\hat{n}$). This is shown in Figure 1-3. Because of the right-hand-rule convention introduced in the definition,

$$\bar{A} \times \bar{B} = -\bar{B} \times \bar{A}. \tag{1-2.19}$$

The cross product between orthogonal unit vectors is written

$$\hat{i}_{(i)} \times \hat{i}_{(j)} = e_{ijk} \hat{i}_{(k)}. \tag{1-2.20}$$

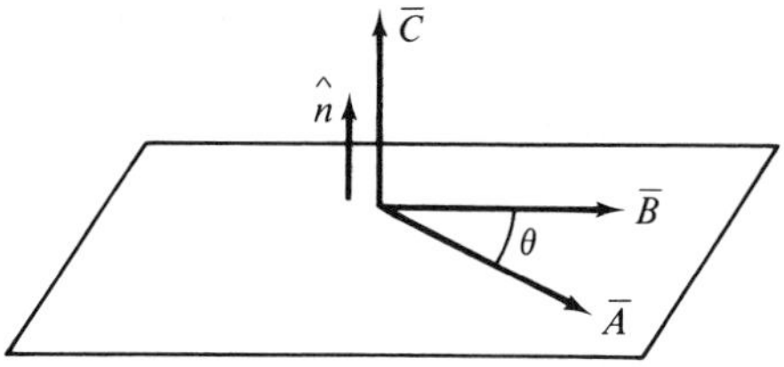

Figure 1-3

Using Gibb's notation, one may write

$$C_k \hat{i}_{(k)} = A_m \hat{i}_{(m)} \times B_n \hat{i}_{(n)} = e_{mnk} A_m B_n \hat{i}_{(k)}. \qquad (1\text{-}2.21)$$

In component notation equation 1-2.21 becomes

$$C_k = e_{mnk} A_m B_n. \qquad (1\text{-}2.22)$$

Note that both m and n are summed on, and k is a free index designating the component being discussed.

The most general type of vector multiplication is a tensorial multiplication that yields nine terms, giving all possible combinations of the components of the two vectors. Symbolically, this may be written as

$$\bar{A}\bar{B} = \bar{\bar{C}}, \qquad (1\text{-}2.23)$$

where $\bar{\bar{C}}$ may be shown to be a second-order tensor. In component notation this becomes

$$C_{ij} = A_i B_j, \qquad (1\text{-}2.24)$$

or in dyadic notation,

$$\hat{i}_{(i)} C_{ij} \hat{i}_{(j)} = A_i \hat{i}_{(i)} B_j \hat{i}_{(j)}. \qquad (1\text{-}2.25)$$

A special second-order tensor that corresponds to the Kronecker delta or the unit tensor may be written as

$$\bar{\bar{I}} = \hat{i}_{(1)}\hat{i}_{(1)} + \hat{i}_{(2)}\hat{i}_{(2)} + \hat{i}_{(3)}\hat{i}_{(3)}. \qquad (1\text{-}2.26)$$

The general tensor product shown in equation 1-2.25 is also called the dyadic product. The two unit vectors carry the directional properties of the second-order tensor.

The continued dyadic product may be used to represent higher-order tensors. For example, a third-order tensor may be written

$$\bar{\bar{\bar{T}}} = \hat{i}_{(i)}\hat{i}_{(j)}\hat{i}_{(k)} T_{ijk}. \qquad (1\text{-}2.27)$$

Some authors separate the elements of a continued dyadic product by semicolons but we shall not use that convention. The order of these elements is important and must be preserved. The greatest value of the dyadic or extended Gibb's notation is as a means of representing the tensors in orthogonal curvilinear coordinates without the necessity of utilizing general curvilinear tensor notation.

We have looked at a few examples of the dyadic product and the continued dyadic product. We may, if we wish, define continued dyadic products of rank 0, 1, 2, etc. This corresponds to the number of vectors in the product and is a way, as has been shown, of introducing higher-order tensors. A continued dyadic product of rank 1 would be a vector, whereas that of rank 0 would be a scalar. We may now look at some of the laws that this type of multiplication is assumed to obey.

1. Associative law. The continued dyadic product may be bracketed in any manner without change of meaning. For example,
$$\bar{A}\bar{B}\bar{C} = (\bar{A}\bar{B})\bar{C} = \bar{A}(\bar{B}\bar{C}). \qquad (1\text{-}2.28)$$
2. Scalar law. A scalar may be placed anywhere in the continued dyadic product without change of the meaning;
$$\lambda\bar{A}\bar{B}\bar{C} = \bar{A}(\lambda\bar{B})\bar{C} = \bar{A}\bar{B}(\lambda\bar{C}). \qquad (1\text{-}2.29)$$
3. Distributive law. Let $\bar{B}$ and $\bar{C}$ denote continued dyadic products of the same rank and let $\bar{A}$ be any continued dyadic product. Then
$$\bar{A}(\bar{B} + \bar{C}) = \bar{A}\bar{B} + \bar{A}\bar{C}. \qquad (1\text{-}2.30)$$

The order of the dyadic product is important, and we may define the conjugate of a dyad or of a continued dyadic product of rank 2 as
$$(\bar{A}\bar{B})_C = (\bar{B}\bar{A}). \qquad (1\text{-}2.31)$$
A symmetric tensor would be one that is equal to its conjugate dyad:
$$(\bar{A}\bar{B})_C = (\bar{A}\bar{B}). \qquad (1\text{-}2.32)$$

3 Transformation relations

Consider two right-handed coordinate systems, as shown in Figure 1-4. Quantities may be mathematically classified by the manner in which their components change as the coordinate system changes. Ignoring translation, there are two types of coordinate transformations that might be considered, rotation and inversion. A rotational transformation carries a right-handed system into a right-handed system (as in Figure 1-4), and an inversional transformation carries a right-handed system into a left-handed system, or vice versa. We shall consider the invariance of quantities under transformations that take them from one coordinate system to another system which is not moving relative to the first.

3.1 Scalars. A scalar is a quantity that does not change its mathematical description when one goes from one coordinate system to another, and will

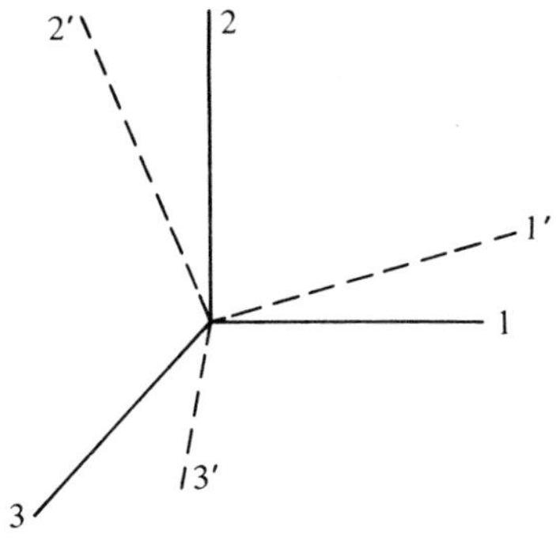

Figure 1-4

change only if the units of measurement are changed:

$$B' = B. \tag{1-3.1.1}$$

Examples of scalar quantities are temperature and mass.

3.2 Vectors. Let us consider the components of a vector in both the primed and the unprimed systems. Note that either the components V'_i or V_i completely describe the vector $\bar{V}$, both in magnitude and direction. Since the quantity is a fixed entity, it must be independent of the coordinate system. There must be a relationship between the components in the two systems that depends only upon the orientation of the two coordinate systems. It is obvious from Figure 1-5 that the components do not equal one another in general.

The relations between the components in the two systems may be written as

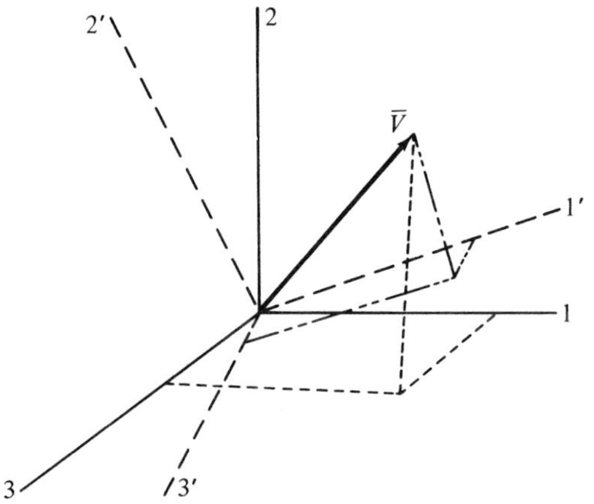

Figure 1-5

$$\begin{aligned} V'_1 &= V_1 \cos(x'_1, x_1) + V_2 \cos(x'_1, x_2) + V_3 \cos(x'_1, x_3), \\ V'_2 &= V_1 \cos(x'_2, x_1) + V_2 \cos(x'_2, x_2) + V_3 \cos(x'_2, x_3), \\ V'_3 &= V_1 \cos(x'_3, x_1) + V_2 \cos(x'_3, x_2) + V_3 \cos(x'_3, x_3). \end{aligned} \qquad (1\text{-}3.2.1)$$

If we define the symbol a_{ij} to be the cosine of the angle between the x'_i axis and the x_j axis and use the summation notation, the transformation equations may be written as

$$V'_i = a_{ij}V_j. \qquad (1\text{-}3.2.2)$$

Equation 1-3.3.3 serves as the definition of a vector or a first-order tensor, In this sense, Equation 1-3.1.1 serves as the definition of a scalar or a zero-order tensor.

3.3 Properties of the transformation matrix. The array of cosines between the coordinate axes may be written as a 3×3 matrix with elements a_{ij}. This matrix is in general not symmetric, but its elements possess certain orthonormal relationships. These may be obtained by observing the transformation of the components of a vector to the primed coordinate system and back again. These transformation equations may be written as

$$V'_i = a_{ij}V_j, \qquad (1\text{-}3.3.1)$$

$$V_i = a_{ji}V'_j. \qquad (1\text{-}3.3.2)$$

These transformations applied in sequence should result in the original components:

$$V_i = a_{ji}a_{jk}V_k, \qquad (1\text{-}3.3.3)$$

$$V'_i = a_{ij}a_{kj}V'_k. \qquad (1\text{-}3.3.4)$$

For these expressions to be true, the transformation matrix elements must satisfy the relations

$$a_{ji}a_{jk} = \delta_{ik}, \qquad (1\text{-}3.3.5)$$

$$a_{ij}a_{kj} = \delta_{ik}. \qquad (1\text{-}3.3.6)$$

These may be written in matrix notation as

$$\mathbf{A}^T\mathbf{A} = \mathbf{I}, \qquad (1\text{-}3.3.7)$$

$$\mathbf{A}\mathbf{A}^T = \mathbf{I}, \qquad (1\text{-}3.3.8)$$

where I is the unit matrix and the T superscript indicates the transpose of the matrix. By the definition of the inverse of a matrix, one may note

$$\mathbf{A}^{-1} = \mathbf{A}^T. \qquad (1\text{-}3.3.9)$$

Equations 1-3.3.5 and 1-3.3.6 are called the orthonormal relationships.

If the determinant is taken of both sides of equation 1-3.3.7, one obtains

$$|\mathbf{A}^T\mathbf{A}| = |\mathbf{I}| = 1. \qquad (1\text{-}3.3.10)$$

However, the determinant of the product of two square matrices is equal to the product of the determinants. Therefore, equation 1-3.3.10 may be written as

$$|\mathbf{A}^T||\mathbf{A}| = 1. \qquad (1\text{-}3.3.11)$$

Since the value of a determinant is unchanged if the rows and columns of the matrix are interchanged, equation 1-3.3.11 becomes

$$|\mathbf{A}|^2 = 1. \qquad (1\text{-}3.3.12)$$

$$|\mathbf{A}| = \pm 1. \qquad (1\text{-}3.3.13)$$

The determinant of the transformation matrix is either plus or minus 1. It can be shown that it will be plus for a rotational transformation and minus for an inversion.*

4 Second-order tensors

The third type of vector multiplication, equation 1-2.23, was defined by

$$A_i B_j. \qquad (1\text{-}4.1)$$

If we examine this in the primed coordinate system, the vector transformation rules yield

$$A'_i B'_j = a_{im} a_{jn} A_m B_n. \qquad (1\text{-}4.2)$$

An array of numbers which transforms in this manner will be defined to be a second-order tensor. The general form of this transformation may be written

$$C'_{ij} = a_{im} a_{jn} C_{mn}, \qquad (1\text{-}4.3)$$

$$C_{ij} = a_{mi} a_{nj} C'_{mn}. \qquad (1\text{-}4.4)$$

A second-order tensor is defined to be symmetric if

$$C_{ij} = C_{ji}, \qquad (1\text{-}4.5)$$

and antisymmetric if

$$C_{ij} = -C_{ji}. \qquad (1\text{-}4.6)$$

Note that all the diagonal elements in an antisymmetric tensor are zero.

Any tensor may be considered to be the sum of a symmetric and an antisymmetric tensor:

$$C_{ij} = \underbrace{\tfrac{1}{2}(C_{ij} + C_{ji})}_{\text{symmetric}} + \underbrace{\tfrac{1}{2}(C_{ij} - C_{ji})}_{\text{antisymmetric}}. \qquad (1\text{-}4.7)$$

* R. Aris, *Vectors, Tensors, and the Basic Equations of Fluid Mechanics*, Prentice-Hall, Inc., Englewood Cliffs, N. J., 1962.

5 Higher-order tensors

Higher-order tensors may be defined by their transformation laws in Cartesian coordinates. In general, this law may be written

$$A'_{ijk\cdots n} = a_{ip}a_{jq}a_{kl}\cdots a_{nm}A_{pql\cdots m}. \qquad (1\text{-}5.1)$$

These tensors may arise from the general tensor product of lower-order tensors. To understand this, it is necessary to examine some of the general tensor operations. Two tensors of the same order may be added or subtracted, yielding a tensor of the same order. This is accomplished by adding or subtracting corresponding components. Symbolically, this may be written

$$\bar{\bar{A}} + \bar{\bar{B}} + \bar{\bar{C}}. \qquad (1\text{-}5.2)$$

In component notation this becomes

$$A_{ij} + B_{ij} = C_{ij}. \qquad (1\text{-}5.3)$$

If a tensor is multiplied by a scalar, the result is a tensor of the same order with each component multiplied by the scalar:

$$\bar{\bar{A}} = \beta\bar{\bar{C}}, \qquad \Longrightarrow \qquad A_{ij} = \beta C_{ij}. \qquad (1\text{-}5.4)$$

General multiplication is the totality of all products made up of combinations of the components of each. The order of the product is the sum of the orders of the two tensor factors:

$$\bar{A}\bar{\bar{B}} = \bar{\bar{\bar{C}}}, \qquad \Longrightarrow \qquad A_iB_{jk} = C_{ijk}. \qquad (1\text{-}5.5)$$

The general vector multiplication shown in Equation 1-2.23 is such a product.

A useful tensorial operation, called *contraction*, is accomplished by making two subscripts of a tensor the same. If the original tensor was of order n, the resulting contracted tensor is of order $n - 2$. The dot product between two vectors may be thought of as the general tensor product followed by a contraction. Symbolically, dots are therefore frequently used to indicate contractions. Multiple dots are appropriate for complex constructions:

$$\bar{\bar{T}}\cdot\cdot\bar{\bar{S}} = \lambda, \qquad \Longrightarrow \qquad T_{ij}S_{ji} = \lambda, \qquad (1\text{-}5.6)$$

$$\bar{\bar{T}}:\bar{\bar{S}} = \beta, \qquad \Longrightarrow \qquad T_{ij}S_{ij} = \beta, \qquad (1\text{-}5.7)$$

$$\bar{V}\cdot\bar{\bar{S}} = \bar{U}, \qquad \Longrightarrow \qquad V_iS_{ij} = U_j. \qquad (1\text{-}5.8)$$

Powers of tensors are examples of contractions:

$$\bar{\bar{T}}^2, \qquad \Longrightarrow \qquad T_{ij}T_{jk} = S_{ik}. \qquad (1\text{-}5.9)$$

The trace of a second-order tensor (sum of the diagonal elements) is a contraction:

$$\bar{\bar{T}}\cdot\cdot\bar{\bar{I}} = T_{ii} = \lambda. \qquad (1\text{-}5.10)$$

Note that the contraction of the Kronecker delta is the sum of the diagonal elements:

$$\delta_{ii} = 3. \qquad (1\text{-}5.11)$$

Any contraction of the alternating tensor is zero:

$$e_{iik} = 0. \qquad (1\text{-}5.12)$$

A very useful relationship between these two symbols or tensors is

$$\begin{aligned} e_{ilm}e_{jpq} = {} & \delta_{ij}\delta_{lp}\delta_{mq} + \delta_{ip}\delta_{lq}\delta_{mj} + \delta_{iq}\delta_{lj}\delta_{mp} \\ & - \delta_{ij}\delta_{lq}\delta_{mp} - \delta_{ip}\delta_{lj}\delta_{mq} - \delta_{iq}\delta_{lp}\delta_{mj}. \end{aligned} \qquad (1\text{-}5.13)$$

Contraction on i and j yields

$$e_{ilm}e_{ipq} = \delta_{lp}\delta_{mq} - \delta_{lq}\delta_{mp}. \qquad (1\text{-}5.14)$$

A second contraction on l and p gives

$$e_{ilm}e_{ilq} = 2\delta_{mq}. \qquad (1\text{-}5.15)$$

Finally, contracting on m and q gives

$$e_{ilm}e_{ilm} = 6. \qquad (1\text{-}5.16)$$

Isomers of tensors are formed by permutation of two subscripts. For example, the isomers of the third-order tensor T_{ijk} are T_{ikj}, T_{jik}, T_{jki}, T_{kij}, and T_{kji}. A tensor of order 2 has only one isomer, its transpose. An important isomeric property of the alternating tensor may be shown as follows:

$$e_{ijk} = e_{jki} = e_{kij} = -e_{jik} = -e_{ikj} = -e_{kji}. \qquad (1\text{-}5.17)$$

The relationships between the alternating tensor and the Kronecker delta may be used to prove vector identities in Cartesian coordinates. Any tensor identity proved in one coordinate system is also valid in all coordinate systems. As an example of one of these proofs, consider the triple vector product identity

$$\bar{A} \times (\bar{B} \times \bar{C}) = \bar{B}(\bar{A} \cdot \bar{C}) - \bar{C}(\bar{A} \cdot \bar{B}). \qquad (1\text{-}5.18)$$

Writing this in component notation gives

$$e_{ijk}A_j e_{klm}B_l C_m = B_i A_j C_j - C_i A_j B_j. \qquad (1\text{-}5.19)$$

Note that this is a vector equation, and i represents the free index or the one designating the component being considered. All other indices are being summed on and may be considered as dummy indices. Using equations 1-5.14 and 1-5.17 yields

$$A_j B_l C_m(\delta_{il}\delta_{jm} - \delta_{im}\delta_{jl}) = B_i A_j C_j - C_i A_j B_j. \qquad (1\text{-}5.20)$$

Since $B_l\delta_{il}$ is zero except for the term $i = l$, the summed index l may be replaced by i. The Kronecker delta is sometimes called the substitution tensor for this reason. Equation 1-5.20 may now be written as

$$B_i A_j C_j - C_i A_j B_j = B_i A_j C_j - C_i A_j B_j, \qquad (1\text{-}5.21)$$

which proves equation 1-5.18.

We may show other examples of tensors involving cross-product and dot-product operations. Symbolically, multiple operations between second-order tensors may be written as

$$\bar{C} = \bar{\bar{T}} \times \cdot \bar{\bar{S}}, \qquad C_i = e_{ijk} T_{jm} C_{mk}, \tag{1-5.22}$$

$$\bar{\bar{C}} = \bar{\bar{T}} \overset{\times}{\times} \bar{\bar{S}}, \qquad C_{ij} = e_{imn} e_{jpq} T_{mp} S_{qn}. \tag{1-5.23}$$

In the last example the upper operational symbol pertains to the outer unit vectors in the dyads and the lower symbol to the inner unit vectors. These are specialized notations and are of less general value than some of those shown earlier.

6 Dual vector of an antisymmetric tensor

Consider a quantity defined by the expression

$$t_i = e_{ijk} T_{jk}. \tag{1-6.1}$$

If this expression is expanded, one obtains

$$\begin{aligned} t_1 &= T_{23} - T_{32}, \\ t_2 &= T_{31} - T_{13}, \\ t_3 &= T_{12} - T_{21}. \end{aligned} \tag{1-6.2}$$

Note that the vector $\bar{t}$ depends only upon the antisymmetric part of T.

Multiplying both sides of equation 1-6.1 by the alternating tensor and contracting yields

$$e_{imn} t_i = e_{imn} e_{ijk} T_{jk}. \tag{1-6.3}$$

Using equation 1-5.14, this may be written

$$\begin{aligned} e_{imn} t_i &= (\delta_{mj} \delta_{nk} - \delta_{mk} \delta_{nj}) T_{jk} \\ &= T_{mn} - T_{nm}. \end{aligned} \tag{1-6.4}$$

If $\bar{\bar{T}}$ is an antisymmetric tensor,

$$T_{mn} = -T_{nm}, \tag{1-6.5}$$

and equation 1-6.4 becomes

$$T_{mn} = \tfrac{1}{2} e_{imn} t_i = -\tfrac{1}{2} e_{min} t_i. \tag{1-6.6}$$

Let $\bar{\omega}$ be defined as

$$\omega_i = -\tfrac{1}{2} t_i = -\tfrac{1}{2} e_{ijk} T_{jk}. \tag{1-6.7}$$

Let us examine the tensor contraction between the antisymmetric tensor $\bar{\bar{T}}$ as a vector $\bar{U}$:

$$T_{mn} U_n = e_{min} \omega_i U_n. \tag{1-6.8}$$

Symbolically, this may be written

$$\bar{\bar{T}} \cdot \bar{U} = \bar{\omega} \times \bar{U}. \tag{1-6.9}$$

The dual vector of the antisymmetric tensor T is $\bar{\omega}$. Operationally, then, one may write

$$\bar{\bar{T}} \cdot \quad \Longrightarrow \quad \bar{\omega} \times . \tag{1-6.10}$$

7 Eigenvalue problem

If a second-order tensor is contracted with a vector, the product will be another vector, generally not parallel to the original vector:

$$T_{ij} v_j = u_i. \tag{1-7.1}$$

Consider now the case where the final vector is parallel to the vector v_j:

$$T_{ij} v_j = \lambda v_i. \tag{1-7.2}$$

This may be written as

$$(T_{ij} - \lambda \delta_{ij}) v_j = 0. \tag{1-7.3}$$

This forms a homogeneous system of equations for v_i, which has a nontrivial solution if the determinant of the coefficients is zero:

$$| T_{ij} - \lambda \delta_{ij} | = 0. \tag{1-7.4}$$

Expanding this determinant yields a cubic equation for λ:

$$\lambda^3 - \mathrm{I}_T \lambda^2 + \mathrm{II}_T \lambda - \mathrm{III}_T = 0, \tag{1-7.5}$$

where

$$\begin{aligned} \mathrm{I}_T &= T_{ii}, \\ \mathrm{II}_T &= \tfrac{1}{2}(T_{ii} T_{jj} - T_{ij} T_{ij}), \\ \mathrm{III}_T &= e_{ijk} T_{i1} T_{j2} T_{k3}. \end{aligned}$$

These are called the invariants of the tensor because it can be shown that they have the same value in all coordinate systems. If the tensor is real and symmetric, the roots of equation 1-7.5, the so-called "eigenvalues," are all real and the three eigenvectors v_i (one corresponding to each eigenvalue) are orthogonal. The magnitude of the eigenvectors is indeterminate and the vectors are usually "normalized" as unit vectors. If a set of primed coordinate axes is taken to correspond with the eigenvector triad, the components of the eigenvectors become

$$v_i^{(m)} = a_{mi}, \tag{1-7.6}$$

where m names the eigenvector and the primed axis. Equation (1-7.2) gives rise to three tensorial equations:

$$T_{ij} v_j^{(m)} = \lambda^{(m)} v_i^{(m)}, \qquad \text{no sum on } m. \tag{1-7.7}$$

Substituting equation 1-7.6 into equation 1-7.7 yields

$$T_{ij} a_{mj} = \lambda^{(m)} a_{mi}, \qquad \text{no sum on } m. \tag{1-7.8}$$

Now consider the general transformation to the primed axes:

$$T'_{nm} = a_{ni} a_{mj} T_{ij}. \tag{1-7.9}$$

Substituting equation (1-7.8) into equation 1-7.9 gives

$$T'_{nm} = a_{ni} a_{mi} \lambda^{(m)}, \qquad \text{no sum on } m. \tag{1-7.10}$$

Using equation 1-3.3.6, this may be written as

$$T'_{nm} = \delta_{nm} \lambda^{(m)}, \qquad \text{no sum on } m. \tag{1-7.11}$$

Equation 1-7.11 shows that the transformation to the coordinate axes corresponding to the eigenvectors "diagonalizes" the tensor; the diagonal elements are the three eigenvalues. The orthogonal eigenvectors are shown in Figure 1-6 and the new coordinate axes are aligned with them.

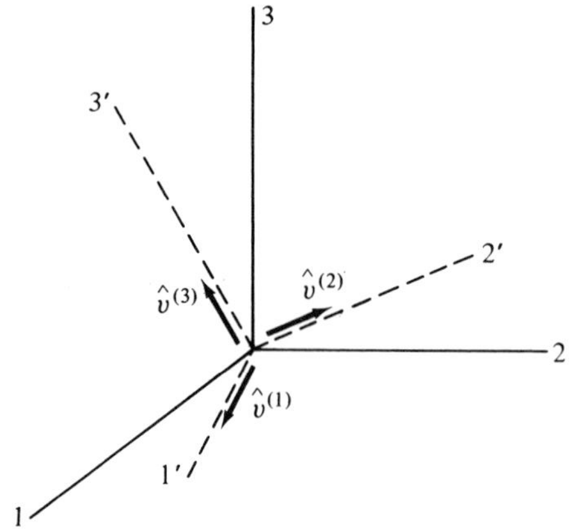

Figure 1-6

8 Isotropic tensors

An isotropic tensor is one whose components are the same in every orthogonal coordinate system. The Kronecker delta is an isotropic second-order tensor, because

$$a_{mi} a_{nj} \delta_{ij} = a_{mi} a_{ni} = \delta'_{mn}. \tag{1-8.1}$$

Every scalar is an isotropic tensor of order zero. There are no isotropic tensors of order 1. It has been shown that δ_{ij} is an isotropic tensor of order 2, and this is the only such tensor of this order. The alternating tensor is the only isotropic tensor of order 3. The product of any two isotropic tensors is also isotropic. Therefore, the following are isotropic tensors of order 4:

$$\begin{aligned} &\delta_{ij}\delta_{pq}, \\ &e_{ijk}e_{pqk}. \end{aligned} \tag{1-8.2}$$

The last tensor may be written

$$e_{ijk}e_{pqk} = \delta_{ip}\delta_{jq} - \delta_{iq}\delta_{jp}. \tag{1-8.3}$$

A tensor defined as follows is also isotropic:

$$T_{ijpq} = a\delta_{ij}\delta_{pq} + b\delta_{ip}\delta_{jq} + c\delta_{iq}\delta_{jp}. \tag{1-8.4}$$

It may be shown* that this is the most general isotropic tensor of order 4. It may be written as the sum of symmetric and antisymmetric tensors as follows:

$$T_{ijpq} = \lambda\delta_{ij}\delta_{pq} + \mu(\delta_{ip}\delta_{jq} + \delta_{iq}\delta_{jp}) + \nu(\delta_{ip}\delta_{jq} - \delta_{iq}\delta_{jp}). \tag{1-8.5}$$

Note that the first two terms are symmetric with respect to i, j and p, q, whereas the last term is antisymmetric with respect to these pairs of indices.

9 Tensor fields

In most applications of continuum mechanics, the components of the tensors depend upon the spatial coordinates and time. The displacement vector, for example, in a deformed body may vary from point to point and with time. We speak of these tensors as tensor fields and denote them as follows:

$$T_{ij} = T_{ij}(x_k, t), \quad \Longrightarrow \quad \bar{\bar{T}} = \bar{\bar{T}}(\bar{x}, t), \tag{1-9.1}$$

$$v_i = v_i(x_j, t), \quad \Longrightarrow \quad \bar{v} = \bar{v}(\bar{x}, t). \tag{1-9.2}$$

If a tensor field is independent of time, it is called a steady field and, conversely, if it depends upon time it is called unsteady.

If we use dyadic notation, we must examine to see if the unit vectors depend upon time or the spatial coordinates. If Cartesian axes are not rotating, the unit vectors are constant vectors:

$$\bar{v} = v_i(x_j, t)\hat{i}_{(i)}. \tag{1-9.3}$$

In polar cylindrical coordinates (r, θ, z), the unit vectors in the r and θ directions are dependent upon the θ coordinate and may therefore have an implicit time dependency:

$$\bar{v} = v_r(r, \theta, z, t)\hat{i}_{(r)}(\theta) + v_\theta(r, \theta, z, t)\hat{i}_{(\theta)}(\theta) + v_z(r, \theta, z, t)\hat{i}_{(z)}. \tag{1-9.4}$$

* Aris, *Vectors, Tensors, and the Basic Equations of Fluid Mechanics.*

If the partial time derivative is taken of the vector field given by equation 1-9.3, we obtain

$$\frac{\partial \bar{v}}{\partial t} = \frac{\partial v_1}{\partial t}\hat{i}_{(1)} + \frac{\partial v_2}{\partial t}\hat{i}_{(2)} + \frac{\partial v_3}{\partial t}\hat{i}_{(3)}. \qquad (1\text{-}9.5)$$

If the vector depends only upon time, this is written

$$\frac{d\bar{v}}{dt} = \frac{dv_k}{dt}\hat{i}_{(k)}. \qquad (1\text{-}9.6)$$

The distributive characteristic of tensor and vector products implies that the derivatives of these products obey the usual calculus rules.

If we express a vector depending only upon time in polar cylindrical coordinates, additional terms are obtained when the time derivative is taken:

$$\bar{v} = v_r(t)\hat{i}_{(r)}(\theta) + v_\theta(t)\hat{i}_{(\theta)}(\theta) + v_z(t)\hat{i}_{(z)}, \qquad (1\text{-}9.7)$$

$$\frac{d\bar{v}}{dt} = \frac{dv_r}{dt}\hat{i}_{(r)} + v_r\frac{d\hat{i}_{(r)}}{d\theta}\frac{d\theta}{dt} + \frac{dv_\theta}{dt}\hat{i}_{(\theta)} + v_\theta\frac{d\hat{i}_{(\theta)}}{d\theta}\frac{d\theta}{dt} + \frac{dv_z}{dt}\hat{i}_{(z)}. \qquad (1\text{-}9.8)$$

Considering the unit vectors as functions of θ, we obtain the following by examination of Figure 1-7:

$$\frac{d\hat{i}_{(r)}}{d\theta} = \hat{i}_{(\theta)}, \qquad (1\text{-}9.9)$$

$$\frac{d\hat{i}_{(\theta)}}{d\theta} = -\hat{i}_{(r)}. \qquad (1\text{-}9.10)$$

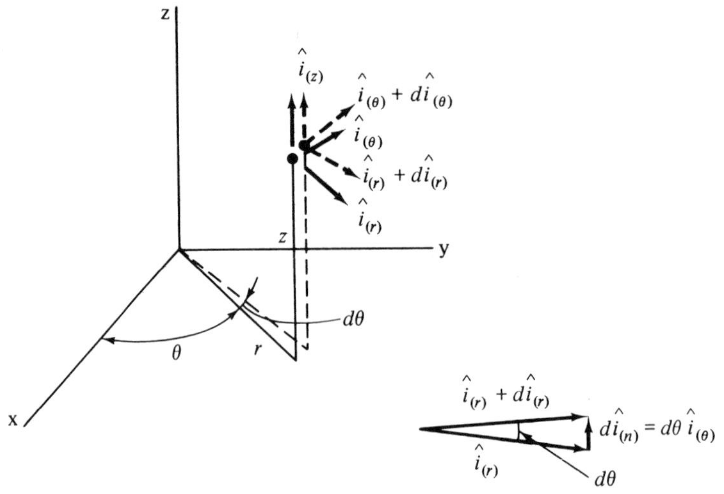

Figure 1-7

THE DEL OPERATOR

A directed rate of change of a tensor field with respect to the coordinate directions may be obtained by use of the del operator. This vector operator is defined as

$$\bar{\nabla} = \hat{i}_{(k)} \frac{\partial}{\partial x_k} = \hat{i}_{(x)} \frac{\partial}{\partial x} + \hat{i}_{(y)} \frac{\partial}{\partial y} + \hat{i}_{(z)} \frac{\partial}{\partial z}. \qquad (1\text{-}9.11)$$

The components of this operator in Cartesian coordinates may also be designated by ∇_k.

If we consider a scalar field $\varphi(x_k)$, the gradient of this field is written

$$\bar{\nabla}\varphi \quad \text{or} \quad \text{grad}\ \varphi. \qquad (1\text{-}9.12)$$

This may be expanded as follows in Cartesian coordinates

$$\bar{\nabla}\varphi = \hat{i}_{(k)} \frac{\partial \varphi}{\partial x_k}. \qquad (1\text{-}9.13)$$

In polar cylindrical coordinates, the gradient may be written

$$\bar{\nabla}\varphi = \hat{i}_{(r)} \frac{\partial \varphi}{\partial r} + \hat{i}_{(\theta)} \frac{1}{r} \frac{\partial \varphi}{\partial \theta} + \hat{i}_{(z)} \frac{\partial \varphi}{\partial z}. \qquad (1\text{-}9.14)$$

The gradient vector itself points in the direction of the greatest rate of change of φ and its magnitude represents this rate of change. Therefore, if we consider surfaces of constant φ, $\bar{\nabla}\varphi$ would be normal to these surfaces (assuming that φ is continuous and single valued). If we want the rate of change of φ in a particular direction designated by the unit vector, $\hat{n}$, the dot product between this vector and the gradient of φ will give this derivative:

$$\frac{\partial \varphi}{\partial n} = \hat{n} \cdot \bar{\nabla}\varphi.$$

GRADIENTS OF VECTORS AND HIGHER-ORDER TENSORS

The directed rate of change of a vector may also be obtained by forming the gradient of each component of that vector. This may be represented symbolically by forming the general tensor or dyadic product between the vector operator $\bar{\nabla}$ and the vector. Since this product does not commute, there are two possible gradient forms, which for lack of better terms will be called the pregradient, $\vec{\nabla}\bar{v}$, and the postgradient, $\bar{v}\overleftarrow{\nabla}$. Expanding each of these yields

$$\vec{\nabla}\bar{v} = \hat{i}_{(i)} \frac{\partial v_j}{\partial x_i} \hat{i}_{(j)}, \qquad (1\text{-}9.15)$$

$$\bar{v}\overleftarrow{\nabla} = \hat{i}_{(i)} \frac{\partial v_i}{\partial x_j} \hat{i}_{(j)}. \qquad (1\text{-}9.16)$$

It is apparent that these are different second-order tensors. The second is the transpose of the first:

$$\bar{v}\overleftarrow{\nabla} = (\vec{\nabla}\bar{v})^T. \qquad (1\text{-}9.17)$$

The symmetric part of either the pregradient or the postgradient is

$$\text{Symmetric part of } \vec{\nabla}\bar{v} = \tfrac{1}{2}(\vec{\nabla}\bar{v} + \bar{v}\overleftarrow{\nabla}). \qquad (1\text{-}9.18)$$

The trace of these two second-order tensors is the same scalar and is called the divergence of the vector. The divergence is written symbolically as

$$\bar{\nabla} \cdot \bar{v} = \text{trace}\,(\vec{\nabla}\bar{v}). \qquad (1\text{-}9.19)$$

The divergence in Cartesian coordinates is

$$\bar{\nabla} \cdot \bar{v} = \hat{i}_{(i)}\frac{\partial}{\partial x_i} \cdot (v_j \hat{i}_{(j)}) = \hat{i}_{(i)} \cdot \hat{i}_{(j)}\frac{\partial v_j}{\partial x_i} = \frac{\partial v_i}{\partial x_i}. \qquad (1\text{-}9.20)$$

In polar cylindrical coordinates, the divergence of the vector $\bar{v}$ takes the form

$$\bar{\nabla} \cdot \bar{v} = \left[\hat{i}_{(r)}\frac{\partial}{\partial r} + \hat{i}_{(\theta)}\frac{1}{r}\frac{\partial}{\partial \theta} + \hat{i}_{(z)}\frac{\partial}{\partial z}\right] \cdot [v_r \hat{i}_{(r)} + v_\theta \hat{i}_{(\theta)} + v_z \hat{i}_{(z)}]. \qquad (1\text{-}9.21)$$

Note that the differentiation involved in the vector operator $\bar{\nabla}$ precedes the dot product. Since $\hat{i}_{(r)}$ and $\hat{i}_{(\theta)}$ are functions of θ, an additional term arises in the divergence:

$$\bar{\nabla} \cdot \bar{v} = \frac{\partial v_r}{\partial r} + \frac{v_r}{r} + \frac{1}{r}\frac{\partial v_\theta}{\partial \theta} + \frac{\partial v_z}{\partial z}. \qquad (1\text{-}9.22)$$

The gradient of a vector in cylindrical coordinates may be handled in a similar manner by dyadic notation,

$$\begin{aligned}\vec{\nabla}\bar{v} = {} & \hat{i}_{(r)}\frac{\partial v_r}{\partial r}\hat{i}_{(r)} + \hat{i}_{(r)}\frac{\partial v_\theta}{\partial r}\hat{i}_{(\theta)} + \hat{i}_{(r)}\frac{\partial v_z}{\partial r}\hat{i}_{(z)} \\ & + \hat{i}_{(\theta)}\frac{1}{r}\frac{\partial v_r}{\partial \theta}\hat{i}_{(r)} + \hat{i}_{(\theta)}\frac{v_r}{r}\hat{i}_{(\theta)} + \hat{i}_{(\theta)}\frac{1}{r}\frac{\partial v_\theta}{\partial \theta}\hat{i}_{(\theta)} - \hat{i}_{(\theta)}\frac{v_\theta}{r}\hat{i}_{(r)} \qquad (1\text{-}9.23) \\ & + \hat{i}_{(\theta)}\frac{1}{r}\frac{\partial v_z}{\partial \theta}\hat{i}_{(z)} + \hat{i}_{(z)}\frac{\partial v_r}{\partial z}\hat{i}_{(r)} + \hat{i}_{(z)}\frac{\partial v_\theta}{\partial z}\hat{i}_{(\theta)} + \hat{i}_{(z)}\frac{\partial v_z}{\partial z}\hat{i}_{(z)}.\end{aligned}$$

Note that the differentiation with respect to θ treats each component times its unit vector as a product and produces two terms. Furthermore, equation 1-9.22 can be obtained from equation 1-9.23 by contraction.

Gradients of higher-order tensors produce tensors one order higher than the original tensors. Therefore, the gradient of a second-order tensor is a third-order tensor:

$$\vec{\nabla}\bar{\bar{T}} = \bar{\bar{\bar{C}}} \quad \Longrightarrow \quad \frac{\partial T_{jk}}{\partial x_i} = C_{ijk}. \qquad (1\text{-}9.24)$$

Contraction of this tensor between the ith and jth components produces a vector that is the divergence of the second-order tensor with respect to the first index, which we may term as a *predivergence*. We also could form a postgradient of this tensor:

$$\bar{\bar{T}}\overleftarrow{\nabla} = \bar{\bar{\bar{D}}}, \qquad \frac{\partial T_{ij}}{\partial x_k} = D_{ijk}. \qquad (1\text{-}9.25)$$

This is an isomer of the pregradient tensor, $\bar{\bar{\bar{C}}}$.

CURL OF A TENSOR FIELD

The cross product between the $\bar{\nabla}$ operator and a tensor is called the curl of that tensor. For example, the curl of a vector produces a vector, this operation may be written

$$\bar{\nabla} \times \bar{v} = \bar{u}, \qquad \Longrightarrow \qquad e_{ijk}\frac{\partial v_k}{\partial x_j} = u_i. \tag{1-9.26}$$

Note that if $\partial v_k/\partial x_j$ is considered to be a second-order tensor, equations 1-6.1 and 1-6.7 imply that $-\frac{1}{2}u_i$ is the component of a dual vector of that tensor. The curl of a vector is related to the dual vector of the second-order tensor formed from the gradient of the vector.

Dyadic notation allows the curl to be represented in other systems orthogonal of coordinates. For example, in polar cylindrical coordinates,

$$\bar{\nabla} \times \bar{v} = \left[\hat{i}_{(r)}\frac{\partial}{\partial r} + \hat{i}_{(\theta)}\frac{1}{r}\frac{\partial}{\partial \theta} + \hat{i}_{(z)}\frac{\partial}{\partial z}\right] \times [v_r\hat{i}_{(r)} + v_\theta\hat{i}_{(\theta)} + v_z\hat{i}_{(z)}] \tag{1-9.27}$$

$$\begin{aligned} &= \hat{i}_{(r)}\left(\frac{1}{r}\frac{\partial v_z}{\partial \theta} - \frac{\partial v_\theta}{\partial z}\right) + \hat{i}_{(\theta)}\left(\frac{\partial v_r}{\partial z} - \frac{\partial v_z}{\partial r}\right) \\ &\quad + \hat{i}_{(z)}\left(\frac{\partial v_\theta}{\partial r} - \frac{1}{r}\frac{\partial v_r}{\partial \theta} + \frac{v_\theta}{r}\right). \end{aligned} \tag{1-9.28}$$

The curl may also be taken of higher-order tensors, yielding a tensor of the same order as the original one.

$$\vec{\nabla} \times \bar{\bar{T}} = \bar{\bar{S}}, \qquad \Longrightarrow \qquad e_{ijk}\frac{\partial T_{km}}{\partial x_j} = S_{im}. \tag{1-9.29}$$

In this case the curl may be applied to the lead vector in the dyadic product or to the last vector. The resulting tensors in these two cases are not the same, and the terminology of *precurl* and *postcurl* may be applied to indicate the difference:

$$(\bar{\bar{T}} \times \overleftarrow{\nabla}) = \bar{\bar{R}}, \qquad e_{ijk}\frac{\partial T_{mj}}{\partial x_k} = R_{mi}. \tag{1-9.30}$$

These may be written in dyadic notation as

$$\begin{aligned} \vec{\nabla} \times \bar{\bar{T}} &= \left(\hat{i}_{(i)}\frac{\partial}{\partial x_i}\right) \times \hat{i}_{(j)}T_{jk}\hat{i}_{(k)} \\ &= (\hat{i}_{(i)} \times \hat{i}_{(j)})\frac{\partial T_{jk}}{\partial x_i}\hat{i}_{(k)}. \end{aligned} \tag{1-9.31}$$

Substituting equation 1-2.20 into equation 1-9.31 yields

$$\vec{\nabla} \times \bar{\bar{T}} = \hat{i}_{(m)}e_{mij}\frac{\partial T_{jk}}{\partial x_i}\hat{i}_{(k)}. \tag{1-9.32}$$

If the del operator and the second-order tensor are written in cylindrical coordinates, extra terms arise upon differentiation because the unit vectors depend upon the coordinates. The second-order tensor must be treated under the general rules of calculus as the product of three terms. This is the case in most coordinate systems with the exception of Cartesian coordinates. The differentiation precedes the vector product operation:

$$\bar{\nabla} \times \bar{\bar{T}} = \left(\hat{i}_{(r)}\frac{\partial}{\partial r} + \hat{i}_{(\theta)}\frac{1}{r}\frac{\partial}{\partial \theta} + \hat{i}_{(z)}\frac{\partial}{\partial z}\right) \times (\hat{i}_{(r)}T_{rr}\hat{i}_{(r)} + \hat{i}_{(r)}T_{r\theta}\hat{i}_{(\theta)} + \hat{i}_{(r)}T_{rz}\hat{i}_{(z)} + \hat{i}_{(\theta)}T_{\theta r}\hat{i}_{(r)} + \hat{i}_{(\theta)}T_{\theta\theta}\hat{i}_{(\theta)} + \hat{i}_{\theta}T_{\theta z}\hat{i}_{(z)} + \hat{i}_{(z)}T_{zr}\hat{i}_{(r)} + \hat{i}_{(z)}T_{z\theta}\hat{i}_{(\theta)} + \hat{i}_{(z)}T_{zz}\hat{i}_{(z)}). \tag{1-9.33}$$

The procedure of obtaining the resulting second-order tensor is lengthy but routine.

Vector and higher-order tensor identities involving the del operator are usually proved in Cartesian coordinates and the results applied to other coordinate systems. For example, the curl of the gradient of any quantity is zero:

$$\bar{\nabla} \times \bar{\nabla}\varphi = 0. \tag{1-9.34}$$

Writing this in Cartesian component notation yields

$$e_{ijk}\frac{\partial}{\partial x_j}\frac{\partial \varphi}{\partial x_k} = 0. \tag{1-9.35}$$

We may prove this by noting the properties of the alternating tensor and a postulated independence of the order of differentiation of the function φ:

$$\frac{1}{2}\left[e_{ijk}\frac{\partial^2 \varphi}{\partial x_j\, \partial x_k} + e_{ijk}\frac{\partial^2 \varphi}{\partial x_j\, \partial x_k}\right] = 0,$$

$$\frac{1}{2}\left[e_{ijk}\frac{\partial^2 \varphi}{\partial x_j\, \partial x_k} - e_{ikj}\frac{\partial^2 \varphi}{\partial x_j\, \partial x_k}\right] = 0.$$

Interchanging the dummy summation indices completes the proof:

$$\frac{1}{2}\left[e_{ijk}\frac{\partial^2 \varphi}{\partial x_j\, \partial x_k} - e_{ijk}\frac{\partial^2 \varphi}{\partial x_k\, \partial x_j}\right] = 0. \tag{1-9.36}$$

Other identities are given in the exercises and may be proved in a similar manner.

One other form involving the del operator deserves special mention. This is the dot product between the $\bar{\nabla}$ operator and itself:

$$\nabla^2 = \bar{\nabla} \cdot \bar{\nabla} = \hat{i}_{(i)}\frac{\partial}{\partial x_i} \cdot \left(\hat{i}_{(j)}\frac{\partial}{\partial x_j}\right) = \frac{\partial^2}{\partial x_i\, \partial x_i}. \tag{1-9.37}$$

This is the Laplacian operator and is a scalar operator. In polar cylindrical coordinates the form of the Laplacian may be obtained by use of dyadic notation:

$$\nabla^2 = \left[\hat{i}_{(r)}\frac{\partial}{\partial r} + \hat{i}_{(\theta)}\frac{1}{r}\frac{\partial}{\partial \theta} + \hat{i}_{(z)}\frac{\partial}{\partial z}\right] \cdot \left[\hat{i}_{(z)}\frac{\partial}{\partial r} + \hat{i}_{(\theta)}\frac{1}{r}\frac{\partial}{\partial \theta} + \hat{i}_{(z)}\frac{\partial}{\partial z}\right] = \frac{\partial^2}{\partial r^2} + \frac{1}{r}\frac{\partial}{\partial r} + \frac{1}{r^2}\frac{\partial^2}{\partial \theta^2} + \frac{\partial^2}{\partial z^2}. \tag{1-9.38}$$

The second term arises from the θ differentiation of the $\hat{i}_{(r)}$ unit vector.

10 Integral theorems

Certain useful integral theorems involving tensor fields will be listed here. No proofs of these theorems will be given; they may be found in most vector calculus books. The first of these is called by many names, such as Gauss's theorem, Green's theorem, or the divergence theorem. Most Russian texts refer to it as Ostrogradsky's theorem.

Consider the volume, V, enclosed by the surface, S, shown in Figure 1-8. Let $\hat{n}$ be a unit vector directed outward and normal to the surface element dS. The divergence theorem for a vector field may be stated as

$$\int_S \hat{n} \cdot \bar{a}\, dS = \int_V \bar{\nabla} \cdot \bar{a}\, dV. \tag{1-10.1}$$

This may be expressed in many other forms with various other operations replacing the indicated dot product. It is vital to observe that the del operator

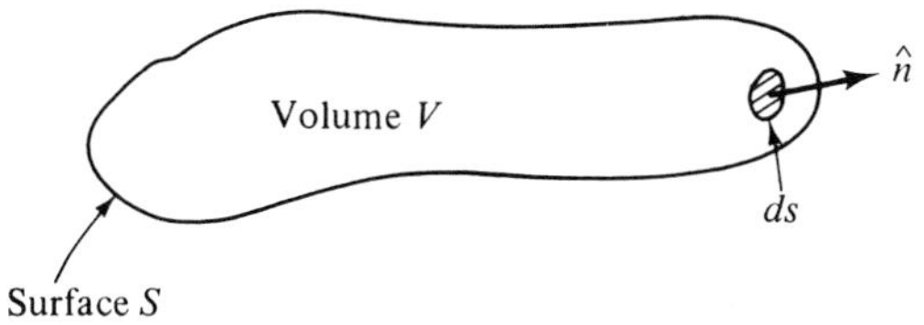

Figure 1-8

always differentiates the entire integrand in the volume integral:

$$\int_S \hat{n} \times \bar{a}\, dS = \int_V \bar{\nabla} \times \bar{a}\, dV, \tag{1-10.2}$$

$$\int_S \hat{n}\varphi\, dS = \int_V \bar{\nabla}\varphi\, dV, \tag{1-10.3}$$

$$\int_S (\hat{n} \cdot \bar{\nabla})\varphi\, dS = \int_V \nabla^2\varphi\, dV, \tag{1-10.4}$$

$$\int_S (\hat{n} \cdot \bar{\nabla})\bar{a}\, dS = \int_V \nabla^2\bar{a}\, dV, \tag{1-10.5}$$

$$\int_S n_i T_{ij}\, dS = \int_V \nabla_i T_{ij}\, dV, \tag{1-10.6}$$

$$\int_S n_j T_{ij}\, dS = \int_V \nabla_j T_{ij}\, dV. \tag{1-10.7}$$

If the vector $\bar{a}$ is a plane vector and $\hat{\nu}$ is a unit vector normal to the line element dL in the plane (Figure 1-9), a two-dimensional counterpart of this theorem may be written:

$$\int_S \bar{\nabla}\varphi\, dS = \oint \hat{\nu}\varphi\, dL, \tag{1-10.8}$$

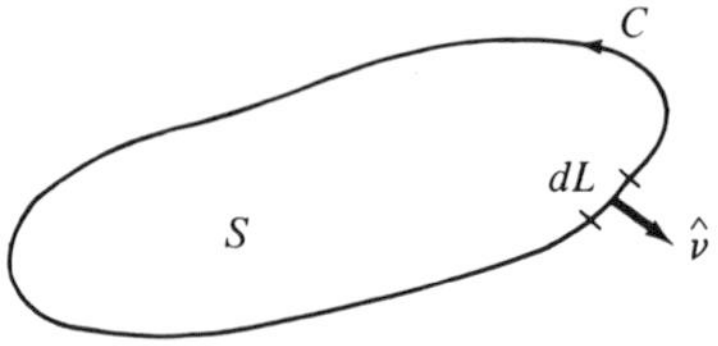

Figure 1-9

$$\int_S \bar{\nabla} \cdot \bar{a}\, dS = \oint \hat{\nu} \cdot \bar{a}\, dL, \tag{1-10.9}$$

$$\int_S \bar{\nabla} \times \bar{a}\, dS = \oint \hat{\nu} \times \bar{a}\, dL. \tag{1-10.10}$$

Another useful theorem is Stoke's theorem, which relates an integral across a surface to an integral around the boundary. In Figure 1-10, let $\hat{t}$

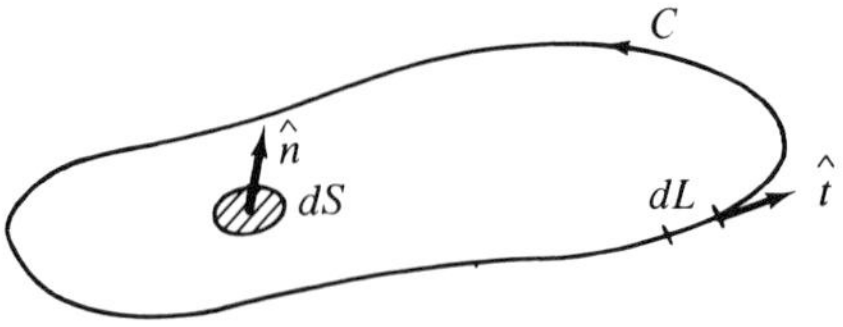

Figure 1-10

be a unit tangential vector to the line element dL. Stoke's theorem states that

$$\int_S (\bar{\nabla} \times \bar{a}) \cdot \hat{n}\, dS = \oint \bar{a} \cdot \hat{t}\, dL. \tag{1-10.11}$$

The integral around the contour is the "circulation" of the vector $\bar{a}$ around the path C.

11 Classification of vector fields

Some vector fields may be classified by operations associated with the del operator. Two such classifications are solenoidal and irrotational. Vector fields that may be classified in one of these manners may be written in simpler mathematical terms.

SOLENOIDAL FIELDS

If a vector field is such that its divergence is everywhere zero, the field is termed solenoidal:

$$\bar{\nabla} \cdot \bar{a} = 0. \tag{1-11.1}$$

If a vector field is solenoidal within any closed volume, the net efflux of that vector quantity through the enclosing surface is zero. This may easily be seen by use of the divergence theorem:

$$\int_S \bar{\nabla} \cdot \bar{a}\, dV = \int_V \hat{n} \cdot \bar{a}\, dS = 0. \tag{1-11.2}$$

It may be shown that a solenoidal field can always be derived from another vector field $\bar{\alpha}$ as follows:

$$\bar{a} = \bar{\nabla} \times \bar{\alpha}, \qquad \bar{\nabla} \cdot \bar{\alpha} = 0, \tag{1-11.3}$$

where $\bar{\alpha}$ may also be chosen to be solenoidal.* The fact that equation 1-11.3 implies equation 1-11.1 is easily shown, but the converse proof is quite difficult.

IRROTATIONAL FIELDS

A field is termed irrotational if its curl is everywhere zero:

$$\bar{\nabla} \times \bar{a} = 0. \tag{1-11.4}$$

An irrotational field may be obtained from a scalar function φ, called the potential of $\bar{a}$;

$$\bar{a} = \bar{\nabla}\varphi. \tag{1-11.5}$$

Again it is easily seen that if $\bar{a}$ is derivable from a potential, the field is irrotational, but demonstrating the converse requires more effort. The irrotational field is frequently encountered in dynamic force fields that are conservative. The gravitational force field is one such example.

HELMHOLTZ'S THEOREM

An extremely useful theorem in mathematics will be given here without proof. The mathematical details of the proof are beyond the scope of this book and may be found in most vector field texts.†

Any finite, continuous vector field that vanishes at infinity may be written as the sum of a solenoidal and an irrotational field:

$$\bar{a} = \bar{a}_S + \bar{a}_{IR}. \tag{1-11.6}$$

This is called Helmholtz's theorem. The vector $\bar{a}$ may be expressed in terms of the appropriate scalar and vector potentials:

$$\bar{a} = \bar{\nabla} \times \bar{\alpha} + \bar{\nabla}\varphi, \tag{1-11.7}$$

where $\bar{\nabla} \cdot \bar{\alpha} = 0$. When a vector field is both solenoidal and irrotational, it may be considered derivable from either a vector potential or a scalar

* Aris, *Vectors, Tensors, and the Basic Equations of Fluid Mechanics.*

† P. M. Morse and H. Feshbach, *Methods of Theoretical Physics*, Part 1, McGraw-Hill Book Company, New York, 1953, pp. 53–54.

potential. If the scalar potential is used, $\bar{\nabla} \times \bar{a} = 0$, we obtain

$$\begin{aligned} \bar{a} &= \bar{\nabla}\varphi, \\ \bar{\nabla} \cdot \bar{a} &= \nabla^2 \varphi = 0. \end{aligned} \tag{1-11.8}$$

The scalar potential is a harmonic function.

As an alternative method, if we first consider the field to be solenoidal, then the vector potential may be used:

$$\begin{aligned} \bar{a} &= \bar{\nabla} \times \bar{\alpha}, \\ \bar{\nabla} \times \bar{a} &= \bar{\nabla} \times (\bar{\nabla} \times \bar{\alpha}) = 0. \end{aligned} \tag{1-11.9}$$

The following vector identity holds for all continuous vector fields:

$$\bar{\nabla} \times (\bar{\nabla} \times \bar{\alpha}) = \bar{\nabla}(\bar{\nabla} \cdot \bar{\alpha}) - \nabla^2 \bar{\alpha}. \tag{1-11.10}$$

Since $\bar{\alpha}$ was chosen to be solenoidal, equation 1.11.9 may be written

$$\bar{\nabla} \times (\bar{\nabla} \times \bar{\alpha}) = -\nabla^2 \bar{\alpha} = 0. \tag{1-11.11}$$

The $\bar{\alpha}$ is a harmonic vector, and in Cartesian coordinates each component is a harmonic function. If the vector field is given in two dimensions, x and y, and is solenoidal and irrotational, only the z component of $\bar{\alpha}$ is required:

$$\bar{\alpha} = \hat{i}_{(z)} \alpha_z(x, y). \tag{1-11.12}$$

The components of $\bar{a}$ become

$$a_x = \frac{\partial \alpha_z}{\partial y}, \qquad a_y = -\frac{\partial \alpha_z}{\partial x}. \tag{1-11.13}$$

One may relate α_z to φ in this case by equating the two different representations for the x and y components of $\bar{a}$:

$$a_x = \frac{\partial \varphi}{\partial x} = \frac{\partial \alpha_z}{\partial y}, \qquad a_y = \frac{\partial \varphi}{\partial y} = -\frac{\partial \alpha_z}{\partial x}. \tag{1-11.14}$$

Since both φ and α_z are harmonic and satisfy the relationships shown in equation 1.11-14, they are called conjugate, and the methods of complex variables may be utilized. Examples of this technique will be shown later.

PROBLEMS

1.1. Construct the arrays a_{ij} for transforming from x_i to x'_j, where axes x'_j are obtained by successively rotating the x_i system by right-hand-rule motions as follows:

(a) 180° about x_3.

(b) 45° about x_1.

(c) 90° about x_1 followed by a 90° rotation about x_2.

1.2. A double transformation of coordinate systems is shown in Figure P. 1-2 in the order a_{ij} (rotation of an angle θ about the 3 axis) and then a'_{ij} (rotation of an angle β about the new 2 axis). Show that the final coordinate system obtained by the reverse order of transformations is not the same unless the rotations may be considered as infinitesimal rotations. (*Hint:* Consider the components of a vector in the two final coordinate systems.)

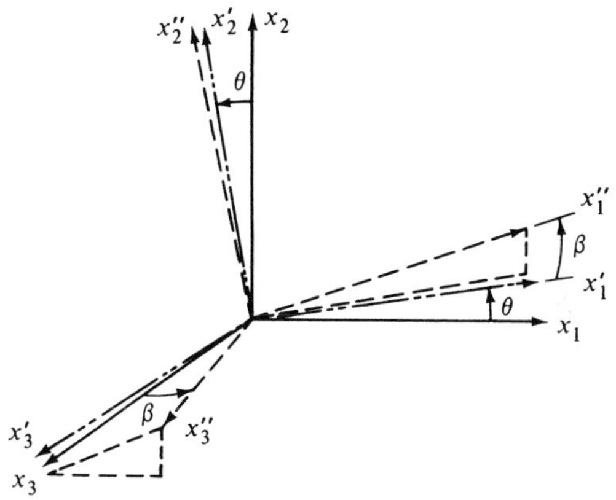

Figure P. 1-2

1.3. Show by enumerating typical cases that

$$e_{ijk}e_{klm} = \delta_{il}\delta_{jm} - \delta_{im}\delta_{jl}.$$

1.4. Show that the condition for $\bar{a}$, $\bar{b}$, and $\bar{c}$ to be coplanar can be written $e_{ijk}a_i b_j c_k = 0$.

1.5. Prove in component notation that

$$(\bar{a} \times \bar{b})\cdot(\bar{a} \times \bar{b}) + (\bar{a}\cdot\bar{b})^2 = |\bar{a}|^2\,|\bar{b}|^2.$$

1.6. Write the following tensor as the sum of a symmetric and an antisymmetric tensor and find the dual vector of the antisymmetric part.

$$\begin{pmatrix} 3 & 6 & 5 \\ -2 & 0 & -2 \\ 1 & 4 & 4 \end{pmatrix}$$

1.7 For the following tensors,
(a) find the eigenvalues.
(b) find the eigenvectors.
(c) show that the eigenvectors are orthogonal.
(d) find the transformation matrix required to go from the original coordinate system to the principal coordinate system.

$$\begin{pmatrix} 1 & 0 & 0 \\ 0 & 3 & -1 \\ 0 & -1 & 3 \end{pmatrix} \qquad \begin{pmatrix} 2 & -1 & 1 \\ -1 & 0 & 1 \\ 1 & 1 & 2 \end{pmatrix}$$

1.8 Three equal mass points are located at coordinates $(a, 0, 0)$, $(0, a, 2a)$, and

(0, 2*a*, *a*). Find the principal moments of inertia about the origin and a set of principal axes. The moment-of-inertia tensor is defined as

$$I_{ij} = \iiint \rho[\delta_{ij} r^2 - x_i x_j]\, dV,$$

where $r^2 = x_k x_k$ and each mass point may be assumed to have mass

$$m = \iiint \rho\, dV.$$

1.9 If symmetric tensors A_{ij} and B_{pq} have a principal direction in common, show that this is also a principal direction of the tensor $(A_{ik}B_{kj} + A_{jk}B_{ki})$.

1.10 If T_{ijpq} is an isotropic fourth-order tensor, show that the tensors A_{ij} and $T_{ijpq}A_{pq}$ have the same principal axes.

1.11 Prove the following vector identities, using indicial notation:
(a) $(\bar{a} \times \bar{b}) \times (\bar{c} \times \bar{d}) = [\bar{c} \cdot (\bar{d} \times \bar{a})]\bar{b} - [\bar{c} \cdot (\bar{d} \times \bar{b})]\bar{a}$.
(b) $[\bar{a} \cdot (\bar{b} \times \bar{c})]\bar{d} = [\bar{d} \cdot (\bar{b} \times \bar{c})]\bar{a} + [\bar{a} \cdot (\bar{d} \times \bar{c})]\bar{b} + [\bar{a} \cdot (\bar{b} \times \bar{d})]\bar{c}$.

1.12 Prove the following identities by component notation ($\bar{r}$ is the position vector having components x_i):
(a) $\bar{\nabla} \cdot \bar{r} = 3$.
(b) $\bar{\nabla} \times \bar{r} = 0$.
(c) $\bar{\nabla} \cdot (\bar{\nabla} \times \bar{A}) = 0$.
(d) $\bar{\nabla} \cdot (\bar{A} \times \bar{B}) = (\bar{\nabla} \times \bar{A}) \cdot \bar{B} - \bar{A} \cdot (\bar{\nabla} \times \bar{B})$.
(e) $\bar{\nabla} \times (\bar{v} \times \bar{w}) = \bar{v}(\bar{\nabla} \cdot \bar{w}) + (\bar{w} \cdot \bar{\nabla})\bar{v} - \bar{w}(\bar{\nabla} \cdot \bar{v}) - (\bar{v} \cdot \bar{\nabla})\bar{w}$.

1.13 If $\bar{u}$ and $\bar{v}$ are vectors, prove that $\bar{\bar{S}}$ is a second-order tensor if it is defined by the equation

$$\bar{u} = \bar{\bar{S}} \cdot \bar{v}.$$

1.14 If T_{ij} is an antisymmetric tensor in one coordinate system, prove that it remains antisymmetric in all coordinate systems.

1.15 By use of dyadic notation, derive the following in spherical coordinates:
(a) $\bar{\nabla} \cdot \bar{v}$.
(b) $\bar{\nabla} \times \bar{v}$.
(c) $\bar{\nabla} \cdot \bar{\bar{T}}$.

CHAPTER

2

KINEMATICS

1 Introduction

A continuum or continuous medium is one without gaps or voids. This is a mathematical model which deals with gross properties that may be thought of as an average of discrete properties over a large number of particles. In contrast, the molecular or particle approach is concerned with the properties of individual particles and their interaction. Although the particle approach is closer to physical reality, it is limited mathematically in its applicability. When a complete understanding of material science is obtained, perhaps particle properties will be used to predict quantitatively the gross phenomenological properties. Until such time we must content ourselves by studying both ends of the spectrum.

In the study of a continuum it is necessary to develop methods of measuring deformations apart from rigid-body motions. Since the material is continuously distributed, we expect these deformations to be measured by tensor fields. The establishment of a coordinate system in which to measure these deformations is our first consideration. It is the choice of reference coordinates that creates much of the mathematical difference between fluid and solid mechanics.

2 Spatial and material coordinates

Consider a body occupying a certain area of space at time $t = 0$, as shown in Figure 2-1. The dashed outline indicates the position of the deformed body at $t = t$. Consider a point P in the undeformed body, which has

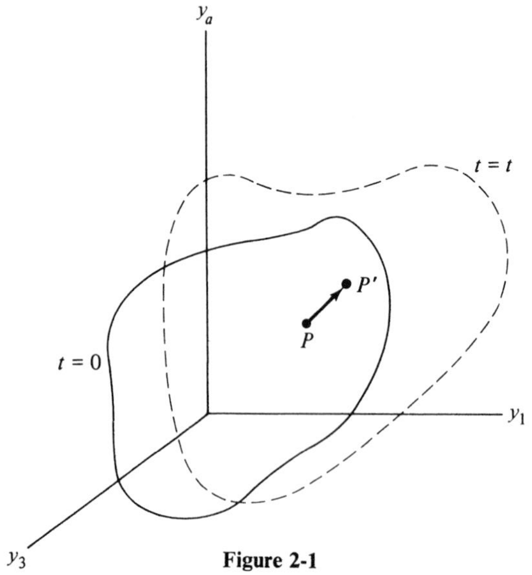

Figure 2-1

coordinates y_i. We may designate these initial coordinates as x_i. In this manner we may label the material point by its initial coordinates. This material point occupies the spatial point P' at time t, having different spatial coordinates. In discussing particle dynamics we generally define the position of a particle by giving its position vector as a function of the material coordinates (that is, which particle) and time. We can then locate a particle with initial coordinates x_i by its new spatial coordinates y_i and time. In this manner we may express the spatial coordinates as a function of the initial or material coordinates and time:

$$y_i = y_i(x_j, t). \tag{2-2.1}$$

The x_i coordinates serve only as a particle-labelling system in this case. Equation 2-2.1 gives the complete kinematic description of the continuum.

We may also describe this motion by looking at a point y_i in space and noting which material point occupies this point at any instant of time:

$$x_i = x_i(y_j, t). \tag{2-2.2}$$

We may use either of the two coordinate systems to describe properties of the continuum. Consider a particular scalar property T, which might be the temperature, mass density, etc. We may express this as a function of the spatial coordinates and time as

$$T = T(y_i, t). \tag{2-2.3}$$

If a point in space is given, we may find the property of the continuum occupying that point at that time. To determine which piece of material we are describing we must also use equation 2-2.2. This type of formulation is called a *spatial description*, or sometimes a *Eulerian description.*

We may also express this property directly in terms of the material coordinates x_i. This would give us a description of the property over this material at any time, and equation 2-2.1 could be used in conjunction with this expression to locate a particular particle:

$$T = T(x_i, t). \tag{2-2.4}$$

This formulation is called a *material description* or a *Lagrangian description.* The names Eulerian and Lagrangian for the spatial and material coordinates, respectively, are historically incorrect, as discussed by C. Truesdell.*

3 Velocity and material time derivative

The velocity of a particle occupying the coordinates y_i at some time t may be found by taking the time derivative and holding the material coordinates constant:

$$v_i = \left.\frac{dy_i}{dt}\right|_{x_j \text{ constant}} = \frac{\partial y_i}{\partial t} \tag{2-3.1}$$

Consider now a scalar property T, which is expressed as a function of the material coordinates x_i and time t:

$$T = T(x_i, t). \tag{2-3.2}$$

If we desire to know how this property of a particular particle changes with time, we must take the material time derivative of T. That is, we must hold x_i constant and differentiate T with respect to t:

$$\frac{\partial T}{\partial t} = \frac{\partial T}{\partial t}(x_i, t). \tag{2-3.3}$$

* "Principles of Classical Mechanics and Field Theory," *Handbuch der Physik*, Vol. III, Springer-Verlag, Berlin, 1960, p. 327.

If T is expressed in terms of spatial coordinates and time, we may still take the material time derivative:

$$\frac{dT}{dt}(y_i, t) = \frac{\partial T}{\partial t} + \frac{\partial T}{\partial y_j}\frac{\partial y_j}{\partial t}. \qquad (2\text{-}3.4)$$

Using equation 2-3.1 this may be written as

$$\frac{dT}{dt} = \frac{\partial T}{\partial t} + v_j \frac{\partial T}{\partial y_j}. \qquad (2\text{-}3.5)$$

The first term gives the rate of change of T due to an increase of this property at the spatial point y_i, and the second yields a "convective" increase due to motion of the particle in a field that varies in space. For example, we may nonuniformly heat a material and the temperature may increase due to the heating of the point it presently occupies ($\partial T/\partial t$). Even if the temperature of this point does not change, the particle may gain heat at a rate $v_j(\partial T/\partial y_j)$ by moving to a warmer point in space.

The material time derivative may be written in operator form as

$$\frac{d}{dt} = \frac{\partial}{\partial t} + \bar{v}\cdot\vec{\nabla}. \qquad (2\text{-}3.6)$$

This operator may be applied to any tensorial quantity expressed in the spatial reference system to obtain the rate of change of this quantity for the particle in question.

The acceleration of a particle is

$$\frac{d\bar{v}(y, t)}{dt} = \frac{\partial \bar{v}}{\partial t} + (\bar{v}\cdot\vec{\nabla})\bar{v}. \qquad (2\text{-}3.7)$$

If $\bar{v}$ has been expressed in material coordinates, the material time derivative would be the simpler expression:

$$\frac{d\bar{v}(x_i, t)}{dt} = \frac{\partial \bar{v}}{\partial t}. \qquad (2\text{-}3.8)$$

The boundary condition on a particular continuum may dictate whether a material or spatial description will be used. In a fluid problem the boundaries are fixed in space and a spatial description is usually used. In the analysis of an elastic solid the original configuration is used as a reference system, and conditions are specified on material boundaries.

4 Volume elements

At any time t, equation 2-2.1 or 2-2.2 gives the relation between the spatial and material coordinate systems. If $\hat{i}_{(j)}$ denote the unit vectors in the spatial system, we may express vectors along coordinate lines, locally, in terms of the spatial coordinates of the current (deformed) configuration of the material, as

$$\hat{I}_{(j)} = \frac{\partial y_i}{\partial x_j} \hat{\imath}_{(i)}. \tag{2-4.1}$$

A directed coordinate arc length may be written as $\hat{I}_1 dx_1$, $\hat{I}_2 dx_2$, or $\hat{I}_3 dx_3$. The triple scalar product gives the volume in the material coordinate system:

$$\begin{aligned} dV &= (\hat{I}_1 \cdot \hat{I}_2 \times \hat{I}_3)\, dx_1\, dx_2\, dx_3 \\ &= e_{mnp} \frac{\partial y_m}{\partial x_1} \frac{\partial y_n}{\partial x_2} \frac{\partial y_p}{\partial x_3}\, dx_1\, dx_2\, dx_3. \end{aligned} \tag{2-4.2}$$

If we introduce the Jacobian J defined as

$$J = e_{mnp} \frac{\partial y_m}{\partial x_1} \frac{\partial y_n}{\partial x_2} \frac{\partial y_p}{\partial x_3}, \tag{2-4.3}$$

and interpret the initial undeformed volume to be

$$dV_0 = dx_1\, dx_2\, dx_3, \tag{2-4.4}$$

we may write the ratio of volumes as

$$\frac{dV}{dV_0} = J. \tag{2-4.5}$$

The Jacobian is usually written as a determinant, as follows:

$$J = \left| \frac{\partial(y_1, y_2, y_3)}{\partial(x_1, x_2, x_3)} \right| = \begin{vmatrix} \dfrac{\partial y_1}{\partial x_1} & \dfrac{\partial y_1}{\partial x_2} & \dfrac{\partial y_1}{\partial x_2} \\ \dfrac{\partial y_2}{\partial x_1} & \dfrac{\partial y_2}{\partial x_2} & \dfrac{\partial y_2}{\partial x_3} \\ \dfrac{\partial y_3}{\partial x_1} & \dfrac{\partial y_3}{\partial x_2} & \dfrac{\partial y_3}{\partial x_3} \end{vmatrix}. \tag{2-4.6}$$

If we consider the integral of some quantity, f, over the deformed volume, we may write the following relation:

$$\int_V f(y_i)\, dV(y_i) = \int_{V^*} F(x_i) J dV_0(x_i), \tag{2-4.7}$$

where V^* represents the material volume corresponding to V and $F(x_i) = f[y_j(x_i)]$. The Jacobian represents an important mathematical and physical quantity, and we must examine it in more detail.

The material time derivative of the Jacobian may be calculated if we note that this determinant may also be written

$$J = e_{mnp} \frac{\partial y_1}{\partial x_m} \frac{\partial y_2}{\partial x_n} \frac{\partial y_3}{\partial x_p}. \tag{2-4.8}$$

The material time derivative of a typical term within this determinant becomes

$$\frac{d}{dt}\left(\frac{\partial y_i}{\partial x_j}\right) = \frac{\partial}{\partial x_j} \frac{dy_i}{dt} = \frac{\partial v_i}{\partial x_j} = \frac{\partial v_j}{\partial y_k} \frac{\partial y_k}{\partial x_j}. \tag{2-4.9}$$

The derivative of a three-by-three determinant can be regarded as the sum of three determinants in which only one row of each is differentiated. This may be seen by differentiating 2-4.8 as a product:

$$\frac{dJ}{dt} = e_{mnp}\left[\frac{\partial v_1}{\partial y_k}\frac{\partial y_k}{\partial x_m}\frac{\partial y_2}{\partial x_n}\frac{\partial y_3}{\partial x_p} + \frac{\partial v_2}{\partial y_k}\frac{\partial y_1}{\partial x_m}\frac{\partial y_k}{\partial x_n}\frac{\partial y_3}{\partial x_p} + \frac{\partial v_3}{\partial y_k}\frac{\partial y_1}{\partial x_m}\frac{\partial y_2}{\partial x_n}\frac{\partial y_k}{\partial x_p}\right]. \tag{2-4.10}$$

Noting that in each of the three sums on k, only one term leads to a nonzero product, Equation 2-4.10 may be written

$$\frac{dJ}{dt} = e_{mnp}\left[\frac{\partial v_1}{\partial y_1} + \frac{\partial v_2}{\partial y_2} + \frac{\partial v_3}{\partial y_3}\right]\frac{\partial y_1}{\partial x_m}\frac{\partial y_2}{\partial x_n}\frac{\partial y_3}{\partial x_p}. \tag{2-4.11}$$

This may be written as

$$\frac{dJ}{dt} = \frac{\partial v_k}{\partial y_k}J = (\bar{\nabla}\cdot\bar{v})J. \tag{2-4.12}$$

5 Reynold's transport theorem

The properties of the Jacobian may be used to derive a transport theorem useful in finding a material time derivative of a spatial volume. Consider a function defined as

$$P = \int_V \mathcal{P}(y_i, t)\, dV. \tag{2-5.1}$$

If the material time derivative of P is desired, we may formally write

$$\frac{dP}{dt} = \frac{d}{dt}\int_V \mathcal{P}(y_i, t)dV. \tag{2-5.2}$$

The differential operator may be taken inside the integral because, in general, the spatial volume is changing in the material sense. This volume may be expressed in the material sense by using equation 2-4.7:

$$\frac{dP}{dt} = \frac{d}{dt}\int_{V^*} \mathcal{P}(y_i[x_j, t])JdV_0(x_j). \tag{2-5.3}$$

This may now be written in a more useful form by differentiating the integrand and using equation 2-4.12:

$$\frac{dP}{dt} = \int_{V^*}\left[\frac{d\mathcal{P}}{dt}J + \mathcal{P}(\bar{\nabla}\cdot\bar{v})J\right]dV_0(x_j). \tag{2-5.4}$$

We may now reconvert this to an expression in terms of the spatial volume:

$$\frac{dP}{dt} = \int_V\left[\frac{d\mathcal{P}}{dt} + (\bar{\nabla}\cdot\bar{v})\mathcal{P}\right]dV. \tag{2-5.5}$$

This is called Reynold's transport theorem after O. Reynolds, who observed it in 1903. Equation 2-5.5 may also be written in an alternative form by use of the definition of the material time derivative:

$$\frac{dP}{dt} = \int_V \left[\frac{\partial \mathcal{P}}{\partial t} + \bar{\nabla}\cdot(\bar{v}\mathcal{P})\right] dV. \tag{2-5.6}$$

6 Displacement vector

The displacement of a particle can be written as the difference between its spatial coordinates at some time t and its initial coordinates:

$$u_i = y_i - x_i. \tag{2-6.1}$$

This displacement can be defined in either material or spatial coordinates by use of equation 2-2.1 or 2-2.2.

The displacement vector changes as one moves from point to point in the continuum. We may express the displacement of the material in the neighborhood of a point in terms of the displacement at that point by use of a Taylor series. Consider two material points P and Q, at time $= 0$, separated by the vector $\overline{dx}$, as shown in Figure 2-2. At time t the two points have moved

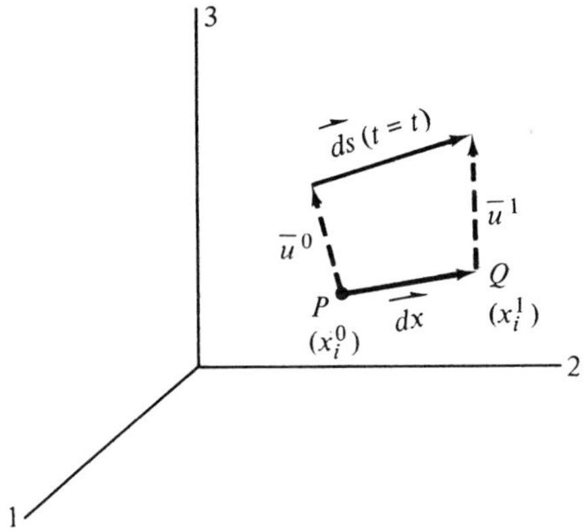

Figure 2-2

through the displacements $\bar{u}^0$ or $\bar{u}^1$. We may express the components of $\bar{u}^1$ in a Taylor series as

$$u_i^1 = u_i^0 + \frac{\partial u_i}{\partial x_j}\bigg|^0 (x_j^1 - x_j^0) + 0(x_j^1 - x_j^0)^2. \tag{2-6.2}$$

Neglecting higher-order terms, this may be written as

$$u_i^1 = u_i^0 + \frac{\partial u_i}{\partial x_j}\bigg|^0 dx_j \Longrightarrow \bar{u}^1 = \bar{u}^0 + (\bar{u}\overleftarrow{\nabla})\cdot\overline{dx}. \tag{2-6.3}$$

The second term is the product of a second-order tensor (the postgradient of $\bar{u}$) and a vector. The second-order tensor may be written as the sum of a symmetric and an antisymmetric tensor:

$$u_i^1 = u_i^0 + \frac{1}{2}\left(\frac{\partial u_i}{\partial x_j} + \frac{\partial u_j}{\partial x_i}\right)_0 dx_j + \frac{1}{2}\left(\frac{\partial u_i}{\partial x_j} - \frac{\partial u_j}{\partial x_i}\right)_0 dx_j. \tag{2-6.4}$$

Let

$$\epsilon_{ij} = \frac{1}{2}\left(\frac{\partial u_i}{\partial x_j} + \frac{\partial u_j}{\partial x_i}\right)_0 \tag{2-6.5}$$

and $\bar{\omega}$ be the dual vector of the antisymmetric tensor. Equation 2-6.4 becomes

$$u_i^1 = u_i^0 + \epsilon_{ij}\,dx_j + e_{ijk}\omega_j\,dx_k. \tag{2-6.6}$$

This may be written in symbolic notation as

$$\bar{u}^1 = \bar{u}_0 + \bar{\bar{\epsilon}}\cdot\overline{dx} + \bar{\omega}\times\overline{dx}. \tag{2-6.7}$$

The first term on the right side represents the rigid-body translation in the neighborhood of the point P, and the last term represents a rigid-body rotation, because it cannot produce extension of the vector $\overline{dx}$. $\bar{\bar{\epsilon}}$ is defined as the Lagrangian strain tensor for the case of small strains. The components of $\bar{\bar{\epsilon}}$ are

$$\bar{\bar{\epsilon}} = \begin{pmatrix} \dfrac{\partial u_1}{\partial x_1} & \dfrac{1}{2}\left(\dfrac{\partial u_1}{\partial x_2} + \dfrac{\partial u_2}{\partial x_1}\right) & \dfrac{1}{2}\left(\dfrac{\partial u_1}{\partial x_3} + \dfrac{\partial u_3}{\partial x_1}\right) \\ & \dfrac{\partial u_2}{\partial x_2} & \dfrac{1}{2}\left(\dfrac{\partial u_2}{\partial x_3} + \dfrac{\partial u_3}{\partial x_2}\right) \\ \text{Sym.} & & \dfrac{\partial u_3}{\partial x_3} \end{pmatrix}. \tag{2-6.8}$$

Symbolically, this may be defined as one half the sum of the pregradient and the postgradient of the displacement vector:

$$\bar{\bar{\epsilon}} = \tfrac{1}{2}(\bar{u}\overleftarrow{\nabla} + \overrightarrow{\nabla}\bar{u}). \tag{2-6.9}$$

The components of this tensor may be related to the elementary engineering definitions of strain by examining an element in two dimensions (Figure 2-3). In elementary treatments the normal strains are defined as the deformation difference along a length divided by the original length. In this manner, the normal strains become

$$\begin{aligned} \epsilon_{11} &= \frac{(\partial u_1/\partial x_1)dx_1}{dx_1} = \frac{\partial u_1}{\partial x_1} \\ \epsilon_{22} &= \frac{(\partial u_2/\partial x_2)dx_2}{dx_2} = \frac{\partial u_2}{\partial x_2}. \end{aligned} \tag{2-6.10}$$

The shear strains may be measured by the differences in the u_1 displacement as one moves in the x_2 direction, divided by the original length $(\partial u_1/\partial x_2)$ or by the difference in the u_2 displacement divided by the x_1 distance $(\partial u_2/\partial x_1)$.

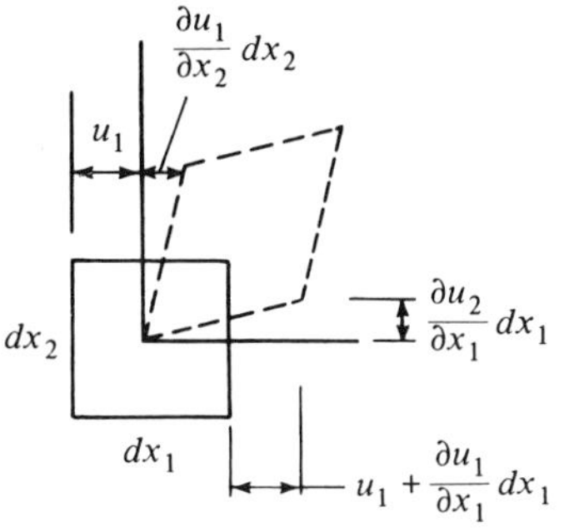

Figure 2-3

The difficulty with these individual definitions is that only the sum $(\partial u_1/\partial x_2 + \partial u_2/\partial x_1)$ is independent of rigid-body rotations of the element so that the sum must be used. If these quantities are small, the sum is the equivalent to the change in the right angle between the dx_1 and dx_2. In order to generate a symmetric tensor that obeys the transformation law, the shear strains are defined to be equal, so one half of the angular change is given to each strain:*

$$\epsilon_{12} = \epsilon_{21} = \frac{1}{2}\left(\frac{\partial u_1}{\partial x_2} + \frac{\partial u_2}{\partial x_1}\right). \tag{2-6.11}$$

Another method of examining strains is to consider an infinitesimal arc ds of a curve C before and after deformation. Let x_i designate the coordinates of the end of a curve before deformation and y_i the coordinates after deformation (Figure 2-4). The squared length of the curve before and after deformation is given by

$$\begin{aligned} dS_0^2 &= dx_2^1 + dx_2^2 + dx_3^2 = dx_i\, dx_i, \\ ds^2 &= dy_i\, dy_i. \end{aligned} \tag{2-6.12}$$

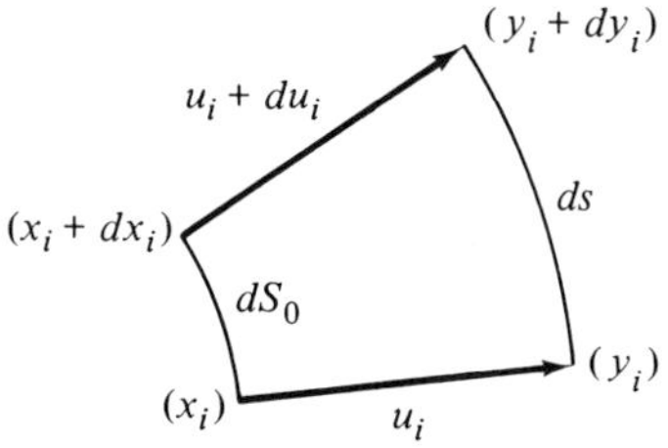

Figure 2-4

* Some books use an "engineering shear strain" defined by

$$\gamma_{12} = \frac{\partial u_1}{\partial x_2} + \frac{\partial u_2}{\partial x_1}.$$

This leads to the necessity of including only one strain component in energy considerations and similar expressions.

We shall choose as a measure of deformation the difference in these lengths squared:

$$ds^2 - dS_0^2 = dy_i dy_i - dx_i dx_i. \tag{2-6.13}$$

If we seek a formulation in material coordinates (Lagrangian development), we may express the spatial coordinates in terms of displacements and the material coordinates:

$$y_i = u_i + x_i. \tag{2-6.14}$$

The differential dy_i may be written

$$dy_i = \frac{\partial y_i}{\partial x_j} dx_j. \tag{2-6.15}$$

Substituting equation 2-6.14 into 2-6.15 yields

$$dy_i = \frac{\partial u_i}{\partial x_j} dx_j + \delta_{ij}\, dx_j. \tag{2-6.16}$$

Equation 2-6.13 becomes

$$ds^2 - dS_0^2 = \left[\frac{\partial y_i}{\partial x_j}\frac{\partial y_i}{\partial x_k} - \delta_{jk}\right] dx_j\, dx_k. \tag{2-6.17}$$

The first term in the bracket is called the Lagrangian deformation tensor. Equation 2-6.17 may be written

$$ds^2 - dS_0^2 = \left[\left(\frac{\partial u_i}{\partial x_j} + \delta_{ij}\right)\left(\frac{\partial u_i}{\partial x_k} + \delta_{ik}\right) - \delta_{jk}\right] dx_j\, dx_k \tag{2-6.18}$$

$$= \left[\frac{\partial u_k}{\partial x_j} + \frac{\partial u_j}{\partial x_k} + \frac{\partial u_i}{\partial x_j}\frac{\partial u_i}{\partial x_k}\right] dx_j\, dx_k. \tag{2-6.19}$$

Let the Lagrangian strain tensor be defined as

$$E_{jk} = \frac{1}{2}\left\{\frac{\partial u_j}{\partial x_k} + \frac{\partial u_k}{\partial x_j} + \frac{\partial u_i}{\partial x_j}\frac{\partial u_i}{\partial x_k}\right\}. \tag{2-6.20}$$

If the deformation gradients are small, the product $(\partial u_i/\partial x_j)(\partial u_i/\partial x_k)$ is negligible and from equation 2-6.5

$$\epsilon_{ij} \approx E_{ij}.$$

The deformation may also be written in terms of spatial coordinates:

$$dx_i = \frac{\partial x_i}{\partial y_j} dy_j. \tag{2-6.21}$$

The deformation measurement becomes

$$ds^2 - dS_0^2 = \left[\delta_{jk} - \frac{\partial x_i}{\partial y_j}\frac{\partial x_i}{\partial y_k}\right] dy_j\, dy_k. \tag{2-6.22}$$

The second term without sign is called the Eulerian deformation tensor. Equation 2-6.22 may be written as

$$ds^2 - dS_0^2 = \left[\delta_{jk} - \left(\delta_{ij} - \frac{\partial u_i}{\partial y_j}\right)\left(\delta_{ik} - \frac{\partial u_i}{\partial y_k}\right)\right] dy_j\, dy_k \tag{2-6.23}$$

$$= \left[\frac{\partial u_j}{\partial y_k} + \frac{\partial u_k}{\partial y_j} - \frac{\partial u_i}{\partial y_j}\frac{\partial u_i}{\partial y_k}\right] dy_j\, dy_k. \tag{2-6.24}$$

A Eulerian strain tensor may be developed in a similar manner to that used for defining the Lagrangian strain tensor, with the result that

$$E_{jk}^* = \frac{1}{2}\left[\frac{\partial u_j}{\partial y_k} + \frac{\partial u_k}{\partial y_j} - \frac{\partial u_i}{\partial y_j}\frac{\partial u_i}{\partial y_k}\right]. \tag{2-6.25}$$

For infinitesimal strains the last term in the bracket may be neglected, and one frequently treats the infinitesimal Lagrangian and Eulerian strains as equivalent.

The change in volume may be obtained from equation 2-4.5 in terms of displacement gradients:

$$\frac{dV}{dV_0} = J = \left|\frac{\partial y_i}{\partial x_j}\right| = \left|\frac{\partial u_i}{\partial x_j} + \delta_{ij}\right|. \tag{2-6.26}$$

This may be written as

$$\frac{dV}{dV_0} = 1 + \Delta, \tag{2-6.27}$$

where Δ is the change in volume per unit volume and is called the cubic dilatation. For small strains only the linear terms are retained, so that

$$\Delta = \frac{\partial u_i}{\partial x_i} = \epsilon_{ii}. \tag{2-6.28}$$

The first invariant of the strain tensor (see equation 1-7.5) is therefore a measure of the volume change per unit volume.

7 Compatibility equations

If elasticity problems are formulated with the displacements as the unknown quantities, it is clear that once these have been obtained strains may be calculated by proper differentiation of these quantities. However, in many problems stresses or strains are taken to be the basic unknowns. Examination of the definition of strain as given by equation 2-6.8 indicates that the six independent strains must be interrelated to serve as an integrability condition leading to a compatible set of only three independent displacements. These interrelations are known as the equations of compatibility, or Saint-Venant's compatibility equations.

We could proceed to obtain the compatibility equations by examining the six definitions of strain in terms of the three displacement components and eliminating these displacements. This in general would produce three independent relations between the six components of the strain tensor. In this case we obtain six relations that are nonindependent. Formally, we may obtain these relations by taking the precurl and postcurl of equation 2-6.9. Noting that the curl of the gradient of any function is zero, we obtain

$$\vec{\nabla} \times \bar{\bar{\epsilon}} \times \overleftarrow{\nabla} = \tfrac{1}{2}\vec{\nabla} \times (\bar{u}\overleftarrow{\nabla} + \vec{\nabla}\bar{u}) \times \overleftarrow{\nabla} = 0. \qquad (2\text{-}7.1)$$

The tensor produced by the cross-product operations on a symmetric tensor is symmetric, and the six components give the Saint-Venant compatibility conditions. In Cartesian component notation, this may be written

$$e_{ilm}e_{jpq}\frac{\partial^2 \epsilon_{mp}}{\partial x_l \partial x_q} = 0. \qquad (2\text{-}7.2)$$

Using equation 1-5.13 gives

$$\delta_{ij}\frac{\partial^2 \epsilon_{ql}}{\partial x_l\,\partial x_q} + \frac{\partial^2 \epsilon_{ij}}{\partial x_l\,\partial x_l} + \frac{\partial^2 \epsilon_{ll}}{\partial x_i\,\partial x_j} - \delta_{ij}\frac{\partial^2 \epsilon_{ll}}{\partial x_p\,\partial x_p} - \frac{\partial^2 \epsilon_{mi}}{\partial x_j\,\partial x_m} - \frac{\partial^2 \epsilon_{jp}}{\partial x_p\,\partial x_i} = 0. \qquad (2\text{-}7.3)$$

The six component equations are

$$\frac{\partial^2 \epsilon_{11}}{\partial x_2^2} + \frac{\partial^2 \epsilon_{22}}{\partial x_1^2} - 2\frac{\partial^2 \epsilon_{12}}{\partial x_1\,\partial x_2} = 0, \qquad (2\text{-}7.4)$$

$$\frac{\partial^2 \epsilon_{22}}{\partial x_3^2} + \frac{\partial^2 \epsilon_{33}}{\partial x_2^2} - 2\frac{\partial^2 \epsilon_{23}}{\partial x_2\,\partial x_3} = 0, \qquad (2\text{-}7.5)$$

$$\frac{\partial^2 \epsilon_{33}}{\partial x_1^2} + \frac{\partial^2 \epsilon_{11}}{\partial x_3^2} - 2\frac{\partial^2 \epsilon_{31}}{\partial x_3\,\partial x_1} = 0, \qquad (2\text{-}7.6)$$

$$-\frac{\partial^2 \epsilon_{11}}{\partial x_2\,\partial x_3} + \frac{\partial}{\partial x_1}\left(-\frac{\partial \epsilon_{23}}{\partial x_1} + \frac{\partial \epsilon_{31}}{\partial x_2} + \frac{\partial \epsilon_{12}}{\partial x_3}\right) = 0, \qquad (2\text{-}7.7)$$

$$-\frac{\partial^2 \epsilon_{22}}{\partial x_3\,\partial x_1} + \frac{\partial}{\partial x_2}\left(-\frac{\partial \epsilon_{31}}{\partial x_2} + \frac{\partial \epsilon_{12}}{\partial x_3} + \frac{\partial \epsilon_{23}}{\partial x_1}\right) = 0, \qquad (2\text{-}7.8)$$

$$-\frac{\partial^2 \epsilon_{33}}{\partial x_1\,\partial x_2} + \frac{\partial}{\partial x_3}\left(-\frac{\partial \epsilon_{12}}{\partial x_3} + \frac{\partial \epsilon_{23}}{\partial x_1} + \frac{\partial \epsilon_{31}}{\partial x_2}\right) = 0. \qquad (2\text{-}7.9)$$

Similar equations may be developed in other orthogonal curvilinear coordinates by use of the dyadic notation. Note that any of these may also be developed by proper differentiation of the six equations defining the strain components. For example, equation 2-7.4 may be obtained as follows:

$$\epsilon_{11} = \frac{\partial u_1}{\partial x_1}, \qquad (2\text{-}7.10)$$

$$\epsilon_{22} = \frac{\partial u_2}{\partial x_2}, \qquad (2\text{-}7.11)$$

$$\epsilon_{12} = \frac{1}{2}\left(\frac{\partial u_1}{\partial x_2} + \frac{\partial u_2}{\partial x_1}\right). \qquad (2\text{-}7.12)$$

By differentiating equation 2-7.10 by x_2 and 2-7.12 by x_1 and substituting, u_1 may be eliminated:

$$\frac{\partial \epsilon_{12}}{\partial x_1} = \frac{1}{2}\left(\frac{\partial \epsilon_{11}}{\partial x_2} + \frac{\partial^2 u_2}{\partial x_1^2}\right). \qquad (2\text{-}7.13)$$

Now u_2 may be eliminated by differentiating equation 2-7.11 twice with respect to x_1 and equation 2-7.13 by x_2 and substituting:

$$\frac{\partial^2 \epsilon_{12}}{\partial x_1\,\partial x_2} = \frac{1}{2}\left(\frac{\partial^2 \epsilon_{11}}{\partial x_2^2} + \frac{\partial^2 \epsilon_{22}}{\partial x_1^2}\right). \qquad (2\text{-}7.14)$$

This is identical to equation 2-7.4 and is, of course, nothing more than the corresponding scalar development. The other equations may be obtained by similar means.

A proof that the compatibility equations are necessary and sufficient conditions for single-valued displacements in a simply connected body was developed by E. Cesaro in 1906 and is presented in Sokolnikoff's and Malvern's texts.*

8 Infinitesimal strain tensor in curvilinear coordinates

Equation 2-6.9 gives the strain tensor in symbolic notation. This expression may be used to obtain the components of the strain tensor in orthogonal curvilinear coordinate systems if dyadic notation is used. Consider the polar cylindrical coordinates (r, θ, z), where the del operator and the displacement vector are

$$\begin{aligned} \bar{\nabla} &= \hat{i}_{(r)}\frac{\partial}{\partial r} + \hat{i}_{(\theta)}\frac{1}{r}\frac{\partial}{\partial \theta} + \hat{i}_{(z)}\frac{\partial}{\partial z}, \\ \bar{u} &= u_r\hat{i}_{(r)} + u_\theta\hat{i}_{(\theta)} + u_z\hat{i}_{(z)}. \end{aligned} \qquad (2\text{-}8.1)$$

Equation 2-6.9 may be written

$$\bar{\bar{\epsilon}} = \tfrac{1}{2}[\vec{\nabla}\bar{u} + (\vec{\nabla}\bar{u})^T]. \qquad (2\text{-}8.2)$$

The pregradient of the displacement vector is

$$\begin{aligned} \vec{\nabla}\bar{u} = {} & \hat{i}_{(r)}\frac{\partial u_r}{\partial r}\hat{i}_{(r)} + \hat{i}_{(r)}\frac{\partial u_\theta}{\partial r}\hat{i}_{(\theta)} + \hat{i}_{(r)}\frac{\partial u_z}{\partial r}\hat{i}_{(z)} \\ & + \hat{i}_{(\theta)}\frac{1}{r}\frac{\partial u_r}{\partial \theta}\hat{i}_{(r)} + \hat{i}_{(\theta)}\frac{u_r}{r}\hat{i}_{(\theta)} + \hat{i}_{(\theta)}\frac{1}{r}\frac{\partial u_\theta}{\partial \theta}\hat{i}_{(\theta)} \\ & - \hat{i}_{(\theta)}\frac{u_\theta}{r}\hat{i}_{(r)} + \hat{i}_{(\theta)}\frac{1}{r}\frac{\partial u_z}{\partial \theta}\hat{i}_{(z)} \\ & + \hat{i}_{(z)}\frac{\partial u_r}{\partial z}\hat{i}_{(r)} + \hat{i}_{(z)}\frac{\partial u_\theta}{\partial z}\hat{i}_{(\theta)} + \hat{i}_{(z)}\frac{du_z}{dz}\hat{i}_{(z)}. \end{aligned} \qquad (2\text{-}8.3)$$

* L. E. Malvern, *Introduction to the Mechanics of a Continuous Medium*, Prentice-Hall, Inc., Englewood Cliffs, N.J., 1969; I. S. Sokolnikoff, *Mathematical Theory of Elasticity*, McGraw-Hill Book Company, New York, 1956.

The strain tensor in polar cylindrical coordinates thus becomes

$$\bar{\bar{\epsilon}} = \begin{pmatrix} \dfrac{\partial u_r}{\partial r} & \dfrac{1}{2}\left[\dfrac{\partial u_\theta}{\partial r} + \dfrac{1}{r}\dfrac{\partial u_r}{\partial \theta} - \dfrac{u_\theta}{r}\right] & \dfrac{1}{2}\left[\dfrac{\partial u_z}{\partial r} + \dfrac{\partial u_r}{\partial z}\right] \\ & \left[\dfrac{1}{r}\dfrac{\partial u_\theta}{\partial \theta} + \dfrac{u_r}{r}\right] & \dfrac{1}{2}\left[\dfrac{1}{r}\dfrac{\partial u_z}{\partial \theta} + \dfrac{\partial u_\theta}{\partial z}\right] \\ & \text{Symmetric} & \dfrac{\partial u_z}{\partial z} \end{pmatrix}. \qquad (2\text{-}8.4)$$

This tensor can be derived by direct geometrical examination but with greater difficulty. This type of derivation is sketched below in two dimensions to illustrate the difficulties of such a derivation. An element in the plane, z equals a constant, is shown in Figure 2-5. The displacement of point O to O'

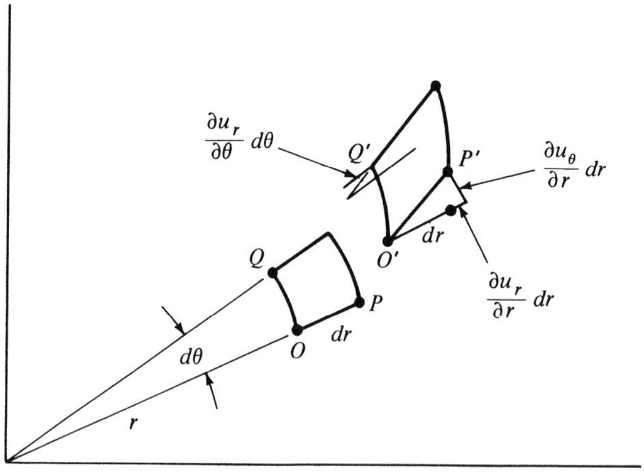

Figure 2-5

will be designated by the components u_r and u_θ. The strain in the radial direction may be obtained by examining the new length of the line OP shown as $O'P'$ in Figure 2-5:

$$\overline{O'P'}^2 = [1 + \epsilon_{rr}]^2 dr^2 = \left(dr + \frac{\partial u_r}{\partial r} dr\right)^2 + \left(\frac{\partial u_\theta}{\partial r}\right)^2. \qquad (2\text{-}8.5)$$

Expanding and dropping higher-order terms yields

$$\epsilon_{rr} = \frac{\partial u_r}{\partial r}. \qquad (2\text{-}8.6)$$

In a similar manner, the length of the line $O'Q'$ may be written

$$\overline{O'Q'}^2 = (1 + \epsilon_{\theta\theta})^2 (r\, d\theta)^2 = \left[(r + u_r)\, d\theta + \frac{\partial u_\theta}{\partial \theta}\partial\theta\right]^2 + \left[\frac{\partial u_r}{\partial \theta}\right]^2. \qquad (2\text{-}8.7)$$

Expanding and neglecting higher-order terms yields

$$\epsilon_{\theta\theta} = \frac{u_r}{r} + \frac{1}{r}\frac{\partial u_\theta}{\partial \theta}. \tag{2-8.8}$$

The shearing strain $\epsilon_{r\theta}$ is equal to one half the change in angle between the lines $\bar{O}\bar{P}$ and $\bar{O}\bar{Q}$:

$$\epsilon_{r\theta} = \left[\frac{(\partial u_\theta/\partial r)\,dr}{(1+\epsilon_{rr})\,dr} + \frac{(\partial u_r/\partial \theta)\,d\theta}{(1+\epsilon_{\theta\theta})r\,d\theta} - \frac{u_\theta}{r+u_r}\right]\frac{1}{2}. \tag{2-8.9}$$

This may be approximated for small strains as

$$\epsilon_{r\theta} = \frac{1}{2}\left[\frac{\partial u_\theta}{\partial r} + \frac{1}{r}\frac{\partial u_r}{\partial \theta} - \frac{u_\theta}{r}\right]. \tag{2-8.10}$$

These expressions agree with those obtained by dyadic development, but great care is required to avoid missing some terms.

The strain tensor in spherical coordinates may also be obtained by either of these methods. Only the results will be presented here, with the development left to the reader. The spherical coordinates are shown in Figure 2-6.

$$\bar{\bar{\epsilon}}(R,\varphi,\theta) = \begin{pmatrix} \dfrac{\partial u_R}{\partial R} & \dfrac{1}{2}\left[\dfrac{1}{R}\dfrac{\partial u_R}{\partial \varphi} + \dfrac{\partial u_\varphi}{\partial R} - \dfrac{u_\varphi}{R}\right] & \dfrac{1}{2}\left[\dfrac{1}{R\sin\varphi}\dfrac{\partial u_R}{\partial \theta} + \dfrac{\partial u_\theta}{\partial R} - \dfrac{u_\theta}{R}\right] \\ & \dfrac{1}{R}\dfrac{\partial u_\varphi}{\partial \varphi} + \dfrac{u_R}{R} & \dfrac{1}{2}\left[\dfrac{1}{R\sin\varphi}\dfrac{\partial u_\varphi}{\partial \theta} + \dfrac{1}{R}\dfrac{\partial u_\theta}{\partial \varphi} - \dfrac{u_\theta}{R}\cot\varphi\right] \\ & & \dfrac{1}{R\sin\varphi}\dfrac{\partial u_\theta}{\partial \theta} + \dfrac{u_\varphi}{R}\cot\varphi + \dfrac{u_R}{R} \end{pmatrix}. \tag{2-8.11}$$

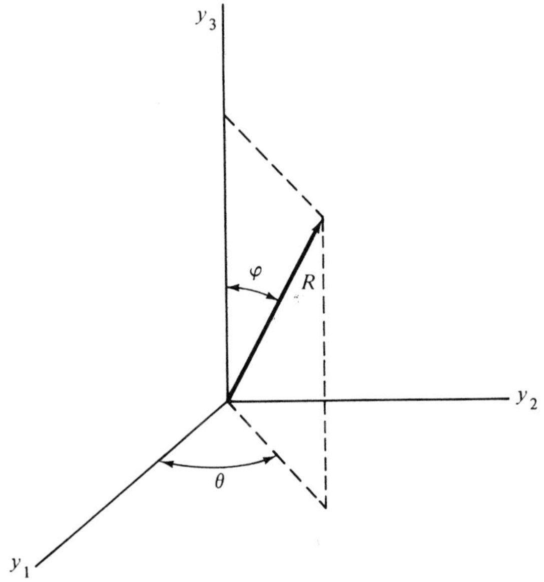

Figure 2-6

9 Spherical and deviatoric strain tensors

It is convenient for some purposes to separate the strain tensor into two parts, called the spherical and the deviatoric parts:

$$\bar{\bar{\epsilon}} = \bar{\bar{\epsilon}}^s + \bar{\bar{\epsilon}}^d. \tag{2-9.1}$$

The spherical part is equal to one third of the first invariant times the unity tensor.

$$\begin{aligned} \epsilon^s_{ij} &= \tfrac{1}{3}I_\epsilon \delta_{ij} \\ &= \tfrac{1}{3}\epsilon_{kk}\delta_{ij}. \end{aligned} \tag{2-9.2}$$

In section 6, the first invariant of the strain tensor was shown to be equal to the change in volume per unit volume. Therefore, the average of normal strains is a measure of the volumetric change. Note that the spherical strain tensor is an isotropic tensor and therefore independent of coordinate orientation.

The deviatoric strain tensor is defined to be the difference between the strain tensor and the spherical strain tensor:

$$\epsilon^d_{ij} = \epsilon_{ij} - \tfrac{1}{3}\epsilon_{kk}\delta_{ij}. \tag{2-9.3}$$

As implied by its name, this tensor gives a measure of how the new shape deviates from the original shape, and carries all the shear strain effects. The principal axes of the deviatoric and the original tensor coincide, although the principal strains differ by the amount of the spherical tensor's components.

PROBLEMS

2.1 In steady flow $[(\partial\bar{v}/\partial t) = 0]$, show that

$$\frac{d^2J}{dt^2} = J\bar{\nabla}\cdot[(\bar{\nabla}\cdot\bar{v})\bar{v}].$$

2.2 Obtain a transport theorem for surface integral in the form

$$\frac{a}{dt}\iint_S \bar{a}\cdot\hat{n}\,dS = \iint_S \left[\frac{d\bar{a}}{dt} + \bar{a}(\vec{\bar{\nabla}}\cdot\bar{v}) - (\bar{a}\cdot\vec{\bar{\nabla}})\bar{v}\right]\cdot\hat{n}\,dS.$$

2.3 At some time t, the relation between the spatial and material coordinates is as follows:

$$\begin{aligned} y_1 &= 2x_1 + x_2 + x_3, \\ y_2 &= x_1 + x_2, \\ y_3 &= x_1 + 2x_3. \end{aligned}$$

Find the finite Lagrangian and Eulerian strain tensors.

2.4 Given the displacement field

$$u_1 = (3x_1^2 x_2 + 6) \times 10^{-2},$$
$$u_2 = (x_2^2 + 6x_1 x_3) \times 10^{-2},$$
$$u_3 = (6x_3^2 + 2x_2 x_3 + 10) \times 10^{-2},$$

calculate the infinitesimal Lagrangian strain at a point (1, 0, 2). What is the extension of a line of initial length and orientation dx_1 at this point? What is the rotation of this line?

2.5 For the strain tensor calculated in problem 2.4, compute the principal strains and the principal directions. Find the invariants of the strain tensor.

2.6 Verify equation 2-8.11.

2.7 Derive the compatibility equations in cylindrical coordinates.

2.8 Show that the eigenvectors of the spherical and deviatoric strain tensors are the same.

2.9 The Green deformation tensor, $\bar{\bar{C}}$, is used when working with large deformations. Defined in terms of the material strains, $\bar{\bar{C}}$ is

$$C_{ij} = 2E_{ij} + \delta_{ij}.$$

Find the first two invariants of $\bar{\bar{C}}$ in terms of the components of $\bar{\bar{E}}$. Note that the third invariant is equal to 1 if the material is incompressible.

CHAPTER 3

STRESS

1 Introduction

When nondeformable bodies are studied, the internal transmission of force is not of great concern. However, elasticians have worried about internal forces since the "stargazer" Galileo examined beams in 1638. Bernoulli, Coulomb, and Euler introduced the concepts of shear and normal internal forces. Cauchy built a general theory on these ideas and described the nature of these internal forces.

One obvious general classification for the forces acting upon a continuum is to divide them into external and internal forces. Further distinctions may be made between body forces and surface forces. *Body forces* are external forces that act upon each element of the continuum. The gravitational force is such a force. These forces are usually expressed as force per unit mass and act throughout the volume. Some authors refer to such forces as *extrinsic* *loads.** *Surface forces* or contact forces act on the bounding surfaces of a body or on the surfaces of an element within the body. They therefore may be considered as either internal or external forces. These forces are usually expressed as force per unit area.

2 Stress tractions

Consider the body shown in Figure 3-1, subjected to some external loading. At a point within the body, an element of the internal surface has a normal unit vector $\hat{n}$ and an area ΔS. Some authors define an area vector,

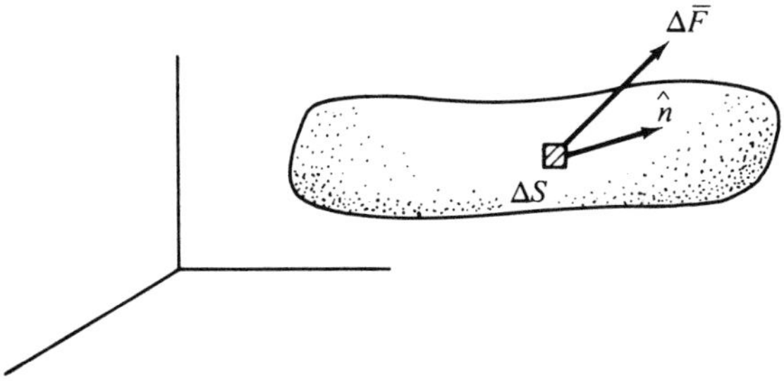

Figure 3-1

$\overline{\Delta A} = \hat{n}\Delta S$, and use this in their discussions. The total force acting on this element of surface may be written

$$\overline{\Delta F} = \bar{t}^{(n)}\,\Delta S. \qquad (3\text{-}2.1)$$

The surface traction is $\bar{t}^{(n)}$, where the superscript n designates the normal to the surface. In the limit, equation 3-2.1 becomes

$$\overline{dF} = \bar{t}^{(n)}\,dS. \qquad (3\text{-}2.2)$$

The surface traction is clearly the surface force per unit area. At any point this traction vector would change, depending upon the orientation of the surfaces, of which there are an infinite number passing through this point. The stress traction defined in the limit is independent of the surface as long as it has the same normal at the point of evaluation.

Of special interest are the stress tractions at a point associated with planes normal to the coordinate axes. These may be written $\bar{t}^{(i)}$, where $i = 1$, 2, 3, and correspond to the coordinate axes y_1, y_2, and y_3. The components

* C. Truesdell, "Principles of Classical Mechanics and Field Theory," *Handbook der Physik*, Vol. III, Springer-Verlag, Berlin, 1960, p. 587.

of these stress tractions will be written

$$T_{ij} = t_j^{(i)}, \tag{3-2.3}$$

where T_{ij} are the components of a stress tensor. The first subscript of T_{ij} designates the direction of the normal of the surface, and the second subscript designates the direction of the force component on that surface. The following sign convention is associated with these components: a positive component is in the positive coordinate direction on the face with a position normal or in the negative direction on the face with a negative normal. This convention is shown on the element in Figure 3-2. The stress tensor will be formally

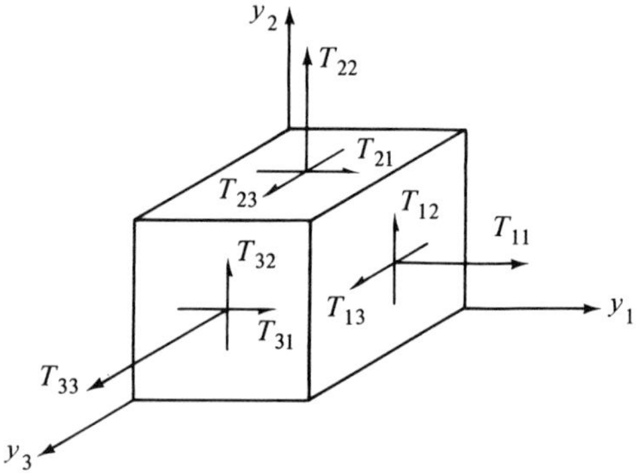

Figure 3-2

related to the stress traction by the tensor equation

$$\bar{t}^{(n)} = \hat{n} \cdot \bar{\bar{T}}. \tag{3-2.4}$$

This may be written

$$t_j^{(n)} = n_i T_{ij}. \tag{3-2.5}$$

We shall show shortly that $\bar{\bar{T}}$ is a second-order tensor by showing that it satisfies the appropriate transformation laws.

Consider a tetrahedron at a point in the continuum shown in Figure 3-3. If the area of the face normal to the $\hat{n}$ vector is S, the areas of the other faces are

$$S_j = \hat{n} \cdot \hat{i}_{(j)} S. \tag{3-2.6}$$

If we consider the volume enclosed by this tetrahedron to be shrunk to a point, body forces and acceleration effects are higher-order infinitesimal effects and Newton's law yields

$$\bar{t}^{(n)} S = -[\bar{t}^{(1)} S_1 + \bar{t}^{(2)} S_2 + \bar{t}^{(3)} S_3]. \tag{3-2.7}$$

Substituting Equation 3-2.6 into Equation 3-2.7 gives

$$\bar{t}^{(n)} = -[\bar{t}^{(1)}(\hat{n} \cdot \hat{i}_{(1)}) + \bar{t}^{(2)}(\hat{n} \cdot \hat{i}_{(2)}) + \bar{t}^{(3)}(\hat{n} \cdot \hat{i}_{(3)})]. \tag{3-2.8}$$

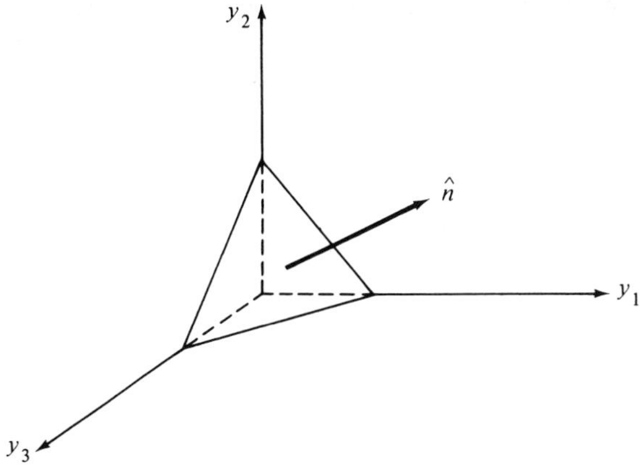

Figure 3-3

Equation 3-2.8 gives the relationship of the stress traction on one plane in terms of those on the coordinate planes.

Let us consider another coordinate system with the y'_i axis coinciding with the normal vector $\hat{n}$ and the other two axes lying in the plane of area S. The components of the stresses on this face may be written

$$T'_{ij} = t^{(i')}_{j'} = \hat{i}'_{(j)} \cdot \bar{t}^{(i')}. \tag{3-2.9}$$

Noting that in equation 3-2.8

$$\bar{t}^{(i')} = \bar{t}^{(n)},$$

we obtain

$$T'_{ij} = -\hat{i}'_{(j)} \cdot [\bar{t}^{(1)}(\hat{i}'_{(i)} \cdot \hat{i}_{(1)}) + \bar{t}^{(2)}(\hat{i}'_{(i)} \cdot \hat{i}_{(2)}) + \bar{t}^{(3)}(\hat{i}'_{(i)} \cdot \hat{i}_{(3)})]. \tag{3-2.10}$$

Using equation 3-2.5 and observing the sign convention, equation 3-2.10 becomes

$$T'_{ij} = a_{im} a_{jn} T_{mn},$$

where $$a_{im} = \cos(y'_i, y_m) = (\hat{i}'_{(i)} \cdot \hat{i}_{(m)}). \tag{3-2.11}$$

The components of the stress tensor thus transform as a second-order tensor and the array $\bar{\bar{T}}$ is correctly termed a second-order tensor.

3 Stress tensor in the material sense

The Cauchy stress tensor defined by equations 3-2.2 and 3-2.3 refers to the deformed body in the spatial coordinates. This means that the surface traction $\bar{t}^{(n)}$ is the force per unit area of deformed surface. We shall now describe a method of relating the force to an original element of undeformed area. Consider the original surface dS_0, which deforms to the surface dS, and the force $d\bar{F}$ on the surface dS to have been oriented originally in the $d\bar{F}_0$ direction (Figure 3-4). The easiest way to define a stress tensor in the material coordinates would be to introduce a stress traction $\bar{t}^{(N_0)}$ in the following manner:

$$\bar{t}^{(N_0)}\, dS_0 = \overline{dF} = \bar{t}^{(n)}\, dS. \tag{3-3.1}$$

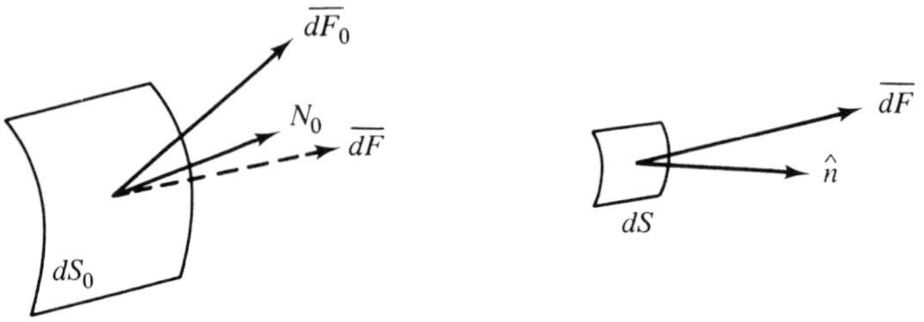

Figure 3-4

This is consistent with the ideas used in simple tension tests where the actual load is measured and divided by the original area. We may now introduce a stress tensor associated with the original stress traction vector:

$$\bar{t}^{(N_0)} = \hat{N}_0 \cdot \bar{\bar{T}}^0. \tag{3-3.2}$$

Although this procedure seems reasonable because of past procedures in elementary calculations, we are actually defining a force in the spatial coordinates in terms of a stress tensor in the material coordinates. We may relate the two stress tensors by studying how an area dS_0 changes to dS.

Let the edges of the area dS be designated by the vectors $\overline{\delta y}$ and $\overline{dy}$ (Figure 3-5):

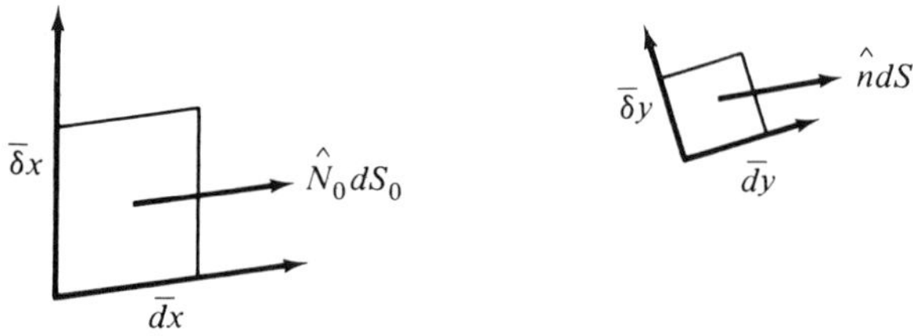

Figure 3-5

$$\begin{aligned} \hat{N}_0\, dS_0 &= \overline{dx} \times \overline{\delta x}, \\ \hat{n}\, dS &= \overline{dy} \times \overline{\delta y}. \end{aligned} \tag{3-3.3}$$

The undeformed vectors may be related to the deformed vectors by the relation

$$\begin{aligned} dx_i &= \frac{\partial x_i}{\partial y_j}\, dy_j, \\ \delta x_i &= \frac{\partial x_i}{\partial y_j}\, \delta y_j. \end{aligned} \tag{3-3.4}$$

Equation 3-3.3 may be written in component notation as

$$dS_0 N_I^0 = e_{IJK}\, dx_J\, \delta x_K = e_{IJK} \frac{\partial x_J}{\partial y_s} \frac{\partial x_K}{\partial y_t}\, dy_s\, \delta y_t. \tag{3-3.5}$$

Capital subscripts are used here to clearly indicate which quantities depend upon the material coordinates and which are in spatial coordinates. Multiplying both sides of equation 3-3.5 by $(\partial x_I/\partial y_r)$ and summing yields

$$dS_0 N_I^0 \frac{\partial x_I}{\partial y_r} = e_{IJK} \frac{\partial x_I}{\partial y_r} \frac{\partial x_J}{\partial y_s} \frac{\partial x_K}{\partial y_t}\, dy_s\, \delta y_t. \tag{3-3.6}$$

The Jacobian was defined in section 2-4 to be

$$J = e_{mnp} \frac{\partial y_1}{\partial x_m} \frac{\partial y_2}{\partial x_n} \frac{\partial y_3}{\partial x_p}. \tag{3-3.7}$$

This may be rewritten as

$$e_{IJK} J = e_{rst} \frac{\partial y_r}{\partial x_I} \frac{\partial y_s}{\partial x_J} \frac{\partial y_t}{\partial x_K}. \tag{3-3.8}$$

Equation 3-3.8 uses these properties of a determinant: if any two rows and columns are equal, the determinant is zero; and the sign of the determinant changes if two adjacent rows or columns are interchanged.

In a similar manner, we may define the inverse Jacobian J^{-1} as

$$J^{-1} = e_{IJK} \frac{\partial x_I}{\partial y_1} \frac{\partial x_J}{\partial y_2} \frac{\partial x_K}{\partial y_3}, \tag{3-3.9}$$

or

$$e_{rst} J^{-1} = e_{IJK} \frac{\partial x_I}{\partial y_r} \frac{\partial x_J}{\partial y_s} \frac{\partial x_K}{\partial y_t}. \tag{3-3.10}$$

Equation 3-3.6 may now be written

$$\frac{\partial x_I}{\partial y_r} N_I^0\, dS_0 = e_{rst} J^{-1}\, dy_s\, \delta y_t. \tag{3-3.11}$$

From equation 3-3.3,

$$n_r\, dS = e_{rst}\, dy_s\, \delta y_t, \tag{3-3.12}$$

so equation 3-3.11 becomes

$$J^{-1} \hat{n}\, dS = \hat{N}^0 \cdot (\bar{x} \overleftarrow{\nabla}_y)\, dS_0 \tag{3-3.13}$$

where $\bar{x}\overleftarrow{\nabla}_y$ is the symbolic representation for a second-order "two-point tensor" having components $\partial x_I/\partial y_j$, and is associated with two coordinate systems, one material and one spatial.

Equation 3-3.13 takes to final form

$$\hat{n}\, dS = J \hat{N}^0 \cdot (\bar{x} \overleftarrow{\nabla}_y)\, dS_0. \tag{3-3.14}$$

Equations 3-3.1, 3-3.2, and 3-2.4 yield the identity

$$\hat{N}^0 \cdot \bar{\bar{T}}^0\, dS_0 = \hat{n} \cdot \bar{\bar{T}}\, dS. \tag{3-3.15}$$

Substituting equation 3-3.14 into equation 3-3.15 yields

$$\hat{N}_0 \cdot [\bar{\bar{T}}^0 - J(\bar{x}\overleftarrow{\nabla}_y) \cdot \bar{\bar{T}}]\, dS_0 = 0, \tag{3-3.16}$$

$$\bar{\bar{T}}^0 = J(\bar{x}\overleftarrow{\nabla}_y) \cdot \bar{\bar{T}}, \tag{3-3.17}$$

$$T_{Ij} = J \frac{\partial x_I}{\partial y_k} T_{kj}. \tag{3-3.18}$$

This is called the first Piola–Kirchhoff stress tensor and is a two-point tensor. Two disadvantages exist in working with this quantity. First, it is a two-point tensor and, second, if T_{kj} is symmetric (as we shall discuss in Chapter 4), T^0_{Ij} need not be. We therefore shall develop a third stress tensor which overcomes these two difficulties and relates the original force to the original area. Let us define force $d\bar{F}_0$ related to $d\bar{F}$ by the equation

$$dF_{0I} = \frac{\partial x_I}{\partial y_j} dF_j \quad \Longrightarrow \quad \overline{dF_0} = (\bar{x}\overleftarrow{\nabla}_y) \cdot \overline{dF}. \tag{3-3.19}$$

Furthermore, let the force $d\bar{F}_0$ be related to a stress tensor $\bar{\bar{\sigma}}$ by the following relation:

$$\overline{dF_0} = \hat{N}_0 \cdot \bar{\bar{\sigma}}\, dS_0 = (\bar{x}\overleftarrow{\nabla}_y) \cdot \overline{dF}. \tag{3-3.20}$$

Substituting equation 3-3.1 and 3-3.2 into 3-3.20 yields

$$\hat{N}_0 \cdot \bar{\bar{\sigma}}\, dS_0 = (\bar{x}\overleftarrow{\nabla}_y) \cdot (\hat{N}_0 \cdot \bar{\bar{T}}^0)\, dS_0, \tag{3-3.21}$$

$$N_0 \cdot \bar{\bar{\sigma}} = \hat{N}_0 \cdot \bar{\bar{T}}^0 \cdot (\overrightarrow{\nabla}_y \bar{x}). \tag{3-3.22}$$

The stress tensor $\bar{\bar{\sigma}}$ is related to the first Piola–Kirchhoff stress tensor by

$$\bar{\bar{\sigma}} = \bar{\bar{T}}^0 \cdot (\overrightarrow{\nabla}_y \bar{x}), \tag{3-3.23}$$

$$\sigma_{IJ} = T^0_{Im} \frac{\partial x_J}{\partial y_m}. \tag{3-3.24}$$

Replacing the first Piola–Kirchhoff tensor by the Cauchy stress tensor gives

$$\begin{aligned} \bar{\bar{\sigma}} &= J(\bar{x}\overleftarrow{\nabla}_y) \cdot \bar{\bar{T}} \cdot (\overrightarrow{\nabla}_y \bar{x}), \\ \sigma_{IJ} &= J \frac{\partial x_I}{\partial y_m} T_{mn} \frac{\partial x_J}{\partial y_n}. \end{aligned} \tag{3-3.25}$$

The second Piola–Kirchhoff stress tensor is $\bar{\bar{\sigma}}$ and will be used in problems formulated in material coordinates. Note that if $\bar{\bar{T}}$ is symmetric, $\bar{\bar{\sigma}}$ is symmetric. Equation 3-3.25 may be inverted as follows:

$$T_{ij} = J^{-1}\left(\frac{\partial y_i}{\partial x_M} \sigma_{MN} \frac{\partial y_j}{\partial x_N}\right). \tag{3-3.26}$$

4 Properties of the stress tensor

It will be shown in Chapter 4 that the Cauchy stress tensor and, therefore, the second Piola–Kirchhoff stress tensor are usually symmetric. If we may borrow on credit this result, we can examine the stress tensor in more detail. In section 1-7, it was shown that for a symmetric, real second-order tensor, there exists a coordinate system that diagonalizes the tensor. These coordinate axes are called the principal axes of stress, and the eigenvalues asscociated with them are called the principal stresses. The eigenvalues and eigenvectors are found by solution of the eigenvalue problem,

$$(T_{ij} - \lambda\delta_{ij})u_j = 0. \tag{3-4.1}$$

The following combinations can be shown to be invariants in all coordinate systems and are called the stress invariants:

$$\begin{aligned} \mathrm{I}_T &= T_{ii}, \\ \mathrm{II}_T &= \tfrac{1}{2}[T_{ii}T_{jj} - T_{ij}T_{ij}], \\ \mathrm{III}_T &= e_{ijk}T_{i1}T_{j2}T_{k3}. \end{aligned} \tag{3-4.2}$$

As an example, consider the stress tensor

$$T_{ij} = \begin{pmatrix} 2000 & 4000 & -6{,}000 \\ 4000 & 2000 & -6{,}000 \\ -6000 & -6000 & -15{,}000 \end{pmatrix}.$$

The principal stresses are obtained by finding the roots of the cubic equation formed by setting the determinant $|T_{ij} - \lambda\delta_{ij}|$ equal to zero.

$$\begin{aligned} T^1 &= 9000, \\ T^2 &= -2000, \\ T^3 &= -18{,}000. \end{aligned}$$

Substituting T^1 into equation 3-4.1 yields

$$\begin{aligned} -7u_1 + 4u_2 - 6u_3 &= 0, \\ 4u_1 - 7u_2 - 6u_3 &= 0, \\ -6u_1 - 6u_2 - 24u_3 &= 0, \end{aligned} \implies \begin{aligned} u_1 &= u_2, \\ u_3 &= -\tfrac{1}{2}u_2. \end{aligned}$$

The unit vector corresponding to this axis is

$$\hat{e}^{(1)} = (\tfrac{2}{3}, \tfrac{2}{3}, -\tfrac{1}{3}).$$

In a similar manner, the other two eigenvectors are

$$\hat{e}^{(2)} = \left(-\frac{1}{\sqrt{2}}, \frac{1}{\sqrt{2}}, 0\right),$$

$$\hat{e}^{(3)} = \left(\frac{1}{\sqrt{18}}, \frac{1}{\sqrt{18}}, \frac{4}{\sqrt{18}}\right).$$

Note that the eigenvectors are orthogonal and form a right-handed coordinate system.

A graphical method of solving this eigenvalue problem generally used for two-dimensional problems was given by 0. Mohr in 1882 and is presented in most texts on the mechanics of materials.

Let p denote the mean normal pressure:

$$p = \tfrac{1}{3} T_{ii}. \tag{3-4.3}$$

Note that this is an invariant quantity. The stress tensor may be split into two parts, one associated with the mean normal pressure and called the spherical stress tensor, and the remainder, called the deviatoric stress tensor:

$$T_{ij} = -p\,\delta_{ij} + T^d_{ij}. \tag{3-4.4}$$

The deviatoric stress tensor is T^d and is equal to

$$T^d_{ij} = T_{ij} - \tfrac{1}{3} T_{kk}\,\delta_{ij}. \tag{3-4.5}$$

This decomposition is used to separate volumetric effects from distortional effects.

PROBLEMS

3.1 If $\bar{t}^{(n)}$ and $\bar{t}^{(m)}$ are stress tractions at a point P corresponding to the unit normal vectors $\hat{n}$ and $\hat{m}$, respectively, show that the component of $\bar{t}^{(n)}$ in the m direction is equal to the component of $\bar{t}^{(m)}$ in the n direction if the stress tensor is assumed to be symmetric.

3.2 If the stress tensor at a point within the body is given by

$$\begin{pmatrix} 1000 & 200 & 0 \\ 200 & -600 & -400 \\ 0 & -400 & 0 \end{pmatrix},$$

find the normal stress in the direction n, where $\hat{n}$ is the unit vector:

$$\hat{n} = \tfrac{2}{3}\hat{i}_1 + \tfrac{2}{3}\hat{i}_2 - \tfrac{1}{3}\hat{i}_3.$$

3.3 Consider a body under a state of biaxial shear:

$$\begin{pmatrix} 0 & 2x_3 & 0 \\ 2x_3 & 0 & 0 \\ 0 & 0 & 0 \end{pmatrix}.$$

Find the principal axes and the principal stresses at a point in the body with coordinates (0, 2, 3).

3.4 For the state of stress shown in problem 3.3, find the stress tensor with reference to a coordinate system rotated 30° about the x_3 axis by the right-hand rule.

3.5 Find the spherical and deviatoric stress tensors from the stress tensor shown in problem 3.2.

3.6 In a state of plane stress ($\sigma_{33} = \sigma_{13} = \sigma_{23} = 0$), the remaining components of the stress tensor may be expressed in terms of a scalar function $\varphi(x_1, x_2)$. Determine an expression for σ_{nn} and σ_{nt}, the tangential and normal stresses along a curve C.

$$\sigma_{11} = \frac{\partial^2 \varphi}{\partial x_2^2},$$

$$\sigma_{22} = \frac{\partial^2 \varphi}{\partial x_1^2},$$

$$\sigma_{12} = -\frac{\partial^2 \varphi}{\partial x_1\, \partial x_2}.$$

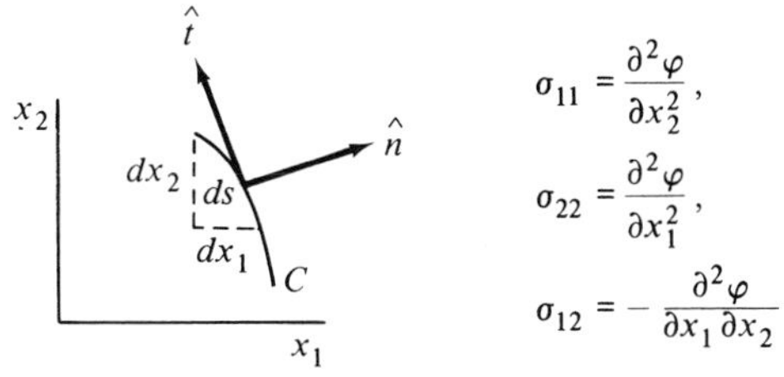

Figure P. 3-6

CHAPTER 4

BASIC EQUATIONS OF CONTINUUM MECHANICS

1 Introduction

Legend says that Sir Isaac Newton was hit by a falling apple and developed the laws of classical mechanics. It might have been better for our purposes had he been hit by applesauce, for then his laws would have been directly applicable to a continuum. Although Newton published his axioms in *Principia* (1687), it was not until 1749 that Euler gave them useful mathematical clothing. Cauchy is, however, credited with writing them as the present equations of motion.

We must extend the concepts and laws governing mass points to laws governing continuously distributed masses and velocities of particles to velocity fields for continua. The total mass of a body will be considered to be the sum of the mass of its parts distributed throughout the continuum. This mass will be assumed to be a conserved quantity, which is consistent with the general assumptions of classical mechanics. In a material sense, this mass may be used in Newton's laws, and linear and angular momentum equations may be written accordingly. Conservation equations for mass and for energy complete the system of basic equations. These equations apply to all continua and will be developed in this chapter.

2 Conservation of mass

Consider the mass enclosed by an arbitrary surface within the continuum at any time t, as shown in Figure 4-1. The mass inside this surface

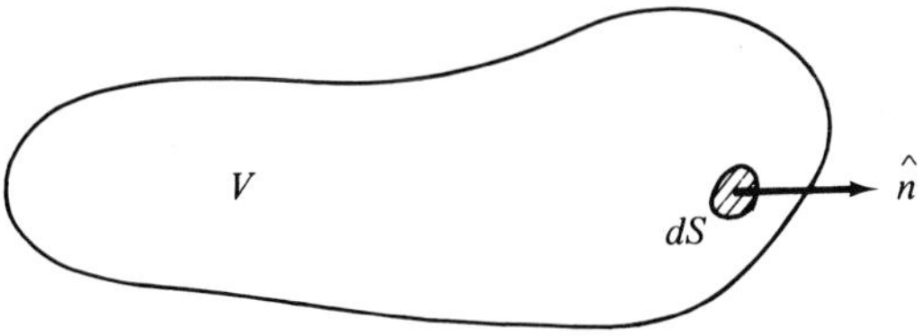

Figure 4-1

may be written

$$M = \int_{V(y_i)} \rho \, dV. \tag{4-2.1}$$

In the material sense this mass is conserved. The material time derivative of equation 4-2.1 is equal to zero:

$$\frac{d}{dt} \int \rho \, dV = 0. \tag{4-2.2}$$

Using Reynold's transport theorem (equation 2-5.5), we obtain

$$\int \left[\frac{d\rho}{dt} + \rho(\bar{\nabla} \cdot \bar{v}) \right] dV = 0. \tag{4-2.3}$$

Since the volume was arbitrarily chosen, the integrand may be set equal to zero:

$$\frac{d\rho}{dt} + \rho \bar{\nabla} \cdot \bar{v} = 0. \tag{4-2.4}$$

Using the definition of the material derivative, this may be written in alternative forms:

$$\frac{\partial \rho}{\partial t} + \bar{v} \cdot \bar{\nabla} \rho + \rho \bar{\nabla} \cdot \bar{v} = 0, \tag{4-2.5}$$

$$\frac{\partial \rho}{\partial t} + \bar{\nabla} \cdot (\rho \bar{v}) = 0. \tag{4-2.6}$$

These equations are each a form of the continuity equation.

It is instructive in this case to use an alternative approach in the derivation of the equation of continuity. In Figure 4-2, consider a volume $V_0(x_i)$, defined in the material coordinate system at time $t = 0$, which occupies the volume $V(y_i)$ at some later time $t = t$. The mass in the material volume V_0 is

$$M = \int_{V(x_i)} \rho_0 \, dV(x_i). \tag{4-2.7}$$

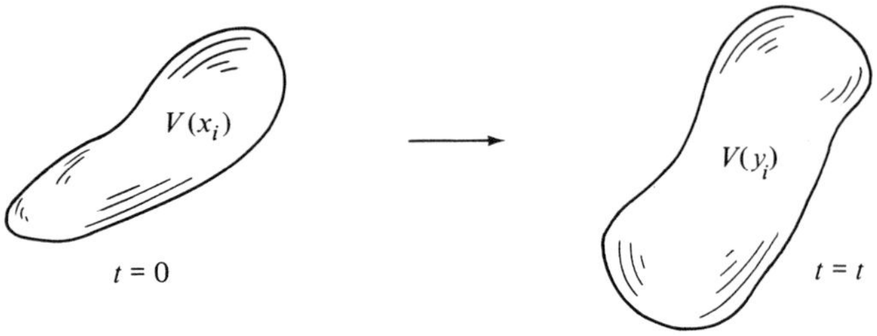

Figure 4-2

At time t, this mass occupies the spatial volume $V(y_i)$ and may be written

$$M = \int_{V(y_i)} \rho \, dV(y_i). \tag{4-2.8}$$

Using equation 2-4.5, we may convert the integration in the spatial sense to a material volume:

$$\int_{V(x_i)} \rho_0 \, dV(x_i) = \int_{V(x_i)} \rho J \, dV(x_i). \tag{4-2.9}$$

Since the material volume including this mass remains constant, the integrands in equation 4-2.9 may be equated:

$$\rho_0 = \rho J. \tag{4-2.10}$$

The initial mass density is ρ_0, and ρ is the mass density at time t. Equation 4-2.10 is another form of the equation of continuity and may be expressed in the same form as equation 4-2.4 if the material time derivative is taken of both sides of equation 4-2.10. Noting that ρ_0 is independent of time and using equation 2-4.12 yields

$$\frac{d}{dt}(\rho J) = \left[\frac{d\rho}{dt} + \rho \bar{\nabla} \cdot \bar{v}\right] J = 0. \tag{4-2.11}$$

Since the Jacobian of the transformation must be greater than zero, we obtain

$$\frac{d\rho}{dt} + \rho \bar{\nabla} \cdot \bar{v} = 0. \tag{4-2.12}$$

Equations 4-2.4, 4-2.5, 4-2.6, and 4-2.10 are all referred to as the *equation of continuity* and are the mathematical expressions for the conservation of mass.

The equation of continuity may be used with Reynold's transport theorem to simplify the material derivative of a spatial volume when the integrand is proportional to the mass density:

$$\frac{d}{dt}\int_{V(y_i)} \rho\mathfrak{F}(y_i, t)\, dV(y) = \int_{V(y_i)} \left[\frac{d}{dt}(\rho\mathfrak{F}) + \rho\mathfrak{F}\bar{\nabla}\cdot\bar{v}\right] dV. \qquad (4\text{-}2.13)$$

The integrand on the right-hand side may be written

$$\frac{d}{dt}(\rho\mathfrak{F}) + \rho\mathfrak{F}\bar{\nabla}\cdot\bar{v} = \rho\frac{d\mathfrak{F}}{dt} + \mathfrak{F}\left(\frac{d\rho}{dt} + \rho\bar{\nabla}\cdot\bar{v}\right)\cdot \qquad (4\text{-}2.14)$$

The second term is zero from the equation of continuity, and equation 4-2.13 reduces to

$$\frac{d}{dt}\int_{V(y)} \rho\mathfrak{F}(y_i, t)\, dV(y_i) = \int_{V(y)} \rho\frac{d\mathfrak{F}}{dt}\, dV(y). \qquad (4\text{-}2.15)$$

This special form of the Reynold's transport theorem will be put to use in the next sections.

3 Cauchy's equations of motion

Newton's laws may now be expressed as equations of linear and angular momentum with the concept of conservation of mass included. Consider our now familiar potato-like volume within the continuum at time t. The forces acting on this volume are shown in Figure 4-3 and have been introduced and classified in Chapter 3.

The surface traction on the element of surface dS is $\bar{t}^n$ and the body force per unit of mass acting upon the element dV is $\bar{b}$. The surface traction may represent an applied force if dS is an element of exterior surface of the continuum, or the surface traction may arise from the transfer of the forces in one area of the continuum to another through hypothetical surfaces. At the

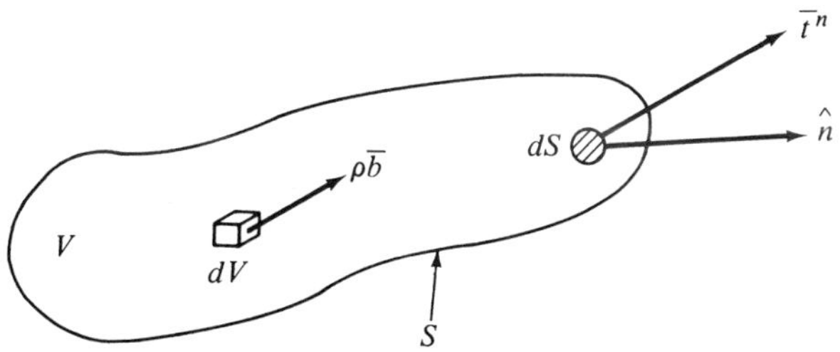

Figure 4-3

time t, the surface encloses some material and the linear momentum must be balanced in accordance with Newton's laws. The total force acting upon the particular volume equals the time rate of change (in the material sense) of the total linear momentum. This is expressed in the following equation:

$$\int_S \bar{t}^n \, dS + \int_V \rho \bar{b} \, dV = \frac{d}{dt} \int_V \rho \bar{v} \, dV, \tag{4-3.1}$$

where $\bar{v}$ represents the velocity field within the element and all quantities are expressed in spatial coordinates. The stress traction vector $\bar{t}^n$ may be related to the Cauchy stress tensor by use of equation 3-2.4 and the surface integral converted to a volume integral by use of the divergence theorem:

$$\int_V \bar{\nabla} \cdot \bar{\bar{T}} \, dV + \int_V \rho \bar{b} \, dV = \frac{d}{dt} \int_V \rho \bar{v} \, dV. \tag{4-3.2}$$

Equation 4-2.15 permits us to bring the material time derivative inside the volume integral, and, noting that the volume chosen is arbitrary, we can write

$$\bar{\nabla} \cdot \bar{\bar{T}} + \rho \bar{b} = \rho \frac{d\bar{v}}{dt}. \tag{4-3.3}$$

This vector equation is Cauchy's equation of motion. Expressing this equation in component notation yields

$$\frac{\partial T_{ij}}{\partial x_i} + \rho b_j = \rho \left(\frac{\partial v_j}{\partial t} + v_i \frac{\partial v_j}{\partial x_i} \right). \tag{4-3.4}$$

Note that the presence of the material time derivative produces a nonlinear term on the right-hand side of equation 4-3.4.

Equation 4-3.4 expresses the balance of linear momentum in spatial coordinates in terms of the Cauchy stress tensor. This form of the equation of motion is used in the development of most fluid mechanics problems, where the boundary conditions are usually given in a spatial sense. The boundary conditions in problems of elastic solids are most easily expressed in material coordinates. We shall therefore develop equivalent equations of motion in Lagrangian coordinates.

Consider the volume V_0 at the time $t = 0$ (Figure 4-4), which becomes

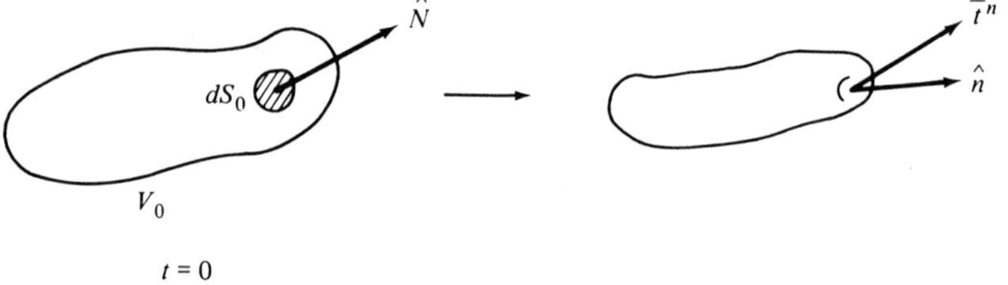

Figure 4-4

the volume V at time t. We shall designate quantities referring to time zero by zero subscripts or capital letters. Using equations 3-3.15 and 2-4.5, we may write the equation of motion in terms of the initial mass and the first Piola–Kirchhoff stress tensor:

$$\int_{S_0} \hat{N}\cdot\bar{\bar{T}}^0\, dS_0 + \int_{V_0} \rho_0\bar{b}_0\, dV_0 = \frac{d}{dt}\int_{V_0} \rho_0\bar{v}\, dV_0, \tag{4-3.5}$$

where $\bar{b}_0$ is the body force in terms of material coordinates:

$$\bar{b}_0 = \bar{b}[y(x)].$$

The order of integration and differentiation may be interchanged in the last term, because it is an integral in the material coordinates. Using the divergence theorem, equation 4-3.5 may be written

$$\bar{\nabla}_x\cdot\bar{\bar{T}}^0 + \rho_0\bar{b}_0 = \rho_0\frac{d}{dt}\bar{v}, \tag{4-3.6}$$

where $\bar{\nabla}_x$ designates the del operator in the material coordinates. We may solve equation 3-3.24 to obtain the first Piola–Kirchhoff stress tensor in terms of the second, obtaining

$$\bar{\bar{T}}^0 = \bar{\bar{\sigma}}\cdot\vec{\bar{\nabla}}_x\bar{y}, \tag{4-3.7}$$

or

$$T^0_{Im} = \sigma_{IJ}\frac{\partial y_m}{\partial x_J}. \tag{4-3.8}$$

We shall write equation 4-3.6 in component notation and replace the velocity term with the material time derivative of the spatial coordinate. This equation now becomes

$$\frac{\partial}{\partial x_I}\left(\sigma_{IJ}\frac{\partial y_m}{\partial x_J}\right) + \rho_0 b_{0m} = \rho_0\frac{d^2}{dt^2}y_m. \tag{4-3.9}$$

The displacement vector was defined as the difference between the spatial and material coordinates, so that

$$y_m = u_M + x_M. \tag{4-3.10}$$

Equation 4-3.9 takes the form

$$\frac{\partial}{\partial x_I}\left[\sigma_{IJ}\left(\delta_{JM} + \frac{\partial u_M}{\partial x_J}\right)\right] + \rho_0 b_{0M} = \rho_0\frac{d^2u_M}{dt^2}. \tag{4-3.11}$$

This equation is in material coordinates. The time derivative is actually a simple partial time derivative, and the term is linear. However, the equation is still nonlinear due to the presence of the product of the stress tensor and the gradient of the displacement. When the displacement gradient can be considered small compared to 1, a linear equation is obtained:

$$\vec{\bar{\nabla}}_x\cdot\bar{\bar{\sigma}} + \rho_0\bar{b}_0 = \rho_0\frac{\partial^2\bar{u}}{\partial t^2}. \tag{4-3.12}$$

This form of Cauchy's equation of motion is used in linear elasticity theory.

4 Considerations of angular momentum

We shall now show that the laws of angular momentum predict the symmetry of the Cauchy stress tensor, and therefore the symmetry of the second Piola–Kirchhoff stress tensor, if a number of other considerations are ignored. Three additional concepts have received some attention recently, any one of which would destroy the much-coveted symmetry of the stress tensor.

If one can conceive of body forces that reach inside a body and act upon each unit of mass, why not postulate body moments producing a tendency for rotation of each element of mass? Furthermore, since we have considered surface stress tractions per unit area of surface on an arbitrary volume, we might also postulate surface couples. These couples will give rise to what are called *couple stresses.* Couple stresses were introduced into continuum mechanics by E. Voigt (1887, 1894) and the F. Cosserats (1909) and have gained popularity in the 1960s.* Finally, we may also postulate that particles in the continuum have a spin angular momentum or an intrinsic angular momentum which may be affected by the applied forces and moments and, therefore, must be included in any angular momentum balance. If we consider each of these in addition to the terms arising from forces and the linear momentum, we obtain

$$\int_S [\bar{y} \times \bar{t}^n + \bar{c}^n]\, dS + \int_V [\bar{y} \times \rho\bar{b} + \rho\bar{m}]\, dV = \frac{d}{dt}\int_V [\bar{y} \times \rho\bar{v} + \rho\bar{l}]\, dV, \tag{4-4.1}$$

where $\bar{c}^n$ is the surface couple vector, $\bar{m}$ is the body moment, and $\bar{l}$ is the intrinsic angular momentum vector.

We may relate the surface couple vector to a couple stress tensor, $\bar{\bar{C}}$, in a manner similar to that used to introduce the Cauchy stress tensor:

$$\bar{c}^n = \hat{n}\cdot\bar{\bar{C}}. \tag{4-4.2}$$

Substituting equation 4-4.2 into equation 4-4.1 and using the divergence theorem and the special Reynolds transport theorem gives

$$\frac{\partial}{\partial y_i}[e_{mnp}y_nT_{ip} + C_{im}] + e_{mnp}y_n(\rho b_p) + \rho m_m$$
$$= \rho\left[\frac{d}{dt}(e_{mnp}y_nv_p + l_m)\right]. \tag{4-4.3}$$

Expanding yields

$$e_{mnp}\,\delta_{in}\,T_{ip} + \frac{\partial C_{im}}{\partial y_i} + \rho m_m + e_{mnp}y_n\left[\frac{\partial T_{ip}}{\partial y_i} + \rho b_q - \rho\frac{dv_p}{dt}\right] = \rho\frac{d}{dt}l_m. \tag{4-4.4}$$

* R. D. Mindlin and H. F. Tiersten, "Effects of couple-stresses in linear elasticity," *Arch. Rat. Mech. Anal.*, **11**, 415–448, 1962.

Using Cauchy's equation of motion, Equation 4-4.4 becomes

$$e_{mnp}T_{np} + \frac{\partial C_{im}}{\partial y_i} + \rho m_m = \rho \frac{d}{dt} l_m. \tag{4-4.5}$$

If the effects of couple stresses, body moments, and spin angular momentum are neglected, the angular momentum balance simplifies

$$e_{mnp}T_{np} = 0. \tag{4-4.6}$$

This dictates that in these cases the Cauchy stress tensor is symmetric. In the balance of this book we shall consider only those cases in which the stress tensor is symmetric. This includes most of the traditional field of elasticity.

In Chapter 3 we assumed that the stress tensor was symmetric and saw how the corresponding eigenvalue problem provided a means of finding principal stresses and directions.

5 Energy-conservation equation

A final field equation will be developed that has great importance in the general field of continuum mechanics. Where we can assume a separation of mechanical and thermal energy in problems not involving thermal gradients, this balance equation is not directly needed in linear elasticity. This, perhaps, is the most drastic and, to some, the most offensive of the assumptions made in the development of a simplified theory. However, in ideal elasticity, a state closely approximated by many materials in some range of their loading, heat transfer is negligible and all work put into the system is stored as recoverable elastic strain energy.

In a fluid continuum, a hydrostatic pressure p is assumed to exist when the fluid is at rest. This variable is interrelated with the other conventional thermodynamic variables in the energy-balance equation, which is a representation of the first law of thermodynamics. This equation will introduce a quantity known as the internal energy, which then requires an additional equation of state to describe particular material characteristics. The equation of state is considered to be a type of constitutive equation, together with the equation relating the stress and the strain or strain rate tensors.

The energy contained in an arbitrary volume of the continuum at any instant of time will be divided into two parts. The kinetic energy may be written as

$$K = \int_V \tfrac{1}{2} \rho v_k v_k \, dV. \tag{4-5.1}$$

If e defines a quantity known as the specific internal energy per unit mass, the total internal energy becomes

$$E = \int_V \rho e \, dV. \tag{4-5.2}$$

Let us introduce a new vector q_k, which is a nonmechanical energy flux vector. It is so defined that the rate of any nonmechanical energy transmitted outward through the surface element dS, with outward directed normal n_k, is $q_k n_k \, dS$.

The power associated with the surface and body forces is

$$P = \int_S T_{ij} n_i v_j \, dS + \int_V \rho b_j v_j \, dV. \tag{4-5.3}$$

The basic energy-balance equation may now be written by equating the rate of change of kinetic and internal energy to the rate at which surface and body forces do work and the rate at which nonmechanical energy is transferred:

$$\frac{d}{dt}[K + E] = P - \int_S q_k n_k \, dS. \tag{4-5.4}$$

This may be rewritten as

$$\frac{d}{dt}\left\{\int_V \frac{1}{2} \rho v_k v_k \, dV + \int_V \rho e \, dV\right\} = \int_S T_{ij} n_i v_j \, dS + \int_V \rho b_j v_j \, dV - \int_S q_k n_k \, dS. \tag{4-5.5}$$

Using the special form of Reynold's transport theorem and the divergence theorem, equation 4-5.5 becomes

$$\int_V \left[\rho \frac{d}{dt}\left\{\frac{v_k v_k}{2} + e\right\} - \nabla_i(T_{ij} v_j) - \rho b_j v_j + \nabla_k q_k\right] dV = 0. \tag{4-5.6}$$

Noting that the volume was arbitrarily chosen, the integrand may be expanded and equated to zero, yielding

$$v_j\left[\rho \frac{dv_j}{dt} - \nabla_i T_{ij} - \rho b_j\right] + \rho \frac{de}{dt} - T_{ij}(\nabla_i v_j) + \nabla_j q_j = 0. \tag{4-5.7}$$

The terms within the first bracket are zero by virtue of Cauchy's equation of motion, and the basic energy-balance equation becomes

$$\rho \frac{de}{dt} - T_{ij} \frac{\partial v_j}{\partial y_i} + \frac{\partial q_j}{\partial y_j} = 0. \tag{4-5.8}$$

Because T_{ij} is symmetric, the second term may be written as

$$T_{ij} \nabla_i v_j = T_{ij} d_{ij}, \tag{4-5.9}$$

where

$$d_{ij} = \tfrac{1}{2}(\nabla_i v_j + \nabla_j v_i). \tag{4-5.10}$$

The rate of strain tensor is d_{ij}.

Equation 4.5-8 is the basic energy-conservation equation. It will be examined when energy methods are considered and must be included in the formulation if thermal effects are considered.

PROBLEMS

4.1 Show that Cauchy's equation of motion can be written

$$\frac{\partial}{\partial t}(\rho v_i) = \rho b_i + \frac{\partial}{\partial y_j}(T_{ji} - \rho v_j v_i).$$

4.2 If u_i is a unit vector and T is a positive scalar, the stress tensor $T_{ij} = Tu_iu_j$ represents the "uniaxial" state of stress whose intensity and direction of axis are given by T and u_i, respectively. Consider such a uniaxial stress field with variable intensity, but constant direction of axis, and show that in the absence of body forces the gradient of the intensity must be perpendicular to the axis.

4.3 Show that if ψ is any function of position and time, then

$$\iint_S \psi T_{ij} r_j \, dS = \iiint_V \left[T_{ij}\frac{\partial \psi}{\partial y_j} + \rho\psi\left(\frac{dv_i}{dt} - b_i\right)\right] dV.$$

This is the theorem of stress means.

4.4 Derive the equation of continuity by considering a fixed volume in space and equating the mass buildup within this volume to the mass influx.

4.5 If a fluid moves radially and the velocity v is a function of R and t only, where R is the distance from a fixed point, prove that the equation of continuity is

$$\frac{\partial \rho}{\partial t} + v\frac{\partial \rho}{\partial R} + \frac{\rho}{R^2}\frac{\partial}{\partial R}(R^2 v) = 0.$$

4.6 Consider a fluid under a state of hydrostatic pressure. It is subjected to gravity and rotating with a constant angular velocity, ω, about an axis that forms the angle α with the vertical. Determine the expression for the pressure at any point.

4.7 Consider a fluid with the equation $\rho = a + bp$, relating the mass density ρ to the pressure p. Assuming that gravity is the only body force acting on this fluid, determine the dependence of the pressure on the depth below the surface of the fluid.

CHAPTER 5

LINEAR ELASTICITY

1 Introduction

Up to this point in our development, we have considered the basic mathematical description of a continuum without focusing on any particular material. We wish now to specialize not to a particular material but to a model that will characterize a large number of materials through some range of their load-deformation history. The mathematical model we shall use is that of a linear elastic material. By elastic, we mean that the material may be deformed and will return to its original configuration upon release of the deforming loads. The qualifying adjective of linear means that the load-deformation law may be represented by a linear relationship.

The concept of linear elasticity was first proposed by Robert Hooke in 1678 when he wrote, "Ut tensio sic vis," or "the power of any springy body is in the same proportion as the extension." Hooke's experiments were crude and related only force to extension. No account was taken of the size and shape of the piece being tested. The law which currently bears his name is that proposing a linear relationship between the stress and strain tensors. Cauchy developed this relation, and it is usually called generalized Hooke's law.

This linear relation is the simplest constitutive equation that may be assumed for an idealized solid continuum. Much has been done in the last two decades to formalize the development and requirements of constitutive equations. Consistency with the basic tensonal equations of continuum mechanics and the concept of coordinate invariance requires constitutive laws to have tensorial character. In addition, these equations must satisfy the principle of *material indifference.* Material indifference requires that the response of the material be independent of the observer. For example, we assume that the spring constant of a spring does not vary when the spring is moved from place to place. This assumption is inherent in many of our experimental pieces of equipment. This does not eliminate a dependency upon temperature and other thermodynamic variables. The linear theory of elasticity satisfies this requirement by relating the stress tensor to the strain tensor. Thus stress is related to differences in displacements and not to the displacements themselves. It is obvious that if the constitutive equation involved displacements themselves, this would cause the stresses to vary just by rigid-body motions and the material properties would become observer dependent. Note that the requirement of material indifference is much stronger than the requirement of coordinate invariance.

2 Generalized Hooke's law

The stress and strain tensors may be related linearly as follows:

$$\sigma_{ij} = c_{ijkl} E_{kl}. \qquad (5\text{-}2.1)$$

Equation 5-2.1 is written such that it relates the second Piola–Kirchhoff stress tensor to the Lagrangian strain tensor. There are 81 coefficients in this equation relating the nine components of the stress tensor to the nine components of the strain tensor. Not all these coefficients are independent, and the number may be reduced to 36 by noting the symmetry of the stress and strain tensors.

The general stress–strain relationships will depend upon the directions of the reference axes. The relations simplify when the material exhibits certain kinds of symmetry. It is necessary to explain the geometrical character of the symmetry that is observed in various materials. The nature of the anisotropy of a material is not completely determined by its elastic behavior alone, because the material may be anisotropic with respect to other physical properties. For example, a material may be anisotropic in its refraction, or the reflection of light. If in a body two lines exist along which the material properties are the same, the lines are sometimes called or said to be equivalent. It is important to note that lines in opposite directions are not always equivalent. Therefore, the properties along the $+z$ direction may not be the same as along the $-z$ direction. When certain crystals undergo changes in temperature, opposite ends of a particular axis become oppositely electrified. This is the phenomenon known as pyroelectricity. Some crystals exhibit the effect that opposite ends of axes become oppositely charged when compressed perpendicular to these axes. This effect is known as piezoelectricity. Additional properties of crystals may be studied by examination of J. F. Nye's book.*

A body that permits certain transformations of reference axes without change in properties along these axes is said to possess certain types of symmetry. If the body allows reflection in a plane, this plane is called a *plane of symmetry.* If its properties are independent of rotation about an axis, this axis is an *axis of symmetry.* Methods of solution of problems dealing with anisotropic elastic bodies are covered in texts by Lekhnitskii and Hearmon.†

If the material being considered is isotropic in nature, we may reduce the number of coefficients to two. Using equation 1-8.5 and dropping the antisymmetric term yields

$$\sigma_{ij} = \{\lambda\delta_{ij}\delta_{kl} + \mu(\delta_{ik}\delta_{jl} + \delta_{il}\delta_{jk})\}E_{kl}. \qquad (5\text{-}2.2)$$

This may be written as

$$\sigma_{ij} = \lambda E_{kk}\delta_{ij} + 2\mu E_{ij}. \qquad (5\text{-}2.3)$$

This form of constitutive equation was proposed in 1821 by Navier with $\lambda = \mu$ and by Cauchy in 1823 in the more general form. The two constants, λ and μ, are called the Lame constants.

Contraction on the indices i and j yields a relationship between the first invariants of the stress and strain tensors:

$$\sigma_{kk} = (3\lambda + 2\mu)E_{kk}. \qquad (5\text{-}2.4)$$

We may now invert equation 5-2.3 to obtain an expression for strains in terms of stresses:

$$E_{ij} = \frac{1}{2\mu}\left\{\sigma_{ij} - \frac{\lambda}{3\lambda + 2\mu}\sigma_{kk}\delta_{ij}\right\}. \qquad (5\text{-}2.5)$$

The constitutive equation for linear elasticity may also be written in two parts, one relating the spherical stress tensor to the spherical strain tensor, and another relating the deviatoric parts of these two tensors. Equation

* J. F. Nye, *Physical Properties of Crystals*, Oxford, Clarendon Press, 1957.

† S. G. Lekhnitskii, *Anisotropic Plates*, Gordon and Breach, Science Publishers, Inc., New York, 1968; R. F. S. Hearmon, *An Introduction to Applied Anisotropic Elasticity*, Oxford University Press, London, 1961.

5-2.4 when divided by three yields

$$\sigma_{ij}^s = (3\lambda + 2\mu)E_{ij}^s,$$

where

$$\sigma_{ij}^s = \frac{\sigma_{kk}}{3}\delta_{ij} \quad \text{and} \quad E_{ij}^s = \frac{E_{kk}}{3}\delta_{ij}. \tag{5-2.6}$$

The bulk modulus is $(3\lambda + 2\mu)$ and relates the average normal stress to the volumetric change. Subtracting equation 5-2.6 from equation 5-2.3 yields a relationship between the deviatoric tensors:

$$\sigma_{ij}^d = 2\mu E_{ij}^d. \tag{5-2.7}$$

These forms of the constitutive equations are most often seen in introductions to the subjects of plasticity or viscoelasticity, where special assumptions are usually made concerning the compressibility of the material.

The reduction to the isotropic case presented here uses the general properties of a fourth-order isotropic tensor, and this reduction is both formal and brief. Although the study of anisotropic materials is not included in this text, it is beneficial to indicate how the reduction to some special anisotropic constitutive equations is obtained. If the x_1x_2 plane is considered a plane of symmetry, the constitutive equation must be independent of a transformation that carries the $+z$ axis into the $-z$ axis. The transformation matrix is

$$a_{ij} = \begin{pmatrix} 1 & 0 & 0 \\ 0 & 1 & 0 \\ 0 & 0 & -1 \end{pmatrix}. \tag{5-2.8}$$

Using the symmetry of the stress and strain tensors, the constitutive equation in the transformed coordinate system is

$$\begin{aligned}
\sigma'_{11} &= c_{1111}\epsilon'_{11} + c_{1122}\epsilon'_{22} + c_{1133}\epsilon'_{33} + c_{1112}\epsilon'_{12} + c_{1113}\epsilon'_{13} + c_{1123}\epsilon'_{23},\\
\sigma'_{22} &= c_{2211}\epsilon'_{11} + c_{2222}\epsilon'_{22} + c_{2233}\epsilon'_{33} + c_{2212}\epsilon'_{12} + c_{2213}\epsilon'_{13} + c_{2223}\epsilon'_{23},\\
\sigma'_{33} &= c_{3311}\epsilon'_{11} + c_{3322}\epsilon'_{22} + c_{3333}\epsilon'_{33} + c_{3312}\epsilon'_{12} + c_{3313}\epsilon'_{13} + c_{3323}\epsilon'_{23},\\
\sigma'_{12} &= c_{1211}\epsilon'_{11} + c_{1222}\epsilon'_{22} + c_{1233}\epsilon'_{33} + c_{1212}\epsilon'_{12} + c_{1213}\epsilon'_{13} + c_{1223}\epsilon'_{23},\\
\sigma'_{13} &= c_{1311}\epsilon'_{11} + c_{1322}\epsilon'_{22} + c_{1333}\epsilon'_{33} + c_{1312}\epsilon'_{12} + c_{1313}\epsilon'_{13} + c_{1323}\epsilon'_{23},\\
\sigma'_{23} &= c_{2311}\epsilon'_{11} + c_{2322}\epsilon'_{22} + c_{2333}\epsilon'_{33} + c_{2312}\epsilon'_{12} + c_{2313}\epsilon'_{13} + c_{2323}\epsilon'_{23}.
\end{aligned} \tag{5-2.9}$$

In this form there are only 36 material constants, because coefficients multiplying symmetrical components of strain have been combined and equations for symmetrical components of stress are the same as those shown.

Transforming each component of the stress and strain tensors to the unprimed coordinate system yields

$$
\begin{aligned}
\sigma_{11} &= c_{1111}\epsilon_{11} + c_{1122}\epsilon_{22} + c_{1133}\epsilon_{33} + c_{1112}\epsilon_{12} - c_{1113}\epsilon_{13} - c_{1123}\epsilon_{23},\\
\sigma_{22} &= c_{2211}\epsilon_{11} + c_{2222}\epsilon_{22} + c_{2233}\epsilon_{33} + c_{2212}\epsilon_{12} - c_{2213}\epsilon_{13} - c_{2223}\epsilon_{23},\\
\sigma_{33} &= c_{3311}\epsilon_{11} + c_{3322}\epsilon_{22} + c_{3333}\epsilon_{33} + c_{3312}\epsilon_{12} - c_{3313}\epsilon_{13} - c_{3323}\epsilon_{23},\\
\sigma_{12} &= c_{1211}\epsilon_{11} + c_{1222}\epsilon_{22} + c_{1233}\epsilon_{33} + c_{1212}\epsilon_{12} - c_{1213}\epsilon_{13} - c_{1223}\epsilon_{23},\\
\sigma_{13} &= -c_{1311}\epsilon_{11} - c_{1322}\epsilon_{22} - c_{1333}\epsilon_{33} - c_{1312}\epsilon_{13} + c_{1313}\epsilon_{13} + c_{1323}\epsilon_{23},\\
\sigma_{23} &= -c_{2311}\epsilon_{11} - c_{2322}\epsilon_{22} - c_{2333}\epsilon_{33} - c_{2312}\epsilon_{12} + c_{2313}\epsilon_{23} + c_{2323}\epsilon_{23}.
\end{aligned}
\tag{5-2.10}
$$

If the x_1x_2 plane is to be a plane of symmetry, the stress–strain relations in equations 5-2.9 and 5-2.10 must be the same. This can be true only if those constants multiplying terms which changed signs are zero. The remaining coefficients displayed as an array appear as follows:

$$
C_{ijkl} = \begin{pmatrix}
c_{1111} & c_{1122} & c_{1133} & c_{1112} & 0 & 0\\
c_{2211} & c_{2222} & c_{2233} & c_{2212} & 0 & 0\\
c_{3311} & c_{3322} & c_{3333} & c_{3312} & 0 & 0\\
c_{1211} & c_{1222} & c_{1233} & c_{1212} & 0 & 0\\
0 & 0 & 0 & 0 & c_{1313} & c_{1323}\\
0 & 0 & 0 & 0 & c_{2313} & c_{2323}
\end{pmatrix}. \tag{5-2.11}
$$

If the material has three orthogonal planes of symmetry (orthotropic symmetry), the coefficient array is

$$
C_{ijkl} = \begin{pmatrix}
c_{1111} & c_{1122} & c_{1133} & 0 & 0 & 0\\
c_{2211} & c_{2222} & c_{2233} & 0 & 0 & 0\\
c_{3311} & c_{3322} & c_{3333} & 0 & 0 & 0\\
0 & 0 & 0 & c_{1212} & 0 & 0\\
0 & 0 & 0 & 0 & c_{1313} & 0\\
0 & 0 & 0 & 0 & 0 & c_{2323}
\end{pmatrix}. \tag{5-2.12}
$$

If it is assumed that the material is the same in any direction, the constitutive equation must be unchanged by any transformation of coordinate axes. This restriction leads directly to the equations for an isotropic material.

3 Summary of the equations of isotropic elasticity

The equations of elasticity may be expressed in either the material or the spatial coordinate systems. In summarizing these equations, the energy-balance equation will not be considered. A discussion of this equation and the general field of thermoelasticity appears in section 8.

The 19 equations of elasticity in the Eulerian coordinates involve 19 unknowns and are

$$E^*_{ij} = \frac{1}{2}\left\{\frac{\partial u_i}{\partial y_j} + \frac{\partial u_j}{\partial y_i} - \frac{\partial u_k}{\partial y_i}\frac{\partial u_k}{\partial y_j}\right\},$$

$$\nabla_j T_{ji} + \rho b_i = \rho\frac{dv_i}{dt} = \rho\left\{\frac{\partial v_i}{\partial t} + v_j \nabla_j v_i\right\},$$

$$T_{ij} = \lambda E^*_{kk}\delta_{ij} + 2\mu E^*_{ij}, \qquad (5\text{-}3.1)$$

$$\frac{\partial \rho}{\partial t} + \rho \nabla_i v_i + v_i \nabla_i \rho = 0,$$

$$v_i = \frac{\partial u_i}{\partial t} + v_j \nabla_j u_i.$$

In the Lagrangian system we have 15 equations and 15 unknowns:

$$E_{ij} = \frac{1}{2}\left\{\frac{\partial u_i}{\partial x_j} + \frac{\partial u_j}{\partial x_i} + \frac{\partial u_k}{\partial x_i}\frac{\partial u_k}{\partial x_j}\right\},$$

$$\frac{\partial}{\partial x_j}\left[\sigma_{jk}\frac{\partial y_i}{\partial x_k}\right] + \rho_0 b_{0i} = \rho_0 \frac{\partial v_i}{\partial t}, \qquad (5\text{-}3.2)$$

$$\frac{\partial}{\partial x_j}\left[\sigma_{jk}\left(\frac{\partial u_i}{\partial x_k} + \delta_{ik}\right)\right] + \rho_0 b_{0i} = \rho_0 \frac{\partial^2 u_i}{\partial t^2},$$

$$\sigma_{ij} = \lambda E_{kk}\delta_{ij} + 2\mu E_{ij}.$$

If the two systems of equations are linearized, they are similar in appearance. In the small deformation (infinitesimal) linear theory of elasticity, the distinction is negligible and we can express the basis equations as

$$\epsilon_{ij} = \frac{1}{2}\left(\frac{\partial u_i}{\partial x_j} + \frac{\partial u_j}{\partial x_i}\right),$$

$$\frac{\partial \sigma_{ij}}{\partial x_i} + \rho b_j = \rho \frac{\partial^2 u_j}{\partial t^2}, \qquad (5\text{-}3.3)$$

$$\sigma_{ij} = \lambda \epsilon_{kk}\delta_{ij} + 2\mu\epsilon_{ij}.$$

These comprise 15 equations in 15 unknowns.*

4 Boundary conditions

If problems of elastostatics are considered, the boundary conditions usually appear as two types, displacements specified and-or surface tractions specified. In general, on all boundaries either the surface traction vector

* If the continuum is at rest, the right-hand side of Cauchy's equations of motion is zero and the resulting equations are called the equilibrium equations. We then have the general field of elastostatics, which is the main topic of this book.

or the displacement vector must be specified. In some cases, called mixed problems, displacements may be specified on some boundaries and stress tractions on others. We may specify components of each vector on the same boundary region. However, the same component of the stress traction vector and the displacement vector cannot be specified at the same point on a boundary.

Where surface tractions are specified, we may write the boundary condition in terms of the ith component of the stress traction on a surface whose normal vector is $\hat{\boldsymbol{n}}$:

$$t_i^n = n_j \sigma_{ji} = \text{given function on } S. \tag{5-4.1}$$

Using the constitutive equation, this may be written in terms of displacements as

$$n_j[\lambda \nabla_k u_k \delta_{ij} + \mu(\nabla_i u_j + \nabla_j u_i)] = \text{given function on } S. \tag{5-4.2}$$

The displacement boundary condition takes the form

$$u_i = \text{given function on } S. \tag{5-4.3}$$

5 Uniqueness and superposition

For a "Green-elastic" material (sometimes called a hyperelastic material), a strain energy function W is assumed to exist, with the property that

$$\sigma_{ij} = \frac{\partial W}{\partial \epsilon_{ij}}. \tag{5-5.1}$$

The existence of this function can be shown to place symmetry constraints on the constants involved in the generalized Hooke's law,

$$\sigma_{ij} = c_{ijrs}\epsilon_{rs}. \tag{5-5.2}$$

If the material obeys Hooke's law, the strain energy function can be written

$$W = \tfrac{1}{2} c_{ijkl}\epsilon_{ij}\epsilon_{kl}. \tag{5-5.3}$$

The symmetry conditions take the form

$$c_{ijrs} = c_{rsij} \tag{5-5.4}$$

and effectively reduce the 36 constants to 21 independent constants. For small strains it may be shown that the function W is positive definite. The positive definiteness of this function allows one to develop uniqueness theorems and minimum potential and complementary energy theorems.*

If either displacements or stress tractions are specified on the boundaries, a uniqueness theorem due to Kirchhoff may be established. This is

* Positive definiteness implies that the function is nonnegative and is zero only in the "natural" (unstrained) state.

based on the positive definiteness of the strain energy function. Consider the body shown in Figure 5-1. The following simplified version of the allowable boundary conditions is assumed:

$$u_i \text{ given on } S' \quad \text{and} \quad t_i^n \text{ given on } S''. \tag{5-5.5}$$

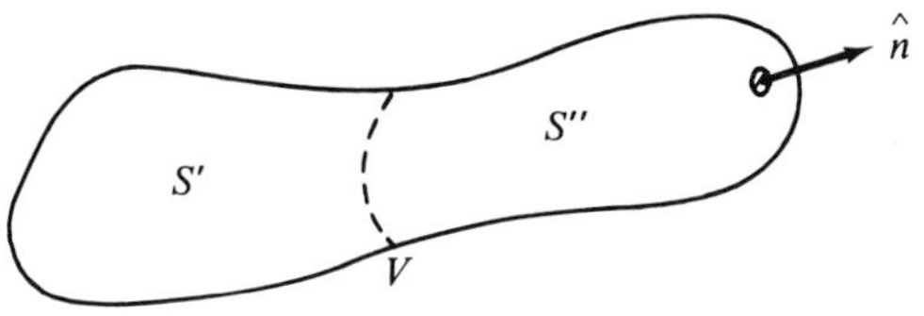

Figure 5-1

Note that the stress traction vector may be written

$$t_i^n = n_j \sigma_{ji} = n_j \frac{\partial W}{\partial \epsilon_{ji}}. \tag{5-5.6}$$

The equilibrium equations for elastostatics are

$$\frac{\partial \sigma_{ij}}{\partial x_i} + \rho b_j = 0. \tag{5-5.7}$$

Uniqueness will be proved by assuming that there exist two distinct solutions, $\bar{u}^1$ and $\bar{u}^2$, which satisfy equation 5-5.7 and the boundary conditions 5-5.5. It will then be shown that the difference, $\bar{u} = \bar{u}^1 - \bar{u}^2$, is zero everywhere in the body. The "difference problem" satisfies the differential equations

$$\frac{\partial \sigma_{ij}}{\partial x_i} = 0, \tag{5-5.8}$$

where $\sigma_{ij} = \sigma_{ij}^1 - \sigma_{ij}^2$.

Equations 5-5.8 may be written

$$\frac{\partial}{\partial x_i}\left(\frac{\partial W}{\partial \epsilon_{ij}}\right) = 0. \tag{5-5.9}$$

The boundary conditions on the difference problem are

$$u_i = 0 \text{ on } S' \quad \text{and} \quad n_j \frac{\partial W}{\partial \epsilon_{ji}} = 0 \text{ on } S''. \tag{5-5.10, 11}$$

Consider the following integral:

$$\int_V u_j \frac{\partial}{\partial x_i}\left(\frac{\partial W}{\partial \epsilon_{ij}}\right) dV = 0. \tag{5-5.12}$$

Note this integral is zero everywhere in the volume due to equation 5-5.9. Equation 5-5.12 may be written

$$\int_V \frac{\partial}{\partial x_i}\left[u_j \frac{\partial W}{\partial \epsilon_{ij}}\right] dV - \int_V \left[\frac{\partial u_j}{\partial x_i} \frac{\partial W}{\partial \epsilon_{ij}}\right] dV = 0. \tag{5-5.13}$$

Using the divergence theorem and the symmetry of the strain tensor,

$$\int_S n_i \frac{\partial W}{\partial \epsilon_{ij}} u_j \, dS - \int_V \frac{\partial W}{\partial \epsilon_{ij}} \epsilon_{ij} \, dV = 0. \qquad (5\text{-}5.14)$$

The first integral is zero on the surface by equations 5-5.10 and 5-5.11.* The second integral may be written by use of equation 5-5.4 as

$$\int_V 2W \, dV = 0. \qquad (5\text{-}5.15)$$

Since W is assumed to be positive definite, the integral can vanish only if W vanishes everywhere in the body. This implies that the difference solution is zero and that $\bar{u}^1$ and $\bar{u}^2$ are the same; thus the solution is unique.

Due to the linearity of the problem and the uniqueness of the solution, a superposition principle may be used. The boundary conditions may be divided between two or more problems and the results combined linearly.

6 Saint-Venant's principle

In 1853 in his "Memoire sur la Torsion des Prismes," Saint-Venant developed solutions for the torsion of prismatic bars which gave the same stress distribution for all cross sections.† He attempted to justify the usefulness of his formulation by the following: "The fact is that the means of application and distribution of the forces toward the extremities of the prisms is immaterial to the perceptible effects produced on the rest of the length, so that one can always, in a sufficiently similar manner, replace the forces applied with equivalent static forces or with those having the same total moments and the same resultant forces . . . ."

In a footnote, Saint-Venant goes on to state that the influence of forces in equilibrium acting on a small portion of a body extend very little beyond parts upon which they act. This principle is actually summarized by one of its more detailed names, the *principle of the elastic equivalence of statically equipollent systems of load.*

M. J. Boussinesq was one of the most distinguished of the pupils of Saint-Venant. In 1885, he published "Application des potentials a l'etude de l'equilibre et du mouvement des solides élastiques." Boussinesq summarized his results of the study of the Saint-Venant principle for the elastic half-space. In 1895, A. E. H. Love‡ stated a more popular version of this prin-

* The surface integral may be written as $\int_S n_i \sigma_{ij} u_j \, ds$. The condition that this integral vanishes indicates that for each value of j, either $n_i \sigma_{ij}$ or u_j can be specified for a unique solution. This demonstates the conditions discussed in Section 4.

† See Chapter 12.

‡ A. E. H. Love, *A Treatise on the Mathematical Theory of Elasticity*, Dover Publications, Inc, New York, 1944, 4th ed. (1926), pp. 131–132.

ciple, which was used by applied mathematicians until 1945; "the strains that are produced in a body by the application, to a small part of its surface, of a system of forces statically equivalent to zero force and zero couple (self-equilibrating), are of negligible magnitude at distances which are large compared with the linear dimensions on the part."

In 1945, R. von Mises* examined tangential forces applied to the boundary and proposed a modification to the principle. He noted that under certain conditions the stress did not decay at a faster rate under simple static equilibrium. He introduced the concept of astatic equilibrium. Astatic equilibrium requires that the surface forces remain in equilibrium even when turned through an arbitrary angle. Another way of imposing this condition is as follows:

$$\begin{aligned} &\sum \bar{F} = 0, && F_i = 0, && \\ &\sum \bar{r}\bar{F} = 0, && x_j F_i = 0, && \text{astatic equilibrium.} \\ &\sum \bar{F} = 0, && F_i = 0, && \\ &\sum \bar{r} \times \bar{F} = 0, && e_{ijk} x_j F_k = 0, && \text{static equilibrium.} \end{aligned}$$

Note that the condition of astatic equilibrium does not come into play unless tangential loads are applied. Von Mises studied examples of the disk to varify this. E. Sternberg† gave a formal proof of this modified principle in 1954.

This principle has been used extensively by elasticians to replace stress distributions on short boundaries by ones that are easier to handle mathematically. Examples of its use will be shown throughout this book.

7 Displacement formulation

One of the boundary conditions that occurs naturally in elasticity problems is the specification of displacements on the boundaries. Even when stresses are specified, these can be stated in terms of derivatives of the displacements. This motivates us to attempt a formulation in terms of displacements only, and the 15 linear equations of elasticity may be reduced to three equations in terms of the displacements. Substituting the definitions of strain in terms of displacements into the stress–strain law yields

$$\sigma_{ij} = \lambda \epsilon_{kk} \delta_{ij} + \mu(\nabla_j u_i + \nabla_i u_j). \tag{5-7.1}$$

Substituting these equations into the equations of motion gives

$$\nabla_i(\lambda \epsilon_{kk})\delta_{ij} + \nabla_i\{\mu(\nabla_j u_i + \nabla_i u_j)\} + \rho b_j = \rho \ddot{u}_j. \tag{5-7.2}$$

* R. von Mises, "On Saint-Venant's principle," *Bull. Amer. Math. Soc.*, **51**, 555, 1945.

† E. Sternberg, "On Saint-Venant's principle," *Quart. Appl. Math.* **11**, 393–402, 1954.

If the material is assumed to be homogeneous, one obtains

$$\lambda \nabla_j \epsilon_{kk} + \mu \nabla_j \nabla_i u_i + \mu \nabla^2 u_j + \rho b_j = \rho \ddot{u}_j, \tag{5-7.3}$$

$$\mu \nabla^2 u_j + (\lambda + \mu) \nabla_j \nabla_i u_i + \rho b_j = \rho \ddot{u}_j. \tag{5-7.4}$$

These are called the Navier equations of elasticity, and they may be rewritten in terms of engineering constants instead of the Lamé constants. The engineering constants are obtained by examining a state of uniaxial tension. Young's modulus, E, is defined to be the ratio of stress to strain in the uniaxial loading direction:

$$\sigma_{11} = E\epsilon_{11} \qquad \text{if } \sigma_{22} = \sigma_{33} = \sigma_{12} = \sigma_{13} = \sigma_{23} = 0. \tag{5-7.5}$$

From Hooke's law, one obtains

$$\begin{aligned} \sigma_{11} &= \lambda(\epsilon_{11} + \epsilon_{22} + \epsilon_{33}) + 2\mu\epsilon_{11}, \\ 0 &= \lambda(\epsilon_{11} + \epsilon_{22} + \epsilon_{33}) + 2\mu\epsilon_{22}, \\ 0 &= \lambda(\epsilon_{11} + \epsilon_{22} + \epsilon_{33}) + 2\mu\epsilon_{33}. \end{aligned} \tag{5-7.6}$$

Adding these equations yields

$$\sigma_{11} = (3\lambda + 2\mu)\epsilon_{ii}, \tag{5-7.7}$$

$$\epsilon_{ii} = \frac{\sigma_{11}}{3\lambda + 2\mu}, \tag{5-7.8}$$

$$\sigma_{11} = \frac{\mu(3\lambda + 2\mu)}{\lambda + \mu}\epsilon_{11},$$

$$E = \frac{\mu(3\lambda + 2\mu)}{\lambda + \mu}. \tag{5-7.9}$$

Poisson's ratio, ν, defined is as follows:

$$\nu = -\frac{\epsilon_{22}}{\epsilon_{11}}, \tag{5-7.10}$$

$$2\mu\epsilon_{22} = -\lambda\epsilon_{ii} = -\frac{\lambda}{3\lambda + 2\mu}\sigma_{11} = -\frac{\lambda\mu}{\lambda + \mu}\epsilon_{11},$$

$$\nu = \frac{\lambda}{2(\lambda + \mu)} \quad \text{or} \quad \lambda = \frac{2\mu\nu}{1 - 2\nu}. \tag{5-7.11}$$

Substituting for λ in the expression for E yields

$$G = \mu = \frac{E}{2(1 + \nu)}^{*} \quad \text{and} \quad \lambda = \frac{E\nu}{(1 + \nu)(1 - 2\nu)}. \tag{5-7.12}$$

The basic constitutive equation may now be written

$$\epsilon_{ij} = \frac{1}{E}\{(1 + \nu)\sigma_{ij} - \nu\sigma_{kk}\delta_{ij}\}, \tag{5-7.13}$$

$$\epsilon_{ii} = \frac{1 - 2\nu}{E}\sigma_{ii}. \tag{5-7.14}$$

* G is frequently used in place of μ and is called the shear modulus of elasticity or the modulus of rigidity.

Using the relations between λ, μ, and ν, we may now rewrite the Navier equations in the form

$$\mu\left\{\nabla^2 u_j + \frac{1}{1-2\nu}\nabla_j(\nabla_i u_i)\right\} + \rho b_j = \rho \ddot{u}_j. \qquad (5\text{-}7.15)$$

Alternatively, using the vector identity

$$\nabla^2 u_j = \nabla_j \nabla_i u_i - e_{jkl} e_{lmn} \nabla_k \nabla_m u_n \qquad (5\text{-}7.16)$$

or

$$\nabla^2 \bar{u} = \bar{\nabla}(\bar{\nabla}\cdot\bar{u}) - \bar{\nabla}\times(\bar{\nabla}\times\bar{u}), \qquad (5\text{-}7.17)$$

still another form is possible.

In symbolic notation, Navier's equations become

$$\begin{aligned} \mu(\bar{\nabla}\cdot\bar{u}) - \mu\bar{\nabla}\times(\bar{\nabla}\times\bar{u}) + (\lambda+\mu)\bar{\nabla}(\bar{\nabla}\cdot\bar{u}) + \rho\bar{b} &= \rho\ddot{\bar{u}}, \\ (\lambda+2\mu)\bar{\nabla}(\bar{\nabla}\cdot\bar{u}) - \mu\bar{\nabla}\times(\bar{\nabla}\times\bar{u}) + \rho\bar{b} &= \rho\ddot{\bar{u}}, \end{aligned} \qquad (5\text{-}7.18)$$

or

$$G\left\{\frac{2(1-\nu)}{1-2\nu}\bar{\nabla}(\bar{\nabla}\cdot\bar{u}) - \bar{\nabla}\times(\bar{\nabla}\times\bar{u})\right\} + \rho\bar{b} = \rho\ddot{\bar{u}}. \qquad (5\text{-}7.19)$$

We may gain a better feel for the variables involved in this equation by examining it in the symbolic form corresponding to equation 5-7.4:

$$\mu\nabla^2\bar{u} + (\lambda+\mu)\bar{\nabla}(\bar{\nabla}\cdot\bar{u}) + \rho\bar{b} = \rho\ddot{\bar{u}}. \qquad (5\text{-}7.20)$$

If problems of elastostatics are considered, $\rho\ddot{\bar{u}} = 0$, and if body forces are not present or are handled separately, the equations become

$$\mu\nabla^2\bar{u} + (\lambda+\mu)\bar{\nabla}(\bar{\nabla}\cdot\bar{u}) = 0. \qquad (5\text{-}7.21)$$

Taking the divergence of this form of the equation yields

$$\mu\bar{\nabla}\cdot(\nabla^2\bar{u}) + (\lambda+\mu)\bar{\nabla}\cdot\bar{\nabla}(\bar{\nabla}\cdot\bar{u}) - 0, \qquad (5\text{-}7.22)$$

$$(\lambda+2\mu)\nabla^2(\bar{\nabla}\cdot\bar{u}) = 0. \qquad (5\text{-}7.23)$$

Now

$$\epsilon_{kk} = \bar{\nabla}\cdot\bar{u}, \qquad (5\text{-}7.24)$$

and

$$\sigma_{ij} = \lambda\epsilon_{kk}\delta_{ij} + 2\mu\epsilon_{ij}, \qquad (5\text{-}7.25)$$

$$\sigma_{kk} = (3\lambda+2\mu)\epsilon_{kk}; \qquad (5\text{-}7.26)$$

thus $\nabla^2(\epsilon_{kk}) = 0$ and the divergence of the displacement is a harmonic function. It may be seen from equation 5-7.26 that in the absence of body forces, the first invariants of both the stress and strain tensors are harmonic functions.

Applying the Laplace operator to the Navier equation yields

$$\mu\nabla^2\nabla^2\bar{\mu} + (\lambda+\mu)\nabla^2\bar{\nabla}(\bar{\nabla}\cdot\bar{u}) = 0. \qquad (5\text{-}7.27)$$

Using the harmonic character of the divergence of the displacement vector, one obtains

$$\nabla^2\nabla^2\bar{u} = 0. \qquad (5\text{-}7.28)$$

Each component of the displacement vector is therefore a biharmonic function in Cartesian coordinates. By examining the stress–strain relations and the definition of strain, it is apparent that each component of the stress and strain tensors is also biharmonic in this coordinate system:

$$\nabla^2\nabla^2\sigma_{ij} = 0, \tag{5-7.29}$$

$$\nabla^2\nabla^2\epsilon_{ij} = 0. \tag{5-7.30}$$

If body forces are present in elastostatic problems, equation 5-7.23 becomes

$$(\lambda + 2\mu)\nabla^2\bar{\nabla}\cdot\bar{u} = -\rho\bar{\nabla}\cdot\bar{b}. \tag{5-7.31}$$

This may be stated as

$$\nabla^2\epsilon_{kk} = -\frac{(1+\nu)(1-2\nu)}{E(1-\nu)}\rho\bar{\nabla}\cdot\bar{b}. \tag{5-7.32}$$

Using the relation between the invariants of the stress and strain tensors,

$$\epsilon_{kk} = \frac{1-2\nu}{E}\sigma_{kk}, \tag{5-7.33}$$

we find that

$$\nabla^2\sigma_{kk} = -\frac{1+\nu}{1-\nu}\rho\bar{\nabla}\cdot\bar{b}. \tag{5-7.34}$$

Equation 5-7.34 will serve as one of the basic points of departure for two-dimensional elasticity; it is, however, fundamentally three dimensional.

8 Thermoelasticity

The effects of a temperature change may be introduced into the theory of elasticity by examining the energy function W introduced in Section 6. If W is assumed to be in the form of a power series expansion in strain, one obtains

$$W = C_0 + C'_{ij}\epsilon_{ij} + \tfrac{1}{2}C_{ijkl}\epsilon_{ij}\epsilon_{kl} + \cdots. \tag{5-8.1}$$

The stress tensor may be related to the strain tensor by the relationship

$$\sigma_{ij} = \frac{\partial W}{\partial \epsilon_{ij}} = C'_{ij} + C_{ijkl}\epsilon_{kl}. \tag{5-8.2}$$

The second term leads to the generalized Hooke's law discussed earlier. The first term implies that stresses may be present in a body which is in a state of zero strain. Such stresses may arise if a linear expansion occurs due to temperature changes. The simplest assumption would be that C'_{ij} is linearly proportional to the temperature change $(T - T_0)$ and equal in all directions for an isotropic material:

$$C'_{ij} = -\alpha(3\lambda + 2\mu)(T - T_0)\delta_{ij}, \tag{5-8.3}$$

where α is the coefficient of linear thermal expansion. The stress–strain relation becomes

$$\sigma_{ij} = \lambda\delta_{ij}\epsilon_{kk} + 2\mu\epsilon_{ij} - \alpha(3\lambda + 2\mu)(T - T_0)\delta_{ij}. \qquad (5\text{-}8.4)$$

This equation is known as the Duhamel–Neumann generalization of Hooke's law.* The thermal strain equation can be obtained by adding a thermal expansion term to the strain equation 5-2.5:

$$\epsilon_{ij} = \frac{1}{2\mu}\left[\sigma_{ij} - \frac{\lambda}{3\lambda + 2\mu}\sigma_{kk}\delta_{ij}\right] + \alpha\delta_{ij}(T - T_0).$$

Equation 5-8.4 is the inverted form of this equation.

For a continuum at rest, the energy-balance equation 4-5.8 reduces to

$$\rho\frac{de}{dt} = -\bar{\nabla}\cdot\bar{q}. \qquad (5\text{-}8.5)$$

The material time derivative of e becomes a simple partial time derivative when this equation is expressed in material coordinates, or the velocity is zero. The specific energy is a function of the strain tensor and the temperature and is related to W as follows:

$$e(\epsilon_{ij}, T) = W(\epsilon_{ij}, T) + T\eta(\epsilon_{ij}, T). \qquad (5\text{-}8.6)$$

The free energy function is W, and η is the specific entropy. If the strain rates, $\partial\epsilon_{ij}/\partial t$, are small, equation 5-8.5 may be written

$$-\frac{\partial q_i}{\partial x_i} = \rho c_E\frac{\partial T}{\partial t}, \qquad (5\text{-}8.7)$$

where c_E is the specific heat at constant volume.

If Fourier's law of heat conduction is assumed, the energy flux vector $\bar{q}$ may be related to the thermal gradient:

$$\bar{q} = -k\bar{\nabla}T, \qquad (5\text{-}8.8)$$

where k is known as the thermal conductivity of the solid.

The energy balance equation may be written as

$$k\nabla^2 T = \rho c_E\frac{\partial T}{\partial t}. \qquad (5\text{-}8.9)$$

If the continuum is not at rest, the energy-balance equation would take the form

$$k\nabla^2 T = \rho c_E\frac{\partial T}{\partial t} + (3\lambda + 2\mu)\alpha\frac{\partial\epsilon_{kk}}{\partial t}. \qquad (5\text{-}8.10)$$

When the last term cannot be neglected, the thermal and mechanical effects are coupled and a simultaneous determination of temperature and deforma-

* For further information on the role of the Duhamel–Neumann equation in thermodynamics, see "Principles of Classical Mechanics and Field Theory," *Handbuch der Physik*, Springer-Verlag, Berlin, 1960, pp. 615–642; or B. A. Boley and J. H. Weiner, *Theory of Thermal Stresses*, John Wiley & Sons, Inc., 1960, pp. 19–30.

tion must be made. When variations in strain occur, the coupled equations imply variations in temperature and a flow of heat. This is known as *thermal dissipation*, and the coupled equations must be examined if this phenomenon is to be studied. If the strain rates are small and are accompanied by small changes in temperature, the coupling effects may often be neglected. The thermal and the displacement problems then separate and may be solved as two independent boundary-value problems. Equation 5-8.6 or 5-8.9 is solved first and the appropriate temperature field introduced in the Duhamel–Neumann equation.

A Navier equation for the decoupled thermal elasticity theory may be developed in a manner similar to that used before, yielding

$$(\lambda + \mu)\bar{\nabla}(\bar{\nabla} \cdot \bar{u}) + \mu\nabla^2\bar{u} - (3\lambda + 2\mu)\alpha\bar{\nabla}T + \rho\bar{b} = 0. \qquad (5\text{-}8.11)$$

The temperature gradient may then be combined with the distributed body force and be considered as a generalized body force. The problem is solved in the same manner as when body forces only are present, and the stresses are obtained from the Duhamel–Neumann equations.

PROBLEMS

5.1. Show that the assumption that an elastic isotropic solid is incompressible is equivalent to the assumption that Poisson's ratio is equal to one half.

5.2. Derive the compatibility equations in terms of stresses (Beltrami–Michell compatibility equations) in the following form:

$$\nabla^2\sigma_{ij} + \frac{1}{(1+\nu)}\nabla_i\nabla_j\sigma_{kk} = -\frac{\nu}{1-\nu}\delta_{ij}\rho\nabla_k b_k - \rho(\nabla_j b_i + \nabla_i b_j).$$

Hint: Start with equation 2-7.2 and use equation 1-5.13 and the stress–strain relations, together with the equations of equilibrium.

5.3. Prove that the principal axes of stress coincide with the principal axes of strain for a linear, isotropic, elastic solid, and relate the principal values to each other.

5.4. Write the stress–strain relations in polar cylindrical coordinates.

5.5. Write Hooke's law for a material having one plane of symmetry.

5.6. Show that the isotropic Hooke's law may be written as

$$\sigma_{ij} = 2\mu\left[\epsilon_{ij} + \frac{\nu}{1-2\nu}\epsilon_{kk}\delta_{ij}\right].$$

5.7. Derive equation 5-8.11.

PART II

TWO-DIMENSIONAL ELASTICITY

CHAPTER 6

GENERAL THEORY OF PLANE ELASTICITY

1 Introduction

The plane problems may be discussed from the classical point of view; (1) the case of *plane stress*, corresponding to a plane plate of small constant thickness, acted upon by forces in the plane of the plate only; (2) the case of *plane deformation* or *plane strain*, corresponding to a long (theoretically infinite) body, acted upon by loads that are uniformly distributed in the infinite direction and have no components normal to the finite planes. The cases of *quasi-plane state of stress*, in which the z variable is ignored, and *generalized plane stress*, in which mean state of stress across the thickness is considered, can be shown to be equivalent to the plane stress when the thickness of the plate is small.*

* For a discussion, see S. P. Timoshenko and J. N. Goodier, *Theory of Elasticity*, 3rd ed., McGraw-Hill Book Company, 1970, p. 274.

Due to the large number of applications, both of the plane problems have received considerable attention. An excellent review of these investigations appears in the article, "One Hundred Years of Investigations on the Plane Problem of the Theory of Elasticity" by P. P. Teodorescu.*

2 Plane deformation or plane strain

Consider the case when the displacement in the z direction is zero, $u_z = 0$, and the other displacements u_x and u_y are independent of z. This leads to a condition of plane strain, since

$$\epsilon_{xz} = \epsilon_{yz} = \epsilon_{zz} = 0. \tag{6-2.1}$$

If the equations of elasticity are examined, it may be noted that σ_{xz} and σ_{yz} are zero and the other stresses are independent of z. The basic equations may then be reduced to the form

$$\nabla_\alpha \sigma_{\alpha\beta} + \rho b_\beta = 0, \tag{6-2.2}$$

$$\epsilon_{\alpha\beta} = \tfrac{1}{2}(\nabla_\alpha u_\beta + \nabla_\beta u_\alpha), \tag{6-2.3}$$

$$\sigma_{\alpha\beta} = \lambda e \delta_{\alpha\beta} + 2\mu\epsilon_{\alpha\beta}, \qquad \text{where } e = \epsilon_{\lambda\lambda}, \tag{6-2.4}$$

$$\epsilon_{\alpha\beta} = \frac{1}{E}\{(1 + \nu)\sigma_{\alpha\beta} - \nu\delta_{\alpha\beta}S\}, \qquad \text{where } S = \sigma_{\lambda\lambda} + \sigma_{zz}. \tag{6-2.5}$$

The Greek subscripts imply the summation convention in two dimensions. These equations may be expanded into scalar notation as follows:

$$\frac{\partial \sigma_{xx}}{\partial x} + \frac{\partial \sigma_{xy}}{\partial y} + \rho b_x = 0,$$

$$\frac{\partial \sigma_{xy}}{\partial x} + \frac{\partial \sigma_{yy}}{\partial y} + \rho b_y = 0,$$

$$\epsilon_{xx} = \frac{\partial u_x}{\partial x}, \qquad \epsilon_{yy} = \frac{\partial u_y}{\partial y}, \qquad \epsilon_{xy} = \frac{1}{2}\left(\frac{\partial u_x}{\partial y} + \frac{\partial u_y}{\partial x}\right),$$

$$\sigma_{xx} = \lambda(\epsilon_{xx} + \epsilon_{yy}) + 2\mu\epsilon_{xx}, \qquad \epsilon_{xx} = \frac{1}{E}[\sigma_{xx} - \nu(\sigma_{yy} + \sigma_{zz})],$$

$$\sigma_{yy} = \lambda(\epsilon_{xx} + \epsilon_{yy}) + 2\mu\epsilon_{yy}, \qquad \epsilon_{yy} = \frac{1}{E}[\sigma_{yy} - \nu(\sigma_{xx} + \sigma_{zz})],$$

$$\sigma_{zz} = \lambda(\epsilon_{xx} + \epsilon_{yy}), \qquad 0 = \frac{1}{E}[\sigma_{zz} - \nu(\sigma_{xx} + \sigma_{yy})],$$

$$\sigma_{xy} = 2\mu\epsilon_{xy}.$$

It is apparent that after σ_{xx} and σ_{yy} have been obtained, σ_{zz} may be easily calculated by observing that ϵ_{zz} is zero and using the relation

* Appl. Mech. Rev., **17**, No. 3, 175–186, 1964.

$$\sigma_{zz} = \frac{\lambda}{2(\lambda + \mu)}(\sigma_{xx} + \sigma_{yy}), \tag{6-2.6}$$

or

$$\sigma_{zz} = \nu(\sigma_{xx} + \sigma_{yy}). \tag{6-2.7}$$

In general, σ_{zz} is not zero but is a function of x and y only. Therefore, some end forces and couples are required to maintain a state of plane deformation. The practical problems normally encountered are those where u_z is not zero even though the loading is such that u_x and u_y are independent of z. Among these are the important cases where the ends are free of stress. The solution obtained by plane strain may be used as one part of the solution of these problems. The rest consists of a body with stress-free boundaries except at the ends, where the negative of the plane strain end stress is applied. If the end stress, σ_{zz}, were linear in its distribution, this second part would be obtained by use of Saint-Venant bending theory (see Chapter 12) plus a uniform stress to remove the simple axial component. The difference between any end-stress distribution arising in plane strain and the linear distribution is a self-equilibrated end traction. If the body is sufficiently long, this remaining self-equilibrated portion may be ignored by Saint-Venant's principle (see Figure 6-1). This indicates that the argument of plane strain will be used when the dimension in the z direction is very large in comparison with the other two dimensions. We shall see that this is quite different from the argument used for plane stress, where it is required that the z dimension be very small.

The problem may now be formulated either in terms of stresses or displacements. To obtain the reduced Navier equations for the displacement problem, substitute the definitions of strain into the stress–strain relations and substitute those equations into the equilibrium equations:

$$\sigma_{xx} = \lambda\left(\frac{\partial u_x}{\partial x} + \frac{\partial u_y}{\partial y}\right) + 2\mu\left(\frac{\partial u_x}{\partial x}\right), \tag{6-2.8}$$

$$\sigma_{yy} = \lambda\left(\frac{\partial u_x}{\partial x} + \frac{\partial u_y}{\partial y}\right) + 2\mu\left(\frac{\partial u_y}{\partial y}\right), \tag{6-2.9}$$

$$\sigma_{xy} = \mu\left(\frac{\partial u_x}{\partial y} + \frac{\partial u_y}{\partial x}\right). \tag{6-2.10}$$

Using the engineering constants G and ν, the reduced Navier equations become

$$G\left[\nabla^2 u_x + \frac{1}{(1 - 2\nu)}\frac{\partial}{\partial x}\left(\frac{\partial u_x}{\partial x} + \frac{\partial u_y}{\partial y}\right)\right] + \rho b_x = 0, \tag{6-2.11}$$

$$G\left[\nabla^2 u_y + \frac{1}{(1 - 2\nu)}\frac{\partial}{\partial y}\left(\frac{\partial u_x}{\partial x} + \frac{\partial u_y}{\partial y}\right)\right] + \rho b_y = 0, \tag{6-2.12}$$

where ∇^2 is the two-dimensional Laplacian operator, $(\partial^2/\partial x^2) + (\partial^2/\partial y^2)$. This yields two coupled equations for the displacements u_x and u_y.

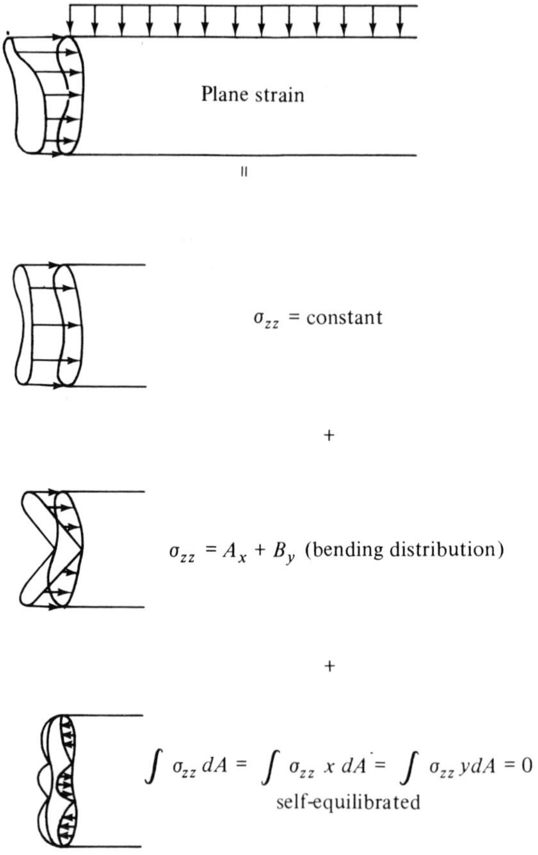

Figure 6-1

To obtain the formulation in terms of stresses, it is necessary to eliminate the displacement components by the following procedure: let P be defined as

$$P = \sigma_{xx} + \sigma_{yy} = (2\lambda + 2\mu)\left(\frac{\partial u_x}{\partial x} + \frac{\partial u_y}{\partial y}\right). \tag{6-2.13}$$

The stress–strain relationships now become

$$\epsilon_{xx} = \frac{\partial u_x}{\partial x} = \frac{1}{2\mu}\left\{\sigma_{xx} - \frac{\lambda}{2(\lambda + \mu)}P\right\}, \tag{6-2.14}$$

$$\epsilon_{yy} = \frac{\partial u_y}{\partial y} = \frac{1}{2\mu}\left\{\sigma_{yy} - \frac{\lambda}{2(\lambda + \mu)}P\right\}. \tag{6-2.15}$$

Differentiation of Equation 6-2.10 yields

$$\frac{\partial^2 \sigma_{xy}}{\partial x\, \partial y} = \mu\left\{\frac{\partial^3 u_x}{\partial x\, \partial y^2} + \frac{\partial^3 u_y}{\partial x^2\, \partial y}\right\}. \tag{6-2.16}$$

Substituting for $\partial u_x/\partial x$ and $\partial u_y/\partial y$, one obtains

$$2\frac{\partial^2 \sigma_{xy}}{\partial x\,\partial y} = \frac{\partial^2 \sigma_{xx}}{\partial y^2} + \frac{\partial^2 \sigma_{yy}}{\partial x^2} - \frac{\lambda}{2(\lambda + \mu)}\nabla^2 P. \qquad (6\text{-}2.17)$$

From the sum of properly differentiated equilibrium equations, we obtain

$$-2\frac{\partial^2 \sigma_{xy}}{\partial x\,\partial y} = \frac{\partial^2 \sigma_{xx}}{\partial x^2} + \frac{\partial^2 \sigma_{yy}}{\partial y^2} + \frac{\partial F_x}{\partial x} + \frac{\partial F_y}{\partial y}, \qquad (6\text{-}2.18)$$

where $F_x = \rho b_x, \qquad F_y = \rho b_y.$

Combining equations 6-2.17 and 6-2.18 yields

$$\nabla^2 P - \frac{\lambda}{2(\lambda + \mu)}\nabla^2 P + \bar{\nabla}\cdot\bar{F} = 0,$$

$$\nabla^2(\sigma_{xx} + \sigma_{yy}) = -\frac{2(\lambda + \mu)}{\lambda + 2\mu}\bar{\nabla}\cdot\bar{F}, \qquad (6\text{-}2.19)$$

$$\nabla^2(\sigma_{xx} + \sigma_{yy}) = -\frac{1}{(1 - \nu)}\bar{\nabla}\cdot\bar{F}.$$

This equation, with the two equilibrium equations, gives a system of three equations for the three unknown stresses, σ_{xx}, σ_{yy}, and σ_{xy}. The last equation is termed the *compatibility equation* because it may be obtained from the only nontrivially satisfied Saint-Venant compatibility equation.

$$\frac{\partial \sigma_{xx}}{\partial x} + \frac{\partial \sigma_{xy}}{\partial y} + F_x = 0,$$

$$\frac{\partial \sigma_{xy}}{\partial x} + \frac{\partial x_{yy}}{\partial y} + F_y = 0, \qquad (6\text{-}2.20)$$

$$\nabla^2(\sigma_{xx} + \sigma_{yy}) = -\frac{1}{(1 - \nu)}\bar{\nabla}\cdot\bar{F}.$$

These are the two desired formulations in terms of either displacements or stresses. Let us introduce two new constants, ν^* and E^*, defined as follows:

$$\nu^* = \frac{\nu}{1 - \nu}, \qquad E^* = \frac{E}{1 - \nu^2}. \qquad (6\text{-}2.21)$$

The equations of plane strain become

$$\boxed{\begin{aligned} &G\left\{\nabla^2 u_x + \frac{1 + \nu^*}{1 - \nu^*}\frac{\partial}{\partial x}\left(\frac{\partial u_x}{\partial x} + \frac{\partial u_y}{\partial y}\right)\right\} + F_x = 0 \\ &G\left\{\nabla^2 u_y + \frac{1 + \nu^*}{1 - \nu^*}\frac{\partial}{\partial y}\left(\frac{\partial u_x}{\partial x} + \frac{\partial u_y}{\partial y}\right)\right\} + F_y = 0 \end{aligned}}, \qquad (6\text{-}2.22)$$

$$\begin{aligned} \epsilon_{xx} &= \frac{1}{E^*}(\sigma_{xx} - \nu^*\sigma_{yy}), & \sigma_{xx} &= \frac{E^*}{(1 - \nu^{*2})}(\epsilon_{xx} + \nu^*\epsilon_{yy}), \\ \epsilon_{yy} &= \frac{1}{E^*}(\sigma_{yy} - \nu^*\sigma_{xx}), & \sigma_{yy} &= \frac{E^*}{(1 - \nu^{*2})}(\epsilon_{yy} + \nu^*\epsilon_{xx}), \qquad (6\text{-}2.23) \\ \epsilon_{xy} &= \frac{1 + \nu^*}{E^*}\sigma_{xy}. \end{aligned}$$

Note that

$$G = \frac{2(1+\nu)}{E} = \frac{2(1+\nu^*)}{E^*}.$$

$$\left.\begin{aligned} \frac{\partial \sigma_{xx}}{\partial x} + \frac{\partial \sigma_{xy}}{\partial y} + F_x = 0, \\ \frac{\partial \sigma_{xy}}{\partial x} + \frac{\partial \sigma_{yy}}{\partial y} + F_y = 0, \end{aligned}\right\} \Longrightarrow \quad \nabla_\alpha \sigma_{\alpha\beta} + F_\beta = 0. \tag{6-2.24}$$

$$\nabla^2(\sigma_{xx} + \sigma_{yy}) = -(1+\nu^*)\bar{\nabla}\cdot\bar{F}. \tag{6-2.25}$$

The reason for introducing the starred constants will become clear in the next section.

3 Plane stress

The condition of plane stress may be defined as

$$\sigma_{zz} = \sigma_{xz} = \sigma_{yz} = 0. \tag{6-3.1}$$

The equations of elasticity become

$$\frac{\partial \sigma_{xx}}{\partial x} + \frac{\partial \sigma_{xy}}{\partial y} + F_x = 0,$$

$$\frac{\partial \sigma_{xy}}{\partial x} + \frac{\partial \sigma_{yy}}{\partial y} + F_y = 0,$$

$$F_z = 0,$$

$$\epsilon_{xx} = \frac{\partial u_x}{\partial x}, \qquad \epsilon_{yy} = \frac{\partial u_y}{\partial y}, \qquad \epsilon_{xy} = \frac{1}{2}\left(\frac{\partial u_x}{\partial y} + \frac{\partial u_y}{\partial x}\right),$$

$$\begin{aligned} \sigma_{xx} &= \lambda(\epsilon_{xx} + \epsilon_{yy} + \epsilon_{zz}) + 2\mu\epsilon_{xx}, & \epsilon_{xx} &= \frac{1}{E}(\sigma_{xx} - \nu\sigma_{yy}), \\ \sigma_{yy} &= \lambda(\epsilon_{xx} + \epsilon_{yy} + \epsilon_{zz}) + 2\mu\epsilon_{yy}, & \epsilon_{yy} &= \frac{1}{E}(\sigma_{yy} - \nu\sigma_{xx}), \\ 0 &= \lambda(\epsilon_{xx} + \epsilon_{yy} + \epsilon_{zz}) + 2\mu\epsilon_{zz}, & \epsilon_{zz} &= -\frac{\nu}{E}(\sigma_{xx} + \sigma_{yy}), \\ \sigma_{xy} &= 2\mu\epsilon_{xy}, \qquad \epsilon_{xy} = \frac{1+\nu}{E}\sigma_{xy}. \end{aligned} \tag{6-3.2}$$

We may eliminate ϵ_{zz} from the equations either in terms of ϵ_{xx} and ϵ_{yy} or in terms of σ_{xx} and σ_{yy}. It should be noted that ϵ_{zz} is not in general zero and that u_x, u_y, and u_z are functions of z. Therefore, plane stress is not truly a two-dimensional problem.

To obtain the displacement formulation, we shall eliminate ϵ_{zz} in terms of the other normal strains and proceed as in the case of plane strain, yielding the following coupled displacement equations:

$$G\left\{\nabla^2 u_x + \frac{1+\nu}{1-\nu}\frac{\partial}{\partial x}\left(\frac{\partial u_x}{\partial x} + \frac{\partial u_y}{\partial y}\right)\right\} + F_x = 0,$$
$$G\left\{\nabla^2 u_y + \frac{1+\nu}{1-\nu}\frac{\partial}{\partial y}\left(\frac{\partial u_x}{\partial x} + \frac{\partial u_y}{\partial y}\right)\right\} + F_y = 0, \qquad (6\text{-}3.3)$$

These equations are in the same form as equation 6-2.22, obtained in the case of plane strain when E^* and ν^* were introduced. The two sets of equations are not entirely equivalent because the Laplacian operator in plane stress includes nonzero z derivatives.

To obtain the equations in terms of stresses requires more arguments since, as has been stated before, this is not truly a two-dimensional set of equations. If we ignore the dependency upon z (quasi-plane stress), we obtain, by considering the definitions of the strains ϵ_{xx}, ϵ_{yy}, and ϵ_{xy}, an equation of compatibility:

$$\nabla^2(\sigma_{xx} + \sigma_{yy}) = -(1+\nu)\vec{\nabla}\cdot\bar{F}. \qquad (6\text{-}3.4)$$

It can be shown that the state of plane stress does not immediately satisfy the conditions of compatibility. A solution that does satisfy compatibility and equilibrium can be shown to depend upon z^2, and hence is applicable for thin plates bounded by $z = \pm h$ (see problems 6.4 and 6.6). The equations for quasi-plane stress and plane strain are identical in form except for the interchange of E^* and E and ν^* and ν.

A conceptually different manner of obtaining the equations for thin plates is called generalized plane stress. Let us consider the case of a thin plate with thickness, $2h$, small compared to the other dimensions (Figure 6-2). Let the surface $z = \pm h$ be free of stress and the body force in the z direction be zero. From the third equilibrium equation,

$$\frac{\partial \sigma_{zx}}{\partial x} + \frac{\partial \sigma_{zy}}{\partial y} + \frac{\partial \sigma_{zz}}{\partial z} = 0, \qquad (6\text{-}3.5)$$

it is immediately seen that at the boundaries

$$\left.\frac{\partial \sigma_{zz}}{\partial z}\right|_{z=\pm h} = 0. \qquad (6\text{-}3.6)$$

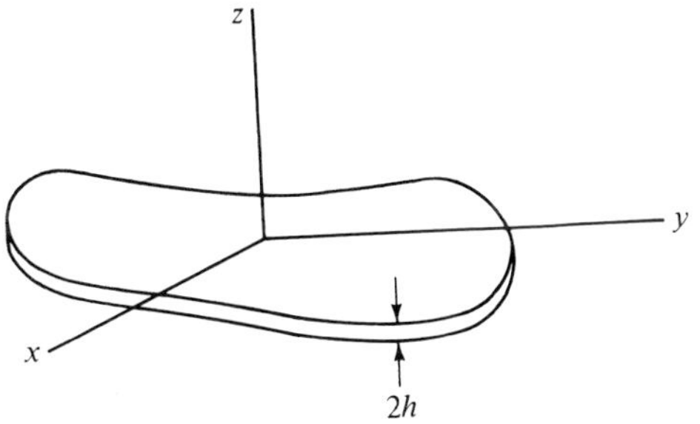

Figure 6-2

Therefore, we may assume that since σ_{zz} and its normal derivative are zero on these boundaries, it does not build to any appreciable value in the interior and may be neglected.

Since the stresses and displacements depend upon z, an average of these properties across the thickness will be taken, and this average will be denoted by an asterisk:

$$u_x^*(x, y) = \frac{1}{2h} \int_{-h}^{+h} u_x(x, y, z)\, dz,$$
$$u_y^*(x, y) = \frac{1}{2h} \int_{-h}^{+h} u_y(x, y, z)\, dz. \tag{6-3.7}$$

The average stresses σ^*_{xx}, σ^*_{yy}, and σ^*_{xy} are defined in a similar manner. Taking the mean value of the equilibrium equations yields

$$\frac{\partial \sigma^*_{xx}}{\partial x} + \frac{\partial \sigma^*_{xy}}{\partial y} + F^*_x = 0,$$
$$\frac{\partial \sigma^*_{xy}}{\partial x} + \frac{\partial \sigma^*_{yy}}{\partial y} + F^*_y = 0. \tag{6-3.8}$$

Note that $1/2h \int_{-h}^{+h} (\partial \sigma_{xz}/\partial z)\, dz$ is zero because σ_{xz} is zero at $y = \pm h$. We may eliminate ϵ_{zz} from the stress–strain relations and take the mean value of these expressions. This yields the same set of plane elasticity equations for the mean displacements and stresses as we obtained for the quasi-plane state. It may be seen that, although the rationale may differ, the form of the results is the same. The complete details of this development are left to the reader.

4 Biharmonic solutions

The plane problem may be reduced to solutions in terms of biharmonic functions defined in terms of either displacements or stresses. In the work that follows, the plane stress formulation will be used but the plane strain problem may be solved by replacing ν and E by ν^* and E^* in the final equations. No distinction need, therefore, be made between the two assumptions during the solution of the problem.

Consider first a solution of Equation 6-3.3 in terms of displacements. As has been noted, these equations are coupled and only in cases where one of the displacements or the divergence of the displacement is zero do the equations decouple. An example of such problems will be discussed later (see Section 8-2). K. Marguerre* in a method basically the same as the three-

* K. Marguerre, "Ebenes und achsensymmetrisches Problem der Elastizitats theorie," *ZAMM*, **13**(6) 437, 1933.

dimensional reduction proposed by B. G. Galerkin* reduced the plane problem to the biharmonic equation when the body forces are zero. The development shown here will be similar to Galerkin's, because the Galerkin vector will be developed in detail in section 13-3 for three-dimensional problems.

It may be noted that equation 6-3.3 may be written in vector form as

$$\nabla^2 \bar{u} + \frac{1+\nu}{1-\nu}\bar{\nabla}(\bar{\nabla}\cdot\bar{u}) + \frac{\bar{F}}{G} = 0. \tag{6-4.1}$$

We now assume that the vector $\bar{u}$ has the form†

$$\bar{u} = a\bar{\nabla}\bar{P} - b\bar{\nabla}(\bar{\nabla}\cdot\bar{P}) \tag{6-4.2}$$

where a and b are constants to be chosen later. Substitution into equation 6-4.1 yields

$$a\nabla^2\nabla^2\bar{P} - b\nabla^2(\bar{\nabla}\bar{\nabla}\cdot\bar{P}) + \frac{1+\nu}{1-\nu}a\bar{\nabla}\bar{\nabla}\cdot(\nabla^2\bar{P}) - b\frac{1+\nu}{1-\nu}\bar{\nabla}\nabla^2(\bar{\nabla}\cdot\bar{P}) = -\frac{\bar{F}}{G}. \tag{6-4.3}$$

The operators ∇^2 and $\bar{\nabla}$ commute, so that

$$\nabla^2(\bar{\nabla}\bar{\nabla}\cdot\bar{P}) = \bar{\nabla}\bar{\nabla}\cdot(\nabla^2\bar{P}) = \bar{\nabla}\nabla^2(\bar{\nabla}\cdot\bar{P}). \tag{6-4.4}$$

The constant b is chosen in terms of a such that the last three terms on the left of equation 6-4.3 cancel:

$$-b + a\frac{1+\nu}{1-\nu} - b\frac{1+\nu}{1-\nu} = 0, \tag{6-4.5}$$

$$b = \frac{1+\nu}{2}a. \tag{6-4.6}$$

The expression governing $\bar{P}$ becomes

$$\nabla^2\nabla^2\bar{P} = -\frac{\bar{F}}{aG}. \tag{6-4.7}$$

Some motivation for the representation of the displacement vector as shown in equation 6-4.2 may be obtained by first considering the general reduction of the displacement vector by the Helmholtz theorem.

$$\bar{u} = \bar{u}_S + \bar{u}_{IR}, \tag{6-4.8}$$

where $\bar{u}_S$ is a solenoidal vector (divergence is zero) and $\bar{u}_{IR}$ is an irrotational vector (curl is zero). The irrotational vector may be considered to be derivable from a potential function and the solenoidal vector may be related to the curl of another vector, which is itself solenoidal (see section 1-11).

$$\bar{u} = \bar{\nabla}\varphi + \bar{\nabla} \times a\bar{\psi}, \tag{6-4.9}$$

* B. G. Galerkin, *Compt. Rend.*, **190**, 1047, 1930; *Compt. Rend. Acad. Sci. U.R.S.S.*, p. 353, 1930, (in Russian).

† See Y. C. Fung, *Foundations of Solid Mechanics*, Prentice-Hall, Inc., Englewood Cliffs, N.J., 1965, for motivation of this form of the displacement vector.

where $\bar{\nabla}\cdot\bar{\psi} = 0$. But if $\bar{\nabla}\cdot\bar{\psi} = 0$, $\bar{\psi}$ may also be generated from another vector $\bar{P}$, with $\bar{\psi} = \bar{\nabla} \times \bar{P}$. The displacement vector now becomes

$$\bar{u} = \bar{\nabla}\varphi + a\bar{\nabla} \times (\bar{\nabla} \times \bar{P}) = \bar{\nabla}\varphi + a\nabla^2\bar{P} - a\bar{\nabla}(\bar{\nabla}\cdot\bar{P}). \qquad (6\text{-}4.10)$$

Whereas $\bar{\nabla}\cdot\bar{P} = 0$ is a common choice, $\bar{\nabla}\cdot\bar{P}$ can still be chosen arbitrarily if desired, and we now take $\bar{\nabla}\cdot\bar{P} = \varphi$. Thus, without loss of generality the first potential function may be combined with the gradient of the divergence of $\bar{P}$, or

$$\bar{u} = a\nabla^2\bar{P} - b\vec{\nabla}(\vec{\nabla}\cdot\bar{P}). \qquad (6\text{-}4.11)$$

This is the form of the displacement assumed by Galerkin, and the rest of the development proceeds as before.

It can be shown that we may choose either the x or y component of $\bar{P}$ to be zero if the body force is zero in either of these directions.* If the entire body force $\bar{F}$ is zero, either component of $\bar{P}$ may be used. Note that the problem is two-dimensional, so that the z component is neglected. In the case of zero body forces, Marguerre proposed that the y component be used, with the choice of a below:

$$a = \frac{2}{1-\nu}, \qquad (6\text{-}4.12)$$

$$\bar{P} = (0, \psi), \qquad (6\text{-}4.13)$$

$$\bar{u} = \frac{2}{1-\nu}\nabla^2\psi\hat{i}_y - \frac{1+\nu}{1-\nu}\bar{\nabla}\frac{\partial\psi}{\partial y}, \qquad (6\text{-}4.14)$$

$$\nabla^2\nabla^2\psi = 0. \qquad (6\text{-}4.15)$$

There is no particular reason to choose a in this manner and it could have been chosen as 1 or $2/(1+\nu)$ so that b would equal 1.

The stresses and the displacements in terms of the Marguerre function become

$$\begin{aligned} u_x &= -\frac{1+\nu}{1-\nu}\frac{\partial^2\psi}{\partial x\,\partial y}, \\ u_y &= \frac{2}{1-\nu}\frac{\partial^2\psi}{\partial x^2} + \frac{\partial^2\psi}{\partial y^2}, \end{aligned} \qquad (6\text{-}4.16)$$

$$\begin{aligned} \sigma_{xx} &= \frac{E}{1-\nu^2}\left\{-\frac{\partial^3\psi}{\partial x^2\,\partial y} + \nu\frac{\partial^3\psi}{\partial y^3}\right\}, \\ \sigma_{yy} &= \frac{E}{1-\nu^2}\left\{(2+\nu)\frac{\partial^3\psi}{\partial x^2\,\partial y} + \frac{\partial^3\psi}{\partial y^3}\right\}, \\ \sigma_{xy} &= \frac{E}{1-\nu^2}\left\{\frac{\partial^3\psi}{\partial x^3} - \nu\frac{\partial^3\psi}{\partial x\,\partial y^2}\right\}. \end{aligned} \qquad (6\text{-}4.17)$$

Examples of problems solved by this formulation will be shown later.

* A detailed discussion of this is presented in section 13-4; see also H. M. Westergaard, *Theory of Elasticity and Plasticity*, Harvard University Press, Cambridge, Mass; John Wiley & Sons, Inc., New Yorks, 1952, pp. 124–125.

A much older biharmonic formulation was proposed by G. B. Airy in 1862.* The stresses were chosen in terms of a function φ, such that the equilibrium equations are satisfied in the case where the body force is derivable from a potential function,

$$\bar{F} = -\vec{\nabla} V. \tag{6-4.18}$$

In symbolic notation the stresses are chosen as

$$\bar{\bar{\sigma}} = \nabla^2 \varphi \bar{\bar{I}} - \vec{\nabla}\vec{\nabla}\varphi + V\bar{\bar{I}}. \tag{6-4.19}$$

In Cartesian component notation, this becomes

$$\sigma_{\alpha\beta} = \nabla^2 \varphi \delta_{\alpha\beta} - \nabla_\alpha \nabla_\beta \varphi + V\delta_{\alpha\beta}. \tag{6-4.20}$$

Substituting into the equilibrium equations, in the form

$$\nabla_\alpha \sigma_{\alpha\beta} - \nabla_\beta V = 0, \tag{6-4.21}$$

yields

$$\nabla_\beta \nabla^2 \varphi - \nabla^2 \nabla_\beta \varphi + \nabla_\beta V - \nabla_\beta V = 0. \tag{6-4.22}$$

The equilibrium equations are satisfied by this relation, and it is the equilibrium equations which imply the existence of the function φ. This may be seen by noting that each of the equilibrium equations,

$$\begin{aligned} \frac{\partial}{\partial x}(\sigma_{xx} - V) + \frac{\partial}{\partial y}(\sigma_{yx}) &= 0, \\ \frac{\partial}{\partial x}(\sigma_{xy}) + \frac{\partial}{\partial y}(\sigma_{yy} - V) &= 0, \end{aligned} \tag{6-4.23}$$

is a condition of integrability for a differential form. The pair 6-4.23 implies the existence of two functions A and B such that

$$\begin{aligned} \sigma_{xx} - V &= \frac{\partial A}{\partial y}, \qquad & \sigma_{yx} &= -\frac{\partial A}{\partial x}, \\ \sigma_{xy} &= \frac{\partial B}{\partial y}, \qquad & \sigma_{yy} - V &= -\frac{\partial B}{\partial x}. \end{aligned} \tag{6-4.24}$$

Since the stress tensor is symmetric, the shear stresses are equal and a further condition of integrability is obtained implying the existence of the stress function φ:

$$\frac{\partial A}{\partial x} + \frac{\partial B}{\partial y} = 0.$$

Therefore

$$A = \frac{\partial \varphi}{\partial y}, \qquad B = -\frac{\partial \varphi}{\partial x}. \tag{6-4.25}$$

Substitution of the expressions for A and B into equation 6-4.24 yields

* G. B. Airy, *Brit. Assoc. Advan. Sci. Rept.*, 1862.

expressions for the stresses:

$$\sigma_{xx} = \frac{\partial^2 \varphi}{\partial y^2} + V,$$

$$\sigma_{yy} = \frac{\partial^2 \varphi}{\partial x^2} + V, \qquad (6\text{-}4.26)$$

$$\sigma_{xy} = -\frac{\partial^2 \varphi}{\partial x\,\partial y}.$$

There are the expanded form of the expressions as shown in equations 6-4.19 and 6-4.20.

Since φ as defined above satisfies equilibrium, the compatibility equation serves as the defining equation for the Airy stress function φ. Substitution into equation 6-3.4 yields

$$\nabla^2\nabla^2\varphi + 2\nabla^2 V - (1 + \nu)\nabla^2 V = 0, \qquad (6\text{-}4.27)$$

$$\nabla^2\nabla^2\varphi = -(1 - \nu)\nabla^2 V. \qquad (6\text{-}4.28)$$

Expanding this equation (when the body forces vanish) shows that φ is governed by the biharmonic equation in this case:

$$\frac{\partial^4 \varphi}{\partial x^4} + 2\frac{\partial^4 \varphi}{\partial x^2\,\partial y^2} + \frac{\partial^4 \varphi}{\partial y^4} = 0. \qquad (6\text{-}4.29)$$

PROBLEMS

6.1. Derive equations 6-2.11 from the basic equations of elasticity expressed in terms of the engineering constants, using two-dimensional Cartesian tensor notation.

6.2. Derive equations 6-2.20 from the basic equations of elasticity expressed in terms of the engineering constants, using two-dimensional Cartesian tensor notation.

6.3. Show that the assumption of plane strain leads to the satisfaction of all but one of the Saint-Venant compatibility equations 2-7.4 to 2-7.9. Derive the third equation of 6-2.20 directly from this one remaining compatibility equation.

6.4. By examination of the Beltrami–Michell compatibility equations (problem 5.2), show that in plane stress, the sum of the normal stresses may be written as follows if the body forces are constant:

$$\sigma_{ii} = \sigma_{\alpha\alpha} = kz + f(x, y),$$

where k is a constant and f is an analytic function of x and y.

6.5. Develop a displacement function similar to the Marguerre function assuming $a = 1$ and $P = (\beta, 0)$.

6.6. Using the results of problem 6.4, show that in plane stress the Airy stress function may be written

$$\varphi = -\frac{1}{2}\frac{\nu}{1+\nu}f(x, y)z^2 + A + Bx + Cy + \varphi_1(x, y)z + \varphi_0(x, y).$$

The stress function (dependent upon z) is φ, and A, B, and C may be functions of z. In problems exhibiting symmetry about the $z = 0$ plane, φ_0 is the quasi-plane Airy stress function.

CHAPTER

7

PROBLEMS IN CARTESIAN COORDINATES

1 Introduction

In this chapter we shall consider problems that may best be solved by use of rectangular coordinates. The choice of a coordinate system is usually dictated by the geometry of the body and, in some cases, the complexity of the differential equations or the boundary conditions involved.

We shall examine problems that may be solved by polynomials first.* This approach is somewhat limited in its applicability but it provides insight into boundary conditions and the utilization of symmetry, or the odd and even properties of a solution. We can attempt an "inverse approach" by examining what problem certain biharmonic polynomials might solve.†

* A. Mesnager, *Compt. Rend.*, **132**, 1475, 1901; and A. Timpe, *Z. Math. Physik*, **52**, 348, 1905.

† See Timoshenko and Goodier, *op. cit.*, p. 35.

Although such an approach is entertaining and builds up a useful library of solutions, a more direct approach will be presented here.

After some familiarization with polynomials has been obtained, solutions involving orthogonal functions are examined in some detail. If one or more dimensions of a body are infinite or semi-infinite, Fourier transforms are employed. Finally, the Saint-Venant principle for semi-infinite strips is examined. The use of nonorthogonal functions is also examined, and numerical techniques are presented to implement this approach.

2 Mathematical preliminaries

Before developing solutions of particular problems using the Marguerre and the Airy functions some mathematical characteristics of functions will be reviewed. In particular, an examination of the even and odd properties of functions helps when one is looking for solutions of problems in two-dimensional elasticity.

An even function is defined by

$$f(x) = f(-x), \tag{7-2.1}$$

whereas an odd function is defined by

$$f(x) = -f(-x). \tag{7-2.2}$$

Examples of even functions are even powers of x, $\cos x$, $x \sin x$, $\cosh x$, etc., examples of odd functions are odd powers of x, $\sin x$, $\sinh x$, $x \cos x$, etc. Not all functions are even or odd in x; for example, e^x and $(x + x^2)$. Any function may, however, be considered to be the sum of an even and an odd function.

$$g(x) = \underbrace{\tfrac{1}{2}[g(x) + g(-x)]}_{\text{even}} + \underbrace{\tfrac{1}{2}[g(x) - g(-x)]}_{\text{odd}}. \tag{7-2.3}$$

As an example, consider the following:

$$e^x = \tfrac{1}{2}(e^x + e^{-x}) + \tfrac{1}{2}(e^x - e^{-x}) = \cosh x + \sinh x. \tag{7-2.4}$$

The use of even and odd functions has particular advantages when one must integrate between symmetric limits:

$$\int_{-a}^{+a} (\text{odd function of } x)dx = 0, \tag{7-2.5}$$

$$\int_{-a}^{+a} (\text{even function of } x)dx = 2\int_{0}^{a} (\text{even function of } x)dx. \tag{7-2.6}$$

It is frequently advantageous to place a coordinate axis as an axis of geometrical symmetry, if such symmetry exists. In this manner one may handle the even and the odd parts independently, thereby reducing the algebraic labor involved.

The product of two even functions is even; so also is the product of two odd functions. The product of an even and an odd function is odd.

The derivative of an odd function is an even one, and the derivative of an even function is an odd one. The second derivatives thus preserve evenness and oddness. The same properties apply to integration, if you omit the constants of integration.

3 Polynomial solutions

It may be noted that in the absence of body forces, low-order polynomials satisfy the defining biharmonic equation term by term. Consider the Airy stress function φ expressed as

$$\varphi = \sum_m \sum_n A_{mn} x^m y^n. \tag{7-3.1}$$

If $m + n \leq 3$, φ satisfies the biharmonic equation for each of these terms. However, the linear function

$$\varphi = A_{00} + A_{10}x + A_{01}y \tag{7-3.2}$$

does not contribute to the stresses and will be omitted.

For simple problems a solution may be constructed from these simple low-order polynomials, which term by term satisfy the biharmonic equation and also satisfy the required boundary conditions in their sum. When $m + n > 3$, combinations of terms must be used to still satisfy the biharmonic equation. For example, when y^4 is operated upon by the biharmonic operator it produces a nonzero constant. The same is true for x^4 and x^2y^2, so that the combination

$$A_{40}x^4 + A_{22}x^2y^2 + A_{04}y^4 \tag{7-3.3}$$

cannot be used unless the condition

$$24A_{40} + 8A_{22} + 24A_{04} = 0 \tag{7-3.4}$$

is met. This condition specifies one constant in terms of the other two, but still leaves two constants to satisfy the boundary conditions.

In general, we do not expect to satisfy all the boundary conditions exactly when we use polynomials, and we often have to resort to the use of the Saint-Venant principle on short boundaries. With this thought in mind, we shall consider problems in which one dimension is very much smaller than the other, for example, long "beams."

C. Y. Neou* formulated a systematic method of examining the polynomial series and obtaining the relations between the coefficients necessary

* C. Y. Neou, "Direct method for determining Airy polynomial stress functions," *J. Appl. Mech.*, **24**(3), 387–390, September 1957.

to satisfy the compatibility equation. In general, infinite polynomial series solutions are too slowly convergent to warrant a detailed examination. We shall use the "semi-inverse" approach to become familiar with the equations and boundary conditions, and shall await the introduction of orthogonal functions before developing a systematic approach.

3.1 Uniaxial tension. Consider the case of a beam subjected to uniform tension T in the x direction, as shown in Figure 7-1. Since the body forces are zero and the boundary conditions are

$$\begin{aligned} x &= \pm l, \quad \sigma_{xx} = T, \quad \sigma_{xy} = 0, \\ y &= \pm h \quad \sigma_{yy} = 0, \quad \sigma_{xy} = 0, \end{aligned} \tag{7-3.1.1}$$

we shall examine solutions in the form of polynomials. We can approach this problem by looking for the stress function that would lead to the known elementary solution, or we can examine the nonhomogeneous boundary condition of the ends, directly. In either case we are led to assume a solution of the form

$$\varphi = A_{02} y^2. \tag{7-3.1.2}$$

The stresses become

$$\sigma_{xx} = 2A_{02}, \tag{7-3.1.3}$$

$$\sigma_{yy} = \sigma_{xy} = 0, \tag{7-3.1.4}$$

so that the boundary conditions are satisfied if $A_{02} = \frac{1}{2}T$. The general approach to any long rectangular shape will be to examine the long boundaries first and to make sure that the assumed stress function satisfies the conditions there. The stress function in Equation 7-3.1.2 does satisfy the long boundary conditions, although the choice was not motivated by considering them in this simple case.

We may now obtain the displacements from the stresses using the stress–strain relations

$$\frac{\partial u_x}{\partial x} = \frac{1}{E}\{\sigma_{xx} - \nu\sigma_{yy}\} = \frac{T}{E}. \tag{7-3.1.5}$$

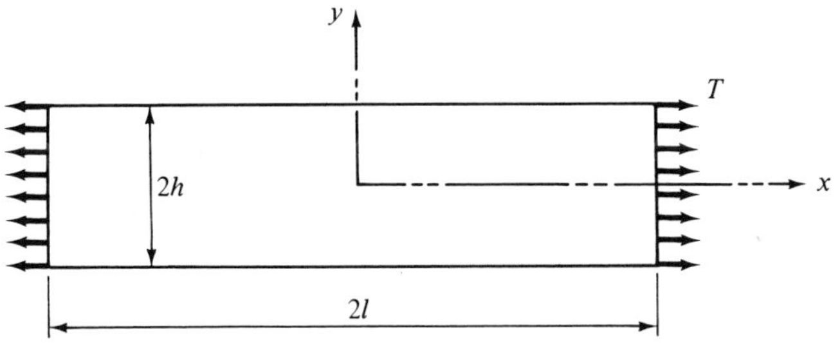

Figure 7-1

therefore,

$$u_x = \frac{T}{E}x + f(y). \tag{7-3.1.6}$$

$$\frac{\partial u_y}{\partial y} = \frac{1}{E}\{\sigma_{yy} - \nu\sigma_{xx}\} = -\frac{\nu T}{E}. \tag{7-3.1.7}$$

$$u_y = -\frac{\nu T}{E}y + g(x). \tag{7-3.1.8}$$

The functions $f(y)$ and $g(x)$ are *functions of integration* and may be determined by substituting equations 7-3.1.6 and 7-3.1.8 into the shear stress–strain equation:

$$\frac{1}{2}\left(\frac{\partial u_x}{\partial y} + \frac{\partial u_y}{\partial x}\right) = \frac{1+\nu}{E}\sigma_{xy}. \tag{7-3.1.9}$$

Therefore,

$$f'(y) + g'(x) = 0, \tag{7-3.1.10}$$

where the prime denotes differentiation.

$$f(y) = -\omega^0 y + u_x^0, \tag{7-3.1.11}$$

$$g(x) = \omega^0 x + u_y^0. \tag{7-3.1.12}$$

These functions involve the rigid-body rotation ω^0, and the translation u_x^0 and u_y^0 in the x and y directions, respectively. If the origin is assumed fixed, the translation constants may be chosen to be zero. Although this problem is simple, it includes most of the elements involved in solutions with polynomials.

This problem could have been solved by the Marguerre function ψ also. Since no body forces are present, ψ is defined by the biharmonic equation:

$$\nabla^2\nabla^2\psi = 0. \tag{7-3.1.13}$$

In general, when using the Marguerre function, the rigid-body motions at the origin will be taken to be zero. They can always be introduced by adding to ψ the terms

$$\psi^0 = -u_x^0\frac{(1-\nu)}{(1+\nu)}xy + u_y^0\frac{y^2}{2} + \frac{(1-\nu)}{2(1+\nu)}\omega^0\left(\frac{\nu x^3}{3} + xy^2\right), \tag{7-3.1.14}$$

where u_x^0, u_y^0, and ω^0 are the rigid-body motions defined previously. These terms lead to the following displacements:

$$u_x = u_x^0 - \omega^0 y, \tag{7-3.1.15}$$

$$u_y = u_y^0 + \omega^0 x. \tag{7-3.1.16}$$

Setting these motions to be zero, assume a Marguerre function for this problem as follows:

$$\psi = B_{21}x^2y + B_{03}y^3. \tag{7-3.1.17}$$

In this case we cannot assume a single term because it produces two normal stresses, one of which must be zero on the boundaries, $y = \pm h$. Two terms were picked, in the hope that the combination would lead to $\sigma_{xx} \neq 0$ and $\sigma_{yy} = 0$. Using Equation 6-4.17, the stresses become

$$\sigma_{xx} = \frac{E}{1 - \nu^2}\{-2B_{21} + 6\nu B_{03}\}, \tag{7-3.1.18}$$

$$\sigma_{yy} = \frac{E}{1 - \nu^2}\{2(2 + \nu)B_{21} + 6B_{03}\}. \tag{7-3.1.19}$$

Setting $\sigma_{yy} = 0$ to meet the boundary conditions along the long edges yields

$$B_{21} = -\frac{3}{2 + \nu}B_{03}. \tag{7-3.1.20}$$

Thus σ_{xx} becomes

$$\sigma_{xx} = \frac{6E(1 + \nu)}{(2 + \nu)(1 - \nu)}B_{03}. \tag{7-3.1.21}$$

To satisfy the final boundary condition. B_{03} becomes

$$B_{03} = \frac{T(2 + \nu)(1 - \nu)}{6E(1 + \nu)}. \tag{7-3.1.22}$$

From equation 7-3.1.20, we get

$$B_{21} = -\frac{T}{2E}\frac{(1 - \nu)}{(1 + \nu)}. \tag{7-3.1.23}$$

Substituting into equation 6-4.16, the displacements are

$$u_x = \frac{Tx}{E}, \tag{7-3.1.24}$$

$$u_y = -\frac{\nu Ty}{E}. \tag{7-3.1.25}$$

This solution agrees, up to the rigid-body motions, with that obtained from the Airy stress function solution.

3.2 Simply supported beam under pure moments. For a comparison of the plane elasticity solution with the Euler–Bernoulli straight beam theory, consider a narrow beam bent by pure moment, M per unit width in the z direction (Figure 7-2). The boundary conditions in this case are

$$y = \pm h, \quad \sigma_{yy} = \sigma_{xy} = 0, \tag{7-3.2.1}$$

$$x = \pm\frac{l}{2}, \quad \sigma_{xy} = 0, \quad \int_{-h}^{+h} \sigma_{xx} y\, dy = -M, \quad \int_{-h}^{+h} \sigma_{xx}\, dy = 0. \tag{7-3.2.2}$$

Note that no attempt is made to specify σ_{xx} exactly on the boundary but only "in the Saint-Venant sense." It is consistent with the statement of the problem in this case. To satisfy the condition that the resultant force on the ends be

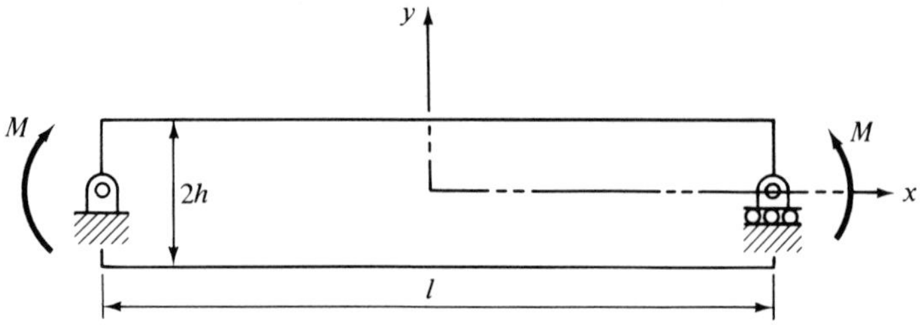

Figure 7-2

zero, we may select σ_{xx} to be an odd function of y. This means that φ is odd in y also, and since the problem is symmetric in the x direction, we shall assume φ to be even in x.

Examining the boundary contidions on the long edge indicates that if φ is taken to be independent of x, σ_{yy} and σ_{xy} will both be zero. The simplest possible choice of φ is therefore

$$\varphi = A_{03}y^3. \tag{7-3.2.3}$$

The stresses become

$$\sigma_{yy} = \sigma_{xy} = 0, \tag{7-3.2.4, 5}$$

$$\sigma_{xx} = 6A_{03}y. \tag{7-3.2.6}$$

Satisfying the moment condition on the short boundary yields

$$\int_{-h}^{+h} 6A_{03}y^2\,dy = -M, \tag{7-3.2.7}$$

$$A_{03} = -\frac{M}{4h^3}, \tag{7-3.2.8}$$

$$\sigma_{xx} = -\frac{3}{2}\frac{My}{h^3}. \tag{7-3.2.9}$$

The displacements may now be calculated:

$$\frac{\partial u_x}{\partial x} = -\frac{3}{2}\frac{My}{Eh^3}, \tag{7-3.2.10}$$

$$u_x = -\frac{3}{2}\frac{Mxy}{Eh^3} + f(y), \tag{7-3.2.11}$$

$$\frac{\partial u_y}{\partial y} = \frac{3}{2}\nu\frac{My}{Eh^3}, \tag{7-3.2.12}$$

$$u_y = \frac{3}{4}\nu\frac{My^2}{Eh^3} + g(x). \tag{7-3.2.13}$$

From the shear stress–strain condition, we obtain

$$\frac{\partial u_x}{\partial y} + \frac{\partial u_y}{\partial x} = 0, \tag{7-3.2.14}$$

$$-\frac{3}{2}\frac{Mx}{Eh^3} + f'(y) + g'(x) = 0, \tag{7-3.2.15}$$

$$f(y) = -\omega^0 y + u_x^0, \tag{7-3.2.16}$$

$$g(x) = \frac{3}{4}\frac{Mx^2}{Eh^3} + \omega^0 x + u_y^0. \tag{7-3.2.17}$$

Therefore, the displacements become

$$u_x = -\frac{3}{2}\frac{Mxy}{Eh^3} - \omega^0 y + u_x^0, \tag{7-3.2.18}$$

$$u_y = \frac{My^2\nu}{Eh^3} + \frac{3}{4}\frac{Mx^2}{Eh^3} + \omega^0 x + u_y^0. \tag{7-3.2.19}$$

If we make the substitution $I = 2h^3/3$, which in elementary strength of materials is the appropiate flexural moment of inertia of a beam of unit width and height $2h$, the expression for stresses and displacements are

$$\sigma_{xx} = -\frac{My}{I}, \qquad \sigma_{xy} = \sigma_{yy} = 0,$$

$$u_x = -\frac{Mxy}{EI} - \omega^0 y + u_x^0, \tag{7-3.2.20}$$

$$u_y = \frac{\nu M}{2EI}y^2 + \frac{Mx^2}{2EI} + \omega^0 x + u_y^0.$$

The rigid-body displacement constants are now chosen such that

$$\begin{aligned} u_x\left(-\frac{l}{2}, 0\right) &= 0,\\ u_y\left(-\frac{l}{2}, 0\right) &= 0,\\ u_y\left(\frac{l}{2}, 0\right) &= 0,\\ u_x^0 &= 0,\\ \omega^0 &= 0,\\ u_y^0 &= -\frac{Ml^2}{8EI}. \end{aligned} \tag{7-3.2.21}$$

The displacements become

$$u_x = -\frac{Mxy}{EI}, \qquad u_y = \frac{M}{EI}\left\{\nu\frac{y^2}{2} + \frac{x^2}{2} - \frac{l^2}{8}\right\}. \tag{7-3.2.22}$$

The elementary solution by Euler–Bernoulli theory gives the following expressions for the stresses and the deflection of the centerline of the beam:

$$\sigma_{xx} = -\frac{My}{I}, \qquad \sigma_{xy} = \sigma_{yy} = 0,$$
$$u_y(x, 0) = \frac{M}{EI}\left\{\frac{x^2}{2} - \frac{l^2}{8}\right\}. \qquad (7\text{-}3.2.23)$$

For this example the solutions are the same and the elementary solution is the exact solution of the equations of elasticity. This is not usually the case for other loadings, because the elementary solution represents certain approximations.

3.3 Beam bent by its own weight. The beam shown in Figure 7-3 has a mass density ρ, and is loaded by its own weight. The body force $\bar{F}$ is

$$\bar{F} = -\rho g \hat{i}_{(y)}. \qquad (7\text{-}3.3.1)$$

The potential function V may now be written as

$$V = \rho g y. \qquad (7\text{-}3.3.2)$$

The defining equation for φ is now

$$\nabla^2\nabla^2\varphi = -(1 - \nu)\nabla^2 V = 0. \qquad (7\text{-}3.3.3)$$

The boundary conditions on the long edges are

$$y = \pm h, \qquad \sigma_{xy} = \sigma_{yy} = 0. \qquad (7\text{-}3.3.4)$$

Using the Saint-Venant assumption on the ends yields

$$x = +l, \qquad \sigma_{xx} = 0, \qquad \int_{-h}^{+h} \sigma_{xy}\, dy = 2\rho g l h, \qquad (7\text{-}3.3.5)$$

$$x = -l, \qquad \sigma_{xx} = 0, \qquad \int_{-h}^{+h} \sigma_{xy}\, dy = -2\rho g l h. \qquad (7\text{-}3.3.6)$$

The stresses are

$$\sigma_{xx} = \frac{\partial^2\varphi}{\partial y^2} + \rho g y, \qquad (7\text{-}3.3.7)$$

$$\sigma_{yy} = \frac{\partial^2\varphi}{\partial x^2} + \rho g y, \qquad (7\text{-}3.3.8)$$

$$\sigma_{xy} = -\frac{\partial^2\varphi}{\partial x\, \partial y}. \qquad (7\text{-}3.3.9)$$

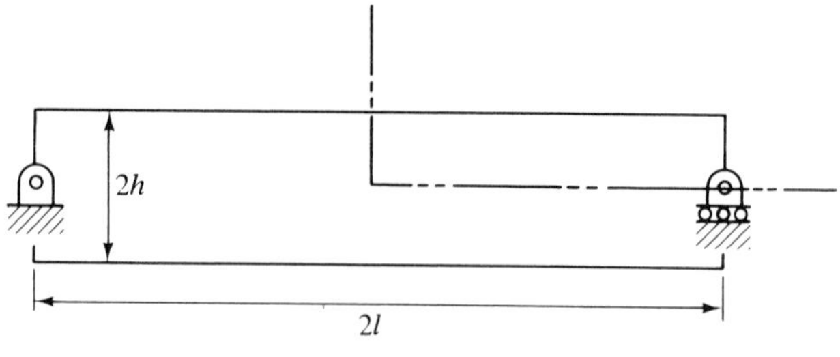

Figure 7-3

Examining the boundary conditions, a first trial at the stress function would be even in x and odd in y with the form

$$\varphi = A_{21}x^2y + A_{23}x^2y^3 + A_{05}y^5. \qquad (7\text{-}3.3.10)$$

The constant A_{05} must be related to A_{23} to satisfy the biharmonic equation. The stresses become

$$\sigma_{xx} = 6A_{23}x^2y + 20A_{05}y^3 + \rho gy, \qquad (7\text{-}3.3.11)$$

$$\sigma_{yy} = 2A_{21}y + 2A_{23}y^3 + \rho gy, \qquad (7\text{-}3.3.12)$$

$$\sigma_{xy} = -2A_{21}x - 6A_{23}xy^2. \qquad (7\text{-}3.3.13)$$

The conditions along the long boundaries and the biharmonic equation may be satisfied by these coefficients, but not the condition that σ_{xx} is zero at the ends. This condition may be replaced by the Saint-Venant equivalent

$$\int_{-h}^{+h} \sigma_{xx}\, dy = 0, \qquad (7\text{-}3.3.14)$$

$$\int_{-h}^{+h} \sigma_{xx} y\, dy = 0. \qquad (7\text{-}3.3.15)$$

The first of these conditions is satisfied because σ_{xx} is an odd function of y, but the second condition is not satisfied. A term $A_{03}y^3$ is added; this will not alter σ_{xy} and σ_{yy}. The Airy stress function is now

$$\varphi = A_{21}x^2y + A_{23}x^2y^3 + A_{03}y^3 + A_{05}y^5. \qquad (7\text{-}3.3.16)$$

All conditions are satisfied by the following values of the coefficients:

$$A_{21} = -\frac{3}{4}\rho g, \qquad A_{23} = \frac{1}{4}\frac{\rho g}{h^2}, \qquad A_{05} = -\frac{1}{20}\frac{\rho g}{h^2},$$
$$A_{03} = -\left[\frac{1}{15}\rho g + \frac{1}{4}\frac{\rho g l^2}{h^2}\right]. \qquad (7\text{-}3.3.17)$$

The stresses become

$$\sigma_{xx} = \left[\frac{3}{5}\rho g + \frac{3}{2}\frac{\rho g}{h^2}(x^2 - l^2)\right]y - \frac{\rho g}{h^2}y^3, \qquad (7\text{-}3.3.18)$$

$$\sigma_{yy} = -\frac{1}{2}\rho g y + \frac{1}{2}\frac{\rho g}{h^2}y^3, \qquad (7\text{-}3.3.19)$$

$$\sigma_{xy} = \frac{3}{2}\rho g x - \frac{3}{2}\frac{\rho g}{h^2}xy^2. \qquad (7\text{-}3.3.20)$$

The displacements may now be calculated with only algebraic difficulties in sight. To relate the results to elementary beam problems, we may introduce the constant I as before:

$$I = 2\frac{h^3}{3}. \qquad (7\text{-}3.3.21)$$

The stresses become

$$\sigma_{xx} = \frac{\rho g h}{I}\left\{\frac{2}{5}h^2 y - \frac{3}{2}y^3 + (x^2 - l^2)y\right\}, \tag{7-3.3.22}$$

$$\sigma_{yy} = \frac{\rho g h}{I}\left\{-\frac{h^2 y}{3} + \frac{y^3}{3}\right\}, \tag{7-3.3.23}$$

$$\sigma_{xy} = \frac{\rho g h}{I}\{h^2 x - y^2 x\}. \tag{7-3.3.24}$$

The strains are

$$\epsilon_{xx} = \frac{\partial u_x}{\partial x} = \frac{\rho g h}{EI}\left\{\frac{2}{5}h^2 y - \frac{2}{3}y^3 + (x^2 - l^2)y + \frac{\nu h^2 y}{3} - \frac{\nu y^3}{3}\right\}, \tag{7-3.3.25}$$

$$\epsilon_{yy} = \frac{\partial u_y}{\partial y} = \frac{\rho g h}{EI}\left\{-\frac{h^2 y}{3} + \frac{y^3}{3} - \frac{2}{5}\nu h^2 y + \frac{2}{3}\nu y^3 - \nu(x^2 - l^2)y\right\}, \tag{7-3.3.26}$$

$$\epsilon_{xy} = \frac{1}{2}\left(\frac{\partial u_x}{\partial y} + \frac{\partial u_y}{\partial x}\right) = \frac{\rho g h}{EI}\{(1 + \nu)(h^2 x - y^2 x)\}. \tag{7-3.3.27}$$

Integrating the two expressions for the normal strains yields

$$u_x = \frac{\rho g h}{EI}\left\{\frac{2}{5}h^2 yx - \frac{2}{3}y^3 x + \left(\frac{x^3}{3} - l^2 x\right)y + \frac{\nu h^2 yx}{3} - \frac{\nu y^3 x}{3} + f(y)\right\}, \tag{7-3.3.28}$$

$$u_y = \frac{\rho g h}{EI}\left\{-\frac{h^2 y^2}{6} + \frac{y^4}{12} - \frac{1}{5}\nu h^2 y^2 + \frac{1}{6}\nu y^4 - \frac{\nu(x^2 - l^2)}{2}y^2 + g(x)\right\}. \tag{7-3.3.29}$$

Substitution of these results into the shear strain expression leads to the relation

$$\begin{aligned}\frac{1}{5}h^2 x - y^2 x + \frac{x^3}{6} - \frac{l^2 x}{2} + \frac{\nu h^2 x}{6} - \frac{\nu y^2 x}{2}& \\ + \frac{f'(y)}{2} - \frac{\nu x y^2}{2} + \frac{g'(x)}{2} &= (1 + \nu)(h^2 x - y^2 x).\end{aligned} \tag{7-3.3.30}$$

Therefore,

$$\begin{aligned} f'(y) &= -\omega^0, \\ g'(x) &= \omega^0 + \frac{8}{5}h^2 x + \frac{5}{3}\nu h^2 x - \frac{x^3}{3} + l^2 x, \end{aligned} \tag{7-3.3.31}$$

$$\begin{aligned} f(y) &= -\omega^0 y + u_x^0, \\ g(x) &= \omega^0 x + \frac{4}{5}h^2 x^2 + \frac{5}{6}\nu h^2 x^2 - \frac{x^4}{12} + \frac{l^2 x^2}{2} + u_y^0. \end{aligned} \tag{7-3.3.32}$$

If we note that $u_y(x, 0) = g(x)$, then $g(x)$ represents the centerline deflection of the beam or the elementary beam theory solution. The constants may be selected to satisfy the boundary conditions

$$u_x(-l, 0) = 0, \qquad u_y(-l, 0) = 0, \qquad u_y(+l, 0) = 0. \tag{7-3.3.33}$$

Using these conditions, the constants become

$$\omega^0 = u_x^0 = 0,$$
$$u_y^0 = -\frac{4}{5}h^2l^2 - \frac{5}{6}\nu h^2 l^2 - \frac{5}{12}l^4. \qquad (7\text{-}3.3.34)$$

The midsurface deflection curve becomes

$$u_y(x, 0) = \frac{\rho g h}{EI}\left\{\frac{l^2x^2}{2} - \frac{x^4}{12} - \frac{5}{12}l^4 + h^2\left(\frac{4}{5} + \frac{5}{6}\nu\right)(x^2 - l^2)\right\}. \qquad (7\text{-}3.3.35)$$

The first three terms represent the corresponding elementary solution from beam theory usually obtained by considering a beam under a uniform external load*.

4 Fourier series solutions

A much more powerful form of solution of the biharmonic equation involves the use of trigonometric functions and Fourier analysis.† Consider the plane elasticity problem when the body force is equal to zero, so that the Airy stress function or the Marguerre displacement function is defined by the biharmonic equation. We shall examine the Airy stress function in this development, but the form of the solution is equally applicable to the Marguerre function:

$$\nabla^2\nabla^2\varphi = 0. \qquad (7\text{-}4.1)$$

Let us seek a solution in the form

$$\varphi = e^{\alpha x}e^{\beta y}. \qquad (7\text{-}4.2)$$

Substitution of equation 7-4.2 into 7-4.1 yields

$$(\alpha^4 + 2\alpha^2\beta^2 + \beta^4)e^{\alpha x}e^{\beta y} = 0. \qquad (7\text{-}4.3)$$

The auxiliary equation for α and β becomes

$$(\alpha^2 + \beta^2)^2 = 0. \qquad (7\text{-}4.4)$$

Therefore, there are double roots of the form

$$\alpha = \pm i\beta. \qquad (7\text{-}4.5)$$

Special consideration must be given to the zero roots of this equation. Note that if $\beta = 0$, there is a fourfold multiplicity of the roots and the solution would take the form

$$\varphi = R_0 + R_1x + R_2x^2 + R_3x^3. \qquad (7\text{-}4.6)$$

* See K. Pearson, *Quart. J. Math.*, **24**, 63, 1889; and T. Von Karman, *Abhandl. Aerodynam. Inst.*, Tech. Hochschule, Aachen, **7**, 3, 1927.

† For the earliest investigation of the use of Fourier series in elasticity, see E. Mathieu, *Théorie de l'élasticité des corps solides*, Gautheir-Villars, Paris, 1890.

In a similar manner, if α is initially considered zero, we obtain a solution of the form

$$\varphi = R_4 y + R_5 y^2 + R_6 y^3 + R_7 xy + R_8 x^2 y + R_9 xy^2. \qquad (7\text{-}4.7)$$

These terms may be recognized as those polynomial solutions which term by term satisfy the biharmonic equation. In considering the general form of the solution of the biharmonic equation they arise naturally due to the zero roots of the characteristic equation.

Examining the general form of the roots shown in equation 7-4.5, one may construct a solution of the form

$$\begin{aligned} \varphi = {} & e^{i\beta x}[Ae^{\beta y} + Be^{-\beta y} + Cye^{\beta y} + Dye^{-\beta y}] \\ & + e^{-i\beta x}[A'e^{\beta y} + B'e^{-\beta y} + C'ye^{\beta y} + D'ye^{-\beta y}]. \end{aligned} \qquad (7\text{-}4.8)$$

Since the double roots may be associated with either the x or the y part of the solution and β may be real or imaginary, the general form of the solution will be taken as the following sum:*

$$\begin{aligned} \varphi = {} & \sin \beta x[A \operatorname{sh} \beta y + B \operatorname{ch} \beta y + C\beta y \operatorname{sh} \beta y + D\beta y \operatorname{ch} \beta y] \\ & + \cos \beta x[A' \operatorname{sh} \beta y + B' \operatorname{ch} \beta y + C'\beta y \operatorname{sh} \beta y + D'\beta y \operatorname{ch} \beta y] \\ & + \sin \alpha y[E \operatorname{sh} \alpha x + F \operatorname{ch} \alpha x + G\alpha x \operatorname{sh} \alpha x + H\alpha x \operatorname{ch} \alpha x] \\ & + \cos \alpha y[E' \operatorname{sh} \alpha x + F' \operatorname{ch} \alpha x + G'\alpha x \operatorname{sh} \alpha x + H'\alpha x \operatorname{ch} \alpha x] \\ & + R_0 + R_1 x + R_2 x^2 + R_3 x^3 + R_4 y + R_5 y^2 + R_6 y^3 \\ & + R_7 xy + R_8 x^2 y + R_9 xy^2. \end{aligned} \qquad (7\text{-}4.9)$$

This form recognizes that the solution must be real and replaces the exponential functions by hyperbolic functions. For problems involving a semi-infinite or infinite dimension, the exponential form is more convenient and will be used. Examples of this type of problem will be shown later. If β is chosen such that it cannot take the value of zero, the polynomial terms will not appear.

Before proceeding to the solution of typical problems, the displacements will be obtained in terms of the general form of the solution given in equation 7-4.2. The stresses are given by

$$\sigma_{xx} = \frac{\partial^2 \varphi}{\partial y^2}, \qquad \sigma_{yy} = \frac{\partial^2 \varphi}{\partial x^2}, \qquad \sigma_{xy} = -\frac{\partial^2 \varphi}{\partial x\, \partial y}. \qquad (7\text{-}4.10)$$

Substituting into the stress–strain relations yields

$$\begin{aligned} \frac{\partial u_x}{\partial x} &= \frac{1}{E}\left[\frac{\partial^2 \varphi}{\partial y^2} - \nu \frac{\partial^2 \varphi}{\partial x^2}\right], \\ \frac{\partial u_y}{\partial y} &= \frac{1}{E}\left[\frac{\partial^2 \varphi}{\partial x^2} - \nu \frac{\partial^2 \varphi}{\partial y^2}\right]. \end{aligned} \qquad (7\text{-}4.11)$$

* The hyperbolic sine will be designated by sh instead of sinh in the remainder of this text. In a similar manner, ch will be used to designate the hyperbolic cosine.

Integrating equation 7-4.11 gives

$$u_x = \frac{1}{E}\left\{\int_x \frac{\partial^2\varphi}{\partial y^2}dx - \nu\frac{\partial\varphi}{\partial x} + f(y)\right\},$$
$$u_y = \frac{1}{E}\left\{\int_y \frac{\partial^2\varphi}{\partial x^2}dy - \nu\frac{\partial\varphi}{\partial y} + g(x)\right\}. \tag{7-4.12}$$

The shear stress condition is

$$\frac{\partial u_x}{\partial y} + \frac{\partial u_y}{\partial x} = -\frac{2(1+\nu)}{E}\frac{\partial^2\varphi}{\partial x\,\partial y}. \tag{7-4.13}$$

Substituting equations 7-4.12 into equation 7-4.13 gives

$$\int_x \frac{\partial^3\varphi}{\partial y^3}dx + 2\frac{\partial^2\varphi}{\partial x\,\partial y} + \int_y \frac{\partial^3\varphi}{\partial x^3}dy + f'(y) + g'(x) = 0. \tag{7-4.14}$$

Using the solution form from equation 7-4.2, equation 7-4.14 becomes

$$\left[\frac{\beta^3}{\alpha} + 2\alpha\beta + \frac{\alpha^3}{\beta}\right]e^{\alpha x}e^{\beta y} + f'(y) + g'(x) = 0. \tag{7-4.15}$$

The expression in the bracket is zero from equation 7-4.4 so that $f(y)$ and $g(x)$ represent only the rigid-body motions, or

$$f(y) = -\omega^0 y + u_x^0,$$
$$g(x) = \omega^0 x + u_y^0. \tag{7-4.16}$$

This form will have to be modified if zero roots are considered, for the polynomial terms will then make a contribution to the functions of integration. The procedure in this case is similar to that used in the discussion of polynomial solutions.

4.1 Beam subjected to sinusoidal load. Consider the beam shown in Figure 7-4. The boundary conditions are

$$y = +h, \quad \sigma_{yy} = -q\sin\frac{\pi x}{l}, \quad \sigma_{xy} = 0, \tag{7-4.1.1}$$

$$y = -h, \quad \sigma_{yy} = 0, \quad \sigma_{xy} = 0, \tag{7-4.1.2}$$

$$x = 0, \quad \sigma_{xx} = 0, \quad \int \sigma_{xy}\,dy = -\frac{1}{2}\int_0^l q\sin\left(\frac{\pi x}{l}\right)dx = -q\frac{l}{\pi}, \tag{7-4.1.3}$$

$$x = l, \quad \sigma_{xx} = 0, \quad \int \sigma_{xy}\,dy = q\frac{l}{\pi}. \tag{7-4.1.4}$$

We have not attempted to specify the exact shear distribution across the ends of the beam. On the presumption that $h/l \ll 1$, we may use the Saint-Venant approximation for this case. Since on the boundary σ_{yy} is a sine function, examination of the general form of the solution in equation 7-4.9 suggests

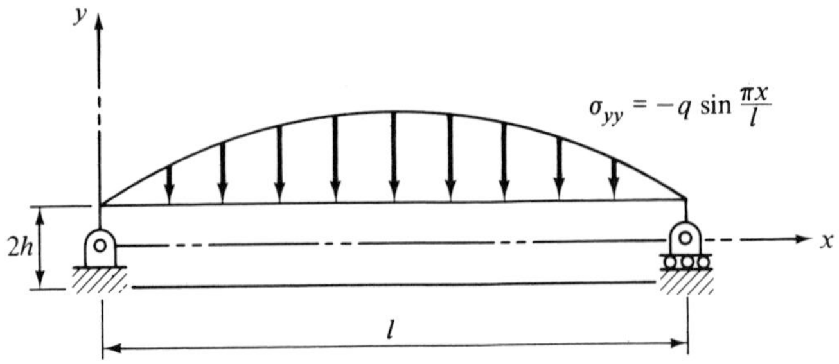

Figure 7-4

that the following form of solution be tried:

$$\varphi = \sin \beta x[A \operatorname{sh} \beta y + B \operatorname{ch} \beta y + C\beta y \operatorname{sh} \beta y + D\beta y \operatorname{ch} \beta y]. \tag{7-4.1.5}$$

The stresses on the long boundary become

$$\sigma_{yy} = -\beta^2 \sin \beta x[A \operatorname{sh} \beta y + B \operatorname{ch} \beta y + C\beta y \operatorname{sh} \beta y + D\beta y \operatorname{ch} \beta y], \tag{7-4.1.6}$$

$$\begin{aligned}\sigma_{xy} = -\beta^2 \cos \beta x[A \operatorname{ch} \beta y + B \operatorname{sh} \beta y + C(\beta y \operatorname{ch} \beta y + \operatorname{sh} \beta y)\\ + D(\beta y \operatorname{sh} \beta y + \operatorname{ch} \beta y)].\end{aligned} \tag{7-4.1.7}$$

To satisfy the condition of zero shearing stress on the upper and lower boundaries, it is necessary that the coefficient of the cos βx term be zero. To satisfy this condition, the even functions and the odd functions of y will be equated independently to zero at the boundaries $y = \pm h$:

$$\begin{aligned}A \operatorname{ch} \beta h + D(\beta h \operatorname{sh} \beta h + \operatorname{ch} \beta h) = 0,\\ B \operatorname{sh} \beta h + C(\beta h \operatorname{ch} \beta h + \operatorname{sh} \beta h) = 0.\end{aligned} \tag{7-4.1.8}$$

The constants A and B become

$$A = -D(\beta h \tanh \beta h + 1), \tag{7-4.1.9}$$

$$B = -C(\beta h \coth \beta h + 1). \tag{7-4.1.10}$$

Substituting these values into equation 7-4.1.6 yields

$$\begin{aligned}\sigma_{yy} = -\beta^2 \sin \beta x[D(\beta y \operatorname{ch} \beta y - \{\beta h \tanh \beta h + 1\} \operatorname{sh} \beta y)\\ + C(\beta y \operatorname{sh} \beta y - \{\beta h \coth \beta h + 1\} \operatorname{ch} \beta y)].\end{aligned} \tag{7-4.1.11}$$

To satisfy the condition that σ_{yy} is zero at $y = -h$, D and C will be related as follows:

$$\begin{aligned}D(\beta h \operatorname{ch} \beta h - \{\beta h \tanh \beta h + 1\} \operatorname{sh} \beta h)\\ = C(\beta h \operatorname{sh} \beta h - \{\beta h \coth \beta h + 1\} \operatorname{ch} \beta h),\\ D\left(\frac{\beta h \operatorname{ch}^2 \beta h - \beta h \operatorname{sh}^2 \beta h}{\operatorname{ch} \beta h} - \operatorname{sh} \beta h\right)\\ = C\left(\frac{\beta h \operatorname{sh}^2 \beta h - \beta h \operatorname{ch}^2 \beta h}{\operatorname{sh} \beta h} - \operatorname{ch} \beta h\right).\end{aligned} \tag{7-4.1.12}$$

Using the identity $\text{ch}^2 \beta h - \text{sh}^2 \beta h = 1$, equation 7.4-1.12 may be written

$$C = -\tanh \beta h \left[\frac{\beta h - \text{sh}\, \beta h \,\text{ch}\, \beta h}{\beta h + \text{sh}\, \beta h \,\text{ch}\, \beta h} \right] D. \qquad (7\text{-}4.1.13)$$

Substituting equation 7-4.1.13 into equation 7-4.1.11 yields

$$\sigma_{yy} = -D\beta^2 \sin \beta x \{[\beta y \,\text{ch}\, \beta y - (\beta h \tanh \beta h + 1)\,\text{sh}\, \beta y] - \tanh \beta h \left[\frac{\beta h - \text{sh}\, \beta h \,\text{ch}\, \beta h}{\beta h + \text{sh}\, \beta h \,\text{ch}\, \beta h} \right] [\beta y \,\text{sh}\, \beta y - (\beta h \coth \beta h + 1)\,\text{ch}\, \beta y]\}. \qquad (7\text{-}4.1.14)$$

The boundary condition at $y = +h$ requires that

$$q \sin \frac{\pi x}{l} = 2\beta^2 D \sin \beta x \left[\frac{\beta h - \text{sh}\, \beta h \,\text{ch}\, \beta h}{\text{ch}\, \beta h} \right]. \qquad (7\text{-}4.1.15)$$

Thus β will be chosen as π/l and D becomes

$$D = \frac{q \,\text{sh}\, \pi h/l}{[2(\pi/l)^2(\pi h/l - \text{sh}\, \pi h/l \,\text{ch}\, \pi h/l)]}. \qquad (7\text{-}4.1.16)$$

From equation 7-4.1.13, C may be found to be

$$C = -\frac{q \,\text{sh}\, \pi h/l}{2(\pi/l)^2[(\pi h/l) + \text{sh}\, \pi h/l \,\text{ch}\, \pi h/l]}. \qquad (7\text{-}4.1.17)$$

It is apparent that since σ_{xx} involves $\sin \pi x/l$, the boundary conditions $\sigma_{xx} = 0$ at $x = 0$ and $x = l$ are satisfied. There is no need to check the final boundary conditions on σ_{xy} at $x = 0$ and $x = l$, because they are automatically satisfied by the equilibrium equations.

The displacement boundary conditions are

$$u_y(0, 0) = 0, \qquad u_x(0, 0) = 0, \qquad u_y(l, 0) = 0. \qquad (7\text{-}4.1.18)$$

Using equations 7-4.12 and 7-4.16, the displacement may be written

$$u_x = -\frac{1}{E}\beta \cos \beta x\{A(1 + \nu)\,\text{sh}\, \beta y + B(1 + \nu)\,\text{ch}\, \beta y + C[(1 + \nu)\beta y \,\text{sh}\, \beta y + 2\,\text{ch}\, \beta y] + D[(1 + \nu)\beta y \,\text{ch}\, \beta y + 2\,\text{sh}\, \beta y]\} - \omega^0 y + u_x^0, \qquad (7\text{-}4.1.19)$$

$$u_y = -\frac{1}{E}\beta \sin \beta x\{A(1 + \nu)\,\text{ch}\, \beta y + B(1 + \nu)\,\text{sh}\, \beta y + C[(1 + \nu)\beta y \,\text{ch}\, \beta y - (1 + \nu)\,\text{sh}\, \beta y] + D[(1 + \nu)\beta y \,\text{sh}\, \beta y - (1 - \nu)\,\text{ch}\, \beta y]\} + \omega^0 x + u_y^0. \qquad (7\text{-}4.1.20)$$

The displacement boundary conditions are satisfied by setting the rigid-body constants equal to the following values:

$$\omega^0 = 0, \qquad u_y^0 = 0, \qquad (7\text{-}4.1.21)$$

$$u_x^0 = -\frac{1}{E}\beta[B(1 + \nu) + 2C]. \qquad (7\text{-}4.1.22)$$

It would be interesting to compare the centerline deflection given by equation 7-4.1.20 with the elementary solution,

$$u_y^{\text{elem}}(x, 0) = -\frac{3}{2} \frac{ql^4}{h^3\pi^4 E} \sin \frac{\pi x}{l}; \tag{7-4.1.23}$$

at $y = 0$, equation 7-4.1.20 becomes

$$u_y(x, 0) = -\frac{\beta}{E} \sin \beta x \{A(1 + \nu) - D(1 - \nu)\}. \tag{7-4.1.24}$$

Using equation 7-4.1.19, $u_y(x, 0)$ becomes

$$u_y(x, 0) = \frac{D\beta}{E} \sin \beta x \{2 + (1 + \nu)\beta h \tanh \beta h\}. \tag{7-4.1.25}$$

From equation 7-4.1.16 D may be written

$$D = \frac{-q \operatorname{ch} \pi h/l}{2(\pi/l)^2(\operatorname{sh} \pi h/l \operatorname{ch} \pi h/l - \pi h/l)}. \tag{7-4.1.26}$$

For the case $h/l \ll 1$, D reduces to

$$D = -\frac{3}{4} \frac{ql^5}{h^3\pi^5}, \tag{7-4.1.27}$$

and the displacement of the centerline may be written

$$u_y(x, 0) = -\frac{3}{2} \frac{ql^4}{Eh^3\pi^4} \sin \frac{\pi x}{l} \left\{1 + \frac{1 + \nu}{2} \frac{\pi h}{l} \tanh \frac{\pi h}{l}\right\}. \tag{7-4.1.28}$$

By the same approximation of $h/l \ll 1$, we may neglect the second term in the brackets. To this degree of approximation, the elementary solution and the elasticity solutions give the same center line deflection.

5 Fourier analysis*

Before looking at solutions of the biharmonic equation, let us consider some of the properties of orthogonal functions. Three mutually perpendicular vectors $\bar{g}_m$ form what is called an orthonormal set if

$$\bar{g}_m \cdot \bar{g}_n = \delta_{mn} = \begin{cases} 1 & m = n \\ 0 & m \neq n \end{cases}. \tag{7-5.1}$$

This is sometimes called an inner product and may also be written

$$(\bar{g}_m, \bar{g}_n) = \delta_{mn}. \tag{7-5.2}$$

* See R. V. Churchill, *Fourier Series and Boundary Value Problems*, McGraw-Hill Book Company, New York, 1941; or P. M. Morse and H. Feshbach, *Methods of Theoretical Physics*, McGraw-Hill Book Company, New York, 1953, for a more detailed presentation.

Every vector $\bar{f}$ in this space may be expressed as a linear combination of the vectors $\bar{g}_1$, $\bar{g}_2$, and $\bar{g}_3$:

$$\begin{aligned} \bar{f} &= c_1\bar{g}_1 + c_2\bar{g}_2 + c_3\bar{g}_3 \\ &= \sum_{m=1}^{3} c_m\bar{g}_m. \end{aligned} \tag{7-5.3}$$

If we take the inner product of both sides of this equation, we obtain

$$(\bar{f}, \bar{g}_n) = c_n(\bar{g}_n, \bar{g}_n) = c_n. \tag{7-5.4}$$

Consider now a set of functions which form a complete orthogonal set over the interval a to b in the sense that

$$\int_a^b \varphi_n(x)\varphi_m(x)\,dx = 0 \qquad \text{if } m \neq n. \tag{7-5.5}$$

A quantity called the "norm" can be defined by

$$N_n = \left[\int_a^b [\varphi_n(x)]^2\,dx\right]^{1/2}. \tag{7-5.6}$$

We can normalize our functions by dividing each φ_n by N_n, so that we obtain an orthonormal set. Given a countable infinite orthonormal set of functions, it is often possible to represent any arbitrary function $f(x)$ in a fundamental interval by a linear combination of these functions:

$$\begin{aligned} f(x) &= c_1\varphi_1(\varphi) + c_2\varphi_2(x) + \cdots + c_n\varphi_n(x) + \cdots, \quad a < x < b, \\ &= \sum_{n=1}^{\infty} c_n\varphi_n(x), \end{aligned} \tag{7-5.7}$$

where

$$c_n = (f, \varphi_n) = \frac{1}{N_n}\int_a^b f(x)\varphi_n(x)\,dx. \tag{7-5.8}$$

The generalized Fourier series corresponding to $f(x)$ is then

$$f(x) \sim \sum_{n=1}^{\infty} \varphi_n(x)\frac{1}{N_n}\int_a^b f(\eta)\varphi_n(\eta)\,d\eta. \tag{7-5.9}$$

5.1 Fourier trigonometric series. The series

$$\frac{1}{2}a_0 + \sum_{n=1}^{\infty}(a_n \cos nx + b_n \sin nx), \quad -\pi < x < \pi, \tag{7-5.1.1}$$

is called the Fourier trigonometric series. These functions form an orthogonal set of functions within the interval $-\pi < x < \pi$:

$$\begin{aligned} \int_{-\pi}^{+\pi} \cos mx \cos nx\,dx &= \frac{1}{2}\int_{-\pi}^{+\pi} [\cos(m+n)x + \cos(m-n)x]\,dx \\ &= \frac{1}{2}\left[\frac{\sin(m+n)x}{m+n} + \frac{\sin(m-n)x}{(m-n)}\right]_{-\pi}^{+\pi} = 0, \quad m \neq n, \end{aligned} \tag{7-5.1.2}$$

$$\int_{-\pi}^{+\pi} \sin mx \sin nx \, dx = 0, \qquad m \neq n, \tag{7-5.1.3}$$

$$\int_{-\pi}^{+\pi} \sin mx \cos nx \, dx = 0 \tag{7-5.1.4}$$

for all m and n, since this is an integral of an odd function across a symmetric interval. Examining the normalization properties,

$$\int_{-\pi}^{+\pi} \cos^2 mx \, dx = \frac{1}{2}\int_{-\pi}^{+\pi} (1 + \cos 2nx) \, dx = \begin{cases} \pi, & m \neq 0 \\ 2\pi, & m = 0 \end{cases}, \tag{7-5.1.5}$$

$$\int_{-\pi}^{+\pi} \sin^2 mx \, dx = \pi. \tag{7-5.1.6}$$

Therefore, a function $f(x)$ may be represented in the interval $-\pi < x < +\pi$ by the expression

$$f(x) \sim \frac{1}{2\pi}\int_{-\pi}^{+\pi} f(\eta) \, d\eta + \sum_{n=1}^{\infty} \frac{1}{\pi}\left[\cos nx \int_{-\pi}^{+\pi} f(\eta) \cos n\eta \, d\eta + \sin nx \int_{-\pi}^{+\pi} f(\eta) \sin n\eta \, d\eta\right]. \tag{7-5.1.7}$$

if this function is piecewise continuous. It is apparent that if $f(x)$ is even, all the sine functions vanish, and if $f(x)$ is odd, all the cosine terms vanish. In equation 7-5.1.7 the following relations were used:

$$a_n = \frac{1}{\pi}\int_{-\pi}^{+\pi} f(\eta) \cos n\eta \, d\eta, \tag{7-5.1.8}$$

$$b_n = \frac{1}{\pi}\int_{-\pi}^{+\pi} f(\eta) \sin n\eta \, d\eta. \tag{7-5.1.9}$$

When $f(x)$ is odd, the above expressions for a_n and b_n become

$$a_n = 0, \tag{7-5.1.10}$$

$$b_n = \frac{2}{\pi}\int_{0}^{\pi} f(\eta) \sin n\eta \, d\eta. \tag{7-5.1.11}$$

and the Fourier series becomes

$$f(x) \sim \frac{2}{\pi}\sum_{n=1}^{\infty} \sin nx \int_{0}^{\pi} f(\eta) \sin n\eta \, d\eta. \tag{7-5.1.12}$$

When $f(x)$ is even these terms and the series become

$$a_n = \frac{2}{\pi}\int_{0}^{\pi} f(\eta) \cos n\eta \, d\eta, \tag{7-5.1.13}$$

$$b_n = 0, \tag{7-5.1.14}$$

$$f(x) \sim \frac{1}{\pi}\int_{0}^{\pi} f(\eta) \, d\eta + \frac{2}{\pi}\sum_{n=1}^{\infty} \cos nx \int_{0}^{\pi} f(\eta) \cos n\eta \, d\eta. \tag{7-5.1.15}$$

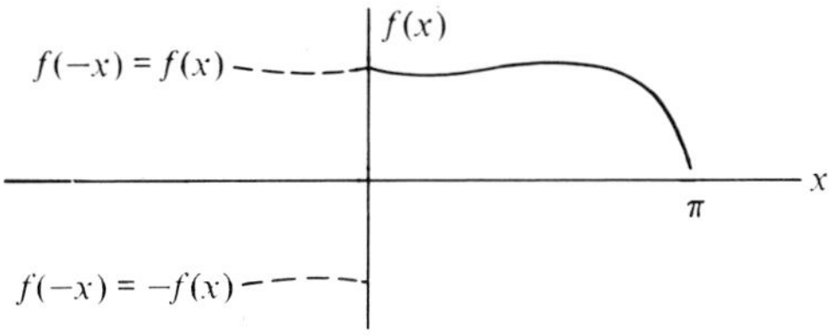

Figure 7-5

When a function is of interest only in the interval $0 < x < \pi$, this function may be thought of as either even or odd and may be expressed as a cosine or sine series.

We may generalize these results to the interval from $-l$ to $+l$ by transforming the series as follows:

$$\frac{1}{2}a_0 + \sum_1^\infty \left(a_n \cos \frac{n\pi x}{l} + b_n \sin \frac{n\pi x}{l}\right), \qquad -l \leq x \leq +l. \tag{7-5.1.16}$$

Let $f(x + 2l) = f(x)$ for all x and let $f(x)$ be piecewise continuous in the interval $(-l, l)$. Then at any point where $f(x)$ has right- and left-hand derivative, it is true that

$$\frac{1}{2}[f(x + 0) + f(x - 0)] = \frac{a_0}{2} + \sum_1^\infty \left(a_n \cos \frac{n\pi x}{l} + b_n \sin \frac{n\pi x}{l}\right), \tag{7-5.1.17}$$

where

$$a_n = \frac{1}{l}\int_{-l}^{+l} f(x) \cos \frac{n\pi x}{l}\, dx, \tag{7-5.1.18}$$

$$b_n = \frac{1}{l}\int_{-l}^{+l} f(x) \sin \frac{n\pi x}{l}\, dx. \tag{7-5.1.19}$$

The previous reductions, when $f(x)$ is either even or odd or defined over the region 0 to l, are also applicable to this series. For more detailed information the reader is referred to one of the many texts on Fourier series. The operations indicated in equations 7-5.1.18 and 7-5.1.19 are often referred to as forming (or taking) "the finite Fourier cosine and sine transform" of $f(x)$.

6 General Fourier solution of elasticity problem

The application of Fourier analysis will first be considered for the case of rectangular bodies with a height–length ratio much less than 1. Thus the Saint-Venant approximation may be applied on the short boundaries, if necessary. Because of the extensive manipulations involved, when using Fourier analysis it is advantageous to separate the problem into combinations of even and odd functions in the x and y directions. As noted before, any function (including boundary stresses) may be separated into the sum of an even and an odd function. The possible combinations are shown in Table 7-1.

Table 7-1.

	φ		σ_{xx}		σ_{yy}		σ_{xy}		u_x		u_x	
Case	x	y	x	y	x	y	x	y	x	y	x	y
1	even	even	even	even	even	even	odd	odd	odd	even	even	odd
2	odd	odd	odd	odd	odd	odd	even	even	even	odd	odd	even
3	even	odd	even	odd	even	odd	odd	even	odd	odd	even	even
4	odd	even	odd	even	odd	even	even	odd	even	even	odd	odd

Due to the linearity of the small-displacement theory of elasticity, we may break any problem into these four cases and superimpose the results. At first glance this may seem to produce more work, but it serves as an efficient bookkeeping procedure. The method recommended here reduces the chance for error and saves work in those problems which naturally have certain symmetry or antisymmetry properties.

Consider the rectangular body shown in Figure 7-6 subjected to arbitrary shear and normal stresses on the upper and lower edges, but free of stresses on the ends. This implies that the system is in equilibrium without end stresses. The normal and shear stresses separate naturally into the four cases mentioned and will have the boundary conditions given by Table 7-2.

***6.1 Case 4—odd in* x *and even in* y.** Case 4 will be selected as an example of the use of Fourier analysis. The Airy stress function will be odd in x and even

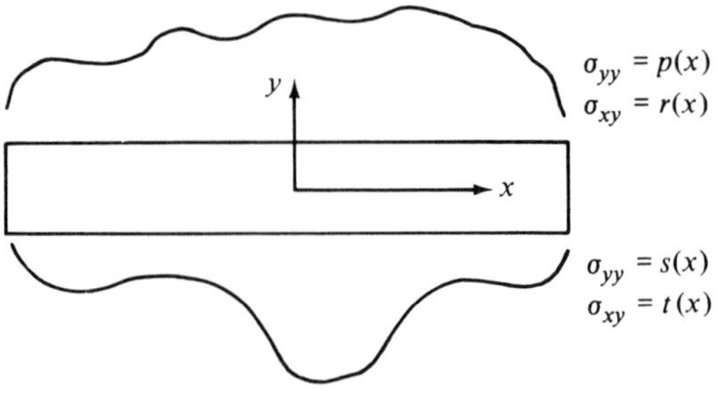

Figure 7-6

Table 7-2

Case	σ_{yy} at $y = +h$	σ_{yy} at $y = -h$	σ_{xy} at $y = +h$	σ_{xy} at $y = -h$
1	$\frac{1}{4}\{p(x)+p(-x)+s(x)+s(-x)\}$	$\frac{1}{4}\{p(x)+p(-x)+s(x)+s(-x)\}$	$\frac{1}{4}\{r(x)-r(-x)-t(x)+t(-x)\}$	$\frac{1}{4}\{-r(x)+r(-x)+t(x)-t(-x)\}$
2	$\frac{1}{4}\{p(x)-p(-x)-s(x)+s(-x)\}$	$-\frac{1}{4}\{p(x)-p(-x)-s(x)+s(-x)\}$	$\frac{1}{4}\{r(x)+r(-x)+t(x)+t(-x)\}$	$\frac{1}{4}\{r(x)+r(-x)+t(x)+t(-x)\}$
3	$\frac{1}{4}\{p(x)+p(-x)-s(x)-s(-x)\}$	$-\frac{1}{4}\{p(x)+p(-x)-s(x)-s(-x)\}$	$\frac{1}{4}\{r(x)-r(-x)+t(x)-t(-x)\}$	$\frac{1}{4}\{r(x)-r(-x)+t(x)-t(-x)\}$
4	$\frac{1}{4}\{p(x)-p(-x)+s(x)-s(-x)\}$	$\frac{1}{4}\{p(x)-p(-x)+s(x)-s(-x)\}$	$\frac{1}{4}\{r(x)+r(-x)-t(x)-t(-x)\}$	$-\frac{1}{4}\{r(x)+r(-x)-t(x)-t(-x)\}$
Σ	$p(x)$	$s(x)$	$r(x)$	$t(x)$

in y in this case, and the boundary conditions may be written

$$y = +h, \qquad \sigma_{yy} = g(x), \qquad \text{odd in } x \tag{7-6.1.1}$$

$$\sigma_{xy} = f(x), \qquad \text{even in } x \tag{7-6.1.2}$$

$$y = -h, \qquad \sigma_{yy} = g(x), \tag{7-6.1.3}$$

$$\sigma_{xy} = -f(x), \tag{7-6.1.4}$$

where

$$g(x) = \tfrac{1}{4}\{p(x) - p(-x) + s(x) - s(-x)\}, \tag{7-6.1.5}$$

$$f(x) = \tfrac{1}{4}\{r(x) + r(-x) - t(x) - t(-x)\}. \tag{7-6.1.6}$$

If possible, the conditions on the short boundaries will be taken to be

$$x = \pm l, \qquad \sigma_{xy} = \sigma_{xx} = 0, \tag{7-6.1.7}$$

Alternatively, we can use the Saint-Venant approximation if necessary:

$$x = \pm l, \qquad \int_{-h}^{+h} \sigma_{xy}\, dy = 0, \tag{7-6.1.8}$$

$$\int_{-h}^{+h} \sigma_{xx}\, dy = 0, \tag{7-6.1.9}$$

$$\int_{-h}^{+h} \sigma_{xx} y\, dy = 0. \tag{7-6.1.10}$$

Since σ_{xx} will be even in y and σ_{xy} is odd equations 7-6.1.10 and 7-7.1.8 are satisfied automatically. The condition of zero axial force, equation 7-6.1.9, must be checked. We can expect to use the Fourier trigonometric series to satisfy the conditions on the long boundaries. The form of the solution will be as follows:

$$\varphi = \sum_n \sin \beta_n x[B_n \operatorname{ch} \beta_n y + C_n \beta_n y \operatorname{sh} \beta_n y]. \tag{7-6.1.11}$$

The stresses become

$$\sigma_{yy} = -\sum_n \beta_n^2 \sin \beta_n x[B_n \operatorname{ch} \beta_n y + C_n \beta_n y \operatorname{sh} \beta_n y], \tag{7-6.1.12}$$

$$\sigma_{xy} = -\sum_n \beta_n^2 \cos \beta_n x[B_n \operatorname{sh} \beta_n y + C_n(\beta_n y \operatorname{ch} \beta_n y + \operatorname{sh} \beta_n y)], \tag{7-6.1.13}$$

$$\sigma_{xx} = \sum_n \beta_n^2 \sin \beta_n x[B_n \operatorname{ch} \beta_n y + C_n(\beta_n y \operatorname{sh} \beta_n y + 2 \operatorname{ch} \beta_n y)]. \tag{7-6.1.14}$$

At first glance, the series for the shear stress may seem incomplete due to the omission of the usual constant term. In many problems it is necessary to include in φ the zero-root polynomials, such as xy, x^2, y^2, xy^2, x^2y, etc., to produce this constant and to maintain the required even and oddness in the stress function. In the present problem, however, we have stated that the ends are to be free of stress and therefore the horizontal forces there must be zero. It is apparent that the average shear must be zero on the top and bottom to maintain equilibrium, and the constant term is not needed. The boundary condition of zero normal stress on the ends may be satisfied by the selection of β_n:

$$\beta_n = \frac{n\pi}{l}. \tag{7-6.1.15}$$

The boundary conditions 7-6.1.1 to 7-6.1.4 become

$$g(x) = -\sum_n \beta_n^2 \sin \beta_n x[B_n \operatorname{ch} \beta_n h + C_n \beta_n h \operatorname{sh} \beta_n h], \tag{7-6.1.16}$$

$$f(x) = -\sum_n \beta_n^2 \cos \beta_n x[B_n \operatorname{sh} \beta_n h + C_n(\beta_n h \operatorname{ch} \beta_n h + \operatorname{sh} \beta_n h)]. \tag{7-6.1.17}$$

Note that by use of the evenness and oddness of the functions we need to consider only one quarter of the geometry.

Taking the finite Fourier sine and cosine transforms yields

$$-\left(\frac{n\pi}{l}\right)^2 l[B_n \operatorname{ch} \beta_n h + C_n \beta_n h \operatorname{sh} \beta_n h] = \int_{-l}^{+l} g(x) \sin \frac{n\pi x}{l}\, dx, \tag{7-6.1.18}$$

$$-\left(\frac{n\pi}{l}\right)^2 l[B_n \operatorname{sh} \beta_n h + C_n(\beta_n h \operatorname{ch} \beta_n h + \operatorname{sh} \beta_n h)] = \int_{-l}^{+l} f(x) \cos \frac{n\pi x}{l}\, dx.$$

$$n = 1, 2, 3, \ldots \tag{7-6.1.19}$$

Given any $f(x)$ and $g(x)$, the Fourier coefficients C_n and B_n can now be determined. Solving the two equations yields

$$B_n = \frac{l}{n^2\pi^2[(n\pi h/l) + \operatorname{sh}(n\pi h/l) \operatorname{ch}(n\pi h/l)]} \int_{-l}^{+l} \left[\frac{n\pi h}{l} \operatorname{sh} \frac{n\pi h}{l} f(x) \cos \frac{n\pi x}{l} - \left(\frac{n\pi h}{l} \operatorname{ch} \frac{n\pi h}{l} + \operatorname{sh} \frac{n\pi h}{l}\right) g(x) \sin \frac{n\pi x}{l}\right] dx, \tag{7-6.1.20}$$

$$C_n = \frac{l}{n^2\pi^2[(n\pi h/l) + \operatorname{sh}(n\pi h/l) \operatorname{ch}(n\pi h/l)]} \int_{-l}^{+l} \left[\operatorname{sh} \frac{n\pi h}{l} g(x) \sin \frac{n\pi x}{l} - \operatorname{ch} \frac{n\pi h}{l} f(x) \cos \frac{n\pi x}{l}\right] dx. \tag{7-6.1.21}$$

A comment may be added here: if the average shear stress had not been zero, the beam would have appeared as shown in Figure 7-7. Since the normal stress σ_{yy} is an odd function of x in this case, its average or resultant value is zero. If the resultant shear were not zero, normal stresses σ_{xx} would have to appear on the ends to preserve equilibrium. A constant term would have to be added to the cosine series in this case. Since the problem being considered

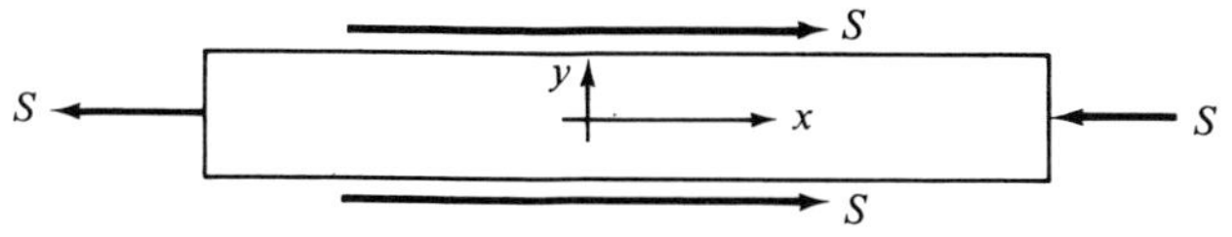

Figure 7-7

has φ odd in x and even in y, the required polynomial would have been xy^2, which would have provided the needed constant in σ_{xy} at the top and bottom for the Fourier analysis and at the same time produced the equilibrating normal stresses at the ends.

If zero roots of β_n are required, they may be formally introduced from the general solution (with the proper oddness and evenness) and the corresponding terms may be evaluated by formal Fourier procedures.

6.2 Displacement solution using Marguerre function. Consider a long plate which is stress free on the short ends and clamped top and bottom, and the long sides are now displaced as shown in Figure 7-8. The boundary conditions are

$$y = +h, \qquad u_x = 0, \qquad u_y = \delta, \tag{7-6.2.1}$$

$$y = -h, \qquad u_x = 0, \qquad u_y = -\delta, \tag{7-6.2.2}$$

$$x = \pm l, \qquad \sigma_{xx} = \sigma_{yx} = 0. \tag{7-6.2.3, 4}$$

If $h/l \ll 1$, a Saint-Venant approximation may be used on the ends. Because u_y must be even in x and odd in y, the following Marguerre solution to the biharmonic equation will be considered:

$$\psi = \sum_n \cos \beta_n x [A_n \operatorname{sh} \beta_n y + D_n \beta_n y \operatorname{ch} \beta_n y]. \tag{7-6.2.5}$$

The stresses and displacements become

$$u_x = \frac{1+\nu}{1-\nu} \sum \beta_n^2 \sin \beta_n x [A_n \operatorname{ch} \beta_n y + D_n(\beta_n y \operatorname{sh} \beta_n y + \operatorname{ch} \beta_n y)], \tag{7-6.2.6}$$

$$u_y = -\frac{1+\nu}{1-\nu} \sum \beta_n^2 \cos \beta_n x \left[A_n \operatorname{sh} \beta_n y + D_n\left(\beta_n y \operatorname{ch} \beta_n y - \frac{2(1-\nu)}{1+\nu} \operatorname{sh} \beta_n y\right)\right], \tag{7-6.2.7}$$

$$\sigma_{xx} = \frac{E}{(1-\nu^2)} \sum \beta_n^3 \cos \beta_n x [A_n(1+\nu) \operatorname{ch} \beta_n y + D_n(\{1+\nu\}\beta_n y \operatorname{sh} \beta_n y + \{1+3\nu\} \operatorname{ch} \beta_n y)], \tag{7-6.2.8}$$

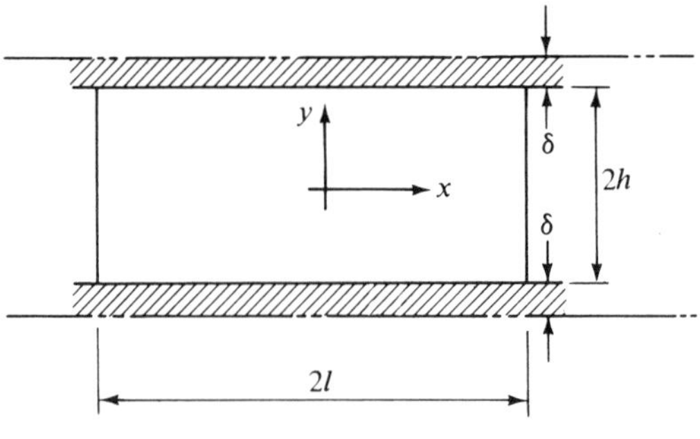

Figure 7-8

$$\sigma_{xy} = \frac{E}{(1-\nu^2)} \sum \beta_n^3 \sin \beta_n x[A_n(1+\nu) \operatorname{sh} \beta_n y + D_n(\{1+\nu\}\beta_n y \operatorname{ch} \beta_n y + 2\nu \operatorname{sh} \beta_n y)], \tag{7-6.2.9}$$

$$\sigma_{yy} = -\frac{E}{(1-\nu^2)} \sum \beta_n^3 \cos \beta_n x[A_n(1+\nu) \operatorname{ch} \beta_n y + D_n(\{1+\nu\}\beta_n y \operatorname{sh} \beta_n y - (1-\nu) \operatorname{ch} \beta_n y)]. \tag{7-6.2.10}$$

To satisfy the condition $u_x = 0$ at $y = \pm h$, A_n will be chosen in terms of D_n as follows:

$$A_n = -D_n[\beta_n h \tanh \beta_n h + 1]. \tag{7-6.2.11}$$

If β_n is chosen such that $\cos \beta_n l = 0$, then the condition that the normal stress is zero on the ends will be satisfied and the boundary functions on the long edge will still form an orthogonal set:

$$\beta_n = \frac{2n+1}{2l}\pi, \qquad n = 0, 1, 2, \ldots \tag{7-6.2.12}$$

Note: $\beta_n = 0$ is not a root and polynomials are not involved. The stress σ_{xy} is an odd function of y, so that the shear boundary condition on $x = \pm l$ is satisfied only in the Saint-Venant sense. The displacement condition on u_y is

$$u_y(h, x) = \delta = -\frac{(1+\nu)}{(1-\nu)} \sum_{n=0}^{\infty} \beta_n^2 \cos \beta_n x \left[A_n \operatorname{sh} \beta_n h + D_n\left(\beta_n h \operatorname{ch} \beta_n h - \frac{2(1-\nu)}{1+\nu} \operatorname{sh} \beta_n h\right)\right]. \tag{7-6.2.13}$$

Substituting equation 7-6.2.11 into equation 7-6.2.13 yields

$$\sum_{n=0} \beta_n^2 \frac{(\beta_n h - (3-\nu)/(1+\nu) \operatorname{sh} \beta_n h \operatorname{ch} \beta_n h)}{\operatorname{ch} \beta_n h} D_n \cos \beta_n x = -\frac{1-\nu}{1+\nu}\delta. \tag{7-6.2.14}$$

Taking the finite Fourier cosine transform of equation 7-6.2.14 gives

$$\beta_n^2 \frac{(1+\nu)\beta_n h - (3-\nu) \operatorname{sh} \beta_n h \operatorname{ch} \beta_n h}{\operatorname{ch} \beta_n h} D_n l = -(1-\nu)\delta \int_{-l}^{+l} \cos \beta_n x \, dx. \tag{7-6.2.15}$$

Therefore,

$$D_n = -\frac{2(1-\nu)\delta(-1)^n \operatorname{ch} \beta_n h}{\beta_n^3 l[(1+\nu)\beta_n h - (3-\nu) \operatorname{sh} \beta_n h \operatorname{ch} \beta_n h]}, \quad n = 0, 1, 2, \ldots \tag{7-6.2.16}$$

$$A_n = \frac{2(1-\nu)\delta(-1)^n(\beta_n h \operatorname{sh} \beta_n h + \operatorname{ch} \beta_n h)}{\beta_n^3 l[(1+\nu)\beta_n h - (3-\nu) \operatorname{sh} \beta_n h \operatorname{ch} \beta_n h]}, \quad n = 0, 1, 2, \ldots \tag{7-6.2.17}$$

The stress in the vertical direction becomes

$$\sigma_{yy} = -\frac{E}{1+\nu}\sum_{n=0}\frac{2}{l}\delta(-1)^n\left[\frac{(1+\nu)(\beta_n h \operatorname{sh}\beta_n h \operatorname{ch}\beta_n y - \beta_n y \operatorname{ch}\beta_n h \operatorname{sh}\beta_n y)}{(1+\nu)\beta_n h - (3-\nu)\operatorname{sh}\beta_n h \operatorname{ch}\beta_n h} + \frac{2\operatorname{ch}\beta_n h \operatorname{ch}\beta_n y}{(1+\nu)\beta_n h - (3-\nu)\operatorname{sh}\beta_n h \operatorname{ch}\beta_n h}\right]\cos\beta_n x. \tag{7-6.2.18}$$

For $h/l \ll 1$, the following approximations hold:

$$\operatorname{ch}\beta_n h \approx 1 + \frac{(\beta_n h)^2}{2}, \qquad \operatorname{sh}\beta_n h \approx \beta_n h + \frac{(\beta_n h)^3}{6} \tag{7-6.2.19}$$

$$\sigma_{yy} = \frac{E}{1+\nu}\frac{2\delta}{l}\sum_{n=0}\frac{(-1)^n}{\beta_n h}\frac{2+(2+\nu)(\beta_n h)^2 - \nu(\beta_n y)^2}{2(1-\nu)+2(3-\nu)(\beta_n h)^2/3}\cos\beta_n x \tag{7-6.2.20}$$

$$\approx \frac{E}{1-\nu^2}\frac{\delta}{h}\sum_{n=0}\frac{2(-1)^n}{\beta_n l}\cos\beta_n x \approx \frac{E\delta}{(1-\nu^2)h}. \tag{7-6.2.21}$$

The solution presented does not examine the singularity in the shear stress at the corners. The nature of this singularity is discussed in Section 8-7.2 of this book and in papers by Benthem, Vorovich and Kopasendo, and Little.*

7 Multiple Fourier analysis

When a rectangular member is short, that is, $h/l \approx 1$, the solution is much more complex because no appeal may be made to the Saint-Venant approximation at the ends. An attempt must be made in these cases to satisfy all boundary conditions. G. Pickett† proposed a method of approaching this problem that results in coupling of the trigonometric series in the x and y directions.

If one made a first attempt at solving a problem satisfying all boundary conditions, superposition might be tried. One could use a Fourier series satisfying one pair of boundary conditions but yielding a remainder on the other pair.‡ Superposition of another Fourier series would remove the remainder, but leaves a new remainder on the original boundaries. Successive iterations might converge to the correct solution. Pickett's method may be thought of as a formalization of this approach. Explicit determination of the Fourier coefficients is not possible in this case, and the solution of a truncated set of infinite equations for infinite unknowns is necessary. Modern

* J. P. Benthem, *Quart. J. Mech. and Appl. Math.*, **16**, pt. 4, 413–429, 1963; I. I. Vorovich and V. V. Kopasenko, *PMM*, **30**(1), 109–115, 1966; R. W. Little, *J. Appl. Mech.*, **36**, 320–324, June 1969.

† G. Pickett, *J. Appl. Mech.*, **11**, 186–72, 1944.

‡ L. N. G. Filon, *Trans. Roy. Soc.* (*London*), A201, 67, 1903; A. Timpe, *Z. Math. Physik*, **55**, 149, 1907; J. N. Goodier, *J. Appl. Mech.*, **54**(18), 173, 1932.

computers allow good approximations in these cases if the boundary functions are such that the series expansions have reasonable rates of convergence.* The details of this method may best be seen by a simple example. As before, to simplify the algebra, all problems will be broken into the four cases shown in Table 7-1. Consider a plate loaded as shown in Figure 7-9. Note that this example has been chosen to correspond to case 1 in Table 7-1. The boundary conditions are

$$y = \pm h, \qquad \sigma_{yy} = \begin{cases} -p, & -c < x < c \\ 0, & -l < x < -c;\, c < x < l \end{cases}, \tag{7-7.1}$$

$$\sigma_{xy} = 0, \tag{7-7.2}$$

$$x = \pm l, \qquad \sigma_{xx} = 0, \tag{7-7.3}$$

$$\sigma_{xy} = 0. \tag{7-7.4}$$

Since $h/l \approx 1$, we cannot replace the boundary conditions in equations 7-7.3 and 7-7.4 by an average condition. Proceeding as before, we can consider the following Airy stress function:

$$\varphi = \cos \beta_x [B \operatorname{ch} \beta y + C\beta y \operatorname{sh} \beta y] + C_0 x^2. \tag{7-7.5}$$

A polynomial to yield the desired constant term has been added to take care of the average stress values. If this solution is examined, it is apparent that B and C may be chosen to satisfy conditions 7-7.1 and 7-7.2 and β may be chosen to satisfy either 7-7.3 or 7-7.4; but one condition on $x = \pm l$ will not be satisfied exactly. Anticipating this difficulty, we therefore assume the more

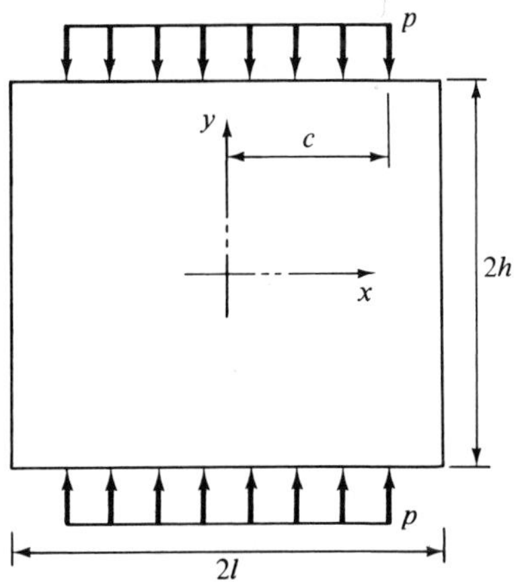

Figure 7-9

* A discussion of the rates of convergence of trigonometric series may be found in L. V. Kantorovich and V. I. Krylov, *Approximate Methods of Higher Analysis*, John Wiley & Sons, Inc. (Interscience Division), New York, p. 77, 1964.

general form:

$$\varphi = \sum_n \cos \beta_n x[B_n \operatorname{ch} \beta_n y + C_n \beta_n y \operatorname{sh} \beta_n y] + C_0 x^2 + \sum_k \cos \alpha_k y[F_k \operatorname{ch} \alpha_k x + G_k \alpha_k x \operatorname{sh} \alpha_k x]. \tag{7-7.6}$$

Using this form of the solution, the stresses are

$$\sigma_{xx} = \sum_n \beta_n^2 \cos \beta_n x[B_n \operatorname{ch} \beta_n y + C_n(\beta_n y \operatorname{sh} \beta_n y + 2 \operatorname{ch} \beta_n y)] - \sum_k \alpha_k^2 \cos \alpha_k y[F_k \operatorname{ch} \alpha_k x + G_k \alpha_k x \operatorname{sh} \alpha_k x], \tag{7-7.7}$$

$$\sigma_{yy} = -\sum_n \beta_n^2 \cos \beta_n x[B_n \operatorname{ch} \beta_n y + C_n \beta_n y \operatorname{sh} \beta_n y] + 2C_0 + \sum_k \alpha_k^2 \cos \alpha_k y[F_k \operatorname{ch} \alpha_k x + G_k(\alpha_k x \operatorname{sh} \alpha_k x + 2 \operatorname{ch} \alpha_k x)], \tag{7-7.8}$$

$$\sigma_{xy} = \sum_n \beta_n^2 \sin \beta_n x[B_n \operatorname{sh} \beta_n y + C_n(\beta_n y \operatorname{ch} \beta_n y + \operatorname{sh} \beta_n y)] + \sum_k \alpha_k^2 \sin \alpha_k y[F_k \operatorname{sh} \alpha_k x + G_k(\alpha_k x \operatorname{ch} \alpha_k x + \operatorname{sh} \alpha_k x)]. \tag{7-7.9}$$

The constants β_n and α_k are chosen to help satisfy conditions 7-7.2 and 7-7.4:

$$\beta_n = \frac{n\pi}{l}, \qquad \alpha_k = \frac{k\pi}{h}. \tag{7-7.10}$$

To satisfy 7-7.2 completely B_n is taken in terms of C_n as follows:

$$y = \pm h; \qquad \sigma_{xy} = 0 = \sum_n \beta_n^2 \sin \beta_n x[B_n \operatorname{sh} \beta_n h + C_n(\beta_n h \operatorname{ch} \beta_n h + \operatorname{sh} \beta_n h)];$$

Therefore,

$$B_n = -C_n[1 + \beta_n h \coth \beta_n h]. \tag{7-7.11}$$

In a similar manner, condition 7-7.4 is satisfied:

$$x = \pm l; \qquad \sigma_{xy} = 0 = \sum_k \alpha_k^2 \sin \alpha_k y[F_k \operatorname{sh} \alpha_k l + G_k(\alpha_k l \operatorname{ch} \alpha_k l + \operatorname{sh} \alpha_k l)];$$

Therefore,

$$F_k = -G_k(1 + \alpha_k l \coth \alpha_k l). \tag{7-7.12}$$

The boundary condition 7-7.3 becomes

$$x = \pm l; \qquad \sigma_{xx} = 0 = \sum_n \beta_n^2 \cos \beta_n l[B_n \operatorname{ch} \beta_n y + C_n(\beta_n y \operatorname{sh} \beta_n y + 2 \operatorname{ch} \beta_n y)] - \sum_k \alpha_k^2 \cos \alpha_k y[F_k \operatorname{ch} \alpha_k l + G_k \alpha_k l \operatorname{sh} \alpha_k l]. \tag{7-7.13}$$

Using equation 7-7.12, equation 7-7.13 may be written

$$\sum_k G_k \alpha_k^2(\alpha_k l \operatorname{sh} \alpha_k l - \alpha_k l \coth \alpha_k l \operatorname{ch} \alpha_k l - \operatorname{ch} \alpha_k l) \cos \alpha_k y = \sum_n \beta_n^2(-1)^n[B_n \operatorname{ch} \beta_n y + C_n(\beta_n y \operatorname{sh} \beta_n y + 2 \operatorname{ch} \beta_n y)]. \tag{7-7.14}$$

Using hyperbolic-function identities, equation 7-7.14 becomes

$$\sum_{k=1} \alpha_k^2 \left(\frac{\alpha_k l + \operatorname{sh} \alpha_k l \operatorname{ch} \alpha_k l}{\operatorname{sh} \alpha_k l}\right) G_k \cos \alpha_k y = \sum_{n=1} \beta_n^2(-1)^{n+1}[B_n \operatorname{ch} \beta_n y + C_n(\beta_n y \operatorname{sh} \beta_n y + 2 \operatorname{ch} \beta_n y)]. \tag{7-7.15}$$

If the constant A_k is introduced as

$$A_k = \alpha_k^2\left(\frac{\alpha_k l + \text{sh}\,\alpha_k l\,\text{ch}\,\alpha_k l}{\text{sh}\,\alpha_k l}\right)G_k, \tag{7-7.16}$$

equation 7-7.15 becomes

$$\sum_{k=1} A_k \cos \alpha_k y = \sum_{n=1} \beta_n^2(-1)^{n+1}[B_n\,\text{ch}\,\beta_n y + C_n(\beta_n y\,\text{sh}\,\beta_n y + 2\,\text{ch}\,\beta_n y)]. \tag{7-7.17}$$

By examination of equation 7-7.11, it may be determined that the average value of the function of y on the right side is zero, so that there is no need to introduce a constant term derived from a polynomial. Taking the finite Fourier cosine transform of equation 7-7.17 yields

$$A_k = \frac{1}{h}\sum_{n=1}\beta_n^2(-1)^{n+1}\int_{-h}^{+h}[B_n\,\text{ch}\,\beta_n y + C_n(\beta_n y\,\text{sh}\,\beta_n y + 2\,\text{ch}\,\beta_n y)]\cos\alpha_k y\,dy. \tag{7-7.18}$$

Using the integrals given in Appendix B and equations 7-7.11 and 7-7.16,

$$G_k = -\frac{\text{sh}\,\alpha_k l}{(\alpha_k l + \text{sh}\,\alpha_k l\,\text{ch}\,\alpha_k l)}\left[\frac{4}{h}\sum_{n=1} C_n\frac{\beta_n^3(-1)^{k+n}\,\text{sh}\,\beta_n h}{(\beta_n^2+\alpha_k^2)^2}\right]. \tag{7-7.19}$$

This gives each of the G_k's in terms of all the C_n's or a single set of infinite equations ($k = 1, \infty$) for a two infinity of unknowns (C_n and G_k).

A second set of equations is obtained by satisfying boundary condition 7-7.1:

$$y = \pm h; \qquad \sigma_{yy} = \begin{cases} -p, & -c < x < c \\ 0, & -l < x < -c, \quad c < x < l \end{cases}$$
$$= -\sum_n \beta_n^2 \cos\beta_n x[B_n\,\text{ch}\,\beta_n h + C_n\beta_n h\,\text{sh}\,\beta_n h] + 2C_0 + \sum_k \alpha_k^2\cos\alpha_k h[F_k\,\text{ch}\,\alpha_k x + G_k(\alpha_k x\,\text{sh}\,\alpha_k x + 2\,\text{ch}\,\alpha_k x)]. \tag{7-7.20}$$

Using equations 7-7.10 and 7-7.11, equation 7-7.20 becomes

$$\sigma_{yy}(x, h) = \sum_n \beta_n^2\cos\beta_n x\left[\frac{\beta_n h + \text{sh}\,\beta h\,\text{ch}\,\beta_n h}{\text{sh}\,\beta_n h}\right]C_n + 2C_0 + \sum_k \alpha_k^2(-1)^k[F_k\,\text{ch}\,\alpha_k x + G_k(\alpha_k x\,\text{sh}\,\alpha_k x + 2\,\text{ch}\,\alpha_k x)]. \tag{7-7.21}$$

This may be written as

$$\sum_{n=0} P_n\cos\beta_n x = \sum_k \alpha_k^2(-1)^{k+1}[F_k\,\text{ch}\,\alpha_k x + G_k(\alpha_k x\,\text{sh}\,\alpha_k x + 2\,\text{ch}\,\alpha_k x)] + f(x), \tag{7-7.22}$$

where

$$P_n = \beta_n^2\left[\frac{\beta_n h + \text{sh}\,\beta_n h\,\text{ch}\,\beta_n h}{\text{sh}\,\beta_n h}\right]C_n, \qquad P_0 = 2C_0, \tag{7-7.23}$$

$$f(x) = \begin{cases} -p, & -c < x < c, \\ 0, & -l < x < -c, c < x < l. \end{cases} \tag{7-7.24}$$

The average value of the first term on the right is zero, but this is not true for the second term. A constant term C_0 is needed in the series to take care of this average. This need was anticipated when the form of φ was selected. The finite Fourier cosine transform is now taken of equation 7-7.22:

$$C_n = -\frac{\operatorname{sh}\beta_n h}{(\beta_n h + \operatorname{sh}\beta_n h \operatorname{ch}\beta_n h)}\left[\frac{4}{l}\sum_k \frac{G_k\alpha_k^3(-1)^{k+n}\operatorname{sh}\alpha_k l}{(\beta_n^2+\alpha_k^2)^2}\right] + \frac{1}{l}\frac{\operatorname{sh}\beta_n h}{\beta_n^2(\beta_n h + \operatorname{sh}\beta_n h \operatorname{ch}\beta_n h)}\int_{-l}^{+l} f(x)\cos\beta_n x\,dx, \quad n > 0, \tag{7-7.25}$$

$$2C_0 = \frac{1}{2l}\int_{-l}^{+l} f(x)\,dx. \tag{7-7.26}$$

Integrating these after using equation 7-7.24 gives

$$C_0 = -\frac{pc}{2l}, \tag{7-7.27}$$

$$C_n = -\frac{\operatorname{sh}\beta_n h}{(\beta_n h + \operatorname{sh}\beta_n h \operatorname{ch}\beta_n h)}\left[\frac{4}{l}\sum_k \frac{G_k\alpha_k^3(-1)^{k+n}\operatorname{sh}\alpha_k h}{(\beta_n^2+\alpha_k^2)^2} - \frac{2p\sin\beta_n c}{\beta_n l\beta_n^2}\right]. \tag{7-7.28}$$

These equations may be written as

$$G_k + \sum_{n=1}^{\infty} R_{kn}C_n = 0, \qquad C_n + \sum_{k=1}^{\infty} S_{nk}G_k = +\frac{2p}{\beta_n l}\frac{\operatorname{sh}\beta_n l\sin\beta_n c}{\beta_n^2(\beta_n h + \operatorname{sh}\beta_n h\operatorname{ch}\beta_n h)}, \tag{7-7.29}$$

where

$$R_{kn} = \frac{4(-1)^{k+n}\operatorname{sh}\alpha_k l\operatorname{sh}\beta_n h}{\beta_n h(\alpha_k l + \operatorname{sh}\alpha_k l\operatorname{ch}\alpha_k l)[1 + (\alpha_k^2/\beta_n^2)]^2}, \tag{7-7.30}$$

$$S_{nk} = \frac{4(-1)^{k+n}\operatorname{sh}\alpha_k l\operatorname{sh}\beta_n h}{\alpha_k l(\beta_n h + \operatorname{sh}\beta_n h\operatorname{ch}\beta_n h)[(\beta_n^2/\alpha_k^2) + 1]^2}. \tag{7-7.31}$$

The C_0 constant need not be included in the system of equations during the inversion process because it is explicitly determined by equation 7-7.27. Equation 7-7.29 represents the desired doubly infinite set of equations in the doubly infinite set of unknowns, which may be truncated and solved to the desired accuracy.

For $h/l \ll 1$, equations 7-7.30 and 7-7.31 become

$$R_{kn} \approx 8(-1)^{k+n}e^{-\alpha_k l}\left(\frac{h}{l}\right)^4\left(\frac{n}{k}\right)^4, \tag{7-7.32}$$

$$S_{nk} \approx (-1)^{k+n}\frac{e^{\alpha_k l}}{\alpha_k l}. \tag{7-7.33}$$

Equation 7-7.29 now may be written

$$C_n \approx +\frac{2p}{\beta_n l}\sin\beta_n c\frac{\operatorname{sh}\beta_n h}{\beta_n^2(\beta_n h + \operatorname{sh}\beta_n h\operatorname{ch}\beta_n h)}, \tag{7-7.34}$$

$$G_k \approx 0. \tag{7-7.35}$$

If we examine the stresses for this case, we obtain

$$\sigma_{xx} = \frac{2p}{\pi}\sum_{n=1}\frac{\sin\beta_n c}{n} \times \left[\frac{\beta_n h \operatorname{ch}\beta_n h \operatorname{ch}\beta_n y - (\beta_n y \operatorname{sh}\beta_n y + \operatorname{ch}\beta_n y)\operatorname{sh}\beta_n h}{\beta_n h + \operatorname{sh}\beta_n h \operatorname{ch}\beta_n h}\right]\cos\beta_n x, \tag{7-7.36}$$

$$\sigma_{yy} = -\frac{pc}{l} - \frac{2p}{\pi}\sum_{n=1}\frac{\sin\beta_n c}{n} \times \left[\frac{\beta_n h \operatorname{ch}\beta_n h \operatorname{ch}\beta_n y - (\beta_n y \operatorname{sh}\beta_n y - \operatorname{ch}\beta_n y)\operatorname{sh}\beta_n h}{\beta_n h + \operatorname{sh}\beta_n h \operatorname{ch}\beta_n h}\right]\cos\beta_n x, \tag{7-7.37}$$

$$\sigma_{xy} = \frac{2p}{\pi}\sum_{n=1}\frac{\sin\beta_n c}{n} \times \left[\frac{\beta_n h \operatorname{ch}\beta_n h \operatorname{sh}\beta_n y - \beta_n y \operatorname{ch}\beta_n y \operatorname{sh}\beta_n h}{\beta_n h + \operatorname{sh}\beta_n h \operatorname{ch}\beta_n h}\right]\sin\beta_n x. \tag{7-7.38}$$

Numerical values were computed for $c = 0.3l$ and $h = 0.2l$ using the single Fourier series.* Even with a strip this thin, $h/l = 0.2$, one may note a significant smoothing in the normal stress σ_{yy}, in going from the boundary to the midline. The boundary stresses at $x = \pm l$ were satisfied to about 0.1 per cent (Figure 7.10).

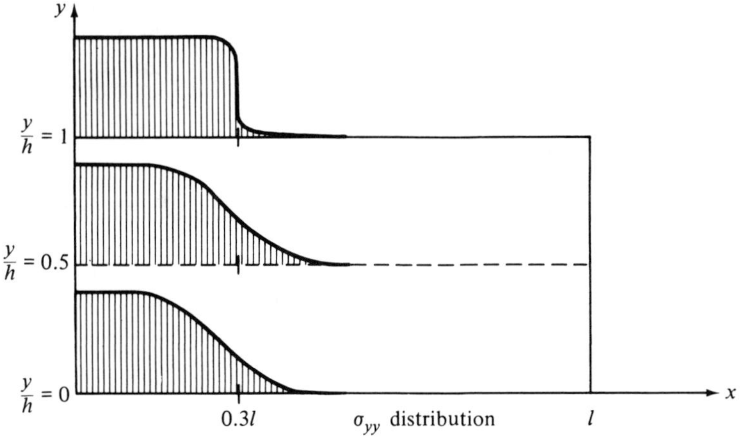

Figure 7-10

8 Problems involving infinite or semi-infinite dimensions

The solution developed for a long strip may be adapted to permit solutions for infinite or semi-infinite strips. If a finite strip utilizing functions

* This numerical work was done by Professor J. L. Lubkin of Michigan State University.

odd in x is considered, the following form of an Airy stress function will be taken as the solution:

$$\varphi = \sum \sin \beta_n x[A_n \operatorname{sh} \beta_n y + B_n \operatorname{ch} \beta_n y + C_n \beta_n y \operatorname{sh} \beta_n y + D_n \beta_n y \operatorname{ch} \beta_n y]. \tag{7-8.1}$$

β_n was usually assigned the discrete values

$$\beta_n = \frac{n\pi}{l}, \qquad n = 1, 2, 3, \ldots \tag{7-8.2}$$

and we obtained solutions in the form of trigonometric series. If, however, β varies continuously from $-\infty$ to $+\infty$, the summation on n goes over into an integral. The coefficients A, B, C, and D are dependent upon the value of n in the finite case, and are functions of the continuously varying parameter β in the infinite case. The sum of the solutions now takes the form of an infinite integral:

$$\begin{aligned}\varphi = \int_{-\infty}^{+\infty} \sin \beta x[&A(\beta) \operatorname{sh} \beta y + B(\beta) \operatorname{ch} \beta y \\ &+ C(\beta)\beta y \operatorname{sh} \beta y + D(\beta)\beta y \operatorname{ch} \beta y]\, d\beta.\end{aligned} \tag{7-8.3}$$

If the order of integration and differentiation is interchanged, it is obvious that this form of solution satisfies the biharmonic equation. It is also instructive to develop the relation between Fourier series and the Fourier integral in more detail. Consider the Fourier sine series for a function $f(x)$ in the region $0 \leq x \leq l$:

$$f(x) = \sum_n C_n \sin \frac{n\pi x}{l}. \tag{7-8.4}$$

The Fourier coefficients may be determined by use of the orthogonality relation

$$C_n = \frac{2}{l} \int_0^l f(\xi) \sin \frac{n\pi\xi}{l}\, d\xi. \tag{7-8.5}$$

Substituting equation 7-8.5 into equation 7-8.4 yields

$$f(x) = \frac{2}{l} \sum_n^{\infty} \int_0^l f(\xi) \sin \frac{n\pi\xi}{l} \sin \frac{n\pi x}{l}\, d\xi. \tag{7-8.6}$$

We shall consider the behavior as $l \to \infty$ and $\beta_n = n\pi/l$ becomes β, which varies continuously. Note that $\Delta\beta_n = \beta_{n+1} - \beta_n = \pi/l$. Equation 7-8.6 may now be written

$$f(x) = \frac{2}{\pi} \sum_n^{\infty} \Delta\beta_n \int_0^l f(\xi) \sin \beta_n \xi \sin \beta_n x\, d\xi. \tag{7-8.7}$$

As $l \to \infty$ and $\Delta\beta_n \to d\beta$, the sum on n is replaced by an integral:

$$f(x) = \frac{2}{\pi} \int_0^{\infty} d\beta \int_0^{\infty} f(\xi) \sin \beta\xi \sin \beta x\, d\xi. \tag{7-8.8}$$

It is clear that the integration may be extended over the whole $\xi\beta$ plane, so that equation 7-8.8 may be written

$$f(x) = \frac{1}{2\pi}\int_{-\infty}^{+\infty} d\beta \int_{-\infty}^{+\infty} f(\xi) \sin \beta\xi \sin \beta x \, d\xi. \tag{7-8.9}$$

This formula holds for odd functions of x only. For even functions the sine functions are replaced by cosine functions:

$$g(x) = \frac{1}{2\pi}\int_{-\infty}^{+\infty} d\beta \int_{-\infty}^{+\infty} g(\xi) \cos \beta\xi \cos \beta x \, d\xi. \tag{7-8.10}$$

The constant term in the cosine series disappears as $l \longrightarrow \infty$, if the term $\int_{-\infty}^{+\infty} g(x)\,dx$ is bounded. This condition is necessary for the Fourier integral to exist. Equations 7-8.8 and 7-8.10 may be written in the symmetrical form:

$$f(x) = \sqrt{\frac{2}{\pi}}\int_0^{\infty} F_s(\beta) \sin \beta x \, d\beta, \tag{7-8.11}$$

where
$$F_s(\beta) = \sqrt{\frac{2}{\pi}}\int_0^{\infty} f(x) \sin \beta x \, dx, \tag{7-8.12}$$

and
$$f(x) = \sqrt{\frac{2}{\pi}}\int_0^{\infty} F_c(\beta) \cos \beta x \, d\beta, \tag{7-8.13}$$

where
$$F_c(\beta) = \sqrt{\frac{2}{\pi}}\int_0^{\infty} f(x) \cos \beta x \, dx. \tag{7-8.14}$$

Equations 7-8.12 and 7-8.14 are referred to as the Fourier sine and cosine transforms of the function $f(x)$, and equations 7-8.11 and 7-8.13 represent the inversion formulas or the inverse transformes. For discussion of the range of validity of these formulas and for more formal derivations, the reader is referred to I. N. Sneddon's *Fourier transforms*,* or similar treatises. If one does not separate a function into its even and odd parts, the general Fourier transform is useful. This may be defined together with its inverse, by

$$\mathfrak{F}[f(x); \beta] = \frac{1}{\sqrt{2\pi}}\int_{-\infty}^{+\infty} f(x)e^{i\beta x} \, dx, \tag{7-8.15}$$

$$f(x) = \frac{1}{\sqrt{2\pi}}\int_{-\infty}^{+\infty} \mathfrak{F}(\beta)e^{-i\beta x} \, d\beta. \tag{7-8.16}$$

Note that if $f(x)$ is either even or odd, this reduces to the transforms given before.

8.1 Infinite strip loaded by uniform pressure. Consider the infinite strip loaded as shown in Figure 7-11. The stress σ_{yy} is symmetric in both the x and the y

* I. N. Sneddon, *Fourier Transforms*, McGraw-Hill Book Company, New York, 1951.

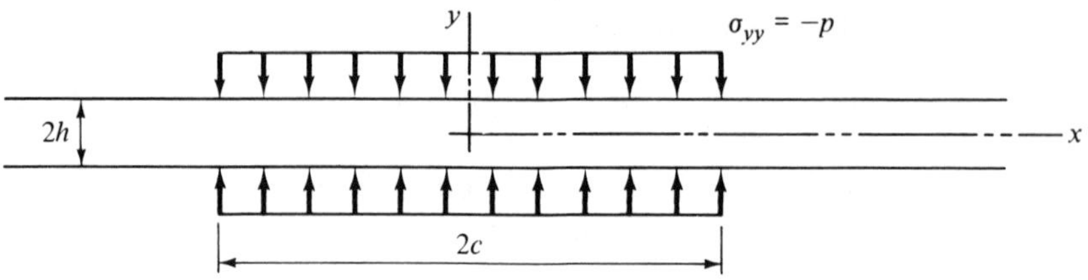

Figure 7-11

directions, so that φ should be taken as even in x and y. The boundary conditions are

$$\sigma_{xy} = 0, \qquad y = \pm h, \tag{7-8.1.1}$$

$$\sigma_{yy} = \begin{cases} -p, & -c < x < c, \quad y = \pm h \\ 0, & |x| > c, \quad y = \pm h \end{cases}. \tag{7-8.1.2}$$

The Airy stress function is taken in the form

$$\varphi = \frac{1}{\sqrt{2\pi}} \int_{-\infty}^{+\infty} [A(\beta) \operatorname{ch} \beta y + C(\beta)\beta y \operatorname{sh} \beta y] \cos \beta x \, d\beta. \tag{7-8.1.3}$$

Boundary condition 7-8.1.1 leads to the following relation:

$$\sigma_{xy} = \frac{1}{\sqrt{2\pi}} \int_{-\infty}^{+\infty} \beta^2 [A(\beta) \operatorname{sh} \beta y + C(\beta)(\beta y \operatorname{ch} \beta y + \operatorname{sh} \beta y)] \sin \beta x \, d\beta. \tag{7-8.1.4}$$

Therefore,

$$A(\beta) = -C(\beta)[1 + \beta h \coth \beta h]. \tag{7-8.1.5}$$

The normal stress σ_{yy} becomes

$$\sigma_{yy} = -\frac{1}{\sqrt{2\pi}} \int_{-\infty}^{+\infty} \beta^2 [A(\beta) \operatorname{ch} \beta y + C(\beta)\beta y \operatorname{sh} \beta y] \cos \beta x \, d\beta. \tag{7-8.1.6}$$

After substitution of equation 7-8.1.5 into equation 7-8.1.6, σ_{yy} becomes, at the boundary,

$$\sigma_{yy}\big|_{y=+h} = \frac{1}{\sqrt{2\pi}} \int_{-\infty}^{+\infty} C(\beta)\beta^2 \left[\frac{\beta h + \operatorname{sh} \beta h \operatorname{ch} \beta h}{\operatorname{sh} \beta h}\right] \cos \beta x \, dx. \tag{7-8.1.7}$$

If we introduce the notation

$$F_c(\beta) = C(\beta)\beta^2 \left[\frac{\beta h + \operatorname{sh} \beta h \operatorname{ch} \beta h}{\operatorname{sh} \beta h}\right], \tag{7-8.1.8}$$

then we recognize $F_c(\beta)$ to be the Fourier cosine transform of σ_{yy} at $y = +h$:

$$F_c(\beta) = \frac{1}{\sqrt{2\pi}} \int_{-\infty}^{+\infty} \sigma_{yy}(x, h) \cos \beta x, dx, \tag{7-8.1.9}$$

$$= \frac{1}{\sqrt{2\pi}} \int_{-\infty}^{+\infty} (-p) \cos \beta x, dx, \tag{7-8.1.10}$$

$$= -\frac{2}{\sqrt{2\pi}} p \frac{\sin \beta c}{\beta}. \tag{7-8.1.11}$$

The solutions for $C(\beta)$ and $A(\beta)$ are thus

$$C(\beta) = -\frac{1}{\sqrt{2\pi}} 2p \frac{\sin \beta c}{\beta^3} \left[\frac{\operatorname{sh} \beta h}{\beta h + \operatorname{sh} \beta h \operatorname{ch} \beta h} \right], \tag{7-8.1.12}$$

$$A(\beta) = \frac{1}{\sqrt{2\pi}} 2p \frac{\sin \beta c}{\beta^3} \left[\frac{\beta h \operatorname{ch} \beta h + \operatorname{sh} \beta h}{\beta h + \operatorname{sh} \beta h \operatorname{ch} \beta h} \right]. \tag{7-8.1.13}$$

The Airy stress function reduces to

$$\varphi = \frac{1}{2\pi} \int_{-\infty}^{+\infty} \frac{2p \sin \beta c}{\beta^3(\beta h + \operatorname{sh} \beta h \operatorname{ch} \beta h)} \times \{(\beta h \operatorname{ch} \beta h + \operatorname{sh} \beta h) \operatorname{ch} \beta y - \beta y \operatorname{sh} \beta h \operatorname{sh} \beta y\} \cos \beta x \, d\beta. \tag{7-8.1.14}$$

This integral may be evaluated numerically, or by means of the calculus of residues.*

8.2 Fourier transform solutions. We may obtain a more direct approach to problems of this type if we examine some properties of Fourier transforms. We need to know the Fourier transforms of derivatives of the unknown function, if we are to use these transforms to solve differential equations. Consider for example, the Fourier cosine transform of the derivative of the function $f(x)$;

$$F_c[f'(x)] = \sqrt{\frac{2}{\pi}} \int_0^{\infty} f'(x) \cos \beta x \, dx. \tag{7-8.2.1}$$

Assuming $f(\infty) = 0$. and integrating by parts produces

$$\begin{aligned} F_c[f'(x)] &= \sqrt{\frac{2}{\pi}} \left\{ -f(0) + \int_0^{\infty} \beta f(x) \sin \beta x \, dx \right\} \\ &= -\sqrt{\frac{2}{\pi}} f(0) + \beta F_s[f(x)] \end{aligned} \tag{7-8.2.2}$$

In a similar manner, we find the Fourier sine transform of a derivative to be

$$F_s[f'(x)] = -\beta F_c[f(x)]. \tag{7-8.2.3}$$

* The integrand is singular at $\beta = 0$, and the evaluation of this integral will be troublesome. However, the corresponding integrals for the stresses do not exhibit this difficulty and these are evaluated numerically.

If we consider second derivatives, we obtain

$$F_c[f''(x)] = -\sqrt{\frac{2}{\pi}}\, f'(0) - \beta^2 F_c[f(x)], \tag{7-8.2.4}$$

$$F_s[f''(x)] = \sqrt{\frac{2}{\pi}}\, \beta f(0) - \beta^2 F_s[f(x)]. \tag{7-8.2.5}$$

Transforms of higher derivatives may be obtained by use of these formulas. For example, the fourth derivative becomes

$$\begin{aligned} F_c[f^{\text{IV}}(x)] &= -\sqrt{\frac{2}{\pi}}\, f'''(0) - \beta^2 F_c[f''(x)], \\ &= -\sqrt{\frac{2}{\pi}}\, f'''(0) + \sqrt{\frac{2}{\pi}}\, \beta^2 f'(0) + \beta^4 F_c[f(x)], \end{aligned} \tag{7-8.2.6}$$

$$F_s[f^{\text{IV}}(x)] = \sqrt{\frac{2}{\pi}}\, \beta f''(0) - \sqrt{\frac{2}{\pi}}\, \beta^3 f(0) + \beta^4 F_s[f(x)]. \tag{7-8.2.7}$$

Similar results hold for the Fourier transforms of functions of several variables when one is dealing with the partial derivatives of the transformed variable. Note that if one is working in infinite regions, the integrals are from $-\infty$ to $+\infty$ and the integration by parts does not produce the boundary terms at zero. As above, f and its derivatives are assumed to be zero at infinity. The transforms of derivatives simplify in this case. For example,

$$\mathfrak{F}[f^{\text{IV}}(x)] = \beta^4 \mathfrak{F}[f(x)]. \tag{7-8.2.8}$$

Consider the biharmonic equation and a transformation of the x variable. We denote the transformed function of φ as Φ, which is a function of β and y:

$$\mathfrak{F}[\nabla^2\nabla^2\varphi] = \frac{d^4\Phi}{dy^4} - 2\beta^2\frac{d^2\Phi}{dy^2} + \beta^4\Phi. \tag{7-8.2.9}$$

Note that the Fourier transform for an infinite region was used, and that it reduces the partial differential equation to an ordinary differential equation, If we deal with semi-infinite regions and take the cosine transform, the equation is more complex:

$$\begin{aligned} F_c[\nabla^2\nabla^2\varphi] &= \frac{d^4\Phi}{dy^4} - 2\beta^2\frac{d^2\Phi}{dy^2} + \beta^4\Phi \\ &+ \sqrt{\frac{2}{\pi}}\left\{-2\frac{\partial^3\varphi}{\partial x\,\partial y^2}(0, y) - \frac{\partial^3\varphi}{\partial x^3}(0, y) + \beta^2\frac{\partial\varphi}{\partial x}(0, y)\right\}. \end{aligned} \tag{7-8.2.10}$$

In a similar manner, the sine transform of the biharmonic operator in a semi-infinite region is

$$\begin{aligned} F_s[\nabla^2\nabla^2\varphi] &= \frac{d^4\Phi}{dy^4} - 2\beta^2\frac{d^2\Phi}{dy^2} + \beta^4\Phi \\ &+ \sqrt{\frac{2}{\pi}}\left\{2\beta\frac{\partial^2\varphi(0, y)}{\partial y^2} + \beta\frac{\partial^2\varphi(0, y)}{\partial x^2} - \beta^3\varphi(0, y)\right\}. \end{aligned} \tag{7-8·2.11}$$

The Fourier convolution theorem is also useful, and the convolution integral is defined as

$$f \circ g = \frac{1}{\sqrt{2\pi}} \int_{-\infty}^{+\infty} f(t - u) g(u)\, du. \qquad (7\text{-}8.2.12)$$

Some of the properties of this convolution integral are

$$f \circ (\lambda g) = (\lambda f) \circ g = \lambda (f \circ g), \qquad \text{where } \lambda \text{ is a constant}, \qquad (7\text{-}8.2.13)$$

$$f \circ (h + g) = f \circ h + f \circ g, \qquad (7\text{-}8.2.14)$$

$$f \circ g = g \circ f, \qquad (7\text{-}8.2.15)$$

$$(f \circ g) \circ h = f \circ (g \circ h), \qquad (7\text{-}8.2.16)$$

$$\mathfrak{F}[f \circ g] = F(\beta) G(\beta), \qquad \text{where } F(\beta) = \mathfrak{F}[f],\ G(\beta) = \mathfrak{F}[g]. \qquad (7\text{-}8.2.17)$$

The last property leads to the convolution theorem,

$$\mathfrak{F}^{-1}[FG] = f \circ g, \qquad (7\text{-}8.2.18)$$

where $\mathfrak{F}^{-1}$ denotes the inverse transform. This may be written

$$\int_{-\infty}^{+\infty} F(\beta) G(\beta) e^{-i\beta x}\, d\beta = \int_{-\infty}^{+\infty} f(u) g(x - u)\, du. \qquad (7\text{-}8.2.19)$$

8.3 Solution of the infinite-strip problem using integral transforms. Consider the strip loaded as in section 7-8.1. The problem is defined by the biharmonic equation, and boundary conditions

$$\nabla^2 \nabla^2 \varphi = 0, \qquad (7\text{-}8.3.1)$$

$$\sigma_{yy} = \frac{\partial^2 \varphi}{\partial x^2} = -p, \qquad -c < x < c, \qquad y = \pm h, \qquad (7\text{-}8.3.2)$$

$$\sigma_{xy} = -\frac{\partial^2 \varphi}{\partial x\, \partial y} = 0, \qquad y = \pm h. \qquad (7\text{-}8.3.3)$$

Taking the Fourier cosine transform of these equations on the x direction leads to the transformed problem

$$\frac{d^4 \Phi}{dy^4} - 2\beta^2 \frac{d^2 \Phi}{dy^2} + \beta^4 \Phi = 0, \qquad (7\text{-}8.3.4)$$

$$-\beta^2 \Phi = -\frac{1}{\sqrt{2\pi}} \int_{-c}^{+c} p \cos \beta x\, dx \Longrightarrow \Phi = \frac{2p}{\sqrt{2\pi}} \frac{\sin \beta c}{\beta^3}, \qquad y = \pm h, \qquad (7\text{-}8.3.5)$$

$$\frac{d\Phi}{dy} = 0, \qquad y = \pm h, \qquad (7\text{-}8.3.6)$$

where

$$\Phi(\beta, y) = \frac{1}{\sqrt{2\pi}} \int_{-\infty}^{+\infty} \varphi(x, y) \cos \beta x\, dx.$$

Solution of equation 7-8.3.4 yields

$$\Phi = A \operatorname{ch} \beta y + B \operatorname{sh} \beta y + C \beta y \operatorname{sh} \beta y + D \beta y \operatorname{ch} \beta y. \qquad (7\text{-}8.3.7)$$

Noting that the problem is even in y, one obtains

$$B = D = 0. \tag{7-8.3.8}$$

Boundary condition 7-8.3.6 may be satisfied by relating A and C as follows:

$$\begin{aligned} \frac{d\Phi}{dy}\bigg|_{y=h} &= \beta\{A \operatorname{sh} \beta h + C(\beta h \operatorname{ch} \beta h + \operatorname{sh} \beta h)\} = 0, \\ A &= -C(1 + \beta h \coth \beta h). \end{aligned} \tag{7-8.3.9}$$

Boundary condition 7-8.3.5 yields

$$-C\left[\frac{\beta h + \operatorname{sh} \beta h \operatorname{ch} \beta h}{\operatorname{sh} \beta h}\right] = \frac{2p}{\sqrt{2\pi}} \frac{\sin \beta c}{\beta^3}. \tag{7-8.3.10}$$

The transformed function becomes

$$\begin{aligned} \Phi = \frac{1}{\sqrt{2\pi}} \frac{2p \sin \beta c}{\beta^3(\beta h + \operatorname{sh} \beta h \operatorname{ch} \beta h)} &\{(\beta h \operatorname{ch} \beta h + \operatorname{sh} \beta h) \operatorname{ch} \beta y \\ &- \beta y \operatorname{sh} \beta h \operatorname{sh} \beta y\}. \end{aligned} \tag{7-8.3.11}$$

The solution to the problem is obtained by taking the inverse Fourier transform:

$$\begin{aligned} \varphi = \frac{1}{2\pi} \int_{-\infty}^{+\infty} \frac{2p \sin \beta c}{\beta^3(\beta h + \operatorname{sh} \beta h \operatorname{ch} \beta h)} &\{(\beta h \operatorname{ch} \beta h + \operatorname{sh} \beta h) \operatorname{ch} \beta y \\ &- \beta y \operatorname{sh} \beta h \operatorname{sh} \beta y\} \cos \beta x \, d\beta. \end{aligned} \tag{7-8.3.12}$$

This is the same solution obtained in section 7-8.1; only the method of reaching it has differed.

8.4 Semi-infinite strip problems. Examination of the Fourier sine and cosine transforms of the biharmonic equation for the semi-infinite case would lead one to believe that, except for particular boundary conditions at $x = 0$, this problem may be solved in a manner similar to that used for infinite strips. Any function may be considered to be either odd or even in x, since it is only defined from $0 \leq x < \infty$. The choice of using the sine or cosine transform is therefore dictated by what we know at $x = 0$. If the sine transform is used, one must specify at this end

$$\varphi(0, y) \quad \text{and} \quad \frac{\partial^2 \varphi}{\partial x^2}(0, y). \tag{7-8.4.1}$$

Note that specifying $\varphi(0, y)$ also specifies $(\partial^2\varphi/\partial y^2)(0, y)$. Suppose that we wish to set these equal to zero on this end. What does this correspond to physically? Setting $\varphi(0, y)$ equal to zero is equivalent to setting $\sigma_{xx} = 0$. Of course, $\partial^2\varphi/\partial x^2$ is equal to σ_{yy}, but this is not a boundary traction and is not specified on the boundary. However, the displacement in the y direction may specified:

$$\frac{\partial u_y}{\partial y} = \frac{1}{E}\{\sigma_{yy} - \nu\sigma_{xx}\}. \tag{7-8.4.2}$$

Therefore specifying these functions is equivalent to specifying the normal stress and the tangential displacement on this boundary. This will not specify the shear stress on this boundary, as is sometimes desired.

If the cosine transform is used, one would like to specify $(\partial\varphi/\partial x)(0, y)$ and $(\partial^3\varphi/\partial x^3)(0, y)$. The first of these is obviously equivalent to the shear stress σ_{xy}. The second may be shown to be equivalent to specifying the normal displacement:

$$\frac{\partial u_x}{\partial y} + \frac{\partial u_y}{\partial x} = \frac{2(1+\nu)}{E}\sigma_{xy}. \tag{7-8.4.3}$$

Differentiating with respect to y yields

$$\frac{\partial^2 u_x}{\partial y^2} + \frac{\partial^2 u_y}{\partial x\,\partial y} = \frac{2(1+\nu)}{E}\frac{\partial\sigma_{xy}}{\partial y}. \tag{7-8.4.4}$$

However, this may be written in a different manner by noting that

$$\frac{\partial^2 u_y}{\partial x\,\partial y} = \frac{1}{E}\left\{\frac{\partial\sigma_{yy}}{\partial x} - \nu\frac{\partial\sigma_{xx}}{\partial x}\right\}, \tag{7-8.4.5}$$

and from the equilibrium equation

$$\frac{\partial\sigma_{xx}}{\partial x} = -\frac{\partial\sigma_{xy}}{\partial y}. \tag{7-8.4.6}$$

Substitution of these two relations into equation 7-8.4.4 yields

$$\frac{\partial^2 u_x}{\partial y^2} = \frac{1}{E}\left\{-\frac{\partial\sigma_{yy}}{\partial x} + (2+\nu)\frac{\partial\sigma_{xy}}{\partial y}\right\}. \tag{7-8.4.7}$$

From the definition of the Airy stress function,

$$\frac{\partial\sigma_{yy}}{\partial x} = \frac{\partial^3\varphi}{\partial x^3}, \tag{7-8.4.8}$$

and we may finally write

$$\frac{\partial^2 u_x}{\partial y^2} = \frac{1}{E}\left\{-\frac{\partial^3\varphi}{\partial x^3} + (2+\nu)\frac{\partial\sigma_{xy}}{\partial y}\right\}. \tag{7-8.4.9}$$

Therefore, if u_x and σ_{xy} are known at $x = 0$, then $(\partial^3\varphi/\partial x^3)$ is also known. This, however, does not specify the normal stress σ_{xx} on this end.

This restricts the type of problem which may be considered by this method in the case of semi-infinite domains. An example will help to illustrate the use.

Consider the problem shown in Figure 7-12. The boundary conditions are

$$y = \pm h, \qquad \sigma_{xy} = 0, \tag{7-8.4.10}$$

$$\sigma_{yy} = g(x). \tag{7-8.4.11}$$

If we use the cosine transform, we set σ_{xy} and u_x equal to zero at $x = 0$. The transformed biharmonic equation and the transformed boundary condi-

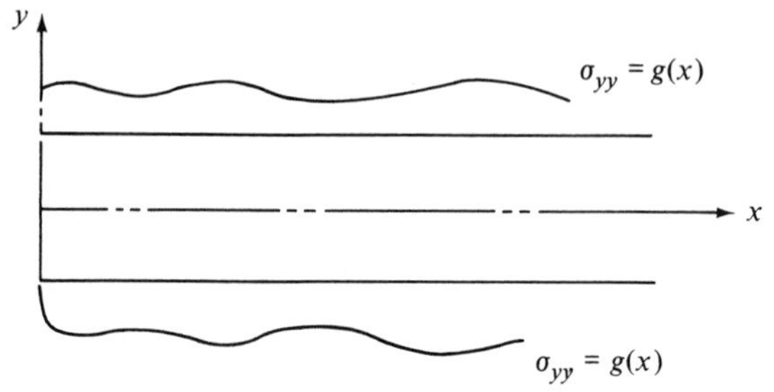

Figure 7-12

tions become

$$\frac{d^4\Phi}{dy^4} - 2\beta^2\frac{d^2\Phi}{dy^2} + \beta^4\Phi = 0, \qquad (7\text{-}8.4.12)$$

$$y = \pm h, \qquad \frac{d\Phi}{dy} = 0, \qquad (7\text{-}8.4.13)$$

$$-\beta^2\Phi = F_c(g) = G(\beta), \qquad (7\text{-}8.4.14)$$

where

$$\Phi = F_c(\varphi). \qquad (7\text{-}8.4.15)$$

The solution of equation 7-8.4.12 may be taken in the form

$$\Phi = A \operatorname{ch} \beta y + C\beta y \operatorname{sh} \beta y. \qquad (7\text{-}8.4.16)$$

As before, the shear boundary condition yields

$$A = -C(1 + \beta h \coth \beta h). \qquad (7\text{-}8.4.17)$$

The transformed Airy stress function now becomes

$$\Phi = C\{\beta y \operatorname{sh} \beta y - \operatorname{ch} \beta y - \beta h \coth \beta h \operatorname{ch} \beta y\}. \qquad (7.8.4.18)$$

Note that $\sigma_{xx} = (\partial^2\varphi/\partial y^2)$, so that when transformed this becomes $d^2\Phi/dy^2$. The integral of this function from $-h$ to $+h$ is zero by examination of equation 7-8.4.13. The end, $x = 0$, is therefore free of force and moment. If stress-free conditions were specified, this problem would be solved within the Saint-Venant approximation. The constant C may be chosen to satisfy the final boundary condition:

$$C = \frac{\operatorname{sh} \beta h}{\beta^2(\beta h + \operatorname{sh} \beta h \operatorname{ch} \beta h)} G(\beta). \qquad (7\text{-}8.4.19)$$

This may be substituted into the equation for Φ and the final solution obtained by taking the inverse Fourier transform.

8.5 Solution for the half-plane. Consider the half-plane defined by $y \geq 0$, and loaded at the boundary $y = 0$ with some normal stress (Figure 7-13).

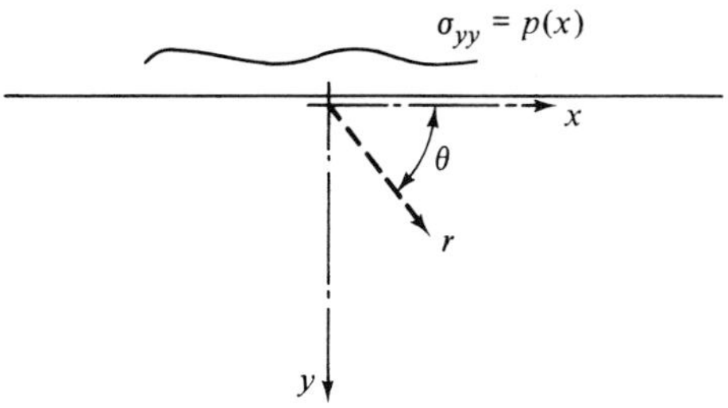

Figure 7-13

The problem may be defined as follows:

$$\nabla^2\nabla^2\varphi = 0, \tag{7-8.5.1}$$

$$\left.\begin{aligned} \sigma_{yy} &= \frac{\partial^2\varphi}{\partial x^2} = p(x), \qquad (7\text{-}8.5.2) \\ \sigma_{xy} &= 0, \qquad (7\text{-}8.5.3) \end{aligned}\right\} y = 0.$$

The integral of $p(x)$ from $-\infty$ to $+\infty$ is assumed to be bounded. Taking the Fourier transform of these equations yields

$$\Phi^{IV} - 2\beta^2\Phi'' + \beta^4\Phi = 0, \tag{7-8.5.4}$$

$$y = 0\left\{\begin{aligned} -\beta^2\Phi &= P(\beta). \qquad (7\text{-}8.5.5) \\ \Phi' &= 0. \qquad (7\text{-}8.5.6) \end{aligned}\right.$$

The solution will be expressed in terms of exponential functions in y. Terms involving the positive exponential will be dropped to maintain boundness of the solution as $y \longrightarrow \infty$. Since β ranges from $-\infty$ to $+\infty$, it is necessary to place absolute value signs on β in the exponential:

$$\Phi = [A + C|\beta|y]e^{-|\beta|y}. \tag{7-8.5.7}$$

The boundary conditions on $y = 0$ may be met by choosing A and C as follows:

$$A = C = -\frac{P(\beta)}{\beta^2}. \tag{7-8.5.8}$$

The transformed function becomes

$$\Phi = -\frac{P(\beta)}{\beta^2}(1 + |\beta|y)e^{-|\beta|y}. \tag{7-8.5.9}$$

The transformed stresses are

$$\mathfrak{F}(\sigma_{xx}) = \frac{d^2\Phi}{dy^2} = P(\beta)[1 - |\beta|y]e^{-|\beta|y}, \tag{7-8.5.10}$$

$$\mathfrak{F}(\sigma_{yy}) = -\beta^2\Phi = P(\beta)[1 + |\beta|y]e^{-|\beta|y}, \tag{7-8.5.11}$$

$$\mathfrak{F}(\sigma_{xy}) = i\beta\frac{d\Phi}{dy} = P(\beta)[i\beta y]e^{-|\beta|y}. \tag{7-8.5.12}$$

Although the first derivative of the general Fourier transform has not been given earlier, it may be obtained as shown in section 7-8.2 and is

$$\mathfrak{F}\left(\frac{\partial \varphi}{\partial x}\right) = -i\beta\Phi.$$

The inverse Fourier transforms of parts of these functions are given as follows:

$$\mathfrak{F}^{-1}[(1 - |\beta|y)e^{-|\beta|y}] = 2^{3/2}\pi^{-1/2}x^2y(x^2 + y^2)^{-2} \tag{7-8.5.13}$$

$$\mathfrak{F}^{-1}[(1 + |\beta|y)e^{-|\beta|y}] = 2^{3/2}\pi^{-1/2}y^3(x^2 + y^2)^{-2}, \tag{7-8.5.14}$$

$$\mathfrak{F}^{-1}[i\beta ye^{-|\beta|y}] = 2^{3/2}\pi^{-1/2}xy^2(x^2 + y^2)^{-2}. \tag{7-8.5.15}$$

We may now apply the convolution theorem of equation 7-8.2.19. The inverses of both functions into which the transformed stresses naturally divide are known, so that the stresses may be written

$$\sigma_{xx} = \frac{2y}{\pi}\int_{-\infty}^{+\infty}\frac{(x - \xi)^2 p(\xi)}{[(x - \xi)^2 + y^2]^2}\,d\xi, \tag{7-8.5.16}$$

$$\sigma_{xy} = \frac{2y^2}{\pi}\int_{-\infty}^{+\infty}\frac{(x - \xi)p(\xi)}{[(x - \xi)^2 + y^2]^2}\,d\xi, \tag{7-8.5.17}$$

$$\sigma_{yy} = \frac{2y^3}{\pi}\int_{-\infty}^{+\infty}\frac{p(\xi)}{[(x - \xi)^2 + y^2]^2}\,d\xi. \tag{7-8.5.18}$$

The details of this solution are taken from "The Use of Transform Methods in Elasticity" by I. N. Sneddon.* If the particular loading considered is a concentrated load $-P$ directed downward at the origin, then

$$p(\xi) = -P\delta(\xi) \tag{7-8.5.19}$$

where $\delta(\xi)$ is the Dirac delta function. The stresses in this case become

$$\sigma_{xx} = -\frac{2Px^2y}{\pi(x^2 + y^2)^2} = -\frac{2P}{\pi}\frac{\cos^2\theta\sin\theta}{r}, \tag{7-8.5.20}$$

$$\sigma_{yy} = -\frac{2P}{\pi}\frac{y^3}{(x^2 + y^2)^2} = -\frac{2P}{\pi}\frac{\sin^3\theta}{r}, \tag{7-8.5.21}$$

$$\sigma_{xy} = -\frac{2P}{\pi}\frac{xy^2}{(x^2 + y^2)^2} = -\frac{2P}{\pi}\frac{\cos\theta\sin^2\theta}{r}. \tag{7-8.5.22}$$

This solution was originally obtained by Flamant in 1892† by modification of the three-dimensional solution found by Boussinesq. It is therefore frequently called the Flamant solution. Note that equations 7-8.5.16 to 7-8.5.18 allow any normal stress distribution to be specified on the boundary, and closed form solutions can be obtained if the infinite integrals can be evaluated.

* *AFORS Tech. Rept. 64–1789.*

† A. Flamant, *Compt. Rend.*, **114**, 1465, 1892.

Related material dealing with loading of the interior of an infinite two-dimensional solid is given at the end of section 8-4.

9 Saint-Venant boundary region in elastic strips

When one uses Fourier methods on semi-infinite strips, one still has to resort often to the Saint-Venant approximation on the end of the strip. The solution for the case of self-equilibrating stresses specified on this end and zero boundary stresses elsewhere is needed to remove this approximation. This solution is also needed in thermal shock and punch problems and in other cases where the stress distribution in the Saint-Venant boundary region is desired. The problem has been termed the fundamental strip problem and has been investigated by P. F. Papkovich, V. K. Prokopov, R. C. T. Smith, G. Horvay, J. P. Benthem, M. W. Johnson and R. W. Little, and others. A brief sketch of one approach* will be given here to show the decay characteristics of this type of stress distribution.

Consider the semi-infinite elastic strip occupying the region $|y| \leq 1$, $0 \leq x < \infty$ (Figure 7-14). The boundary conditions on the edges $y = \pm 1$ are

$$\sigma_{xy} = \sigma_{yy} = 0. \tag{7-9.1}$$

On the edge $x = 0$, one of the following conditions hold:

$$\sigma_{xx} = \sigma^0_{xx}(y), \qquad u_y = u^0_y(y), \tag{7-9.2a}$$

$$\sigma_{xy} = \sigma^0_{xy}(y), \qquad u_x = u^0_x(y), \tag{7-9.2b}$$

$$\sigma_{xx} = \sigma^0_{xx}(y), \qquad \sigma_{xy} = \sigma^0_{xy}(y), \tag{7-9.2c}$$

$$u_x = u^0_x(y), \qquad u_y = u^0_y(y). \tag{7-9.2d}$$

We also shall require that the solution approach zero as x approaches infinity.

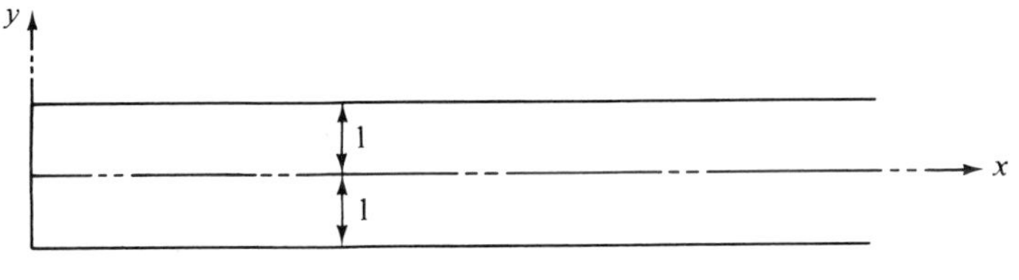

Figure 7-14

* M. W. Johnson and R. W. Little, "The semi-infinite elastic strip," *Quart. Appl. Math.*, **23**, 335–344, 1965. Other references may be found in this paper.

The equations of equilibrium and compatibility are

$$\frac{\partial \sigma_{xx}}{\partial x} + \frac{\partial \sigma_{xy}}{\partial y} = 0, \tag{7-9.3}$$

$$\frac{\partial \sigma_{xy}}{\partial x} + \frac{\partial \sigma_{yy}}{\partial y} = 0, \tag{7-9.4}$$

$$\nabla^2(\sigma_{xx} + \sigma_{yy}) = 0. \tag{7-9.5}$$

A new function $p(x, y)$ will be introduced, defined by the equation

$$\frac{\partial \sigma_{xx}}{\partial y} - \frac{\partial p}{\partial x} = 0. \tag{7-9.6}$$

Equation 7-9.5 may be written

$$\frac{\partial^2}{\partial x^2}(2\sigma_{xx} + \sigma_{yy}) + \frac{\partial^2 p}{\partial x\, \partial y} = 0. \tag{7-9.7}$$

Integrating with respect to x and setting the function of integration equal to zero (since the solution approaches zero as x approaches infinity) yields

$$\frac{\partial p}{\partial y} + \frac{\partial}{\partial x}(2\sigma_{xx} + \sigma_{yy}) = 0. \tag{7-9.8}$$

Equations 7-9.3, 7-9.4, 7-9.6, and 7-9.8 may be written in a matrix-vector form as

$$\frac{\partial \bar{f}}{\partial y} + U\frac{\partial \bar{f}}{\partial x} = 0, \tag{7-9.9}$$

where

$$\bar{f} = \begin{bmatrix} \sigma_{yy} \\ -\sigma_{xy} \\ \sigma_{xx} \\ p \end{bmatrix}, \qquad U = \begin{bmatrix} 0 & -0 & 0 & 0 \\ 0 & 0 & -1 & 0 \\ 0 & 0 & 0 & -1 \\ 1 & 0 & 2 & 0 \end{bmatrix}.$$

Note that p and σ_{xx} are related to the displacements as follows:

$$\frac{\partial u_y}{\partial y} = \frac{1}{E}\{\sigma_{yy} - \nu\sigma_{xx}\}, \tag{7-9.10}$$

$$\frac{\partial u_x}{\partial y} = \frac{1}{E}\{p + \nu\sigma_{xy}\}. \tag{7-9.11}$$

On the boundaries $y = \pm 1, f^1 = f^2 = 0$. We assume a solution in the form

$$\bar{f} = \sum_n C_n \bar{Z}_n(y) e^{+i\alpha_n x}. \tag{7-9.12}$$

The equation for $\bar{Z}_n$ becomes

$$\frac{d}{dy}(\bar{Z}_n) + i\alpha_n U \bar{Z}_n = 0, \tag{7-9.13}$$

$\bar{Z}_n$ may be separated into parts corresponding to σ_{yy} odd and even in y:

$$ {}_0\bar{Z}_n = {}_0\alpha_n^2 \begin{Bmatrix} -[{}_0\alpha_n y \operatorname{sh} {}_0\alpha_n \operatorname{ch} {}_0\alpha_n y - {}_0\alpha_n \operatorname{ch} {}_0\alpha_n \operatorname{sh} {}_0\alpha_n y] \\ i[{}_0\alpha_n y \operatorname{sh} {}_0\alpha_n \operatorname{sh} {}_0\alpha_n y - ({}_0\alpha_n \operatorname{ch} {}_0\alpha_n - \operatorname{sh} {}_0\alpha_n) \operatorname{ch} {}_0\alpha_n y \\ [{}_0\alpha_n y \operatorname{sh} {}_0\alpha_n \operatorname{ch} {}_0\alpha_n y - ({}_0\alpha_n \operatorname{ch} {}_0\alpha_n - 2 \operatorname{sh} {}_0\alpha_n) \operatorname{sh} {}_0\alpha_n y] \\ -i[{}_0\alpha_n y \operatorname{sh} {}_0\alpha_n \operatorname{sh} {}_0\alpha_n y - ({}_0\alpha_n \operatorname{ch} {}_0\alpha_n - 3 \operatorname{sh} {}_0\alpha_n) \operatorname{ch} {}_0\alpha_n y] \end{Bmatrix}, \tag{7-9.14a} $$

$$ {}_e\bar{Z}_n = {}_e\alpha_n^2 \begin{Bmatrix} -[{}_e\alpha_n y \operatorname{ch} {}_e\alpha_n \operatorname{sh} {}_e\alpha_n y - {}_e\alpha_n \operatorname{sh} {}_e\alpha_n \operatorname{ch} {}_e\alpha_n y] \\ i[{}_e\alpha_n y \operatorname{ch} {}_e\alpha_n \operatorname{ch} {}_e\alpha_n y - ({}_e\alpha_n \operatorname{sh} {}_e\alpha_n - \operatorname{ch} {}_e\alpha_n) \operatorname{sh} {}_e\alpha_n y] \\ [{}_e\alpha_n y \operatorname{ch} {}_e\alpha_n \operatorname{sh} {}_e\alpha_n y - ({}_e\alpha_n \operatorname{sh} {}_e\alpha_n - 2 \operatorname{ch} {}_e\alpha_n) \operatorname{ch} {}_e\alpha_n y] \\ -i[{}_e\alpha_n y \operatorname{ch} {}_e\alpha_n \operatorname{ch} {}_e\alpha_n y - ({}_e\alpha_n \operatorname{sh} {}_e\alpha_n - 3 \operatorname{ch} {}_e\alpha_n) \operatorname{sh} {}_e\alpha_n y] \end{Bmatrix}. \tag{7-9.14b} $$

The eigenvalues are defined by the following transcendental equations:

$$ \operatorname{sh}(2\,{}_0\alpha_n) = 2\,{}_0\alpha_n, \qquad \operatorname{sh}(2\,{}_e\alpha_n) = -2\,{}_e\alpha_n. \tag{7-9.15} $$

The lowest eigenvalues are

$$ \begin{aligned} {}_0\alpha_1 &= 1.38 + 3.74i, \\ {}_e\alpha_1 &= 1.13 + 2.11i. \end{aligned} \tag{7-9.16} $$

The minimum rate of decay of the stresses and displacements is given as $e^{-2.11x}$. A biorthogonal vector may be developed by use of the adjoint equation of equation 7-9.13, leading to the condition

$$ \int_{-1}^{+1} W_k^+ U Z_n \, dy = 0, \qquad k \neq n \tag{7-9.17} $$

where $(\ \ )^+$ in the complex conjugate transposed.

$$ {}_0W_n = \begin{Bmatrix} -i[{}_0\alpha_n^* y \operatorname{sh} {}_0\alpha_n^* \operatorname{sh} {}_0\alpha_n^* y - ({}_0\alpha_n^* \operatorname{ch} {}_0\alpha_n^* + \operatorname{sh} {}_0\alpha_n^*) \operatorname{ch} {}_0\alpha_n^* y] \\ [{}_0\alpha_n^* y \operatorname{sh} {}_0\alpha_n^* \operatorname{ch} {}_0\alpha_n^* y - ({}_0\alpha_n^* \operatorname{ch} {}_0\alpha_n^* + 2 \operatorname{sh} {}_0\alpha_n^*) \operatorname{sh} {}_0\alpha_n^* y] \\ -i[{}_0\alpha_n^* y \operatorname{sh} {}_0\alpha_n^* \operatorname{sh} {}_0\alpha_n^* y - ({}_0\alpha_n^* \operatorname{ch} {}_0\alpha_n^* - \operatorname{sh} {}_0\alpha_n^*) \operatorname{ch} {}_0\alpha_n^* y] \\ [{}_0\alpha_n^* y \operatorname{sh} {}_0\alpha_n^* \operatorname{ch} {}_0\alpha_n^* y - {}_0\alpha_n^* \operatorname{ch} {}_0\alpha_n^* \operatorname{sh} {}_0\alpha_n^* y] \end{Bmatrix}, \tag{7-9.18a} $$

where ${}_0\alpha_n^*$ is the complex conjugate of ${}_0\alpha_n$;

$$ {}_eW_n = \begin{Bmatrix} -i[{}_e\alpha_n^* y \operatorname{ch} {}_e\alpha_n^* \operatorname{ch} {}_e\alpha_n^* y - ({}_e\alpha_n^* \operatorname{sh} {}_e\alpha_n^* + \operatorname{ch} {}_e\alpha_n^*) \operatorname{sh} {}_e\alpha_n^* y] \\ [{}_e\alpha_n^* y \operatorname{ch} {}_e\alpha_n^* \operatorname{sh} {}_e\alpha_n^* y - ({}_e\alpha_n^* \operatorname{sh} {}_e\alpha_n^* + 2 \operatorname{ch} {}_e\alpha_n^*) \operatorname{ch} {}_e\alpha_n^* y] \\ -i[{}_e\alpha_n^* y \operatorname{ch} {}_e\alpha_n^* \operatorname{ch} {}_e\alpha_n^* y - ({}_e\alpha_n^* \operatorname{sh} {}_e\alpha_n^* - \operatorname{ch} {}_e\alpha_n^*) \operatorname{sh} {}_e\alpha_n^* y] \\ [{}_e\alpha_n^* y \operatorname{ch} {}_e\alpha_n^* \operatorname{sh} {}_e\alpha_n^* y - {}_e\alpha_n^* \operatorname{sh} {}_e\alpha_n^* \operatorname{ch} {}_e\alpha_n^* y] \end{Bmatrix}. \tag{7-9.18b} $$

Using $\bar{W}$ as a biorthogonal transform on equation 7-8.12 at* $x = 0$ yields

$$C_n = K_n^{-1} \int_{-1}^{+1} \bar{W}_n^+ U \bar{f}_b \, dy, \tag{7-9.19}$$

$$K_n = \int_{-1}^{+1} \bar{W}_n^+ U \bar{Z}_n \, dy. \tag{7-9.20}$$

For the important case 7-9.2c the boundary conditions are specified as follows:

$$\begin{aligned} f_b^1 &= \sum_n C_n Z_n^{(1)}, & f^2 &= -\sigma_{xy}^0 \\ f_b^3 &= \sigma_{xx}^0, & f^4 &= \sum C_n Z_n^4. \end{aligned} \tag{7-9.21}$$

The equation for C_n becomes

$$C_n = K_n^{-1} \sum_j C_j \int_{-1}^{+1} [W_n^{*4} Z_j^1 - W_n^{*3} Z_j^4] \, dy + F_n, \tag{7-9.22}$$

where $$F_n = K_n^{-1} \int_{-1}^{+1} [W_n^{*1} \sigma_{xy}^0 + (2W_n^{*4} - W_n^{*2}) \sigma_{xx}^0] \, dy. \tag{7-9.23}$$

L. D. Power and S. B Childs† have adapted this eigenfunction solution to finite circular bars. This solution may be superimposed to remove the Saint-Venant approximation on the ends. This approach is used instead of the multiple Fourier analysis proposed by Pickett (section 7).

The eigenfunctions of Papkovich do not give stresses on the boundaries $y = \pm h$. As a result, the Fourier series and the Papkovich series decouple. The Papkovich eigenfunctions and the corresponding eigenvalues are more difficult to handle numerically.

Power's and Childs' method considers the finite problem to be the sum of the solution for the region $y \in [-1, 1]$, $x \in [0, \infty]$ and the solution for the region $[-1, 1]$, $x \in [-\infty, l]$. Only the eigenvalues in the upper half of the complex plane $(\alpha_j, -\alpha_j)$ are used for the first of these solutions and the eigenvalues in the lower half-plane $(\alpha_j^*, -\alpha_j^*)$ for the second part. The solution may be written incorporating this fact and using only the eigenvalues in the upper half-plane in the following manner:

$$\bar{f} = \sum_n \{A_n \bar{Z}_n(\alpha_n y) e^{i\alpha_n x} + B_n Z_n(\alpha_n^* y) e^{i\alpha_n^*(x-l)}\}. \tag{7-9.24}$$

The biorthogonality relationship 7-9.19 may be used to evaluate the coefficients in the following manner:

$$A_n = K_n^{-1} \int_{-1}^{+1} \bar{W}_n^+(\alpha_n y) U \bar{f}_b(0, y) \, dy, \tag{7-9.25}$$

$$B_n = K_n^{-1} \int_{-1}^{+1} \bar{W}_n^+(\alpha_n^* y) U \bar{f}_b(l, y) \, dy. \tag{7-9.26}$$

* There is an infinite set of eigenvalues in each quaadrant of the complex plane: α, α^*, $-\alpha$, $-\alpha^*$. Only those in the upper half-plane will ensure decaying solutions.

† L. D. Power and S. B. Childs, "Axisymmetric stresses and displacements in a finite circular bar," *Int. J. Engng. Sci.*, **9**, pp. 241–255, 1971.

Convergence characteristics of these expansions have been examined by Smith* and Gusein-Zade.†

10 Nonorthogonal boundary function expansions

Frequently we are able to obtain solutions to problems of solid mechanics in the form

$$\psi = \sum_m A_m \varphi_m(x, y), \tag{7-10.1}$$

where each term satisfies the defining differential equation term by term. On the boundary the required conditions can usually be expressed in terms of a function of the variables:

$$f(\zeta) = \sum_m A_m \bar{\varphi}_m(\zeta). \tag{7-10.2}$$

Here, $\bar{\varphi}_m(\zeta)$ may be the original function or derivatives of it, or linear combinations of these, depending upon the required boundary condition. In the following discussion the bar will be omitted. Assuming that equation 7-10.1 forms a complete set of functions, we shall examine methods of satisfying equation 7-10.2.‡

10.1 Point matching. The simplest method of satisfying equation 7-10.2 approximately is to force satisfaction at n points, taken along the boundary (usually evenly spaced), with n coefficients A_i.

This condition may be expressed by the following equations:

$$\begin{aligned}
A_1\varphi_1(\zeta_1) + A_2\varphi_2(\zeta_1) + \cdots + A_n\varphi_n(\zeta_1) &= f(\zeta_1) \\
A_1\varphi_1(\zeta_2) + A_2\varphi_2(\zeta_2) + \cdots + A_n\varphi_n(\zeta_2) &= f(\zeta_2) \\
\cdot \qquad\qquad\qquad\qquad\qquad \cdot \qquad & \quad \cdot \\
\cdot \qquad\qquad\qquad\qquad\qquad \cdot \qquad & \quad \cdot \\
\cdot \qquad\qquad\qquad\qquad\qquad \cdot \qquad & \quad \cdot \\
A_1\varphi_1(\zeta_n) + A_2\varphi_2(\zeta_n) + \cdots + A_n\varphi_n(\zeta_n) &= f(\zeta_n).
\end{aligned} \tag{7-10.1.1}$$

We note that there is a one-to-one correspondence between that number of

* R. C. T. Smith, "The bending of a semi-infinite strip," *Austral. J. Sci. Res.*, **5** 227–237, 1952.

† M. J. Gusein-Zade, "On necessary and sufficient conditions for the existence of decaying solutions of the plane problem of the theory of elasticity for a semi-strip," *Appl. Math. Mech.*, **29**, 892–901, 1965.

‡ A detailed discussion of numerical analysis may be found in the many texts on this subject. For example, see F. B. Hildebrand, *Introduction to Numerical Analysis*, McGraw-Hill Book Company, New York, 1956.

points and the number of terms in the series. This is called a point-matching or semicollocation procedure.

10.2 Least squares.* A more difficult but much more effective way to satisfy equation 7-10.2 involves the minimization of the square of the error on the boundary:

$$\int_B \{f(\zeta) - \sum_m A_m\varphi_m(\zeta)\}^2\, d\zeta = \text{min.} \tag{7-10.2.1}$$

Expanding gives

$$\int_B \{f^2(\zeta) - 2f(\zeta)\sum_m A_m\varphi_m(\zeta) + \sum_m\sum_n A_mA_n\varphi_m(\zeta)\varphi_n(\zeta)\}\, d\zeta = \text{min.} \tag{7-10.2.2}$$

Minimizing with respect to each coefficient A_j yields

$$\frac{\partial}{\partial A_j}\int_B \{f(\zeta) - \sum_m A_m\varphi_m(\zeta)\}^2\, d\zeta = 0. \tag{7-10.2.3}$$

Interchanging the order of differentiation and integration and expanding yields

$$\sum_m A_m \int_B \varphi_m(\zeta)\varphi_n(\zeta)\, d\zeta = \int_B f(\zeta)\varphi_n(\zeta)\, d\zeta. \tag{7-10.2.4}$$

If the φ's form an orthonormal set of functions on the boundary

$$\int_B \varphi_m(\zeta)\varphi_n(\zeta)\, d\zeta = \delta_{mn}, \tag{7-10.2.5}$$

and the solution for the A's is obtained quickly,

$$A_n = \int_B f(\zeta)\varphi_n(\zeta)\, d\zeta. \tag{7-10.2.6}$$

All too frequently the set of φ's are not orthogonal on the boundary and one must resort to numerical techniques. The simplest form of numerical integration is to represent the integral by a sum of the values of the integrand evaluated at evenly spaced points, times the spacing interval:

$$\sum_{m=1}^{N} A_m \int_B \varphi_m(\zeta)\varphi_n(\zeta)\, d\zeta = \sum_{m=1}^{N} A_m \sum_{k=1}^{K\geq N} \varphi_m(\zeta_k)\varphi_n(\zeta_k)\,\Delta\zeta, \tag{7-10.2.7}$$

$$\int_B f(\zeta)\varphi_n(\zeta)\, d\zeta = \sum_{k=1}^{K\geq N} f(\zeta_k)\varphi(\zeta_k)\,\Delta\zeta. \tag{7-10.2.8}$$

Therefore, the boundary condition becomes

$$\sum_{m=1}^{N} A_m \sum_{k=1}^{K\geq N} \varphi_m(\zeta_k)\varphi_n(\zeta_k) = \sum_{k=1}^{K\geq N} f(\zeta_k)\varphi_n(\zeta_k). \tag{7-10.2.9}$$

Note that the sum in the numerical integration must involve an equal or greater number of points than the number of terms in the series.

* See C. Lanczos, *Linear Differential Operators*, 1st ed., Van Nostrand Reinhold Company, New York, 1961.

Writing in expanded form yields

$$\begin{bmatrix} \sum_k \varphi_1^2(\zeta_k) & \sum_k \varphi_1(\zeta_k)\varphi_2(\zeta_k) & \cdots & \sum_k \varphi_1(\zeta_k)\varphi_N(\zeta_k) \\ \sum_k \varphi_1(\zeta_k)\varphi_2(\zeta_k) & \sum_k \varphi_2^2(\zeta_k) & \cdots & \sum_k \varphi_2(\zeta_k)\varphi_N(\zeta_k) \\ \cdot & \cdot & & \cdot \\ \cdot & \cdot & & \cdot \\ \sum_k \varphi_1(\zeta_k)\varphi_N(\zeta_k) & \sum_k \varphi_2(\zeta_k)\varphi_N(\zeta_k) & \cdots & \sum_k \varphi_N^2(\zeta_k) \end{bmatrix} \begin{bmatrix} A_1 \\ A_2 \\ \cdot \\ \cdot \\ A_N \end{bmatrix} = \begin{bmatrix} \sum_k f(\zeta_k)\varphi_1(\zeta_k) \\ \sum_k f(\zeta_k)\varphi_2(\zeta_k) \\ \cdot \\ \cdot \\ \sum_k f(\zeta_k)\varphi_N(\zeta_k) \end{bmatrix}. \qquad (7\text{-}10.2.10)$$

Note that the coefficient matrix in equation 7-10.2.10 is symmetric, making the solution easier.

10.3 Iterative improvements to point-matching techniques. * It is possible to obtain improvements to the point-matching solution outlined in equation 7-10.1.1 by use of iterative techniques. Two such methods discussed in the paper by Ojalvo and Linzer will be reviewed here. (1) Several evenly spaced points are selected and a solution is obtained. The calculated boundary values f_0 may be compared to the true values f. The negatives of the maximum errors in between matched points are then imposed as the right-hand side of equation 7-10.1.1. A correction solution for ψ is then obtained called ψ_1. The new solution is

$$\psi_1 = \psi_0 + \psi_1. \qquad (7\text{-}10.3.1)$$

This procedure may be repeated m times to give

$$\psi_m = \psi_0 + \psi_1 + \psi_2 + \cdots + \psi_m. \qquad (7\text{-}10.3.2)$$

(2) Several spaced points are again selected as in (1) and a solution for ψ obtained. A comparison between f and f_0 is made on the boundary. The next solution is obtained by using the same points as before plus the points in between, where $(f - f_0)$ has its maximum value. This results in doubling the number of equations with each iteration.

Error bounds for the solution of the Poisson equation are given by Ojalvo and Linzer together with an excellent discussion of each of these methods. Practical experience has shown that the least-squares technique is usually the best.

* Ojalvo and Linzer, "Improved point-matching techniques," *Quart. J. Mech. and Appl. Math*, **18**, 1, 1965.

11 Plane elasticity problems using nonorthogonal functions

Consider a body occupying the region $-l < x < l$ and $f(x) < y < g(x)$, as shown in Figure 7-15. We may assume a solution of the form

$$\varphi = \sum_n \sin \beta_n x[A_n \operatorname{sh} \beta_n y + B_n \operatorname{ch} \beta_n y + C_n \beta_n y \operatorname{sh} \beta_n y + D_n \beta_n y \operatorname{ch} \beta_n y]$$
$$+ \sum_n \cos \beta_n x[E_n \operatorname{sh} \beta_n y + F_n \operatorname{ch} \beta_n y + G_n \beta_n y \operatorname{sh} \beta_n y + H_n \beta_n y \operatorname{ch} \beta_n y]. \tag{7-11.1}$$

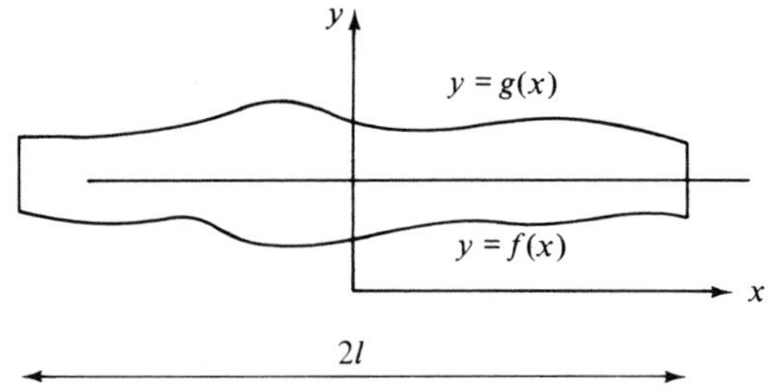

Figure 7-15

In general, the geometry prohibits the separation of the problem into even and odd parts, and the complete solution in the form of equation 7-11.1 is needed. Depending upon the choice of β, zero roots in the form of polynomials may have to be added. On the upper and lower boundaries, the solution is in the form of nonorthogonal series and the methods of section 10 are applicable.

The boundary geometries are substituted into equation 7-11.1, or the appropriate derivatives and the series is equated to the specified boundary stresses or displacements. The two sets of four constants are required to satisfy the four boundary conditions on the long boundaries. If the boundaties $x = \pm l$ are to be treated without the Saint-Venant approximation, or have specified nonvanishing tractions, two additional series will be needed. These extra terms are generated by interchanging x and y in the equations above. The example that follows avoids the Saint-Venant approximation, and has the general character described.

11.1 Examples requiring functions nonorthogonal on the boundaries. Consider the body shown in Figure 7-16. The boundary conditions are

$$x = 0, \quad 0 < y < b, \quad \begin{aligned} u_x &= 0, \\ \sigma_{xy} &= 0, \end{aligned} \tag{7-11.1.1}$$

$$x = l, \quad 0 < y < a, \quad \begin{aligned} \sigma_{xx} &= T, \\ \sigma_{xy} &= 0, \end{aligned} \tag{7-11.1.2}$$

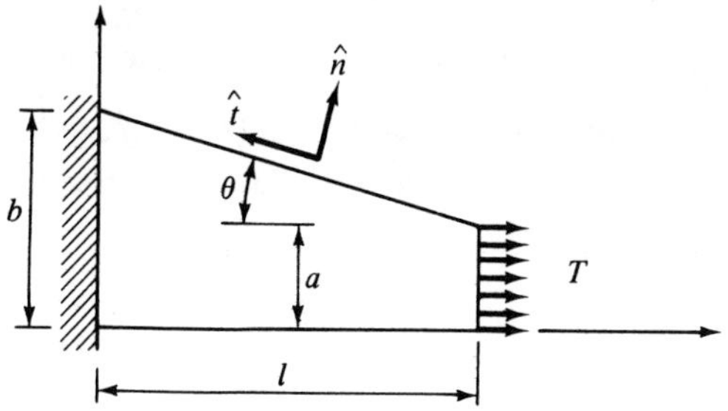

Figure 7-16

$$y = 0, \qquad 0 < x < l, \qquad \begin{array}{l}\sigma_{yy} = 0,\\ \sigma_{xy} = 0,\end{array} \tag{7-11.1.3}$$

$$\begin{array}{l} y = b - \dfrac{b-a}{l}x = b - x\tan\theta \\ y = a + (l-x)\tan\theta \end{array} \qquad \begin{cases}\sigma_{nn} = 0 \\ \sigma_{nt} = 0\end{cases}. \tag{7-11.1.4}$$

We shall assume a solution in the form

$$\varphi = \sum_n \cos\beta_n x[A_n \operatorname{sh}\beta_n y + B_n \operatorname{ch}\beta_n y + C_n\beta_n y \operatorname{sh}\beta_n y + D_n\beta_n y \operatorname{ch}\beta_n y] + \sum_k \cos\alpha_k y[E_k \operatorname{sh}\alpha_k x + F_k \operatorname{ch}\alpha_k x + G_k\alpha_k x \operatorname{sh}\alpha_k x + H_k\alpha_k x \operatorname{ch}\alpha_k x]. \tag{7-11.1.5}$$

Note that the solution is neither even nor odd in x or y. The cosine terms were used so that σ_{xy} can be set to zero at $x = 0$ and $y = 0$. Calculating stresses yields

$$\sigma_{xx} = \sum_n \beta_n^2 \cos\beta_n x[A_n \operatorname{sh}\beta_n y + B_n \operatorname{ch}\beta_n y + C_n(2\operatorname{ch}\beta_n y + \beta_n y \operatorname{sh}\beta_n y) + D_n(2\operatorname{sh}\beta_n y + \beta_n y \operatorname{ch}\beta_n y)] - \sum_k \alpha_k^2 \cos\alpha_k y[E_k \operatorname{sh}\alpha_k x + F_k \operatorname{ch}\alpha_k x + G_k\alpha_k x \operatorname{sh}\alpha_k x + H_k\alpha_k x \operatorname{ch}\alpha_k x], \tag{7-11.1.6}$$

$$\sigma_{yy} = -\sum_n \beta_n^2 \cos\beta_n x[A_n \operatorname{sh}\beta_n y + B_n \operatorname{ch}\beta_n y + C_n\beta_n y \operatorname{sh}\beta_n y + D_n\beta_n y \operatorname{ch}\beta_n y] + \sum_k \alpha_k^2 \cos\alpha_k y[E_k \operatorname{sh}\alpha_k x + F_k \operatorname{ch}\alpha_k x + G_k(2\operatorname{ch}\alpha_k x + \alpha_k x \operatorname{sh}\alpha_k x) + H_k(2\operatorname{sh}\alpha_k x + \alpha_k x \operatorname{ch}\alpha_k x)], \tag{7-11.1.7}$$

$$\sigma_{xy} = \sum_n \beta_n^2 \sin\beta_n x[A_n \operatorname{ch}\beta_n y + B_n \operatorname{sh}\beta_n y + C_n(\operatorname{sh}\beta_n y + \beta_n y \operatorname{ch}\beta_n y) + D_n(\operatorname{ch}\beta_n y + \beta_n y \operatorname{sh}\beta_n y)] + \sum_k \alpha_k^2 \sin\alpha_k y[E_k \operatorname{ch}\alpha_k x + F_k \operatorname{sh}\alpha_k x + G_k(\operatorname{sh}\alpha_k x + \alpha_k x \operatorname{ch}\alpha_k x) + H_k(\operatorname{ch}\alpha_k x + \alpha_k x \operatorname{sh}\alpha_k x)]. \tag{7-11.1.8}$$

If we fix the rigid-body displacements at the origin, the displacement u_x becomes

$$Eu_x = \int_x \frac{\partial^2\varphi}{\partial y^2}dx - \nu\frac{\partial\varphi}{\partial x}. \tag{7-11.1.9}$$

$$Eu_x = \sum_n \beta_n \sin \beta_n x[A_n(1 + \nu) \operatorname{sh} \beta_n y + B_n(1 + \nu) \operatorname{ch} \beta_n y + C_n(2 \operatorname{ch} \beta_n y + [1 + \nu]\beta_n y \operatorname{sh} \beta_n y) + D_n(2 \operatorname{sh} \beta_n y + [1 + \nu]\beta_n y \operatorname{ch} \beta_n y)] - \sum_k \alpha_k \cos \alpha_k y[E_k(1 + \nu) \operatorname{ch} \alpha_k x + F_k(1 + \nu) \operatorname{sh} \alpha_k x + G_k\{(1 + \nu)\alpha_k x \operatorname{ch} \alpha_k x - (1 + \nu) \operatorname{sh} \alpha_k x\} + H_k\{(1 + \nu)\alpha_k x \operatorname{sh} \alpha_k x - (1 - \nu) \operatorname{ch} \alpha_k x\}]. \quad (7\text{-}11.1.10)$$

Examining the boundary $0 < y < b$, $x = 0$,

$$\sigma_{xy}|_{x=0} = 0 = \sum \alpha_k^2 \sin \alpha_k y(E_k + H_k), \quad (7\text{-}11.1.11)$$

$$u_x|_{x=0} = 0 = \sum \alpha_k \cos \alpha_k y[E_k(1 + \nu) - H_k(1 - \nu)]. \quad (7\text{-}11.1.12)$$

Therefore,

$$E_k = H_k = 0. \quad (7\text{-}11.1.13)$$

For the boundary $0 < x < l$, $y = 0$,

$$\sigma_{xy}|_{y=0} = 0 = \sum \beta_n^2 \sin \beta_n x[A_n + D_n] \Longrightarrow A_n = -D_n, \quad (7\text{-}11.1.14)$$

$$\sigma_{yy}|_{y=0} = 0 = -\sum \beta_n^2 \cos \beta_n x[B_n] + \sum_k \alpha_k^2[F_k \operatorname{ch} \alpha_k x + G_k(2 \operatorname{ch} \alpha_k x + \alpha_k x \operatorname{sh} \alpha_k x)]. \quad (7\text{-}11.1.15)$$

We may use Fourier analysis to satisfy this condition, but must specify β_n first. Note that we can have orthogonal series with either $\beta_n = n\pi/l$ or $\beta_n = (2n + 1)(\pi/l)/2$. To make the proper selection, examine the conditions at $x = l$:

$$\sigma_{xy}|_{x=l} = \sum_n \beta_n^2 \sin \beta_n l[B_n \operatorname{sh} \beta_n y + C_n(\operatorname{sh} \beta_n y + \beta_n y \operatorname{ch} \beta_n y) + D_n\beta_n y \operatorname{sh} \beta_n y] + \sum_k \alpha_k^2 \sin \alpha_k y[F_k \operatorname{sh} \alpha_k l + G_k(\operatorname{sh} \alpha_k l + \alpha_k l \operatorname{ch} \alpha_k l)] = 0. \quad (7\text{-}11.1.16)$$

Now choose

$$\beta_n = n\pi/l \quad \text{and} \quad F_k = -G_k\left[\frac{\operatorname{sh} \alpha_k l + \alpha_k l \operatorname{ch} \alpha_k l}{\operatorname{sh} \alpha_k l}\right]. \quad (7\text{-}11.1.17)$$

Returning to examine the $\sigma_{yy}|_{y=0}$ condition, the Fourier series is

$$\sum_n B_n\beta_n^2 \cos \frac{n\pi x}{l} = \sum_k \alpha_k^2[F_k \operatorname{ch} \alpha_k x + G_k(2 \operatorname{ch} \alpha_k x + \alpha_k x \operatorname{sh} \alpha_k x)]. \quad (7\text{-}11.1.18)$$

We see that we do not need the constant term for this series: the integral of the right side from 0 to l is zero. Taking the finite Fourier cosine transform yields equations of the form (using the relation between F_k and G_k)

$$B_n = \sum_k \mathfrak{F}_{nk} G_k. \quad (7\text{-}11.1.19)$$

Examining the remaining condition at $x = l$,

$$\sigma_{xx}|_{x=l} = T = \sum_n \beta_n^2 \cos \beta_n l\{B_n \operatorname{ch} \beta_n y + C_n(2 \operatorname{ch} \beta_n y + \beta_n y \operatorname{sh} \beta_n y) + D_n(\operatorname{sh} \beta_n y + \beta_n y \operatorname{ch} \beta_n y)\} - \sum_k \alpha_k^2 \cos \alpha_k y\{F_k \operatorname{ch} \alpha_k l + G_k\alpha_k l \operatorname{sh} \alpha_k l\}. \quad (7\text{-}11.1.20)$$

Eliminating F_k yields the condition

$$\sum_k G_k\alpha_k^2\left[\frac{\alpha_k l + \operatorname{sh}\alpha_k l \operatorname{ch}\alpha_k l}{\operatorname{sh}\alpha_k l}\right]\cos\alpha_k y = T - \sum_n \beta_n^2\cos\beta_n l\{B_n \operatorname{ch}\beta_n y + C_n(2\operatorname{ch}\beta_n y + \beta_n y \operatorname{sh}\beta_n y) + D_n(\operatorname{sh}\beta_n y + \beta_n y\operatorname{ch}\beta_n y)\}. \tag{7-11.1.21}$$

α may be chosen as $\alpha_k = k\pi/a$ or $\alpha_k = [(2k+1)/2](\pi/a)$. Using $\alpha_k = [(2k+1)/2](\pi/a)$, this equation takes the form, after taking the transform,

$$G_k = \mathfrak{J}_{(k)} - \sum_n \mathfrak{R}_{kn}(B_n, C_n, D_n). \tag{7-11.1.22}$$

Note that at this point we have selected α_k and β_n and have the following relations for our eight original constants:

$$\begin{aligned} E_k &= 0,\\ H_k &= 0,\\ A_n &= -D_n,\\ F_k &= -G_k\left[\frac{\operatorname{sh}\alpha_k l + \alpha_k l\operatorname{ch}\alpha_k l}{\operatorname{sh}\alpha_k l}\right],\\ B_n &= \sum_k \mathfrak{F}_{nk}G_k,\\ G_k &= \mathfrak{J}_{(k)} - \sum_n \mathfrak{R}_{kn}(B_n, C_n, D_n). \end{aligned} \tag{7-11.1.23}$$

We have six relations for eight unknowns and have two more conditions to satisfy. To examine these two conditions, let us note the following transformation equations: $\sigma'_{ij} = a_{ik}a_{jm}\sigma_{km}$. Let $\cos\theta = \gamma$ and $\sin\theta = \lambda$.

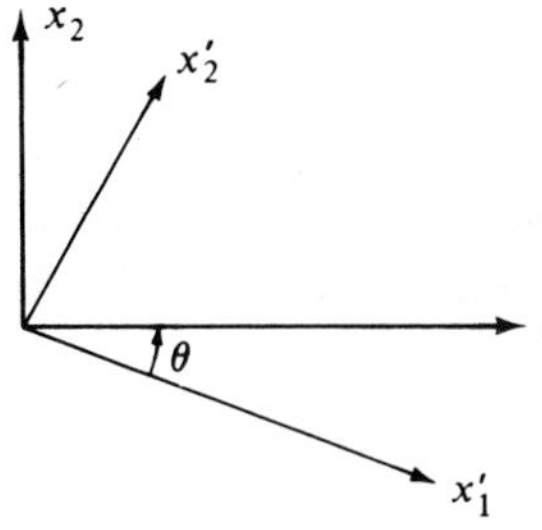

Figure 7-17

$$\sigma'_{22} = \lambda^2\sigma_{11} + 2\lambda\gamma\sigma_{12} + \gamma^2\sigma_{22} \tag{7-11.1.24}$$

or

$$\sigma_{nn} = \lambda^2\frac{\partial^2\varphi}{\partial y^2} + \gamma^2\frac{\partial^2\varphi}{\partial x^2} - 2\gamma\lambda\frac{\partial^2\varphi}{\partial x\,\partial y}.$$

Now we shall prove that the relation $\sigma_{nn} = \sigma''_{22} = \partial^2\varphi/\partial t^2$ if t stands for the tangential direction (see Figure 7-16).

$$\frac{\partial\varphi}{\partial x_1'} = \frac{\partial\varphi}{\partial x_1}\frac{\partial x_1}{\partial x_1'} + \frac{\partial\varphi}{\partial x_2}\frac{\partial x_2}{\partial x_1'}, \tag{7-11.1.25}$$

$$\begin{aligned} \frac{\partial^2\varphi}{\partial x_1'\,\partial x_1'} &= \frac{\partial^2\varphi}{\partial x_1^2}\left(\frac{\partial x_1}{\partial x_1'}\right)^2 + 2\frac{\partial^2\varphi}{\partial x_1\,\partial x_2}\left(\frac{\partial x_1}{\partial x_1'}\frac{\partial x_2}{\partial x_2'}\right) + \frac{\partial^2\varphi}{\partial x_2^2}\left(\frac{\partial x_2}{\partial x_1'}\right)^2\\ &= \lambda^2\frac{\partial^2\varphi}{\partial y^2} + \gamma^2\frac{\partial^2\varphi}{\partial x^2} - 2\gamma\lambda\frac{\partial^2\varphi}{\partial x\,\partial y} \end{aligned} \tag{7-11.1.26}$$

Therefore,

$$\sigma_{nn} = \frac{\partial^2 \varphi}{\partial t^2}. \tag{7-11.1.27}$$

In a similar manner,

$$\sigma_{nt} = -\frac{\partial^2 \varphi}{\partial n \partial t}. \tag{7-11.1.28}$$

We may note that the boundary conditions, $\sigma_{nn} = \sigma_{nt} = 0$, are equivalent to

$$\varphi|_{\text{boundary}} = 0 \quad \text{and} \quad \left.\frac{\partial \varphi}{\partial n}\right|_{\text{boundary}} = 0, \tag{7-11.1.29}$$

where $$\frac{\partial \varphi}{\partial n} = \frac{\partial \varphi}{\partial x}\frac{\partial x}{\partial n} + \frac{\partial \varphi}{\partial y}\frac{\partial y}{\partial x} = \frac{\partial \varphi}{\partial x}\sin\theta + \frac{\partial \varphi}{\partial y}\cos\theta$$

on the inclined boundary. The conditions are thus

$$\begin{aligned}\varphi = 0 = \sum_n \cos\beta_n x[A_n \operatorname{sh}\beta_n y + B_n \operatorname{ch}\beta_n y + C_n\beta_n y \operatorname{sh}\beta_n y + D_n\beta_n y \operatorname{ch}\beta_n y]\\ + \sum_k \cos\alpha_k y[F_k \operatorname{ch}\alpha_k x + G_k\alpha_k x \operatorname{sh}\alpha_k x], \qquad y = a + (l - x)\tan\theta,\end{aligned} \tag{7-11.1.30}$$

and $\partial\varphi/\partial n = 0$. Therefore,

$$\frac{\partial \varphi}{\partial x}\tan\theta + \frac{\partial \varphi}{\partial y} = 0 \qquad \text{at } y = a + (l - x)\tan\theta. \tag{7-11.1.31}$$

These can be viewed as conditions to determine D_n and C_n. These two equations relate all the constants to each other and may be solved by point-matching or least squares. We now have a system of linear equations to determine G, B, C, and D. If, for example, we took two B's, two C's, two D's, and three G's, then

$$\begin{array}{ccccccccc} B_1 & B_2 & C_1 & C_2 & D_1 & D_2 & G_1 & G_2 & G_3 \end{array}$$

$$\begin{bmatrix} x & 0 & 0 & 0 & 0 & 0 & x & x & x \\ 0 & x & 0 & 0 & 0 & 0 & x & x & x \\ x & x & x & x & x & x & x & 0 & 0 \\ x & x & x & x & x & x & 0 & x & 0 \\ x & x & x & x & x & x & 0 & 0 & x \\ x & x & x & x & x & x & x & x & x \\ x & x & x & x & x & x & x & x & x \\ x & x & x & x & x & x & x & x & x \\ x & x & x & x & x & x & x & x & x \end{bmatrix} \begin{bmatrix} 0 \\ 0 \\ T_1 \\ T_2 \\ T_3 \\ 0 \\ 0 \\ 0 \\ 0 \end{bmatrix} \begin{matrix} \left.\begin{matrix} \\ \\ \end{matrix}\right\} \sigma_{yy}|_{y=0} = 0 \\ \left.\begin{matrix} \\ \\ \\ \end{matrix}\right\} \sigma_{xx}|_{x=l} = T \\ \left.\begin{matrix} \\ \\ \end{matrix}\right\} \varphi = 0 \\ \left.\begin{matrix} \\ \\ \end{matrix}\right\} \dfrac{\partial \varphi}{\partial n} = 0 \end{matrix}$$

Here, x's denote nonzero matrix entries; a solution may be obtained by use of digital computers.

PROBLEMS

7.1. Find the stress and displacement distributions in the beam shown and compare the elasticity solution to the elementary beam solution. Use polynomials and a Saint-Venant approximation on the short boundaries (if necessary).

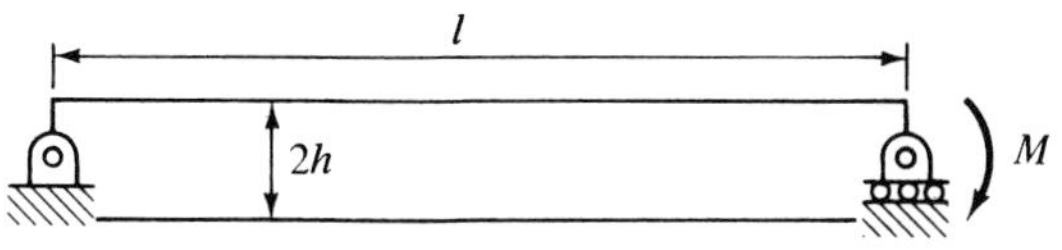

Figure P7-1

7.2. Find the stress and displacement distributions in the beam shown by use of polynomials and compare to the elementary solution.

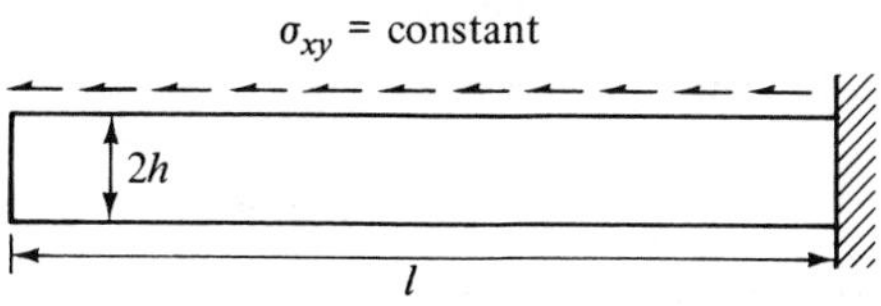

Figure P7-2

7.3. Determine the stress distribution in the beam shown by use of polynomials.

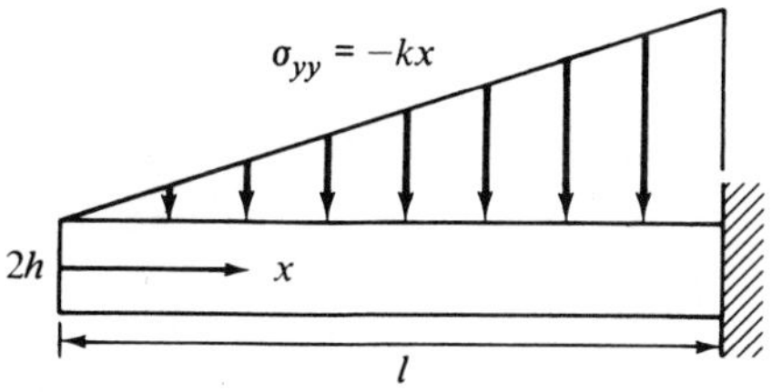

Figure P7-3

7.4. Use Fourier trigonometric series to approximate the solution to the beam problem shown below.

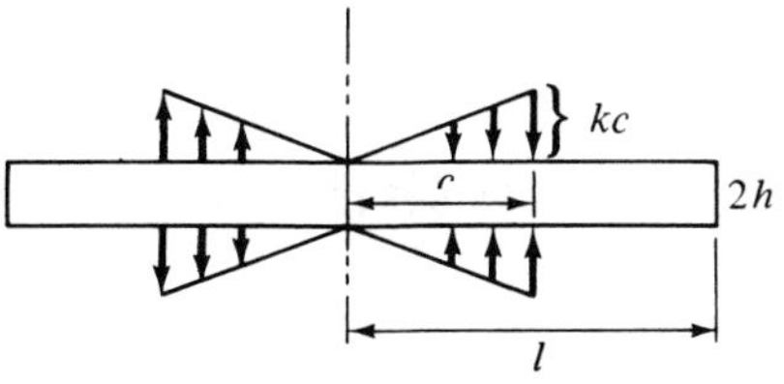

Figure P7-4

7.5. Set up the equations for Problem 7.4 if $h/l \approx 1$ and a Saint-Venant approximation is not permissible on the ends.

7.6. Set up the solution for the following problem of a square plate compressed as shown.

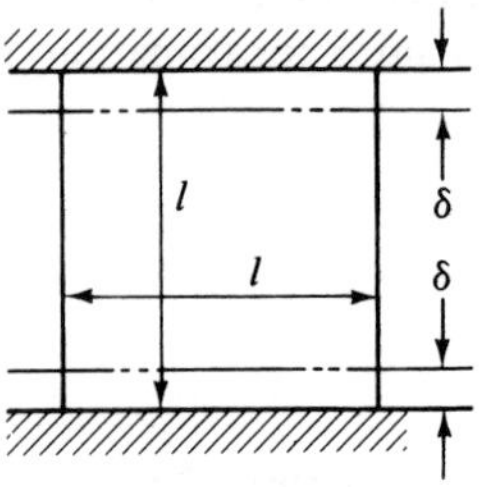

Figure P7-6

Each end of the plate undergoes a deflection of amount δ in the y direction and no deflection in the x direction (see Section 7-6.2).

7.7. Derive the following integral.

$$\int \cosh au \sin bu \, du.$$

7.8. Find the stresses for the half-plane loaded as shown.
The following inverse transform may be useful:

$$\mathfrak{F}^{-1}[i(\text{sign } \beta)(2 - |\beta| y)e^{-|\beta| y}] = 2^{3/2}\pi^{-1/2}x^3(x^2 + y^2)^{-2}.$$

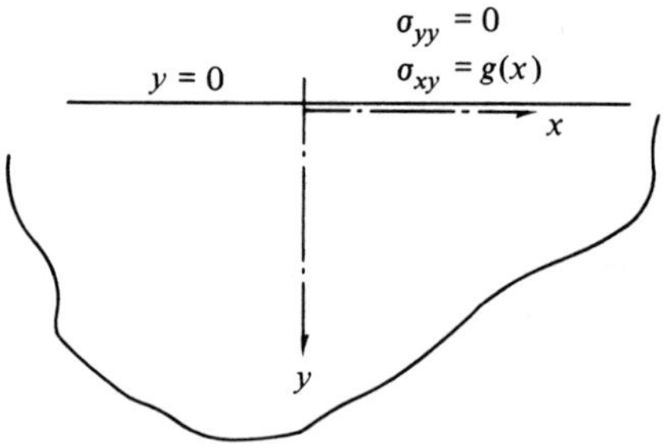

Figure P7-8

7.9.   Use transform methods to set up the problem of an infinite quadrant, subjected to the following boundary conditions:

$$x = 0 \quad \begin{cases} \sigma_{xy} = 0 \\ u_x = 0 \end{cases}$$

$$y = 0 \quad \begin{cases} \sigma_{yy} = f(x) \\ u_x = 0 \end{cases}$$

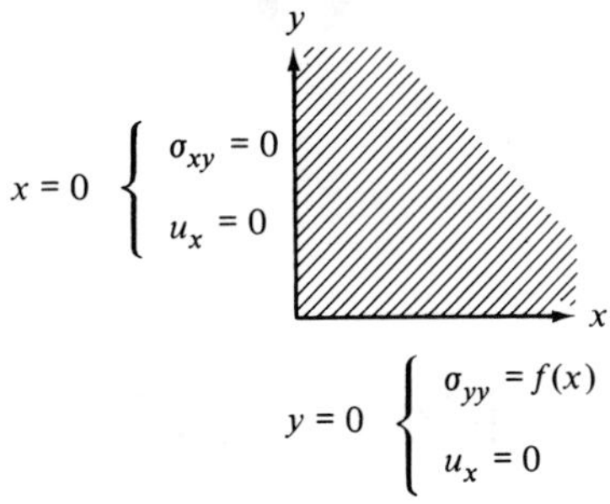

Figure P7-9

7.10.   Separate the problem shown into four parts so that an Airy stress function may use odd and even functions. Show each by a sketch and indicate the properties of the corresponding stress function.

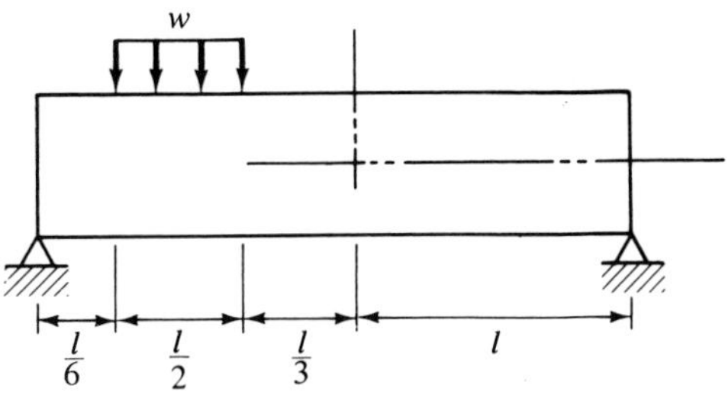

Figure P7-10

7.11.   Indicate how you would solve the following problem by use of the Flamant solution.

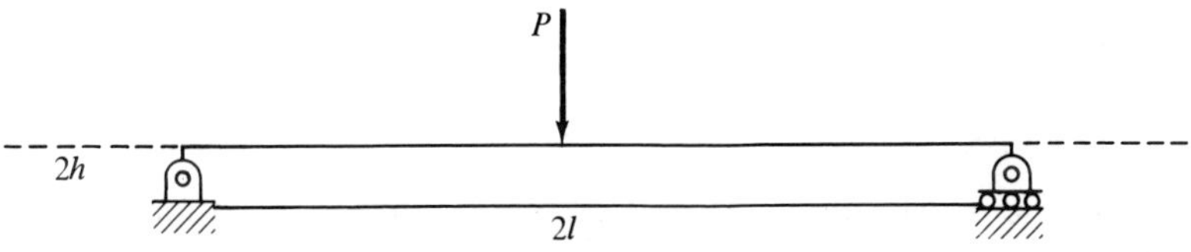

Figure P7-11

CHAPTER 8

PROBLEMS IN POLAR COORDINATES

1 Introduction

We now consider problems in which the geometry affords a simpler formulation in polar coordinates than in Cartesian coordinates. Before consideration of the general form of solution is undertaken, we display for future reference the equations of elasticity in polar coordinates. In symbolic notation the equations of equilibrium are

$$\bar{\nabla}\cdot\bar{\bar{\sigma}} + \bar{F} = 0. \tag{8-1.1}$$

Using dyadic expansions, the corresponding scalar equations in polar coordinates become

$$\begin{aligned} \frac{\partial \sigma_{rr}}{\partial r} + \frac{\sigma_{rr} - \sigma_{\theta\theta}}{r} + \frac{1}{r}\frac{\partial \sigma_{r\theta}}{\partial \theta} + F_r = 0, \\ \frac{\partial \sigma_{r\theta}}{\partial r} + \frac{2\sigma_{r\theta}}{r} + \frac{1}{r}\frac{\partial \sigma_{\theta\theta}}{\partial \theta} + F_\theta = 0. \end{aligned} \tag{8-1.2}$$

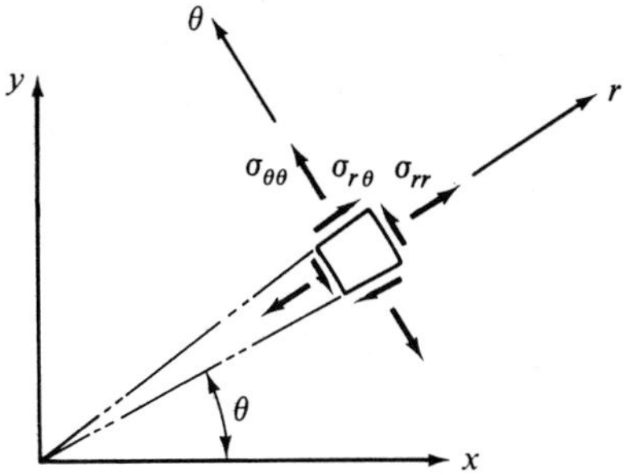

Figure 8-1

For the case when the body forces are zero, a symbolic form for the relation between the stress tensor and the Airy stress function is

$$\bar{\bar{\sigma}} = \nabla^2\varphi\bar{\bar{I}} - \bar{\nabla}\bar{\nabla}\varphi. \tag{8-1.3}$$

Expanding produces

$$\begin{aligned} \sigma_{rr} &= \frac{1}{r}\frac{\partial\varphi}{\partial r} + \frac{1}{r^2}\frac{\partial^2\varphi}{\partial\theta^2}, \\ \sigma_{\theta\theta} &= \frac{\partial^2\varphi}{\partial r^2}, \\ \sigma_{r\theta} &= -\frac{\partial}{\partial r}\left(\frac{1}{r}\frac{\partial\varphi}{\partial\theta}\right). \end{aligned} \tag{8-1.4}$$

The Laplacian operator in polar coordinates is

$$\nabla^2 = \frac{\partial^2}{\partial r^2} + \frac{1}{r}\frac{\partial}{\partial r} + \frac{1}{r^2}\frac{\partial^2}{\partial\theta^2}. \tag{8-1.5}$$

The biharmonic equation defining φ in this coordinate system takes the form

$$\left(\frac{\partial^2}{\partial r^2} + \frac{1}{r}\frac{\partial}{\partial r} + \frac{1}{r^2}\frac{\partial^2}{\partial\theta^2}\right)^2\varphi = 0. \tag{8-1.6}$$

The definitions of strain become (see section 2-8)

$$\begin{aligned} \epsilon_{rr} &= \frac{\partial u_r}{\partial r}, \\ \epsilon_{\theta\theta} &= \frac{u_r}{r} + \frac{1}{r}\frac{\partial u_\theta}{\partial\theta}, \\ \epsilon_{r\theta} &= \frac{1}{2}\left[\frac{\partial u_\theta}{\partial r} - \frac{u_\theta}{r} + \frac{1}{r}\frac{\partial u_r}{\partial\theta}\right]. \end{aligned} \tag{8-1.7}$$

These may be obtained from the symbolic equation, using dyads:

$$\bar{\bar{\epsilon}} = \text{sym. part } (\vec{\nabla}\bar{u}) = \tfrac{1}{2}[\vec{\nabla}\bar{u} + \bar{u}\overleftarrow{\nabla}]. \tag{8-1.8}$$

Hooke's stress–strain relations for an isotropic material in plane stress are

$$\epsilon_{rr} = \frac{1}{E}\{\sigma_{rr} - \nu\sigma_{\theta\theta}\},$$

$$\epsilon_{\theta\theta} = \frac{1}{E}\{\sigma_{\theta\theta} - \nu\sigma_{rr}\}, \qquad (8\text{-}1.9)$$

$$\epsilon_{r\theta} = \frac{1+\nu}{E}\sigma_{r\theta}.$$

2 Axially symmetric problems

Consider the class of problems in which φ does not depend upon θ. The biharmonic equation for this case becomes an ordinary "equidimensional" equation:

$$\frac{d^4\varphi}{dr^4} + \frac{2}{r}\frac{d^3\varphi}{dr^3} - \frac{1}{r^2}\frac{d^2\varphi}{dr^2} + \frac{1}{r^3}\frac{d\varphi}{dr} = 0. \qquad (8\text{-}2.1)$$

This equation is known to have a solution in terms of integer powers of r:

$$\varphi = r^n. \qquad (8\text{-}2.2)$$

The characteristic equation has two sets of multiple roots leading to a general solution of the form

$$\varphi = A \ln r + Br^2 \ln r + Cr^2 + D. \qquad (8\text{-}2.3)$$

The stresses become

$$\sigma_{rr} = \frac{A}{r^2} + 2B \ln r + B + 2C, \qquad (8\text{-}2.4)$$

$$\sigma_{\theta\theta} = -\frac{A}{r^2} + 2B \ln r + 3B + 2C, \qquad (8\text{-}2.5)$$

$$\sigma_{r\theta} = 0. \qquad (8\text{-}2.6)$$

The zero shear stress is consistent with the assumption of axially symmetric conditions. When $B = 0$, inspection of the solution shows that the constant $2C$ is equivalent to the negative of the hydrostatic pressure. The terms multiplied by the constants A and B become singular at the origin and may be used in problems involving an annulus or the region external to a given radius. The problem of a disk under axisymmetric loading contains only the term C, and is equivalent to hydrostatic tension or compression.

We may calculate the displacements from the stresses for the general problem, as follows:

$$\epsilon_{rr} = \frac{du_r}{dr} = \frac{1}{E}\left\{\frac{A(1+\nu)}{r^2} + 2C(1-\nu) + B(1-\nu)[2\ln r + 1] - 2\nu B\right\}, \tag{8-2.7}$$

$$\epsilon_{\theta\theta} = \frac{u_r}{r} = \frac{1}{E}\left\{-\frac{A(1+\nu)}{r^2} + 2C(1-\nu) + B(1-\nu)[2\ln r + 1] + 2B\right\}. \tag{8-2.8}$$

These give two expressions for u_r which must be consistent. Integrating the first yields

$$u_r = \frac{1}{E}\left\{-\frac{A(1+\nu)}{r} + 2C(1-\nu)r + B(1-\nu)[2r\ln r - r] - 2\nu Br + \text{const}\right\}. \tag{8-2.9}$$

From the second expression we obtain

$$u_r = \frac{1}{E}\left\{-\frac{A(1+\nu)}{r} + 2C(1-\nu)r + B(1-\nu)[2r\ln r + r] + 2Br\right\}, \tag{8-2.10}$$

if the displacements are to be truly independent of θ. Now to obtain consistent results, the constant B and the constant in 8-2.9 must be zero. We could have obtained this result directly from the Navier equations because they can be solved easily in this case. The vector Navier equation is of the form

$$G\left\{\nabla^2\bar{u} + \frac{1+\nu}{1-\nu}\vec{\nabla}(\vec{\nabla}\cdot\bar{u})\right\} = 0. \tag{8-2.11}$$

where

$$\bar{u} = u_r\hat{i}_{(r)}. \tag{8-2.12}$$

The divergence of $\bar{u}$ becomes

$$\vec{\nabla}\cdot\bar{u} = \frac{du_r}{dr} + \frac{u_r}{r}. \tag{8-2.13}$$

The Laplacian of $\bar{u}$ is

$$\frac{d^2u_r}{dr} + \frac{1}{r}\frac{du_r}{dr} - \frac{1}{r^2}u_r = \frac{d}{dr}\left\{\frac{du_r}{dr} + \frac{u_r}{r}\right\}. \tag{8-2.14}$$

The first and second terms of equation 8-2.11 thus yield exactly the same expression, so that the Navier equation reduces to

$$\frac{d}{dr}\left\{\frac{du_r}{dr} + \frac{u_r}{r}\right\} = 0. \tag{8-2.15}$$

This may be simplified to the following form:

$$\frac{d}{dr}\left\{\frac{1}{r}\frac{d}{dr}(ru_r)\right\} = 0. \tag{8-2.16}$$

Integrating once yields

$$\frac{1}{r}\frac{d}{dr}(ru_r) = C_1. \tag{8-2.17}$$

The second integration gives the following expression for u_r:

$$u_r = \frac{C_1 r}{2} + \frac{C_2}{r}. \tag{8-2.18}$$

The form of this solution is the same as that obtained from the Airy stress function approach after the constant B has been taken to be zero.

If, on the other hand, we assume that the stresses are independent of θ but that the displacements may depend upon θ, we obtain much different results. This may be considered as the quasi-axisymmetric case. The strain displacement relations of the Airy stress function approach now provide

$$\epsilon_{rr} = \frac{\partial u_r}{\partial r} = \frac{1}{E}\left\{\frac{A(1+\nu)}{r^2} + 2C(1-\nu) + B(1-\nu)[2\ln r + 1] - 2\nu B\right\}, \tag{8-2.19}$$

$$\epsilon_{\theta\theta} = \frac{u_r}{r} + \frac{1}{r}\frac{\partial u_\theta}{\partial \theta} = \frac{1}{E}\left\{-\frac{A(1+\nu)}{r^2} + 2C(1-\nu) + B(1-\nu)[2\ln r + 1] + 2B\right\}, \tag{8-2.20}$$

$$\epsilon_{r\theta} = \frac{1}{2}\left\{\frac{\partial u_\theta}{\partial r} - \frac{u_\theta}{r} + \frac{1}{r}\frac{\partial u_r}{\partial \theta}\right\}. \tag{8-2.21}$$

Integrating equation 8-2.19 yields

$$u_r = \frac{1}{E}\left\{-\frac{A(1+\nu)}{r} + 2C(1-\nu)r + B(1-\nu)[2r\ln r - r] - 2\nu Br + f(\theta)\right\}. \tag{8-2.22}$$

Substituting this into equation 8-2.20,

$$\frac{1}{r}\frac{\partial u_\theta}{\partial \theta} = \frac{1}{E}\left\{4B - \frac{f(\theta)}{r}\right\}, \tag{8-2.23}$$

$$u_\theta = \frac{1}{E}\left\{4B\theta r - \int f(\theta)\, d\theta + g(r)\right\}. \tag{8-2.24}$$

Substituting u_r and u_θ into the shear strain expression yields

$$g'(r) - \frac{g(r)}{r} + \frac{1}{r}\int f(\theta)\, d\theta + \frac{1}{r} f'(\theta) = 0, \tag{8-2.25}$$

$$rg'(r) - g(r) = \text{const.} = D_1, \tag{8-2.26}$$

$$f'(\theta) + \int f(\theta)\, d\theta = -D_1. \tag{8-2.27}$$

From equation 8-2.26, one obtains

$$g(r) = -D_1 + D_2 r, \tag{8-2.28}$$

$$f(\theta) = F \sin\theta + G \cos\theta, \tag{8-2.29}$$

$$\int f(\theta)\, d\theta = -D_1 - F\cos\theta + G\sin\theta. \tag{8-2.30}$$

Substitution of these results into the expressions for u_r and u_θ yields

$$u_r = \frac{1}{E}\left\{-\frac{A(1+\nu)}{r} + 2C(1-\nu)r + B[(1-\nu)(2r\ln r - r) - 2\nu r] + F\sin\theta + G\cos\theta\right\}, \tag{8-2.31}$$

$$u_\theta = \frac{1}{E}\{4Br\theta + F\cos\theta - G\sin\theta + D_2 r\}, \tag{8-2.32}$$

where the constants A, B, and C appear in the stress equations and the constants F, G, and D_2 represent the rigid-body motions. It should be noted that we could not obtain these results from the Navier equations without some difficulty, because in this case we would have to face the task of solving coupled differential equations. For a problem which is truly axisymmetric, that is, both the stresses and the displacements are independent of angle, we would have to choose the constants B, F, G, and D_2 to be zero. This was noted before when u_θ was assumed to be identically zero. It is important to also note that the first term in the expression for u_θ is not single valued, and cannot be used in doubly connected regions (such as an annulus) except in those special problems involving "dislocations."

2.1 Lamé problem.* One of the most useful axisymmetric problems is the case of an annulus subjected to internal and external pressure, as shown in Figure 8-2. The boundary conditions are

$$\sigma_{rr} = -p_i, \qquad \sigma_{r\theta} = 0, \qquad r = a, \tag{8-2.1.1}$$

$$\sigma_{rr} = -p_0, \qquad \sigma_{r\theta} = 0, \qquad r = b, \tag{8-2.1.2}$$

Using equations 8-2.3 to 8-2.4, after setting B equal to zero to guarantee single-valued displacements, leads to the following conditions for boundary stresses:

$$\sigma_{rr}(a) = -p_i = \frac{A}{a^2} + 2C, \tag{8-2.1.3}$$

$$\sigma_{rr}(b) = -p_0 = \frac{A}{b^2} + 2C. \tag{8-2.1.4}$$

* This solution is credited to G. Lamé, *Leçons sur la théorie. . . de l'élasticité*, Gautheir-Villars, Paris, 1852.

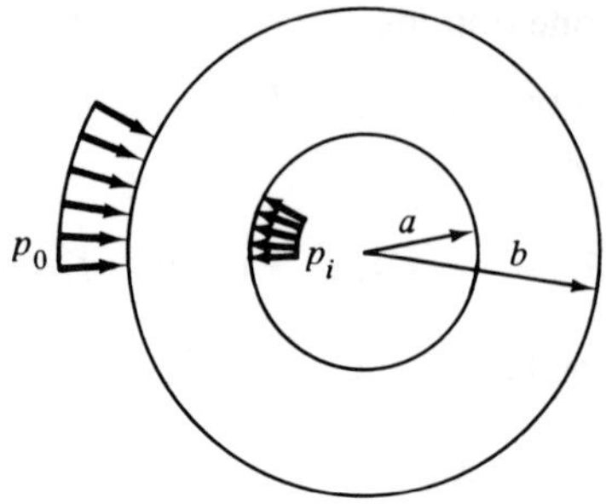

Figure 8-2

Solving for A and C yields

$$\sigma_{rr} = \frac{a^2b^2}{b^2 - a^2}\frac{p_0 - p_i}{r^2} + \frac{a^2p_i - b^2p_0}{b^2 - a^2}, \qquad (8\text{-}2.1.5)$$

$$\sigma_{\theta\theta} = -\frac{a^2b^2}{b^2 - a^2}\frac{p_0 - p_i}{r^2} + \frac{a^2p_i - b^2p_0}{b^2 - a^2}. \qquad (8\text{-}2.1.6)$$

If the mean radius and the thickness of the annulus are introduced,

$$t = b - a, \qquad R_m = \frac{b + a}{2}, \qquad (8\text{-}2.1.7)$$

then the internal and external radii may be written as follows:

$$b = R_m + \tfrac{1}{2}t, \qquad a = R_m - \tfrac{1}{2}t, \qquad b^2 - a^2 = 2tR_m. \quad (8\text{-}2.1.8)$$

For the case where the external pressure p_0 is zero and terms of the the order t^2/R_m^2 may be neglected, the expression for the hoop stress $\sigma_{\theta\theta}$ becomes

$$\sigma_{\theta\theta} = \frac{p_iR_m}{2t}\left[1 + \frac{R_m^2}{r^2} - \frac{t}{R_m}\right]. \qquad (8\text{-}2.1.9)$$

The elementary membrane approximation is $\sigma_{\theta\theta} = p_iR_m/t$, and is equivalent to both neglecting the t/R_m term and evaluating the stress when $r = R_m$; this is consistent with the assumptions of the membrane solution.*

The solution of the problem of a pressure applied inside a hole in an infinite plane may be obtained by setting $p_0 = 0$ and letting b approach infinity. For this case the stresses become

$$\sigma_{rr} = -p_i\frac{a^2}{r^2}, \qquad \sigma_{\theta\theta} = p_i\frac{a^2}{r^2}. \qquad (8\text{-}2.1.10)$$

* The problem for an eccentric bore was solved by G. B. Jeffery, *Trans. Roy. Soc. (London)*, **A221**, 265, 1921.

2.2 *Pure bending of a curved beam.* Consider the problem of a curved beam subjected to bending moments at each end.* The beam is shown in Figure 8-3. The boundary conditions are

$$\sigma_{rr} = \sigma_{r\theta} = 0, \qquad \text{at } r = a \text{ and } r = b, \tag{8-2.2.1}$$

$$\left.\begin{aligned} \int \sigma_{\theta\theta}\, dr &= 0 \\ \int \sigma_{\theta\theta} r\, dr &= -M \end{aligned}\right\} \theta = \pm\beta. \tag{8-2.2.2}$$

The last condition introduces the Saint-Venant approximation. This solution may be developed using equation 8-2.3 because there is no requirement

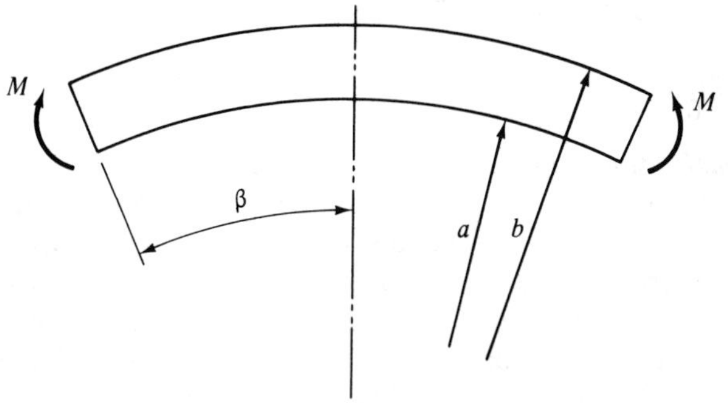

Figure 8-3

that B be taken to be zero. The boundary conditions yield four equations:

$$\frac{A}{a^2} + 2B \ln a + B + 2C = 0, \tag{8-2.2.3}$$

$$\frac{A}{b^2} + 2B \ln b + B + 2C = 0, \tag{8-2.2.4}$$

$$b\left(\frac{A}{b^2} + 2B \ln b + B + 2C\right) - a\left(\frac{A}{a^2} + 2B \ln a + B + 2C\right) = 0, \tag{8-2.2.5}$$

$$-A \ln \frac{b}{a} + B(b^2 \ln b - a^2 \ln a) + C(b^2 - a^2) = -M. \tag{8-2.2.6}$$

The third equation is automatically satisfied if the first two are, so that there are only three independent equations to determine the three constants. Solv-

* H. Golovin, *Trans. Inst. Tech.*, St. Petersburg, 1881. Also see M. C. Ribiere, *Compt. Rend.*, **108**, 1889; A. Föppl, *Vorlesungen über Technische Mechanik*, **5**, 72, 1907; and A. Timpe, *Z. Math. Physik*, **52**, 348, 1905.

ing for the constants yields

$$A = -\frac{4M}{N}a^2b^2 \ln\frac{b}{a}, \qquad B = -\frac{2M}{N}(b^2 - a^2), \tag{8-2.2.7}$$

$$C = \frac{M}{N}[b^2 - a^2 + 2(b^2 \ln b - a^2 \ln a)], \tag{8-2.2.8}$$

where

$$N = (b^2 - a^2)^2 - 4a^2b^2\left(\ln\frac{b}{a}\right)^2. \tag{8-2.2.9}$$

The stresses become

$$\sigma_{rr} = -\frac{4M}{N}\left\{\frac{a^2b^2}{r^2}\ln\frac{b}{a} + b^2\ln\frac{r}{b} + a^2\ln\frac{a}{r}\right\}, \tag{8-2.2.10}$$

$$\sigma_{\theta\theta} = -\frac{4M}{N}\left\{-\frac{a^2b^2}{r^2}\ln\frac{b}{a} + b^2\ln\frac{r}{b} + a^2\ln\frac{a}{r} + b^2 - a^2\right\}, \tag{8-2.2.11}$$

$$\sigma_{r\theta} = 0. \tag{8-2.2.12}$$

2.3 Rotational dislocation. It was mentioned earlier that for a complete annulus the term in equation 8-2.32 containing the constant B would be multivalued. If we consider a rotational dislocation of particular type, this term is essential. The term dislocation was used by A. E. H. Love in his discussions of the work of G. Weingarten, A. Timpe, V. Volterra, and E. Cesaro. Volterra used the term "distorisioni" in discussing the physical interpretations of multivalued displacements. Love* gives a complete discussion of the nature of the allowable discontinuities in the displacements and the conditions that must be imposed upon the strains.

Consider a ring that is opened and separated by small angle α, as shown in Figure 8-4. If the ring is now forced back together to form an annulus and

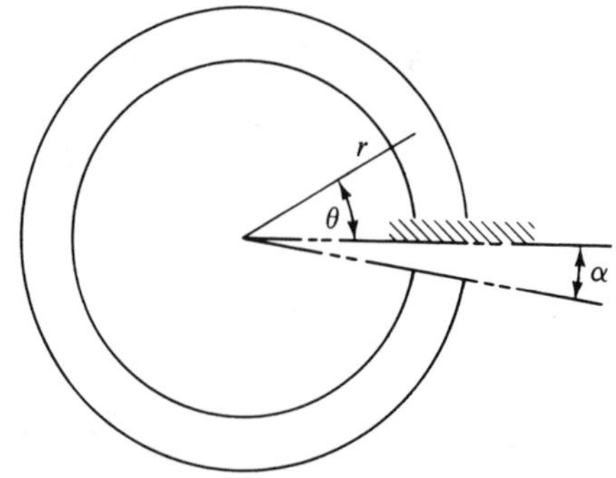

Figure 8.4

* A. E. H. Love, *A Treatise on the Mathematical Theory of Elasticity*, Dever Publications Inc. New York, 1944, 4th ed. (1926).

held together by welding, gluing, etc., we may calculate the internal stresses due to this action. The boundary conditions on the outer and inner edges are that the body is stress free. At the interface, the displacement condition is

$$u_\theta = r\alpha, \qquad \text{at } \theta = 2\pi. \tag{8-2.3.1}$$

Examining equation 8-2.32 and setting the rigid-body motions equal to zero yields

$$\frac{4B}{E}2\pi = \alpha, \qquad B = \frac{E\alpha}{8\pi}. \tag{8-2.3.2}$$

Substituting equation 8-2.3.2 into the equations obtained in section 8-2.2, we obtain an expression for the internal bending moment:

$$M = -\frac{E\alpha}{16\pi}\frac{N}{b^2 - a^2} = -\frac{E\alpha}{16\pi}\frac{(b^2 - a^2)^2 - 4a^2b^2(\ln b/a)^2}{(b^2 - a^2)}. \tag{8-2.3.3}$$

The stresses may now be calculated by use of the equations of section 8-2.2.

3 Solution of axisymmetric problems using the Navier equation

The Navier equation may be written as

$$G\left\{\nabla^2\bar{u} + \frac{1+\nu}{1-\nu}\vec{\nabla}(\vec{\nabla}\cdot\bar{u})\right\} + \bar{F} = 0. \tag{8-3.1}$$

We have considered the reduced equation previously in our discussions of displacements, but without the effects of body forces. For axisymmetric problems, equation 8-3.1 reduces to a single ordinary differential equation:

$$\frac{2G}{1-\nu}\left\{\frac{d}{dr}\left[\frac{du_r}{dr} + \frac{u_r}{r}\right]\right\} + F_r = 0. \tag{8-3.2}$$

Consider a rotating disk where the body force may be taken to be the inertia force:

$$F_r = \rho r\omega^2, \tag{8-3.3}$$

where ρ is the mass density and ω is the angular velocity*

Substituting equation 8-3.3 into equation 8-3.2 yields

$$\frac{d}{dr}\left[\frac{du_r}{dr} + \frac{u_r}{r}\right] = -\frac{(1-\nu)}{2G}\rho\omega^2 r. \tag{8-3.4}$$

* See the discussion of this problem in K. Loffler, *Die Berechnung von Rotierenden Scheiben und Schalen*, Springer-Verlag, Gottingen, Germany, 1961.

This may be written

$$\frac{d}{dr}\left\{\frac{1}{r}\frac{d}{dr}(ru_r)\right\} = -\frac{1-\nu}{2G}\rho\omega^2 r. \tag{8-3.5}$$

Integrating yields

$$u_r = -\frac{1-\nu}{16G}\rho\omega^2 r^3 + C_1 r + \frac{C_2}{r}. \tag{8-3.6}$$

The strains are

$$\begin{aligned} \epsilon_{rr} &= -\frac{3}{16}\frac{(1-\nu)}{G}\rho\omega^2 r^2 + C_1 - \frac{C_2}{r^2}, \\ \epsilon_{\theta\theta} &= -\frac{(1-\nu)}{16G}\rho\omega^2 r^2 + C_1 + \frac{C_2}{r^2}. \end{aligned} \tag{8-3.7}$$

The stresses may now be obtained and are equal to

$$\begin{aligned} \sigma_{rr} &= \frac{1-\nu^2}{E}\{\epsilon_{rr} + \nu\epsilon_{\theta\theta}\} \\ &= -\left\{\frac{3+\nu}{8}\rho\omega^2 r^2 - \frac{EC_1(1+\nu)}{1-\nu^2} + \frac{EC_2(1-\nu)}{r^2(1-\nu^2)}\right\}, \\ \sigma_{\theta\theta} &= -\left\{\frac{1+3\nu}{8}\rho\omega^2 r^2 - \frac{EC_1(1+\nu)}{1-\nu^2} - \frac{EC_2(1-\nu)}{r^2(1-\nu^2)}\right\}. \end{aligned} \tag{8-3.8}$$

For solutions bounded at the origin,

$$C_2 = 0. \tag{8-3.9}$$

The constant C_1 may be evaluated by satisfying the final boundary condition

$$\sigma_{rr}|_{r=b} = 0, \tag{8-3.10}$$

$$C_1 = \left(\frac{1-\nu}{E}\right)\frac{3+\nu}{8}\rho\omega^2 b^2. \tag{8-3.11}$$

The stresses become

$$\begin{aligned} \sigma_{rr} &= \frac{3+\nu}{8}\rho\omega^2(b^2 - r^2), \\ \sigma_{\theta\theta} &= \frac{3+\nu}{8}\rho\omega^2 b^2 - \frac{1+3\nu}{8}\rho\omega^2 r^2. \end{aligned} \tag{8-3.12}$$

For the state of stress in a long shaft under rotation the constant ν must be replaced by $\nu^* = \nu/(1-\nu)$ for plane strain and the stress σ_{zz} must be removed.

4 Michell solution

J. H. Michell* sought a general solution of the biharmonic equation in two dimensions that included the theta dependence and was periodic in

* J. H. Michell, *Proc. London Math. Soc.*, **31**, 100, 1899.

nature. We shall develop this solution by considering the biharmonic equation in polar coordinates:

$$\nabla^2\nabla^2\varphi = 0, \tag{8-4.1}$$

$$\left(\frac{\partial^2}{\partial r^2} + \frac{1}{r}\frac{\partial}{\partial r} + \frac{1}{r^2}\frac{\partial^2}{\partial \theta^2}\right)^2 \varphi = 0. \tag{8-4.2}$$

We seek solutions in the following form:

$$\varphi = f(r)e^{\beta\theta}. \tag{8-4.3}$$

Substituting equation 8-4.3 into equation 8-4.2 yields

$$\left[f^{\text{IV}} + 2\frac{f'''}{r} - (1 - 2\beta^2)\frac{f''}{r^2} + (1 - 2\beta^2)\frac{f'}{r^3} + \beta^2(4 + \beta^2)\frac{f}{r^4}\right]e^{\beta\theta} = 0. \tag{8-4.4}$$

Making a change of variables, $r = e^{\xi}$, equation 8-4.4 becomes

$$\left\{\frac{d^4F}{d\xi^4} - 4\frac{d^3F}{d\xi^3} + (4 + 2\beta^2)\frac{d^2F}{d\xi^2} - 4\beta^2\frac{dF}{d\xi} + \beta^2(4 + \beta^2)F\right\}e^{\beta\theta} = 0. \tag{8-4.5}$$

The solution of F can be taken in the form $F = e^{a\xi}$, because the fourth-order differential equation contained in the brackets in equation 8-4.5 has constant coefficients. The characteristic equation for a is

$$a^4 - 4a^3 + (4 + 2\beta^2)a^2 - 4\beta^2 a + \beta^2(4 + \beta^2) = 0. \tag{8-4.6}$$

Factoring yields

$$(a^2 + \beta^2)(a^2 - 4a + 4 + \beta^2) = 0. \tag{8-4.7}$$

The relationships between a and β may be written

$$\begin{aligned} a &= i\beta & & & \beta &= ia \\ a &= -i\beta & & & \beta &= -ia \\ & & &\text{or} & & \\ a &= 2 + \beta i & & & \beta &= i(a - 2) \\ a &= 2 - \beta i & & & \beta &= -i(a - 2) \end{aligned} \tag{8-4.8}$$

Periodic solutions of this form may be obtained by choosing $\beta = in$, with n an integer. It is apparent that a will also be an integer. The only difficulty in obtaining the solution is the necessity to use care in handling multiple roots. We list the values of $n = 0, 1, 2$, and look for repeated roots. The list also shows the values of β obtained from these values of a.

	$n = 0$	$n = 1$	$n = 2$		$a = 0$	$a = 1$	$a = 2$
$a = n$	0	1	2	$\beta = ia$	0	i	$2i$
$a = -n$	0*	-1	-2	$\beta = -ia$	0*	$-i$	$-2i$
$a = 2 - n$	2	1*	0	$\beta = i(a - 2)$	$-2i$	$-i$*	0
$a = 2 + n$	2*	3	4	$\beta = -i(a - 2)$	$2i$	i*	0*

* Repeated roots.

The solution is

$$
\begin{aligned}
\varphi = {} & a_0 + b_0 \ln r + c_0 r^2 + d_0 r^2 \ln r \\
& + (A_0 + B_0 \ln r + C_0 r^2 + D_0 r^2 \ln r)\theta \\
& + \left(a_1 r + b_1 r \ln r + \frac{c_1}{r} + d_1 r^3\right)^{\sin\theta}_{\cos\theta} \\
& + (A_1 r + B_1 r \ln r)\theta \, {}^{\sin\theta}_{\cos\theta} \\
& + \sum_{n=2}^{\infty} (a_n r^n + b_n r^{2+n} + c_n r^{-n} + d_n r^{2-n}) {}^{\sin n\theta}_{\cos n\theta}.
\end{aligned} \qquad (8\text{-}4.9)
$$

The notation for $\sin n\theta$ and $\cos n\theta$ is meant to imply that $b_n r^{2+n} \sin n\theta$ is one term and $b_n r^{2+n} \cos n\theta$ is another independent term available to satisfy boundary conditions. Later, b_n and b'_n are distinguished by "upper" and "lower" superscripts.

This is essentially the solution obtained by J. H. Michell, except that he omitted some of the repeated roots. It should be pointed out that this is not a general solution because a may take any noninteger value. These, however, do not lead to periodic solutions or allow the use of orthogonal functions on the boundaries and are not of practical value in many problems. Particular examples using these will be shown later.

Using the stress definitions in terms of the Airy stress function produces the following stresses:

$$
\begin{aligned}
\sigma_{rr} = {} & \frac{b_0}{r^2} + 2c_0 + d_0(2\ln r + 1) + B_0\frac{\theta}{r^2} + 2C_0\theta + D_0(2\ln r + 1)\theta \\
& + \left(\frac{b_1}{r} - \frac{2c_1}{r^3} + 2d_1 r\right)^{\sin\theta}_{\cos\theta} + \left(\frac{2A_1}{r}\right)^{\cos\theta}_{-\sin\theta} \\
& + \left(\frac{B_1}{r}\theta\right)^{\sin\theta}_{\cos\theta} + \left(\frac{2B_1}{r}\ln r\right)^{\cos\theta}_{-\sin\theta} \\
& - \sum_{n=2}^{\infty} \{a_n n(n-1)r^{n-2} + b_n(n+1)(n-2)r^n + c_n n(n+1)r^{-(n+2)} \\
& + d_n(n-1)(n+2)r^{-n}\} {}^{\sin n\theta}_{\cos n\theta},
\end{aligned} \qquad (8\text{-}4.10)
$$

$$
\begin{aligned}
\sigma_{\theta\theta} = {} & -\frac{b_0}{r^2} + 2c_0 + d_0(2\ln r + 3) - \frac{B_0\theta}{r^2} + 2C_0\theta + D_0(2\ln r + 3)\theta \\
& + \left(\frac{b_1}{r} + \frac{2c_1}{r^3} + 6d_1 r\right)^{\sin\theta}_{\cos\theta} + \left(\frac{B_1\theta}{r}\right)^{\sin\theta}_{\cos\theta} \\
& + \sum_{n=2}^{\infty} \{a_n n(n-1)r^{n-2} + b_n(n+1)(n+2)r^n + c_n n(n+1)r^{-(n+2)} \\
& + d_n(n-2)(n-1)r^{-n}\} {}^{\sin n\theta}_{\cos n\theta},
\end{aligned} \qquad (8\text{-}4.11)
$$

$$
\begin{aligned}
\sigma_{r\theta} = {} & \frac{A_0}{r^2} + B_0\frac{\ln r - 1}{r^2} - C_0 - D_0(\ln r + 1) + \left(-\frac{b_1}{r} + \frac{2c_1}{r} - 2d_1 r\right)^{\cos\theta}_{-\sin\theta} \\
& - \left(\frac{B_1}{r}\right)^{\sin\theta}_{\cos\theta} - \left(\frac{B_1}{r}\theta\right)^{\cos\theta}_{-\sin\theta} \\
& - \sum_{n=2}^{\infty} \{a_n n(n-1)r^{n-2} + b_n n(n+1)r^n - c_n n(n+1)r^{-(n+2)} \\
& - d_n n(n-1)r^{-n}\} {}^{\cos n\theta}_{-\sin n\theta}.
\end{aligned} \qquad (8\text{-}4.12)
$$

The displacements may be obtained from the following relations:

$$\frac{\partial u_r}{\partial r} = \frac{1}{E}\{\sigma_{rr} - \nu\sigma_{\theta\theta}\}, \tag{8-4.13}$$

$$\frac{1}{r}\frac{\partial u_\theta}{\partial \theta} = \frac{1}{E}\{\sigma_{\theta\theta} - \nu\sigma_{rr}\} - \frac{u_r}{r}, \tag{8-4.14}$$

$$\frac{1+\nu}{E}\sigma_{r\theta} = \frac{1}{2}\left\{\frac{\partial u_\theta}{\partial r} + \frac{1}{r}\frac{\partial u_r}{\partial \theta} - \frac{u_\theta}{r}\right\}. \tag{8-4.15}$$

Substituting equations 8-4.10 and 8-4.11 into 8-4.13 and integrating yields

$$\begin{aligned} u_r = \frac{1}{E}\Big\{ & -\frac{b_0}{r}(1+\nu) + 2c_0(1-\nu)r + d_0(1-\nu)(2r\ln r - r) - 2d_0\nu r \\ & + \left[-\frac{B_0}{r}(1+\nu) + 2C_0(1-\nu)r + D_0(1-\nu)(2r\ln r - r) - 2D_0\nu r\right]\theta \\ & + \left[b_1(1-\nu)\ln r + \frac{c_1}{r^2}(1+\nu) + d_1 r^2(1-\nu) - 2d_1\nu r^2\right]^{\sin\theta}_{\cos\theta} \\ & + [2A_1\ln r]^{\cos\theta}_{-\sin\theta} + [B_1\ln r(1-\nu)]^{\theta\sin\theta}_{\theta\cos\theta} + [B_1\ln^2 r]^{\cos\theta}_{-\sin\theta} \\ & + \sum_{n=2}^{\infty}[a_n n(1+\nu)r^{n-1} + b_n\{(n-2)+\nu(n+2)\}r^{n+1} \\ & - c_n n(1+\nu)r^{-n+1} - d_n\{(n+2)+\nu(n-2)\}r^{-(n-1)}]^{\sin n\theta}_{\cos n\theta} + g(\theta)\Big\}, \end{aligned} \tag{8-4.16}$$

where $g(\theta)$ is an arbitrary function of integration. From equation 8-4.14, we may now obtain

$$\begin{aligned} u_\theta = \frac{1}{E}\Big\{ & 4d_0\theta r + 2D_0\theta^2 r + \Big[b_1(1-\nu)(1-\ln r) + \frac{c_1(1+\nu)}{r^2} \\ & + d_1(5+\nu)r^2\Big]^{-\cos\theta}_{\sin\theta} - [2A_1(\ln r + \nu)]^{\sin\theta}_{\cos\theta} \\ & + B_1[(1-\nu) - (1+\nu)\ln r - \ln^2 r]^{\sin\theta}_{\cos\theta} \\ & - B_1[(1-\nu) - \ln r(1-\nu)]^{\theta\cos\theta}_{-\theta\sin\theta} + \sum_{n=2}^{\infty}[a_n n(1+\nu)r^{n-1} \\ & + b_n\{n(1+\nu)+4\}r^{n+1} + c_n n(1+\nu)r^{-(n+1)} \\ & + d_n\{n(1+\nu)-4\}r^{-(n-1)}]^{-\cos n\theta}_{\sin n\theta} - \int_\theta g(\theta)\,d\theta + f(r)\Big\}. \end{aligned} \tag{8-4.17}$$

Substituting equations 8-4.12, 8-4.16, and 8-4.17 into equation 8-4.15 yields

$$\begin{aligned} & -2(1+\nu)\frac{A_0}{r^2} + (1-\nu)B_0\left[\frac{-2\ln r + 1}{r^2}\right] + 4C_0 + D_0[4\ln r + 1 + 3\nu] \\ & + \left[\frac{4b_1}{r}\right]^{\cos\theta}_{-\sin\theta} - \left[\frac{2A_1}{r}(1-\nu)\right]^{\sin\theta}_{\cos\theta} + B_1\left[\frac{4}{r}\right]^{\theta\cos\theta}_{-\theta\sin\theta} + \left[B_1\frac{2\nu}{r}\right]^{\sin\theta}_{\cos\theta} \\ & + f'(r) + \frac{1}{r}\left\{g'(\theta) + \int_\theta g(\theta)\,d\theta\right\} - \frac{f(r)}{r} = 0, \end{aligned} \tag{8-4.18}$$

This leads to the following differential equations for the functions of integration:

$$rf'(r) - f(r) = 2(1+\nu)\frac{A_0}{r} + (1+\nu)B_0\left[\frac{2\ln r - 1}{r}\right] - 4C_0 r \qquad (8\text{-}4.19)$$

$$- D_0[4r\ln r + (1+\nu)r] + K$$

$$g'(\theta) + \int_\theta g(\theta)\,d\theta = -[4b_1]^{\cos\theta}_{-\sin\theta} + [2A_1(1-\nu) - B_1 2\nu]^{\sin\theta}_{\cos\theta} - [B_1 4]^{\theta\cos\theta}_{-\theta\sin\theta} - K. \qquad (8\text{-}4.20)$$

To solve equation 8-4.19, we introduce the change of variable $\xi = \ln r$,

$$\frac{df}{dr} = \frac{1}{r}\frac{df}{d\xi}$$

Therefore, equation 8-4.19 becomes

$$f'(\xi) - f(\xi) = 2(1+\nu)A_0 e^{-\xi} + (1+\nu)B_0(2\xi - 1)e^{-\xi} - (4C_0 + (1+\nu)D_0)e^{\xi} - 4D_0\xi e^{\xi} + K.$$

Solving yields

$$f(r) = R_1 r - (1+\nu)B_0\frac{\ln r}{r^2} - 4C_0 r\ln r - (1+\nu)D_0 r\ln r - \frac{(1+\nu)A_0}{r} - 2D_0 r\ln^2 r - K, \qquad (8\text{-}4.21)$$

where R_1 is an arbitrary constant.

Treating $\int_\theta g(\theta)\,d\theta$ as the unknown function, the solution of 8-4.20 is

$$\int_\theta g(\theta)\,d\theta = -[2b_1]^{\theta\sin\theta}_{\theta\cos\theta} - [B_1]^{\theta^2\sin\theta}_{\theta^2\cos\theta} - [A_1(1-\nu) + B_1(1+\nu)]^{\theta\cos\theta}_{-\theta\sin\theta} + S_1\sin\theta + S_2\cos\theta - K, \qquad (8\text{-}4.22)$$

where S_1 and S_2 are arbitrary constants.

Differentiating equation 8-4.22 yields

$$g(\theta) = -[2b_1]^{\theta\cos\theta}_{-\theta\sin\theta} - [2b_1]^{\sin\theta}_{\cos\theta} - [B_1]^{(\theta^2\cos\theta + 2\theta\sin\theta)}_{(-\theta^2\sin\theta + 2\theta\cos\theta)} - [A_1(1-\nu) + B_1(1+\nu)]^{(\cos\theta - \theta\sin\theta)}_{(-\sin\theta - \theta\cos\theta)} + S_1\cos\theta - S_2\sin\theta. \qquad (8\text{-}4.23)$$

We now examine terms that lead to multivalued displacements or stresses. Examining equations 8-4.10, 8-4.11, and 8-4.12, we note that the coefficients B_0, C_0, D_0, and B_1 must be selected to be zero if we wish single-valued stresses, that is, for cases where the origin of the system lies within the body under consideration. In a simply connected region, these terms are single valued if the origin is placed outside the body or on its boundary. In this case they should be included for a complete set of integer solutions. Examining equations 8-4.16, 8-4.17, 8-4.21, 8-4.22, and 8-4.23 indicates that the coefficients d_0, B_0, C_0, D_0, A_1, B_1, and b_1 multiply multivalued displacement terms. For a problem requiring single-valued stresses and displacements, the following conditions must hold:

$$d_0 = B_0 = C_0 = D_0 = B_1 = 0, \qquad (8\text{-}4.24)$$

and b_1 and A_1 must be related in the following manner:

$$b_1^{\text{upper}} = \frac{(1-\nu)}{2} A_1^{\text{lower}}, \qquad (8\text{-}4.25)$$

$$b_1^{\text{lower}} = -\frac{(1-\nu)}{2} A_1^{\text{upper}}, \qquad (8\text{-}4.26)$$

where "upper" and "lower" indicate the upper or lower multiplier of the coefficient, as is given in the solution shown in equation 8-4.9.

The physical significance of some of the terms may be better understood if we consider a small circle of radius a, including the origin (Figure 8-5). The resultant of the traction from the material inside a acting on the

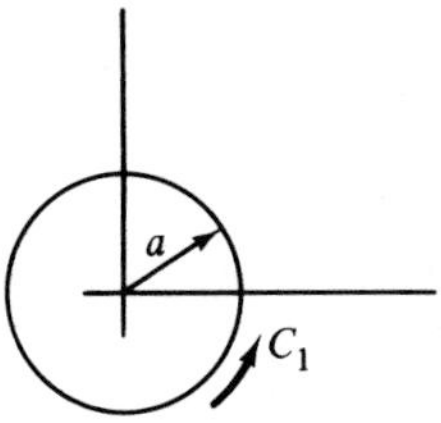

Figure 8-5

material exterior to the contour C_1 is

$$\oint_{C_1} \hat{n}\cdot\bar{\bar{\sigma}}\, ds = \bar{R}, \qquad (8\text{-}4.27)$$

where $\quad \hat{n} = -\hat{\imath}_{(r)}, \qquad ds = a\, d\theta,$

$$\bar{\bar{\sigma}} = \hat{\imath}_{(r)}\sigma_{rr}\hat{\imath}_{(r)} + \hat{\imath}_{(r)}\sigma_{r\theta}\hat{\imath}_{(\theta)} + \hat{\imath}_{(\theta)}\sigma_{\theta\theta}\hat{\imath}_{(\theta)} + \hat{\imath}_{(\theta)}\sigma_{\theta r}\hat{\imath}_{(r)}.$$

Therefore, the integral becomes

$$-\oint_{C_1} (\sigma_{rr}\hat{\imath}_{(r)} + \sigma_{r\theta}\hat{\imath}_{(\theta)})a\, d\theta = \bar{R}. \qquad (8\text{-}4.28)$$

Since $\hat{\imath}_{(r)}$ and $\hat{\imath}_{(\theta)}$ are functions of θ, it is easier to convert these to Cartesian coordinates:

$$\hat{\imath}_{(r)} = \hat{\imath}_{(x)} \cos\theta + \hat{\imath}_{(y)} \sin\theta,$$
$$\hat{\imath}_{(\theta)} = -\hat{\imath}_{(x)} \sin\theta + \hat{\imath}_{(y)} \cos\theta.$$

Therefore,

$$\bar{R} = -\oint_{C_1} [\{\sigma_{rr}\cos\theta - \sigma_{r\theta}\sin\theta\}\hat{\imath}_{(x)} + \{\sigma_{rr}\sin\theta + \sigma_{r\theta}\cos\theta\}\hat{\imath}_{(y)}]a\, d\theta. \qquad (8\text{-}4.29)$$

Using equations 8-4.10 and 8-4.11 evaluated at $r = a$, and omitting terms that are not single valued in stresses, yields

$$\bar{R} = -2\pi A_1^{\text{upper}} \hat{i}_{(x)} + 2\pi A_1^{\text{lower}} \hat{i}_{(y)},$$

$$A_1^{\text{upper}} = -\frac{R_x}{2\pi}, \qquad (8\text{-}4.30)$$

$$A_1^{\text{lower}} = \frac{R_y}{2\pi}.$$

We now see that the two A_1 terms correspond to concentrated forces at the origin. We have seen that the A_1 terms lead to non-single-valued displacements unless the b_1 terms are also included. The stress function corresponding to concentrated forces at the origin or a distribution of stresses around an internal contour about the origin are

$$\varphi = -\frac{R_x}{2\pi}\left[r\theta \sin\theta - \frac{(1-\nu)}{2} r \ln r \cos\theta\right], \qquad (8\text{-}4.31)$$

$$\varphi = \frac{R_y}{2\pi}\left[r\theta \cos\theta + \frac{(1-\nu)}{2} r \ln r \sin\theta\right]. \qquad (8\text{-}4.32)$$

If the resultant force at the origin is zero, then the terms A_1 and b_1 must be taken to be zero. If the origin is included in the body, the stresses and displacements are infinite at the origin due to concentrated loads there. Thus, to preserve elasticity, we can think of R_y as the resultant of a distribution of tractions applied to a very small hole of radius a. When R_x and R_y are infinitesimals, dR_x and dR_y, they may be integrated over infinitesimal lines or areas. The latter removes their singular nature, and we no longer need the device of excluding the singularity with a small hole. This point of view can be rationalized by Kelvin's trick of considering R_x and R_y to be the resultants of distributions of body forces over a small region (circle of radius a) that is not a hole. In other words, the hole-exclusion device is needed to stay within the framework of elasticity theory for finite R_x and R_y, but for distributed tractions (or body forces) of intensity per unit area $p_x(x, y)$ and $p_y(x, y)$, or $p_x(r, \theta)$ and $p_y(r, \theta)$, we are not tied to the hole. The product $p_x\, dA$ ($dA = dx\, dy$ or $r\, dr\, d\theta$ is an area element) is still an infinitesimal, and vanishes with dA, despite the nominal local singularity. When integrated over the region to which such distributions are applied, the corresponding stresses and displacements are well behaved, and acceptable finite resultants are obtained.

Written in xy terms, the R_y stress function 8-4.32 becomes

$$\varphi = \frac{R_y}{2\pi}\left\{x \tan^{-1}\left(\frac{y}{x}\right) + \frac{(1-\nu)}{2} y \ln\sqrt{x^2+y^2}\right\}. \qquad (8\text{-}4.33)$$

As an example of the integration of this to generate a line loading parallel to y, of variation $f(x)$; $-a \leq x \leq b$, the corresponding stress function for the "loadette" $f(\xi)\, d\xi$ at $x = \xi$ is

$$d\varphi = \frac{f(\xi)\,d\xi}{2\pi}\left\{(x-\xi)\tan^{-1}\left(\frac{y}{x-\xi}\right) + \frac{1-\nu}{2}y\ln[(x-\xi)^2+y^2]\right\}.$$

and

$$\varphi = \frac{1}{2\pi}\int_{-a}^{b} f(\xi)\left\{(x-\xi)\tan^{-1}\left(\frac{y}{x-\xi}\right) + \frac{1-\nu}{2}\ln[(x-\xi)^2+y^2]\right\}d\xi.$$

This retains a degree of singularity in the stresses, as a line loading.

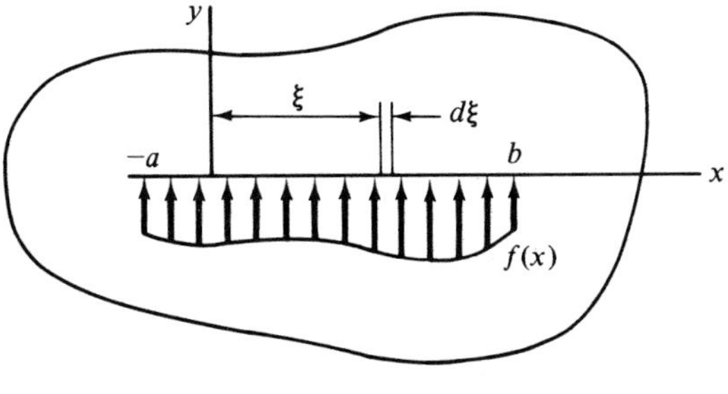

Figure 8-6

An areal loading of intensity $g(x, y)$ over the rectangle $-a \le x \le b$, $-c \le y \le d$ does not have this singularity:

$$\varphi = \frac{1}{2\pi}\int_{-c}^{d} d\eta \int_{-a}^{b} d\xi\left\{g(\xi,\eta)\left[(x-\xi)\tan^{-1}\left(\frac{y-\eta}{x-\xi}\right) + \frac{1-\nu}{2}(y-\eta)\ln[(x-\xi)^2+(y-\eta)^2]\right]\right\}.$$

Similar integrations can be carried out in polar coordinates.

5 Examples using the Michell solution*

5.1 Interior problem—stresses distributed around the edge of a solid disk. Consider the problem of a disk loaded by some stress distribution on the boundary $r = r_0$ (Figure 8-7). The boundary conditions are

$$\sigma_{rr} = p(\theta), \qquad r = r_0, \tag{8-5.1.1}$$

$$\sigma_{r\theta} = q(\theta), \qquad r = r_0. \tag{8-5.1.2}$$

Since the solution must be bounded at the origin, the coefficients of all terms singular at this point must be equated to zero. The stresses and displacements must also be single valued. Therefore, φ reduces to

$$\varphi = c_0 r^2 + [d_1 r^3]^{\sin\theta}_{\cos\theta} + \sum_{n=2}^{\infty}[a_n r^n + b_n r^{(2+n)}]^{\sin n\theta}_{\cos n\theta}. \tag{8-5.1.3}$$

* Analogous solutions will be developed by use of complex variable theory in Chapter 9.

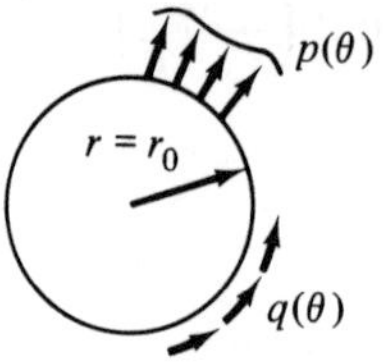

Figure 8-7

The stresses σ_{rr} and $\sigma_{r\theta}$ become

$$\sigma_{rr} = 2c_0 + [2d_1 r]^{\sin\theta}_{\cos\theta} - \sum_{n=2}^{\infty} \{a_n n(n-1)r^{n-2} + b_n(n+1)(n+2)r^n\}^{\sin n\theta}_{\cos n\theta}, \tag{8-5.1.4}$$

$$\sigma_{r\theta} = -[2d_1 r]^{\cos\theta}_{-\sin\theta} - \sum_{n=2}^{\infty} \{a_n n(n-1)r^{n-2} + b_n n(n+1)r^n\}^{\cos n\theta}_{-\sin n\theta}. \tag{8-5.1.5}$$

The boundary functions $p(\theta)$ and $q(\theta)$ must have some restrictions placed upon them by equilibrium, and these restrictions may be seen in the form of the stresses of 8-5.1.4 and 8-5.1.5. Note there is no constant term in the shear stress expansion. The term would correspond to an average or a constant shear around the disk and would produce an unbalancing moment. Equilibrium prohibits such a moment, so that the average value of $q(\theta)$ must be zero for the problem to be one of elastostatics. Another equilibrium restriction is that the resultant forces in the x and y directions must be zero. This is a restriction which interlocks $p(\theta)$ and $q(\theta)$. If we examine the resultants of the series expressions on $r = r_0$ (see section 4), we obtain

$$\begin{aligned} R_x &= \int_0^{2\pi} (\sigma_{rr} \cos\theta - \sigma_{r\theta} \sin\theta) r \, d\theta, \\ &= \int_0^{2\pi} 2d_1' r_0^2 (\cos^2\theta - \sin^2\theta) \, d\theta = 0, \end{aligned} \tag{8-5.1.6}$$

and

$$R_y = \int_0^{2\pi} (\sigma_{rr} \sin\theta + \sigma_{r\theta} \cos\theta) r_0 \, d\theta = 0. \tag{8-5.1.7}$$

This explains why the d_1 term is the same in both the σ_{rr} and $\sigma_{r\theta}$ expressions while there are two independent constants multiplying every other θ-dependent term. σ_{rr} may be written as a Fourier series as

$$\sigma_{rr} = \frac{A_0}{2} + \sum A_n \cos n\theta + \sum B_n \sin n\theta, \tag{8-5.1.8}$$

whence, using primes for cosine terms, and no primes for sine terms,

$$A_0 = 4c_0, \tag{8-5.1.9}$$

$$A_1 = 2d_1' r_0, \tag{8-5.1.10}$$

$$A_n = -\{a_n' n(n-1)r_0^{n-2} + b_n'(n+1)(n-2)r_0^n\}, \qquad n \geq 2, \tag{8-5.1.11}$$

$$B_1 = 2d_1 r_0, \tag{8-5.1.12}$$

$$B_n = -\{a_n n(n-1)r_0^{n-2} + b_n(n+1)(n-2)r_0^n\}, \qquad n \geq 2. \tag{8-5.1.13}$$

The constants a'_n and a_n are different constants, the first designating the coefficient of the cosine terms and the second the coefficient of the sine terms. The same identification holds for d'_1, d_1, b'_n, and b_n. Applying the first boundary condition yields

$$4c_0 = \frac{1}{\pi}\int_0^{2\pi} p(\theta)\,d\theta, \tag{8-5.1.14}$$

$$2d'_1 r_0 = \frac{1}{\pi}\int_0^{2\pi} p(\theta)\cos\theta\,d\theta, \tag{8-5.1.15}$$

$$-\{a'_n n(n-1)r_0^{n-2} + b'_n(n+1)(n-2)r_0^n\} = \frac{1}{\pi}\int_0^{2\pi} p(\theta)\cos n\theta\,d\theta, \tag{8-5.1.16}$$

$$2d_1 r_0 = \frac{1}{\pi}\int_0^{2\pi} p(\theta)\sin\theta\,d\theta, \tag{8-5.1.17}$$

$$-\{a_n n(n-1)r_0^{n-2} + b_n(n+1)(n-2)r_0^n\} = \frac{1}{\pi}\int_0^{2\pi} p(\theta)\sin n\theta\,d\theta. \tag{8-5.1.18}$$

In a similar manner, the shear boundary condition yields

$$2d'_1 r_0 = \frac{1}{\pi}\int_0^{2\pi} q(\theta)\sin\theta\,d\theta, \tag{8-5.1.19}$$

$$\{a'_n n(n-1)r_0^{n-2} + b'_n n(n+1)r_0^n\} = \frac{1}{\pi}\int_0^{2\pi} q(\theta)\sin n\theta\,d\theta, \tag{8-5.1.20}$$

$$-2d_1 r_0 = \frac{1}{\pi}\int_0^{2\pi} q(\theta)\cos\theta\,d\theta, \tag{8-5.1.21}$$

$$\{a_n n(n-1)r_0^{n-2} + b_n n(n+1)r_0^n\} = \frac{1}{\pi}\int_0^{2\pi} q(\theta)\cos\theta\,d\theta. \tag{8-5.1.22}$$

Examining equations 8-5.1.15 and 8-5.1.19 gives

$$d'_1 = \frac{1}{2\pi r_0}\int_0^{2\pi} p(\theta)\cos\theta\,d\theta = \frac{1}{2\pi r_0}\int_0^{2\pi} q(\theta)\sin\theta\,d\theta. \tag{8-5.1.23}$$

This evaluates the constant d'_1 and establishes the equilibrium constraint on $p(\theta)$ and $q(\theta)$. In a similar manner,

$$d_1 = \frac{1}{2\pi r_0}\int_0^{2\pi} p(\theta)\sin\theta\,d\theta = -\frac{1}{2\pi r_0}\int_0^{2\pi} q(\theta)\cos\theta\,d\theta. \tag{8-5.1.24}$$

From equation 8-5.1.14, we obtain

$$c_0 = \frac{1}{4\pi}\int_0^{2\pi} p(\theta)\,d\theta. \tag{8-5.1.25}$$

Equations 8-5.1.16 and 8-5.1.20 yield

$$a'_n = -\frac{1}{\pi 2n(n-1)r_0^{n-2}}\int_0^{2\pi}\{np(\theta)\cos n\theta + (n-2)q(\theta)\sin n\theta\}\,d\theta, \tag{8-5.1.26}$$

$$b'_n = \frac{1}{\pi 2(n+1)r_0^n}\int_0^{2\pi}\{p(\theta)\cos n\theta + q(\theta)\sin n\theta\}\,d\theta. \tag{8-5.1.27}$$

Equations 8-5.1.18 and 8-5.1.22 yield

$$a_n = -\frac{1}{\pi 2n(n-1)r_0^{n-2}}\int_0^{2\pi}\{np(\theta)\sin n\theta - (n-2)q(\theta)\cos n\theta\}\,d\theta, \tag{8-5.1.28}$$

$$b_n = \frac{1}{\pi 2(n+1)r_0^n}\int_0^{2\pi}\{p(\theta)\sin n\theta - q(\theta)\cos n\theta\}\,d\theta. \tag{8-5.1.29}$$

This evaluates all the coefficients. The stresses and displacements may be calculated for any specified conditions $p(\theta)$ and $q(\theta)$.

5.2 Exterior problem—infinite plane with circular hole. We have considered the interior problem and have found that we can represent stress distributions which are sectionally continuous by Fourier trigonometric expansions. Many terms in the Michell solution were dropped due to a boundedness condition at the origin. In a similar manner, we impose a boundedness condition at infinity in the exterior problem. We also require that the stresses and displacements be single-valued functions. If we require the stresses to be bounded at infinity, the Airy stress function takes the form

$$\varphi = b_0 \ln r + c_0 r^2 + A_0\theta + \left(b_1 r \ln r + \frac{c_1}{r}\right)_{\cos\theta}^{\sin\theta} + A_1\theta_{\cos\theta}^{\sin\theta}$$
$$+ \sum_{n=2}^{\infty}(c_n r^{-n} + d_n r^{2-n})_{\cos n\theta}^{\sin n\theta} + (a_2 r^2)_{\cos 2\theta}^{\sin 2\theta}. \tag{8-5.2.1}$$

We have noted that b_1 and A_1 must be related for single-valued displacements. We have also observed that these represent the resultant forces in the x and y directions, on the interior of the hole. If these resultants are nonzero, the displacements are not bounded at infinity (which might be predicted by physical considerations). The stresses may be written

$$\sigma_{rr} = 2c_0 - (2a_2)_{\cos 2\theta}^{\sin 2\theta} - \frac{R_y}{2\pi}\frac{(3+\nu)}{2}\frac{\sin\theta}{r} - \frac{R_x}{2\pi}\frac{(3+\nu)}{2}\frac{\cos\theta}{r} + \frac{b_0}{r^2}$$
$$- \left(\frac{2c_1}{r^3}\right)_{\cos\theta}^{\sin\theta} - \sum_{n=2}^{\infty}[c_n n(n+1)r^{-(n+2)} + d_n(n-1)(n+2)r^{-n}]_{\cos n\theta}^{\sin n\theta}, \tag{8-5.2.2}$$

$$\sigma_{\theta\theta} = 2c_0 + (2a_2)_{\cos 2\theta}^{\sin 2\theta} + \frac{R_y}{2\pi}\frac{(1-\nu)}{2}\frac{\sin\theta}{r} + \frac{R_x}{2\pi}\frac{(1-\nu)}{2}\frac{\cos\theta}{r} - \frac{b_0}{r^2}$$
$$+ \left(\frac{2c_1}{r^3}\right)_{\cos\theta}^{\sin\theta} + \sum_{n=2}^{\infty}[c_n n(n+1)r^{-(n+2)} + d_n(n-2)(n-1)r^{-n}]_{\cos n\theta}^{\sin n\theta}, \tag{8-5.2.3}$$

$$\sigma_{r\theta} = -(2a_2)^{\cos 2\theta}_{-\sin 2\theta} - \frac{R_y}{2\pi}\frac{(1-\nu)}{2}\frac{\cos\theta}{r} - \frac{R_x}{2\pi}\frac{(1-\nu)}{2}\frac{\sin\theta}{r} + \frac{A_0}{r^2}$$

$$+ \left(\frac{2c_1}{r^3}\right)^{\cos\theta}_{-\sin\theta} + \sum_{n=2}^{\infty}[c_n n(n+1)r^{-(n+2)} + d_n n(n-1)r^{-n}]^{\cos n\theta}_{-\sin n\theta}. \tag{8-5.2.4}$$

For large values of r, the solution approaches

$$\sigma_{rr} = 2\{c_0 - a_2 \sin 2\theta - a_2' \cos 2\theta\}, \tag{8-5.2.5}$$

$$\sigma_{\theta\theta} = 2\{c_0 + a_2 \sin 2\theta + a_2' \cos 2\theta\}, \tag{8-5.2.6}$$

$$\sigma_{r\theta} = -2\{a_2 \cos 2\theta - a_2' \sin 2\theta\}. \tag{8-5.2.7}$$

If XX, YY, and XY represent the Cartesian components of the state of stress at infinity, these stresses can be written by using the stress transformation equations and noting that

$$c_0 = \frac{XX + YY}{4}, \qquad a_2' = \frac{YY - XX}{4}, \qquad a_2 = -\frac{XY}{2}. \tag{8-5.2.8}$$

The stresses become

$$\begin{aligned}\sigma_{rr} &= \tfrac{1}{2}\{XX + YY - (YY - XX)\cos 2\theta + 2XY \sin 2\theta\},\\ \sigma_{\theta\theta} &= \tfrac{1}{2}\{XX + YY + (YY - XX)\cos 2\theta - 2XY \sin 2\theta\},\\ \sigma_{r\theta} &= \tfrac{1}{2}\{(YY - XX)\sin 2\theta + 2XY \cos 2\theta\}.\end{aligned} \tag{8-5.2.9}$$

The stresses σ_{rr} and $\sigma_{r\theta}$ are now

$$\sigma_{rr} = \frac{XX+YY}{2} - \frac{YY-XX}{2}\cos 2\theta + XY \sin 2\theta - \frac{3+\nu}{4\pi r}[R_y \sin\theta + R_x \cos\theta]$$

$$+ \frac{b_0}{r^2} - \frac{2c_1'}{r^3}\cos\theta - \sum_{n=2}^{\infty}[c_n' n(n+1)r^{-(n+2)} + d_n'(n-1)(n+2)r^{-n}]\cos n\theta$$

$$- \frac{2c_1}{r^3}\sin\theta - \sum_{n=2}^{\infty}[c_n n(n+1)r^{-(n+2)} + d_n(n-1)(n+2)r^{-n}]\sin n\theta, \tag{8-5.2.10}$$

$$\sigma_{r\theta} = \frac{YY - XX}{2}\sin 2\theta + XY\cos 2\theta - \frac{(1-\nu)}{4\pi r}[R_y \cos\theta + R_x \sin\theta]$$

$$+ \frac{A_0}{r^2} + \frac{2c_1}{r^3}\cos\theta + \sum_{n=2}^{\infty}[c_n n(n+1)r^{-(n+2)} + d_n n(n-1)r^{-n}]\cos n\theta$$

$$- \frac{2c_1'}{r^3}\sin\theta - \sum_{n=2}^{\infty}[c_n' n(n+1)r^{-(n+2)} + d_n' n(n-1)r^{-n}]\sin n\theta. \tag{8-5.2.11}$$

On surfaces $r =$ constant these stresses have the form of a Fourier series plus added terms representing the resultant forces of the boundary stresses on the external material and the conditions at infinity.

As an example of the use of this form of solution, consider a very large plate in uniform tension in the x direction with a small hole at its center

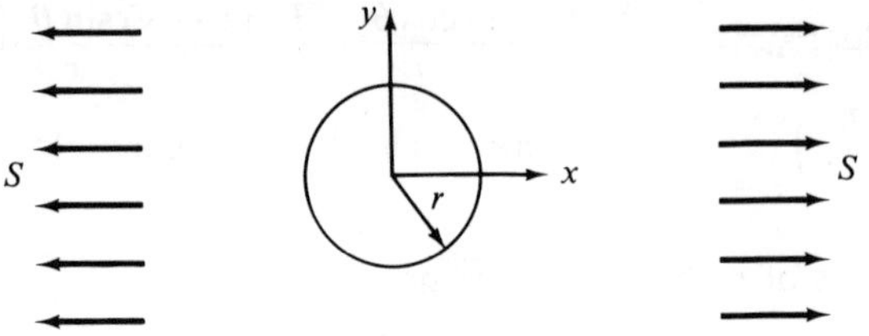

Figure 8-8

of radius r_0 (Figure 8-8).* We can consider this to be an infinite plate with the following boundary conditions:

$$XX = S, \qquad YY = XY = 0$$

$$\sigma_{rr} = 0, \sigma_{r\theta} = 0, \qquad \text{at } r = r_0.$$

Therefore, the resultant forces R_x and R_y are zero and the stresses at $r = r_0$ take the form

$$\sigma_{rr} = \frac{S}{2}(1 + \cos 2\theta) + \text{series},$$

$$\sigma_{r\theta} = -\frac{S}{2}(\sin 2\theta) + \text{series}.$$

It is apparent that we only need two terms from the Fourier trigonometric series and the solution may be written as

$$\sigma_{rr} = \frac{S}{2}(1 + \cos 2\theta) + \frac{b_0}{r^2} - (6c_2' r^{-4} + 4d_2' r^{-2}) \cos 2\theta,$$

$$\sigma_{r\theta} = -\frac{S}{2} \sin 2\theta - (6c_2' r^{-4} + 2d_2' r^{-2}) \sin 2\theta.$$

The boundary conditions at $r = r_0$ can be satisfied by choosing b_0, c_2', and d_2' as follows:

$$b_0 = -\frac{Sr_0^2}{2}, \qquad d_2' = \frac{Sr_0^2}{2}, \qquad c_2' = -\frac{Sr_0^4}{4}.$$

The stresses become

$$\sigma_{rr} = \frac{S}{2}\left(1 - \frac{r_0^2}{r^2}\right) + \frac{S}{2}\left(1 + \frac{3r_0^4}{r^4} - \frac{4r_0^2}{r^2}\right) \cos 2\theta,$$

$$\sigma_{\theta\theta} = \frac{S}{2}\left(1 + \frac{r_0^2}{r^2}\right) - \frac{S}{2}\left(1 + \frac{3r_0^4}{r^4}\right) \cos 2\theta,$$

$$\sigma_{r\theta} = -\frac{S}{2}\left(1 - \frac{3r_0^4}{r^4} + 2\frac{r_0^2}{r^2}\right) \sin 2\theta.$$

Note that at $r = r_0$, the $\sigma_{\theta\theta}$ stress is

$$\sigma_{\theta\theta} = S(1 - 2 \cos 2\theta).$$

* This solution is credited to G. Kirsch, *VDI*, **42**, 1898.

The maximum value of the stress on the edge of the hole occurs at $\theta = 90°$, where

$$\sigma_{\theta\theta} = 3S;$$

at $\theta = 0°$

$$\sigma_{\theta\theta} = -S.$$

For an infinite plate without a hole the stress at $\theta = 90°$ would be

$$\sigma_{\theta\theta} = S.$$

We therefore refer to a stress concentration of 3 due to the presence of the hole.

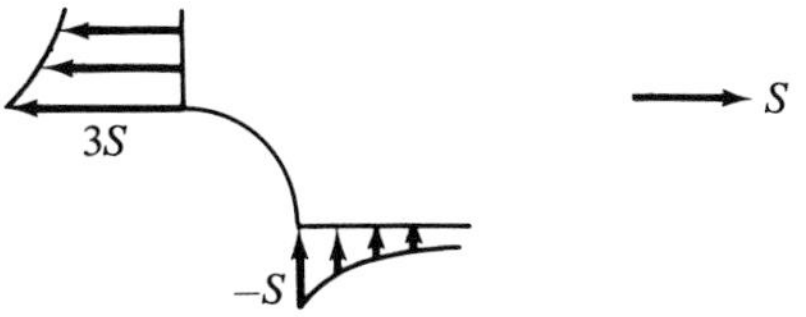

Figure 8-9

We may obtain other solutions by use of the principle of linear superposition. If we wish the solution for uniform tension in the y direction, we replace θ by $(\theta - 90)$ into the results. By this method, the cases of biaxial tension and shear at infinity are easily obtained. These solutions could also be obtained from the general solution.*

5.3 Annulus problem. If we consider problems involving an annulus bounded by the radii r_1 and r_2, we have no conditions of regularity either at the origin or at infinity. If we do not consider dislocations in the displacements, the solution must also be single valued. The stress function (equation 8-4.9) becomes

$$\varphi = a_0 + b_0 \ln r + c_0 r^2 + A_0\theta + \left(a_1 r + \frac{c_1}{r} + d_1 r^3\right)_{\cos\theta}^{\sin\theta}$$
$$+ (b_1 r \ln r + A_1 r\theta)_{\cos\theta}^{\sin\theta} + \sum_{n=2}^{\infty} (a_n r^n + b_n r^{2+n} + c_n r^{-n} + d_n r^{2-n})_{\cos n\theta}^{\sin n\theta}. \tag{8-5.3.1}$$

The terms a_0 and $(a_1 r)_{\cos\theta}^{\sin\theta}$ correspond to a linear function and do not enter into the stresses. b_1 and A_1 are still related to the resultant of the force acting on the inside of the annulus and need not be zero. The body will be in equilibrium if the negative of this resultant acts on the outer boundary.

* See S. P. Timoshenko and J. N. Goodier, *Theory of Elasticity*, 3rd ed., McGraw-Hill Book Company, New York, 1970. p. 93.

The stresses become

$$
\begin{aligned}
\sigma_{rr} = {} & \frac{b_0}{r^2} + 2c_0 + \left(-\frac{2c_1}{r^3} + 2d_1 r\right)^{\sin\theta}_{\cos\theta} - \frac{R_y}{2\pi}\frac{(3+\nu)}{2}\frac{\sin\theta}{\pi} \\
& - \frac{R_x}{2\pi}\frac{(3+\nu)}{2}\frac{\cos\theta}{r} - \sum_{n=2}^{\infty}\{a_n n(n-1)r^{n-2} + b_n(n+1)(n-2)r^n \\
& + c_n n(n+1)r^{-(n+2)} + d_n(n-1)(n+2)r^{-n}\}^{\sin n\theta}_{\cos n\theta},
\end{aligned}
\tag{8-5.3.2}
$$

$$
\begin{aligned}
\sigma_{\theta\theta} = {} & -\frac{b_0}{r^2} + 2c_0 + \left(\frac{2c_1}{r^3} + 6d_1 r\right)^{\sin\theta}_{\cos\theta} + \frac{R_y}{2\pi}\frac{(1-\nu)}{2}\frac{\sin\theta}{r} \\
& + \frac{R_x}{2\pi}\frac{(1-\nu)}{2}\frac{\cos\theta}{r} + \sum_{n=2}^{\infty}\{a_n n(n-1)r^{n-2} + b_n(n+1)(n+2)r^n \\
& + c_n n(n+1)r^{-(n+2)} + d_n(n-2)(n-1)r^{-n}\}^{\sin n\theta}_{\cos n\theta},
\end{aligned}
\tag{8-5.3.3}
$$

$$
\begin{aligned}
\sigma_{r\theta} = {} & \frac{A_0}{r^2} + \left(\frac{2c_1}{r^3} - 2d_1 r\right)^{\cos\theta}_{-\sin\theta} - \frac{R_y}{2\pi}\frac{(1-\nu)}{2}\frac{\cos\theta}{r} \\
& - \frac{R_x}{2\pi}\frac{(1-\nu)}{2}\frac{\sin\theta}{r} - \sum_{n=2}^{\infty}\{a_n n(n-1)r^{n-2} + b_n n(n+1)r^n \\
& - c_n n(n+1)r^{-(n+2)} - d_n n(n-1)r^{-n}\}^{\cos n\theta}_{-\sin n\theta}.
\end{aligned}
\tag{8-5.3.4}
$$

The σ_{rr} and $\sigma_{r\theta}$ expressions may be visualized as two Fourier series for each boundary stress on each boundary. The solution may be algebraically complex but is conceptually simple.

5.4 Symmetry conditions. Frequently, geometric and loading symmetries may be used to simplify the solution. This is best shown by examples. We shall consider a few solid disk problems using the solution from section 8-5.1 to illustrate these points. Consider a disk loaded as shown in Figure 8-10. It may be easily seen that the radial stress at some angle $\theta = +\beta$ would

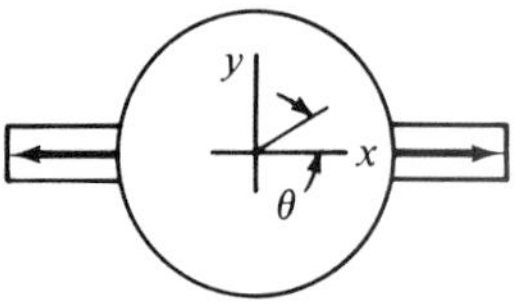

Figure 8-10

be the same as that at the angle $\theta = -\beta$. We can therefore write the stress function as

$$\varphi(r, \theta) = \varphi(r, -\theta). \tag{8-5.4.1}$$

This indicates that φ must be an even function of θ. We would use only the cosine series in our Fourier expansion. We can also note that there is symmetry about the y axis, expressed as follows:

$$\varphi(r, \theta) = \varphi(r, \pi - \theta). \tag{8-5.4.2}$$

Since the theta-dependency involves cosine terms, this indicates that only those terms which admit the relation

$$\cos n\theta = \cos n(\pi - \theta) \tag{8-5.4.3}$$

will enter the solution. For even integers of n, this relations is satisfied. Our solution in this case becomes (equation 8-5.1.3)

$$\varphi = c_0 r^2 + \sum_{n=2,4,6} [a_n r^n + b_n r^{(2+n)}] \cos n\theta. \tag{8-5.4.4}$$

We need consider only the first quadrant and the balance of the solution will proceed as before.

If the disk is loaded as shown in Figure 8-11, we obtain additional symmetry. All the previous statements are valid, but, in addition, there is sym-

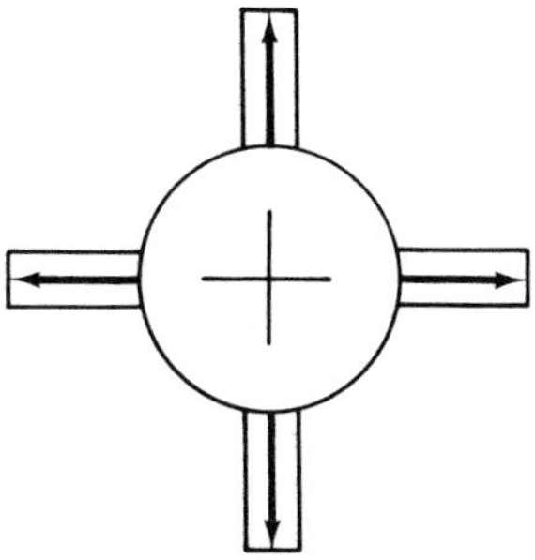

Figure 8-11

metry about the $\theta = 45°$ line. This may be expressed as

$$\varphi(r, \theta) = \varphi\left(r, \frac{\pi}{2} - \theta\right), \tag{8-5.4.5}$$

or, in terms of the cosine functions,

$$\cos n\theta = \cos n\left(\frac{\pi}{2} - \theta\right). \tag{8-5.4.6}$$

This is true for $n = 0, 4, 8, 12, \ldots$, and the solution is simplified to the consideration of the first half-quadrant, or $0 \leq \theta \leq 45°$. If one recognizes that symmetry conditions frequently exist, the mathematical labor of solution of many problems can be reduced.

6 General solutions not involving orthogonal functions

An illustration of a more general problem can be shown by considering the stress concentration existing in a rectangular plate loaded in uniaxial tension with a hole in the center (Figure 8-12). The boundary conditions at the edge of the hole are

$$\sigma_{rr}(c, \theta) = 0, \tag{8-6.1}$$

$$\sigma_{r\theta}(c, \theta) = 0. \tag{8-6.2}$$

On the outer boundary, the conditions become

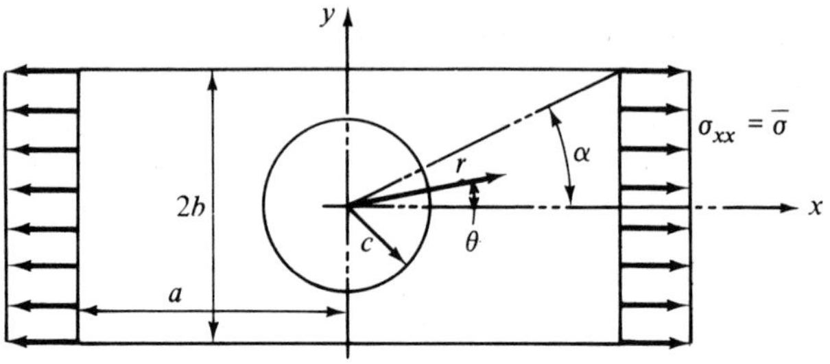

Figure 8-12

(a) for $0 < \theta < \alpha$,

$$\sigma_{xx}(a, y) = f(y) = \bar{\sigma}, \tag{8-6.3}$$

$$\sigma_{xy}(a, y) = 0; \tag{8-6.4}$$

(b) for $\alpha < \theta < \pi/2$,

$$\sigma_{yy}(x, b) = 0, \tag{8-6.5}$$

$$\sigma_{xy}(x, b) = 0, \tag{8-6.6}$$

Due to the symmetry of the problem, only one quadrant will be considered, and terms in the Michell solution corresponding to $n = 0, 2, 4, \ldots,$ will be used. The boundary conditions at the edge of the hole may be written in terms of the Airy stress function as

$$\varphi(c, \theta) = 0, \tag{8-6.7}$$

$$\frac{\partial \varphi}{\partial r}(c, \theta) = 0. \tag{8-6.8}$$

Imposing these conditions on the Michell solution after single-valued terms have been omitted yields

$$\varphi = a_0 + b_0 \ln\left(\frac{r}{b}\right) + c_0\left(\frac{r}{b}\right)^2 + \sum_{n=2,4,6}\left\{a_n\left(\frac{r}{b}\right)^n + b_n\left(\frac{r}{b}\right)^{n+2} + c_n\left(\frac{r}{b}\right)^{-n} + d_n\left(\frac{r}{b}\right)^{2-n}\right\}\cos n\theta. \tag{8-6.9}$$

The constants are related as follows:

$$a_0 = b_0\{\tfrac{1}{2} - \ln\gamma\}, \tag{8-6.10}$$

$$c_0 = -b_0/(2\gamma^2), \tag{8-6.11}$$

and, for $n \geq 2$,

$$a_n = -c_n(n+1)\gamma^{-2n} - d_n n\gamma^{-2(n-1)}, \tag{8-6.12}$$

$$b_n = c_n n\gamma^{-2(n+1)} + d_n(n-1)\gamma^{-2n}, \tag{8-6.13}$$

where the stress function has been nondimensionalized for convenience and

$$\gamma = \frac{c}{b}. \tag{8-6.14}$$

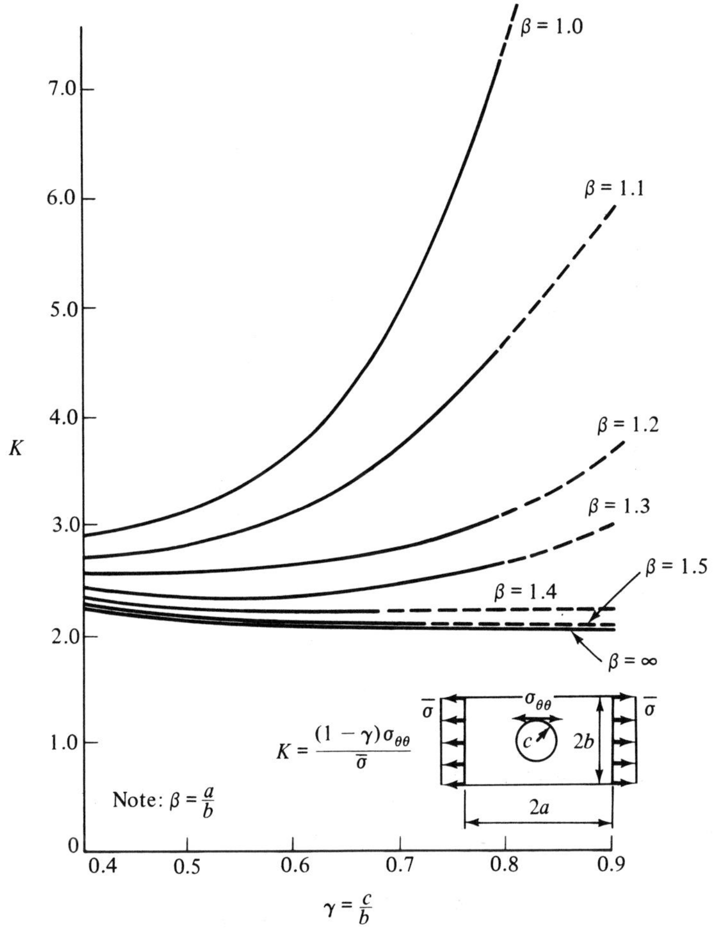

Stress concentration factor
at edge of hole for $\theta = 90°$

Figure 8-13

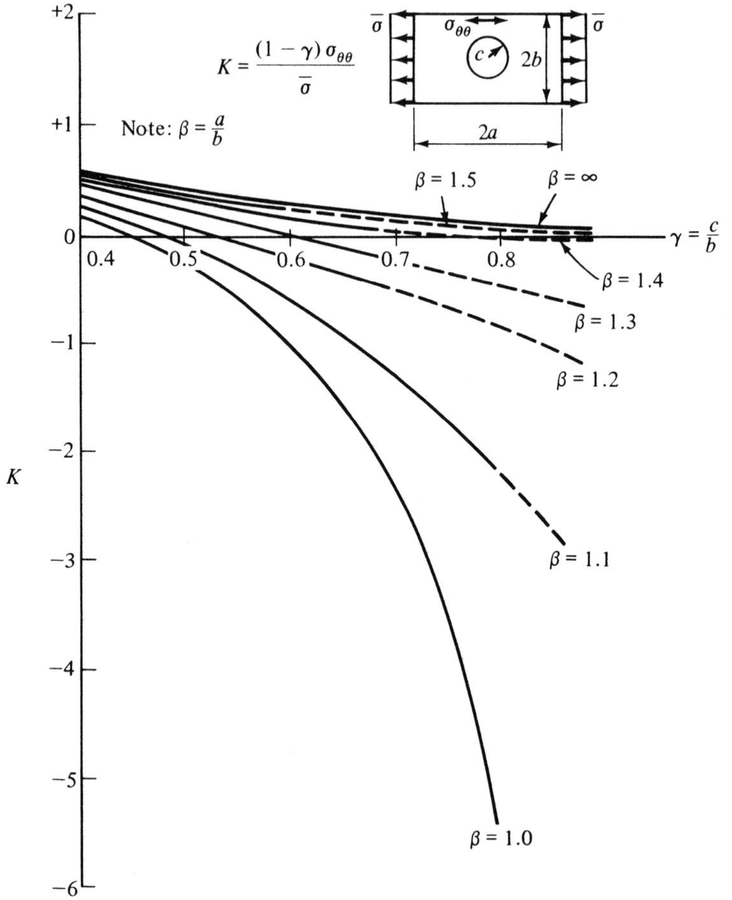

Stress concentration factor at outer boundary for $\theta = 90°$

Figure 8-14

The outer boundaries may be described by the relations

$$0 \leq \theta \leq \alpha, \qquad r = \frac{a}{\cos\theta}, \tag{8-6.15}$$

and

$$\alpha \leq \theta \leq 90°, \qquad r = \frac{b}{\sin\theta}. \tag{8-6.16}$$

Using the stress transformation equations, σ_{xx}, σ_{yy}, and σ_{xy} may be calculated and the outer boundaries may be satisfied in the least-squares sense:

$$\int_0^{\alpha}\left\{\left[\bar{\sigma}-\sigma_{xx}\left(\frac{a}{\cos\theta},\theta\right)\right]^2+\left[\sigma_{xy}\left(\frac{a}{\cos\theta}\right)\right]^2\right\}d\theta$$
$$+\int_{\alpha}^{\pi/2}\left\{\left[\sigma_{yy}\left(\frac{b}{\sin\theta},\theta\right)\right]^2+\left[\sigma_{yx}\left(\frac{b}{\sin\theta},\theta\right)\right]^2\right\}d\theta=\text{min.}\quad (8\text{-}6.17)$$

This problem has been solved by T. R. Thompson* and the solution is summarized in Figures 8-13 and 8-14.

7 Wedge problems

Another geometric configuration that is best approached by use of polar coordinates is a wedge-shaped body. The condition of periodicity is not always applicable in this case although integer value solutions may still be desirable so that boundaries, $\theta=$ constant, have power series in r as the boundary functions. Many problems of this type may be solved by careful selection of terms from the Michell solution. If the boundary functions may be expanded in a Taylor's series in r, these boundary conditions can be approximated by a least-squares analysis.

7.1 Wedge under uniform side load. Consider the wedge shown in Figure 8-15.

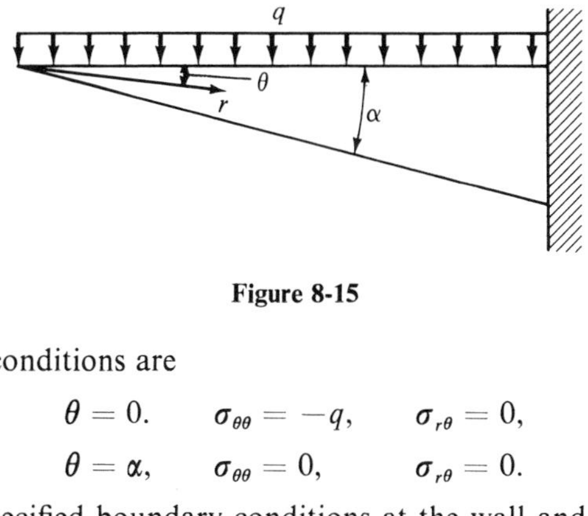

Figure 8-15

The boundary conditions are

$$\theta=0.\qquad \sigma_{\theta\theta}=-q,\qquad \sigma_{r\theta}=0,\qquad (8\text{-}7.1.1)$$

$$\theta=\alpha,\qquad \sigma_{\theta\theta}=0,\qquad \sigma_{r\theta}=0.\qquad (8\text{-}7.1.2)$$

We have not specified boundary conditions at the wall and in this example will accept the equilibrating stresses as a Saint-Venant approximation. $\sigma_{\theta\theta}$

* T. R. Thompson, "Elastostatic problem of a rectangular plate with a circular hole," M.S. dissertation, Oklahoma State University, May 1965. Also see A. L. Schlack and R. W. Little, "Elastostatic problem of a perforated square plate," *Proc. ASCE*, Eng. Mch. Div., October 1964.

See also R.C.T. Howland and Stevenson, "Bi-harmonic analysis in a Perforated strip," Phil. Trans. Roy. Soc. of London, Ser. A, 232, pp. 155–222, 1933 and G.B. Jefferey, Phil. Trans. Roy. Soc. of London, Ser. A., 221, pp. 265–293 1921.

is a constant on the boundaries, and we shall seek solutions that have no r dependence in $\sigma_{\theta\theta}$. Examining the general solution, we can initially pick a solution of the form

$$\sigma_{\theta\theta} = 2c_0 + 2C_0\theta + 2a_2 \sin 2\theta + 2a_2' \cos 2\theta. \qquad (8\text{-}7.1.3)$$

The stress function and the other stresses are

$$\varphi = c_0 r^2 + C_0 r^2\theta + a_2 r^2 \sin 2\theta + a_2' r^2 \cos 2\theta, \qquad (8\text{-}7.1.4)$$

$$\sigma_{r\theta} = -C_0 - 2a_2 \cos 2\theta + 2a_2' \sin 2\theta, \qquad (8\text{-}7.1.5)$$

$$\sigma_{rr} = 2c_0 + 2C_0\theta - 2a_2 \sin 2\theta - 2a_2' \cos 2\theta. \qquad (8\text{-}7.1.6)$$

We have four constants to satisfy the four boundary conditions, and after algebraic manipulations we obtain

$$\varphi = \frac{qr^2}{2}\left[\frac{1}{\tan\alpha - \alpha}\right]\{\sin\theta\cos\theta - \cos^2\theta\tan\alpha + (\alpha - \theta)\}. \qquad (8\text{-}7.1.7)$$

If other stress distributions were specified on the boundaries, we would have had to use more terms in the series. For each power of r, there are four terms multiplying it.*

7.2 Stress singularities at the tip of a wedge (M. L. Williams' soultion).† Consider the problem of a body loaded with some self-equilibrating stress distribution on the curved boundary r = constant. We shall investigate the nature of the singularities that might occur at the tip of the wedge (Figure 8-16). We have considered integer solutions in the development of the Michell

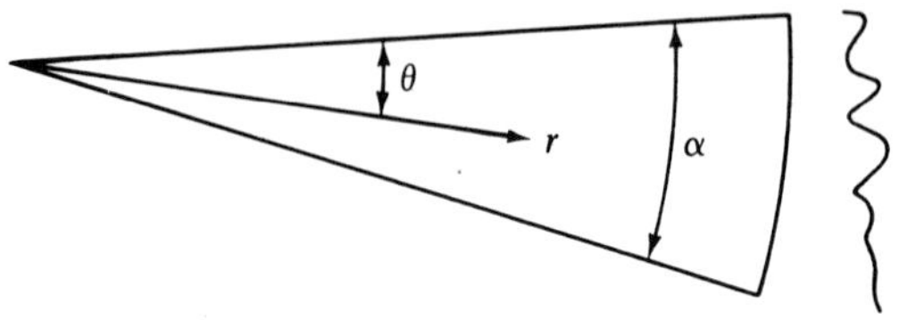

Figure 8-16

solution. If we examine equations 8-4.1 to 8-4.8, we see the solution of φ can be taken:

$$\varphi = r^{\lambda}[A \sin\lambda\theta + B\cos\lambda\theta + C\sin(\lambda - 2)\theta + D\cos(\lambda - 2)\theta]. \qquad (8\text{-}7.2.1)$$

This is obtained by setting $a = \lambda$ in equation 8-4.8.

If the wedge is stress free on the boundaries $\theta = 0$ and $\theta = \alpha$, the

* See Timoshenko and Goodier, *op. cit.*, p. 139.

† M. L. Williams, *J. Appl. Mech.*, **19**, 526, 1952.

constants A, B, C, and D must be interrelated. The boundary stresses are

$$\sigma_{\theta\theta} = \lambda(\lambda - 1)r^{(\lambda-2)}[A \sin \lambda\theta + B \cos \lambda\theta + C \sin (\lambda - 2)\theta + D \cos (\lambda - 2)\theta], \tag{8-7.2.2}$$

$$\sigma_{r\theta} = -(\lambda - 1)r^{(\lambda-2)}[A\lambda \cos \lambda\theta - B\lambda \sin \lambda\theta + C(\lambda - 2) \cos (\lambda - 2)\theta - D(\lambda - 2) \sin (\lambda - 2)\theta]. \tag{8-7.2.3}$$

To satisfy the condition at $\theta = 0$,

$$B = -D, \tag{8-7.2.4}$$

$$\lambda A = -(\lambda - 2)C. \tag{8-7.2.5}$$

The condition at $\theta = \alpha$ becomes

$$C\left[\sin (\lambda - 2)\alpha - \frac{\lambda - 2}{\lambda} \sin \lambda\alpha\right] + D [\cos (\lambda - 2)\alpha - \cos \lambda\alpha] = 0, \tag{8-7.2.6}$$

$$C[(\lambda - 2) \cos (\lambda - 2)\alpha - (\lambda - 2) \cos \lambda\alpha] - D[(\lambda - 2) \sin (\lambda - 2)\alpha - \lambda \sin \lambda\alpha] = 0. \tag{8-7.2.7}$$

To satisfy these homogeneous equations, λ must be the roots of the transcendental equation obtained by expanding the determinant of the coefficients and equating it to zero:

$$\sin (\lambda - 1)\alpha = \frac{\pm \sin \alpha}{\alpha}(\lambda - 1)\alpha. \tag{8-7.2.8}$$

If we had specified a clamped–clamped boundary, λ would be the roots of

$$\sin (\lambda - 1)\alpha = \pm\left[\frac{1 + \nu}{3 - \nu} \frac{\sin \alpha}{\alpha}\right](\lambda - 1)\alpha. \tag{8-7.2.9}$$

For the clamped-free case, the transcendental equation is

$$\sin^2 (\lambda - 1)\alpha = \left[\frac{4}{(1 + \nu)(3 - \nu)}\right] - \left[\frac{\sin^2 \alpha}{\alpha^2} \frac{(1 + \nu)}{(3 - \nu)}\right](\lambda - 1)^2\alpha^2. \tag{8-7.2.10}$$

The stresses all involve terms such as $r^{(\lambda-2)}$, and if λ is less than 2, the stresses will be singular at the origin. If we require that the displacements be bounded at the origin, λ must be greater than 1. Therefore, values of λ having a real part between 1 and 2 will produce singularities at the origin. Each of the transcendental equations has an infinite number of complex roots, but we shall examine only the lowest root and, in particular, its real part. The results from Williams's paper are summarized in Figure 8-17. Note that in the first two cases, singularities do not occur unless the angle is greater than 180°, in which case a reentrant corner exists. However, in the clamped-free case singularities occur for wedge angles of approximately 53° or more.

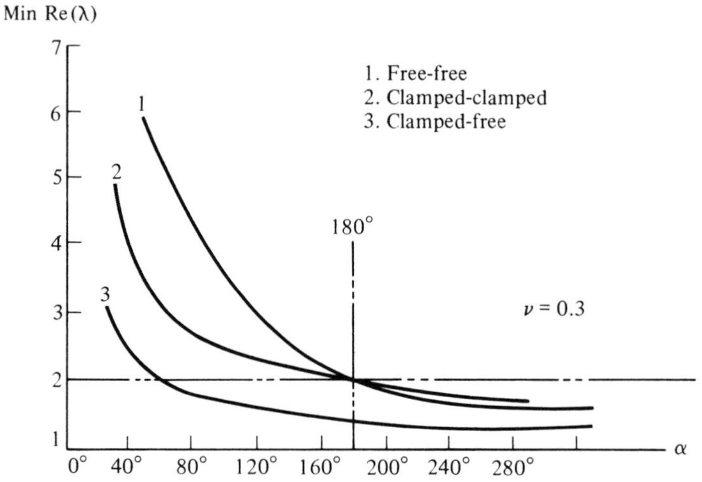

Figure 8-17

The free–free case with $\alpha \approx 2\pi$ is the important case covering the singularity at the root of a crack. One can determine the nature of the singularity and the crack intensity factors that are of interest in fracture mechanics.* For this case, equation 8-7.2.8 becomes

$$\sin(\lambda - 1)2\pi = 0,$$
$$2(\lambda - 1) = n, \qquad (8\text{-}7.2.11)$$
$$\lambda = \frac{n}{2} + 1.$$

The only root of interest is $\lambda = \frac{3}{2}$, and the singularity is of the form $1/\sqrt{r}$.

7.3 Truncated semi-infinite wedge. The Williams solution uses the eigenvalues whose real part was greater than 1. If we examine the truncated semi-infinite wedge shown in Figure 8-18, we obtain a similar formulation involving eigenvalues whose real parts are equal or less than 1.

If self-equilibrated stresses are applied to the boundary $r = r_0$, we have the opportunity to examine the Saint-Venant boundary region for this geometry. T. R. Thompson† examined this problem for various wedge angles and end loading conditions. Thompson assumed a solution of the form shown in equation 8-7.2.1, but with two additional terms to correspond to the eigen-

* M. L. Williams, "On the stress distribution at the base of a stationary crack," ASME Ann. Meet., New York, November 1956, Paper 56-A-16.

† T. R. Thompson, "End effects in a truncated semi-infinite wedge and cone," Ph.D. thesis, Michigan State University, 1968.

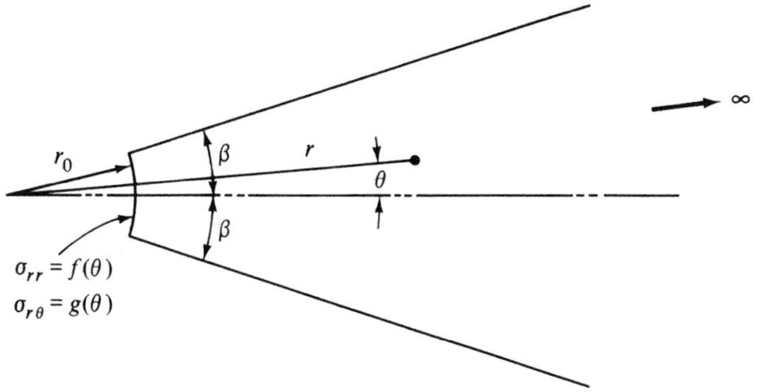

Figure 8-18

value $\lambda = 1$. The solution takes the following form:

$$\varphi = r^{\lambda}[A \sin \lambda\theta + B \cos \lambda\theta + C \sin (\lambda - 2)\theta + D \cos (\lambda - 2)\theta] + F_1 r\theta \cos \theta + F_2 r\theta \sin \theta. \quad (8\text{-}7.3.1)$$

The stress-free boundary conditions at $\theta = \pm\beta$ may be written as

$$A \sin \lambda\beta + C \sin (\lambda - 2)\beta = 0, \qquad A\lambda \cos \lambda\beta + C(\lambda - 2) \cos (\lambda - 2)\beta = 0, \quad (8\text{-}7.3.2)$$

$$B \cos \lambda\beta + D \cos (\lambda - 2)\beta = 0, \qquad B\lambda \sin \lambda\beta + D(\lambda - 2) \sin (\lambda - 2)\beta = 0. \quad (8\text{-}7.3.3)$$

The odd and even theta dependency separates and is handled independently if the boundary functions on the truncated end are written as the sum of odd and even functions. The transcendental equations for the odd and even eigenvalues* are

$$\lambda_n^o \sin 2\beta - 2 \sin (\lambda_n^o\beta) \cos (\lambda_n^o - 2)\beta = 0, \qquad \lambda_n^e \sin 2\beta + 2 \cos (\lambda_n^e\beta) \sin (\lambda_n^e - 2)\beta = 0. \quad (8\text{-}7.3.4)$$

These eigenvalues agree with those given in equation 8-7.2.8 if the wedge angles are equated. The roots of these equations are complex and are grouped in the left and right half-sides of the complex plane. If λ_n is a solution, so also is its complex conjugate, $\bar{\lambda}_n$. The solutions in the right half-plane may be found from those in the left half-plane by the relationship

$$\lambda_n^{(\text{right plane})} = -\lambda_n^{(\text{left plane})} + 2. \quad (8\text{-}7.3.5)$$

Thompson satisfied prescribed stress boundary conditions on the edge, $r = r_0$, by use of a least-squares technique. He examined the decay of stresses from this boundary for the four cases shown in Table 8-1. The lowest

* The odd eigenfunctions correspond to the A, C, and F_1 terms, and the even eignefunctions are the B, D, and F_2 terms.

eigenvalue dominates the decay for the first three cases and this value is given for various wedge angles in Table 8-2.

Table 8-1. End Stress Loadings

	$\sigma_{rr}\|_{r=r_0}$	$\sigma_{r\theta}\|_{r=r_0}$	Principal Decay for Stresses
a.	$A + B\theta^2$	0	$r^{\lambda-2}$
b.	0	$A \sin\frac{\pi\theta}{\beta} + B(\theta^3 - \beta^2\theta)$	$r^{\lambda-2}$
c.	A	$B(\theta^3 - \beta^2\theta)$	$r^{\lambda-2}$
d.	1.0	0	r^{-1}

A and B are related so that the stresses are self-equilibrated.

Table 8-2. Lowest Eigenvalues

β	λ_1^e		λ_1^o	
10°	−11.07	−6.38i	−20.48	−7.87i
20	− 5.06	−3.10i	−9.75	−3.84i
30	−3.06	−1.95i	−6.18	−2.46i
45	−1.74	−1.12i	−3.81	−1.46i
60	−1.09	−0.60i	−2.63	−0.88i
75	−0.91	−0.00i	−1.94	−0.36i

Plots of the stress distribution for two of the loading conditions are shown in Figures 10-8 and 10-9 (see Chap. 10) where this method is compared to a finite element solution.*

8 Special problems using the Flamant solution

Consider a disk subjected to concentrated loads as shown in Figure 8-19.† We can approach this problem by considering the effect of two different Flamant solutions (section 7-8.5), thereby giving the form of the singularities on the disk. We shall convert the previous solution in Cartesian coordinates to polar coordinates by considering the stress transformation equations.

* For other considerations of the truncatee wedge, see S. D. Carothero, *Proc. Roy. Soc. Edinburgh*, **A23**, 292–306, 1912; and E. Sternberg and W. T. Koiter, *J. Appl. Mech.*, **25**, 575–581, 1958.

† H. Hertz, *Z. Math. Physik*, **28**, 1883; and J. H. Michell, *Proc. London Math. Soc.*, **32**, 44, 1900, and **34**, 134, 1901.

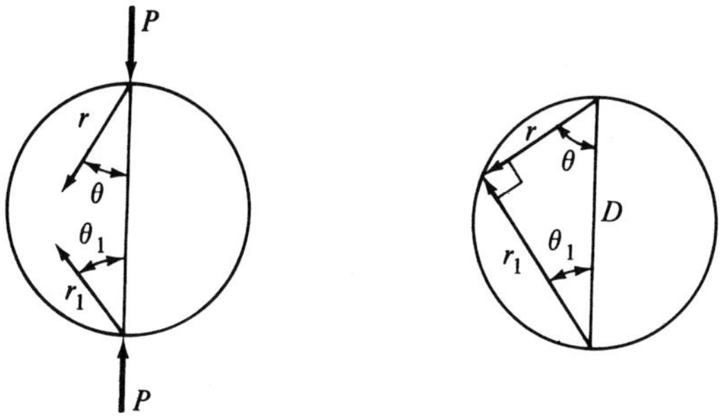

Figure 8-19

The solution in polar coordinates becomes

$$\sigma_{rr} = -\frac{2P}{\pi r}\cos\theta, \tag{8-8.1}$$

$$\sigma_{r\theta} = \sigma_{\theta\theta} = 0. \tag{8-8.2}$$

Examining the geometry of the problem, we obtain the relation on the boundary:

$$D\cos\theta = r, \qquad D\cos\theta_1 = r_1.$$

Therefore,

$$\frac{\cos\theta}{r} = \frac{\cos\theta_1}{r_1} = \frac{1}{D}. \tag{8-8.3}$$

Therefore, on the boundaries the state of stress corresponds to $-(2P/\pi D)$ or biaxial compression. We may remove this unwanted distribution by adding the case of biaxial tension:

$$\sigma_{rr} = \sigma_{\theta\theta} = \frac{2P}{\pi D}, \tag{8-8.4}$$

and the complete solution is

$$\sigma_{rr} = \frac{2P}{\pi}\left\{\frac{1}{D} - \frac{\cos\theta}{r} + [\sigma'_{rr}]_{\text{portion of solution}}\right\}. \tag{8-8.5}$$

Note that this solution is in terms of two different coordinate systems, but this presents no difficulties in its use. The stress function becomes

$$\varphi = \frac{P}{\pi}\left\{\frac{r^2}{D} - r\theta\sin\theta - r_1\theta_1\sin\theta_1\right\}. \tag{8-8.6}$$

8.1 Concentrated load in hole in infinite plate. Consider the problem shown in Figure 8-20. Again we can use the Flamant solution to obtain a start to this problem. We shall introduce a dual coordinate system to facilitate the formulation. As in the last case, some care must be used in calculating the stresses from the final stress function, as the one coordinate must be expressed

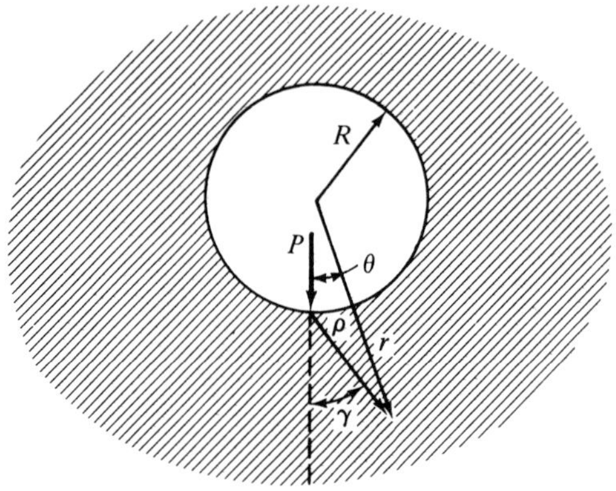

Figure 8-20

in terms of the other before differentiating to obtain stresses. The two sets of stresses may be independently calculated and the results superimposed after transformation to align the coordinates. The Flamant solution is

$$\sigma_{\rho\rho} = \frac{2A\cos\gamma}{\rho}, \tag{8-8.1.1}$$

$$\sigma_{\gamma\gamma} = 0, \qquad A = -\frac{P}{\pi}, \tag{8-8.1.2}$$

$$\sigma_{\rho\gamma} = 0. \tag{8-8.1.3}$$

The stresses in the central coordinates are related to these stresses by the following equations:

$$\sigma_{rr} = \cos^2(\gamma - \theta)\sigma_{\rho\rho}, \tag{8-8.1.4}$$

$$\sigma_{\theta\theta} = \sin^2(\gamma - \theta)\sigma_{\rho\rho}, \tag{8-8.1.5}$$

$$\sigma_{r\theta} = \sin(\gamma - \theta)\cos(\gamma - \theta)\sigma_{\rho\rho}. \tag{8-8.1.6}$$

The relations between ρ, γ and r, θ are

$$\rho\sin\gamma = r\sin\theta, \tag{8-8.1.7}$$

$$\rho\cos\gamma + R = r\cos\theta. \tag{8-8.1.8}$$

On the boundary $r = R$,

$$(180 - \gamma) = (\gamma - \theta),$$

$$\gamma = 90 + \frac{\theta}{2}, \tag{8-8.1.9}$$

$$(\gamma - \theta) = 90 - \frac{\theta}{2},$$

$$\sigma_{\rho\rho} = \frac{2A\cos\gamma\sin\gamma}{R\sin\theta} = -\frac{A}{R}, \tag{8-8.1.10}$$

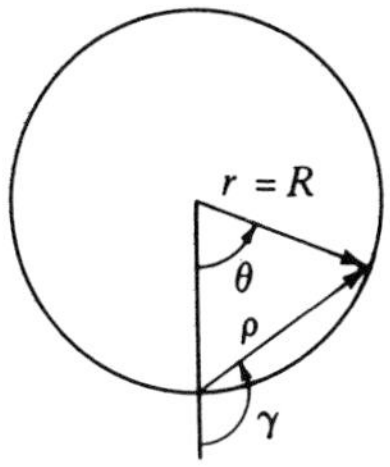

Figure 8-21

$$\sigma_{rr} = -\frac{A}{R}\cos^2(\gamma - \theta) = -\frac{A}{2R}\{1 - \cos\theta\}, \qquad (8\text{-}8.1.11)$$

$$\sigma_{r\theta} = -\frac{A}{R}\frac{\sin 2(\gamma - \theta)}{2} = -\frac{A}{2R}\sin\theta. \qquad (8\text{-}8.1.12)$$

These are the expressions for all points on the boundary except $\theta = 0$, where replacement of ρ would have involved division by zero. To satisfy the boundary conditions at $r = R$, we must pick additional stress terms of the form $\sigma_{rr} = a + b\cos\theta$ and $c\sin\theta$. Note that these solutions must also decay to zero as $r \longrightarrow \infty$. From the general solution, we use

$$\varphi \Longrightarrow b_0 \ln r, \quad A_1 r\theta \sin\theta, \quad b_1 r \ln r \cos\theta, \quad \frac{c_1}{r}\cos\theta. \qquad (8\text{-}8.1.13)$$

Three of these terms are picked to satisfy stress boundary conditions, and A_1 and b_1 are related to give single-valued displacements. Timoshenko* gives the final solution in the form

$$\varphi = -\frac{P}{\pi}\left\{\gamma r\sin\theta - \frac{1}{4}(1-\nu)r\ln r\cos\theta - \frac{1}{2}r\theta\sin\theta + \frac{R}{2}\ln r - \frac{R^2}{8}(3-\nu)\frac{\cos\theta}{r}\right\}. \qquad (8\text{-}8.1.14)$$

PROBLEMS

8.1. Determine the state of stress in a hollow disk rotating at an angular velocity ω.† Examine the state of stress in the disk as the hole radius approaches zero, that is, the case of a stress concentration for a small hole in the disk.

8.2. Determine the state of stress in the composite disk shown below, when rotated at an angular velcoity ω. The material properties of the composite are as follows: $E_A \neq E_B$, $\nu_A \neq \nu_B$, and $\rho_A > \rho_B$.

* Timoshenko and Goodier, *op. cit.*, p. 148.

† Ta-Cheng Ku solved this problem for an eccentric hole; see *J. Appl. Mech.*, **27**, 359–360, 1960.

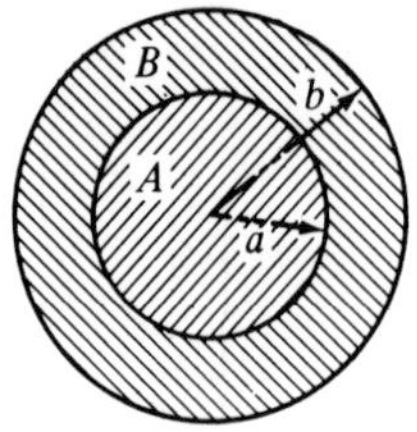

P. 8-2

8.3. Determine the stress distribution in an infinite plate with a hole of radius R if the edges of the hole are given a constant displacement $u_\theta = \Delta$, and the normal stress σ_{rr} on this edge is zero.

8.4. Determine the stresses around the edge of the hole ($\sigma_{\theta\theta}$) in a very large shear panel (considered infinite).

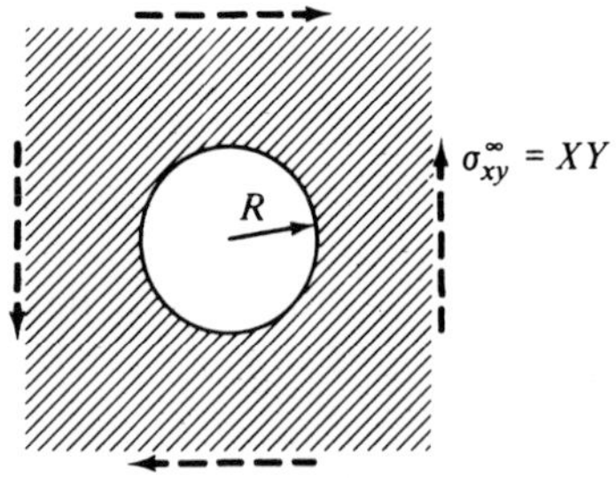

P. 8-4

8.5. Determine the stresses in the ring loaded as shown.

$$t_x^r = ky, \qquad -\alpha < \theta < \alpha.$$

$$t_x^r = -ky, \quad \pi - \alpha < \theta < \pi + \alpha.$$

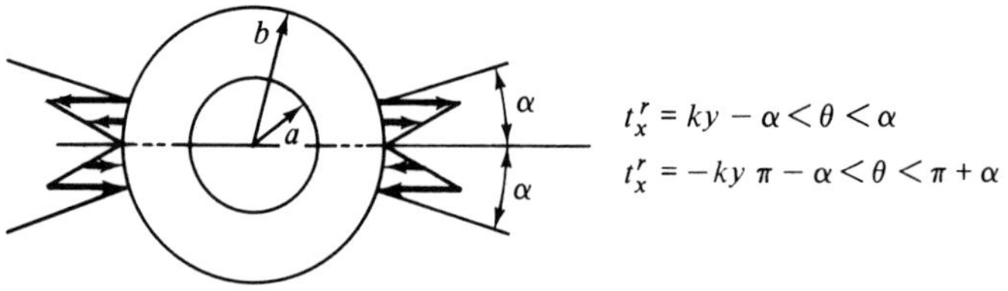

P. 8-5

8.6. Find the solution for the problem shown below.

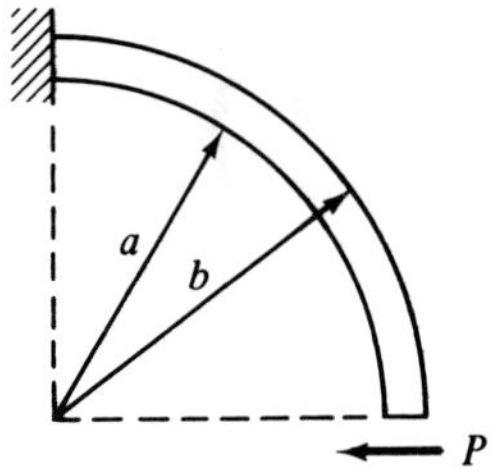

P. 8-6

8.7. Verify the transcendental equation for the clamped–clamped case in the Williams solution given in section 8-7.2.

CHAPTER 9

COMPLEX VARIABLE SOLUTIONS

1 Introduction

The application of complex function theory to two-dimensional elasticity problems yields a powerful tool for solution of certain classes of problems. This method can be used to generate the Fourier-series-type solutions discussed in Chapter 8, but the real value appears when Cauchy's integral formula and conformal mappings are employed. A brief review of basic complex variable theory will be presented before the elasticity formulation is developed, and other concepts of complex function theory will be introduced when needed. Comprehensive texts devoted to complex variable solutions of elasticity are available for the student who wishes a more detailed development.*

* N. I. Muskhelishvili, "*Some Basic Problems of the Mathematical Theory of Elasticity*, Wolters-Noordhoff Publishing, Groningen, 1963; L. M. Milne-Thompson, *Plane Elastic Systems*, Springer-Verlag, Berlin, 1960.

2 Complex variables

Let z represent the complex variable defined as

$$z = x + iy \qquad \text{where } i = \sqrt{-1}. \tag{9-2.1}$$

This can be expressed in polar coordinates as

$$z = r(\cos\theta + i\sin\theta) = re^{i\theta}. \tag{9-2.2}$$

This point can be plotted in the complex plane as shown in Figure 9-1. The

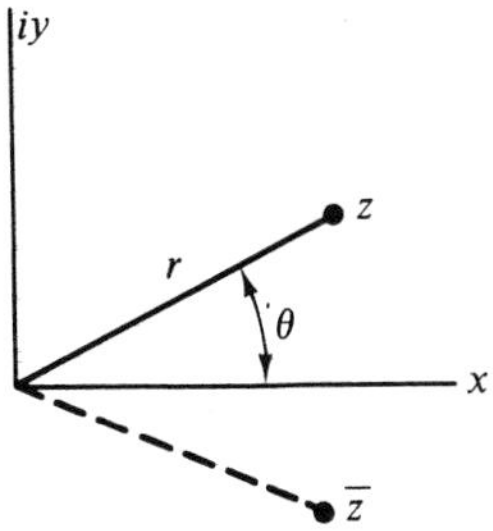

Figure 9-1

complex conjugate $\bar{z}$ is defined as

$$\bar{z} = x - iy = re^{-i\theta}, \tag{9-2.3}$$

and is the reflection of z about the real axis.

The radius r or the modulus of the variable is obtained by taking the square root of the product of the complex variable and its conjugate:

$$r^2 = z\bar{z}. \tag{9-2.4}$$

This form uses the product of two complex numbers, which is defined as follows:

$$z_1z_2 = (x_1 + iy_1)(x_2 + iy_2) = x_1x_2 - y_1y_2 + i(y_1x_2 + x_1y_2). \tag{9-2.5}$$

A function of a complex variable can be written as follows and is called a complex function:

$$f(z) = f(x + iy) = u(x, y) + iv(x, y). \tag{9-2.6}$$

For example, $f(z) = z^2$,

$$u(x, y) = x^2 - y^2 \quad \text{and} \quad v(x, y) = 2xy.$$

$f(z)$ may involve i directly; for example,

$$f(z) = z^2 + iz.$$

The complex conjugate of a complex function is usually designated by barring the function. $\overline{f(z)}$ is the complex conjugate function of $f(z)$:

$$\overline{f(z)} = \bar{f}(\bar{z}). \tag{9-2.7}$$

In the first example

$$f(z) = z^2, \qquad \overline{f(z)} = \bar{z}\bar{z} = u(x, y) - iv(x, y),$$

or in the second example

$$f(z) = z^2 + iz, \qquad \overline{f(z)} = \bar{z}^2 - i\bar{z}.$$

Let $f(z)$ be a single-valued continuous function of z in a domain D. We say that $f(z)$ is differentiable at a point z_0 of D if

$$\lim_{\Delta z \to 0} \frac{1}{\Delta z}[f(\Delta z + z_0) - f(z_0)] \tag{9-2.8}$$

exists as a finite number and is independent of how the complex increment Δz tends to zero. The limit, when it exists, will be denoted by $f'(z_0)$ and is called the derivative of $f(z)$ at z_0.

The function $f(z)$ is differentiable in D if it is differentiable at all points of D, and $f(z)$ is then said to be a holomorhpic function of z in D. Finally, $f(z)$ is holomorphic at $z = z_0$ if it is holomorphic in some ϵ neighborhood of z_0. If $f(z)$ possesses a derivative in the neighborhood of z_0, then it is sometimes called analytic or regular at z_0. A point where the function is not regular is called a singular point.

Let us consider the definition of the complex derivative in a little more detail:

$$w(z) = \frac{df(z)}{dz} = \lim_{\Delta z \to 0} \frac{f(z + \Delta z) - f(z)}{\Delta z}. \tag{9-2.9}$$

Let $f(z) = u + iv$

$$w(z) = \lim_{\substack{\Delta x \to 0 \\ \Delta y \to 0}} \frac{u + \Delta u + iv + i\,\Delta v - u - iv}{\Delta x + i\,\Delta y}.$$

Now let the limit be approached along a line $\Delta y = 0$:

$$w(z) = \lim_{\Delta x \to 0} \left(\frac{\Delta u}{\Delta x} + i\frac{\Delta v}{\Delta x}\right) = \frac{\partial u}{\partial x} + i\frac{\partial v}{\partial x}. \tag{9-2.10}$$

In a similar manner, if the limit is approached along a line $\Delta x = 0$,

$$w(z) = \lim_{\Delta y \to 0} \left(-i\frac{\Delta u}{\Delta y} + \frac{\Delta v}{\Delta y}\right) = \frac{\partial v}{\partial y} - i\frac{\partial u}{\partial y}. \tag{9-2.11}$$

Since the limit must be the same approached in any direction,

$$\frac{\partial u}{\partial x} = \frac{\partial v}{\partial y}, \qquad \frac{\partial u}{\partial y} = -\frac{\partial v}{\partial x}. \tag{9-2.12}$$

These are called the Cauchy–Riemann equations.

By differentiation of these equations we can easily show that

$$\nabla^2 u = 0, \qquad \nabla^2 v = 0. \tag{9-2.13}$$

Note that

$$du = \frac{\partial u}{\partial x}\,dx + \frac{\partial u}{\partial y}\,dy = \frac{\partial v}{\partial y}\,dx - \frac{\partial v}{\partial x}\,dy,$$

so if we know u or v, the other may be found. For example,

$$v = xy, \qquad (\nabla^2 v = 0).$$

$$du = x\,dx - y\,dy.$$

Therefore,

$$u = \frac{x^2}{2} - \frac{y^2}{2} + \mathbb{C}.$$

u and v are called conjugate functions.

The characteristics of holomorphic functions are most important to us because they arise in elasticity theory. An indication that this occurs is seen in equation 9-2.13. If we have harmonic functions, we can define a conjugate function and develop holomorphic functions in the complex plane. In the absence of body forces, the compatibility equation states that the sum of the normal stresses is a harmonic function. This, then, indicates that a complex variable formulation is available. We shall continue to develop properties of complex functions and then show their applications.

LINE INTEGRALS

Let us examine two points in the complex plane connected by a curve C, as shown in Figure 9-2. The line integral of a function $f(z)$ can be written

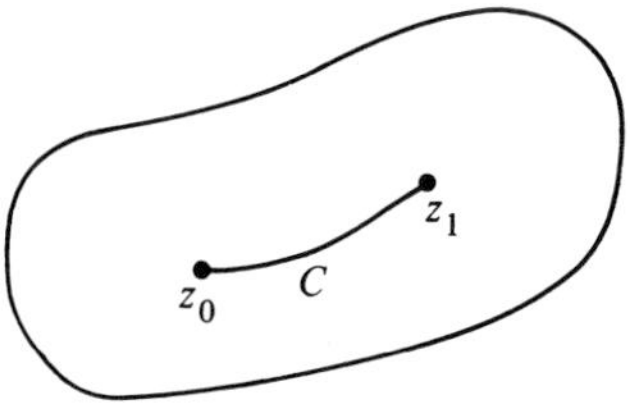

Figure 9-2

$$\begin{aligned}\int_C f(z)\,dz &= \int_C (u + iv)(dx + i\,dy) \\ &= \int_C (u\,dx - v\,dy) + i\int_C (u\,dy + v\,dx).\end{aligned} \tag{9-2.14}$$

This will be an exact differential of some function *if*

$$d\varphi = \frac{\partial \varphi}{\partial x}\,dx + \frac{\partial \varphi}{\partial y}\,dy \tag{9-2.15}$$

for both the real and the imaginary parts. If this is an exact differential (independent of the path of integration),

$$u = \frac{\partial \varphi}{\partial x}, \qquad v = -\frac{\partial \varphi}{\partial y}, \qquad u = \frac{\partial \psi}{\partial y}, \qquad v = \frac{\partial \psi}{\partial x},$$

$$\frac{\partial u}{\partial y} = -\frac{\partial v}{\partial x}, \qquad \frac{\partial u}{\partial x} = \frac{\partial v}{\partial y},$$

These are the Cauchy–Riemann equations. Therefore, the line integral $\int_C f(z)\,dz$ is independent of the path of integration if C can be enclosed in a simply connected region R inside which $f(z)$ is analytic.

We note that $f(z) = u + iv$,

$$\int f(z)\,dz = \int (u\,dx - v\,dy) + i \int (u\,dy + v\,dx), \qquad (9\text{-}2.16)$$

and that $\int f(z)\,dz$ is not a simple line integral. It is related to the sum of four line integrals of real variables. This difference leads to a distinction in name—

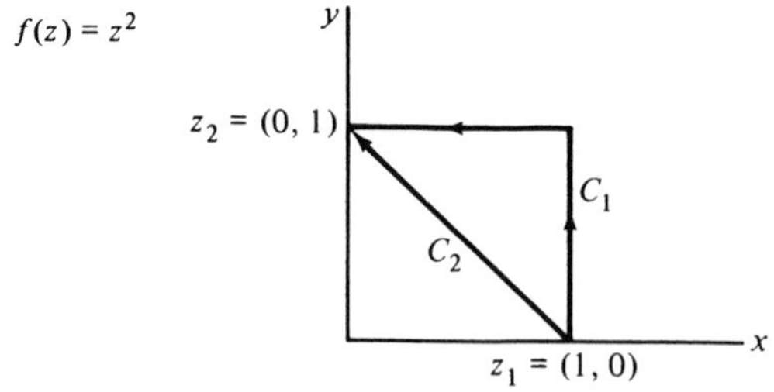

Figure 9-3

contour integrals. Let us consider a few examples:

$$f(z) = z^2.$$

Examining the integral of this function between z_1 and z_2 along contours C_1 and C_2 yields

$$z^2 = (x + iy)^2, \qquad u = x^2 - y^2, \qquad v = 2xy.$$

Along C_1,

$$\int_{C_1} f(z)\,dz = -\int_0^1 2y\,dy + i\int_0^1 (1 - y^2)\,dy + \int_1^0 (x^2 - 1)\,dx + i\int_1^0 2x\,dx$$
$$= -1 + \tfrac{2}{3}i + \tfrac{2}{3} - i = -\tfrac{1}{3}(1 + i).$$

Along C_2,

$$y = 1 - x,$$
$$u = 2x - 1 = -2y + 1,$$
$$v = -2x^2 + 2x = -2y^2 + 2y,$$

$$\int_{C_2} f(z)\,dz = \int_1^0 (2x-1)\,dx - \int_0^1 (-2y^2+2y)\,dy$$
$$+ i\int_0^1 (-2y+1)\,dy + i\int_1^0 (-2x^2+2x)\,dx$$
$$= -\tfrac{1}{3}(1+i).$$

Since $f(z)$ is analytic and holomorphic along C_1 and C_2,

$$\int f(z)\,dz = F(z_2) - F(z_1), \qquad \text{where } F(z) = \frac{z^3}{3}$$
$$= -\tfrac{1}{3}(1+i).$$

Note that z^n is holomorphic in the whole domain with the exception of the point at infinity.

CAUCHY'S INTEGRAL THEOREM

If $f(z)$ is holomorphic on and in some closed curve C,

$$\oint_C f(z)\,dz = 0, \text{ (positive direction keeps enclosed area on left).} \qquad (9\text{-}2.17)$$

If u and v are expressed in polar coordinates

$$dz = d(re^{i\theta}),$$

$$\begin{aligned}\int_C f(z)\,dz &= \int_C (u+iv)(e^{i\theta}\,dr + ire^{i\theta}\,d\theta)\\ &= \int_C (u\cos\theta - v\sin\theta)\,dr - (v\cos\theta + u\sin\theta)r\,d\theta\\ &\quad + \int_C (v\cos\theta + u\sin\theta)\,dr + (u\cos\theta - v\sin\theta)r\,d\theta.\end{aligned} \qquad (9\text{-}2.18)$$

The required condition that the two quantities inside the integrals be exact differentials leads to the equations

$$\frac{\partial u}{\partial r} = \frac{1}{r}\frac{\partial v}{\partial \theta}, \qquad \frac{1}{r}\frac{\partial u}{\partial \theta} = -\frac{\partial v}{\partial r}. \qquad (9\text{-}2.19)$$

Note that these are the exact form of the Cauchy–Riemann conditions one would expect in polar coordinates.

Let us now consider some special integrals about closed paths. If $f(z) = 1/z$, we can calculate its derivative; $f'(z) = -1/z^2$. This function exists everywhere except at the point $z = 0$. We use Cauchy's integral theorem and state that any closed contour not enclosing the origin will lead to the condition

$$\oint \frac{1}{z}\,dz = 0. \qquad (9\text{-}2.20)$$

Next consider the unit circle about the origin:

$$\oint \frac{dz}{z} = \int_0^{2\pi} e^{-i\theta} ie^{i\theta}\,d\theta = \int_0^{2\pi} i\,d\theta = 2\pi i. \qquad (9\text{-}2.21)$$

We can obtain this result by a different way by noting that dz/z is the derivative of $\log z$:

$$w = \log z, \qquad w = u + iv,$$

$$e^w = z,$$

$$z = e^u(\cos v + i \sin v)$$

Therefore,

$$\log z = \log \underset{r}{|z|} + i\theta.$$

Integrating around a unit circle about the origin would produce the result obtained before.

Now let us consider some other curve C_1 that encloses the origin, C and C_1 are simple, closed rectifiable curves. Now suppose that we cut between C and C_1, making a simply connected region as shown in the Figure 9-4. Now $f(z) = 1/z$ is analytic throughout the region R enclosed by the

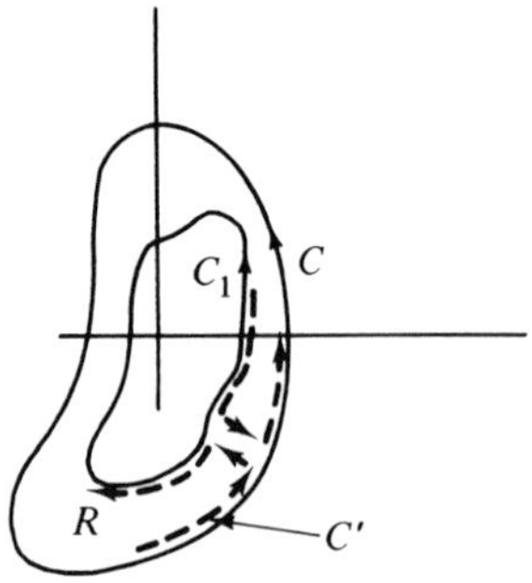

Figure 9-4

contour denoted by C' and indicated on the figure by dashed lines. By Cauchy's integral theorem we may write

$$\oint_{C'} f(z)\, dz = 0. \tag{9-2.22}$$

Now let the cut edges approach one another and note that the integrals along these edges cancel one another:

$$\oint_C f(z)\, dz - \oint_{C_1} f(z)\, dz = 0.$$

Therefore,

$$\oint_C f(z)\, dz = \oint_{C_1} f(z)\, dz. \tag{9-2.23}$$

For any closed path enclosing the origin the following relation holds:

$$\oint_C \frac{dz}{z} = 2\pi i. \tag{9-2.24}$$

This result may be generalized in the case when $f(z)$ is holomorphic in the region D as follows:

$$\oint_{C_0} f(z)\,dz = \sum_{k=1}^{n} \oint_{C_k} f(z)\,dz. \tag{9-2.25}$$

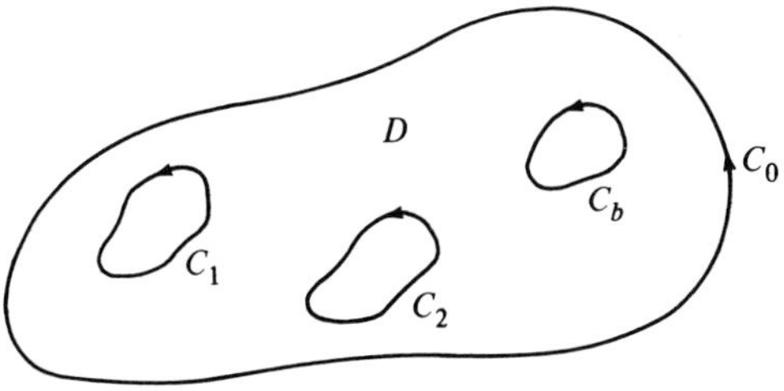

Figure 9-5

Now let us consider the integral of z^n about the contour coinciding with the unit circle:

$$\begin{aligned}\oint_{C_1} z^n\,dz &= i\int_0^{2\pi} e^{i(n+1)\theta}\,d\theta \\ &= i\int_0^{2\pi} [\cos(n+1)\theta + i\sin(n+1)\theta]\,d\theta = 0, \\ &\qquad (n \text{ is an integer and } n \neq -1).\end{aligned} \tag{9-2.26}$$

When n is zero or positive, this coincides with the prediction by Cauchy's integral theorm, because z^n is analytic throughout the unit circle. For n negative but not equal to -1, the integral is still zero, even though z^n is not analytic at the origin. Therefore, we conclude that for any path not passing through the origin

$$\oint_C z^n\,dz = 0, \qquad n \neq -1. \tag{9-2.27}$$

if $n = -1$ and C includes the origin, the integral is equal to $2\pi i$.

We may generalize these results as follows:

Cauchy's Integral Theorem

$$\begin{aligned}\oint_C (z-a)^n\,dz &= 0, \qquad n \neq -1, \\ \oint \frac{dz}{(z-a)} &= 2\pi i,\end{aligned} \tag{9-2.28}$$

where C includes but does not pass through $z = a$ and n is an integer.

The importance of Cauchy's integral theorem may be understood when one considers that an integration contour C_1 may be changed into any number of other more convenient contours as long as C_1 and the new contours bound a closed domain in which the function is holomorphic.

CAUCHY'S INTEGRAL FORMULA

Let C be a closed contour inside of which and along which $f(z)$ is analytic. Let α be a point inside C and let C_ϵ be a small circle of radius ϵ about α (Figure 9-6). $f(z)/(z-\alpha)$ is holormorphic between C_ϵ and C. On C_ϵ the values of z and dz are

$$z = \alpha + \epsilon e^{i\theta}, \qquad dz = i\epsilon e^{i\theta}\, d\theta,$$

$$\oint \frac{f(z)}{z-\alpha}\, dz = \oint_0^{2\pi} f(\alpha + \epsilon e^{i\theta}) i\, d\theta. \tag{9-2.29}$$

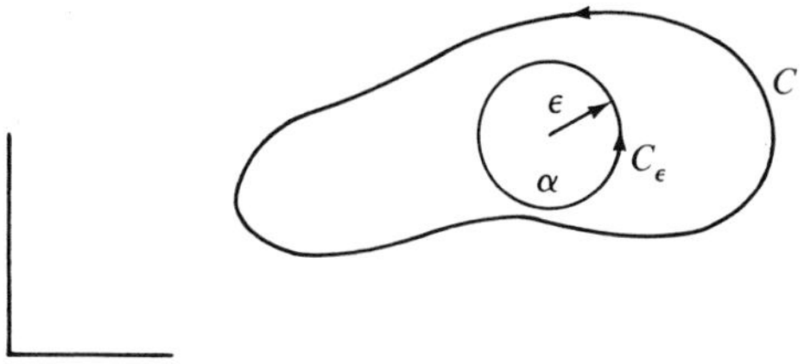

Figure 9-6

Now if we let $\epsilon \longrightarrow 0$, this integral approaches $f(\alpha)2\pi i$:

$$f(\alpha) = \frac{1}{2\pi i} \oint_C \frac{f(z)}{z-\alpha}\, dz. \tag{9-2.30}$$

Changing definitions of z and α yields

$$f(z) = \frac{1}{2\pi i} \oint \frac{f(\alpha)}{\alpha - z}\, d\alpha, \qquad z \text{ inside } C. \tag{9-2.31}$$

Note that if z is outside C, then $f(\alpha)/(\alpha - z)$ is holomorphic throughout C, and by Cauchy's theorem

$$\frac{1}{2\pi i} \oint \frac{f(\alpha)}{\alpha - z}\, dz = 0, \qquad z \text{ outside } C. \tag{9-2.32}$$

Cauchy's integral formula gives a method of specifying the value of holomorphic functions on the interior if their value is known on the boundary. The elasticity problems we have considered involve just this type of condition. Displacements and stresses are specified on the boundaries, and the values of these functions are required in the interior of the elastic body.

TAYLOR SERIES

We use Cauchy's integral formula to develop a Taylor series expansion for complex functions. Note the following:

$$\begin{aligned} \frac{1}{\alpha - z} &= \frac{1}{(\alpha - a) - (z - a)} = \frac{1}{\alpha - a} \cdot \frac{1}{1 - (z-\alpha)/(\alpha - a)} \\ &= \frac{1}{(\alpha - a)} \sum_{n=0}^{\infty} \left(\frac{z-a}{\alpha - a}\right)^n, \qquad (|z - a| < |\alpha - a|). \end{aligned} \tag{9-2.33}$$

We take C to be any circle with center $z = a$ inside which and on which $f(z)$ is analytic. For all points inside the circle, $|z - a| < |\alpha - a| = R$, and the power series converges. This is called the circle of convergence (Figure 9-7).

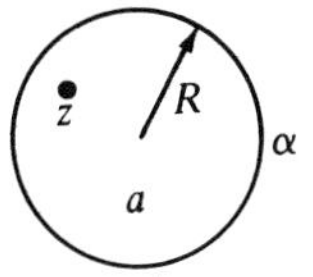

Figure 9-7

Substituting this result into Cauchy's integral formula yields

$$f(z) = \frac{1}{2\pi i} \oint_C \frac{f(\alpha)}{\alpha - z}\, d\alpha, \qquad z \text{ inside } C,$$

$$= \frac{1}{2\pi i} \oint \sum_{n=0}^{\infty} \frac{f(\alpha)}{(\alpha - a)^{n+1}} (z - a)^n\, d\alpha \tag{9-2.34}$$

$$= \sum_{n=0}^{\infty} \left[\frac{1}{2\pi i} \oint \frac{f(\alpha)}{(\alpha - a)^{n+1}}\, d\alpha \right] (z - a)^n,$$

or

$$f(z) = \sum_{n=0}^{\infty} A_n (z - a)^n, \qquad (|z - a| < R), \tag{9-2.35}$$

where

$$A_n = \frac{1}{2\pi i} \oint \frac{f(\alpha)}{(\alpha - a)^{n+1}}\, d\alpha.$$

If we differentiate $f(z)$, we obtain

$$f'(z) = \sum_{n=1}^{\infty} \frac{n}{2\pi i} \left[\oint_C \frac{f(\alpha)\, d\alpha}{(\alpha - a)^{n+1}} \right] (z - a)^{n-1}. \tag{9-2.36}$$

At $z = a$, this becomes

$$f'(a) = \frac{1}{2\pi i} \oint_C \frac{f(z)}{(\alpha - a)^2}\, d\alpha. \tag{9-2.37}$$

In a similar manner, we obtain

$$f^k(a) = \frac{k!}{2\pi i} \oint_C \frac{f(\alpha)}{(\alpha - a)^{k+1}}\, d\alpha, \tag{9-2.38}$$

or, changing notation,

$$f^k(z) = \frac{k!}{2\pi i} \oint_C \frac{f(\alpha)}{(\alpha - z)^{k+1}}\, d\alpha. \tag{9-2.39}$$

Finally, this can be written as

$$f(z) = \sum_{n=0}^{\infty} \frac{f^n(a)}{n!} (z - a)^n. \tag{9-2.40}$$

This is the desired Taylor series. The circle of convergence is the largest circle whose center is at $z = a$ and in which $f(z)$ is analytic.

Without proof, let us present the Laurent series, which is a generalization of the Taylor series.

Let $f(z)$ be analytic in the region R, (Figure 9-8).

$$f(z) = \sum_{n=-\infty}^{\infty} a_n(z-a)^n, \qquad (R_1 < |z-a| < R_2), \qquad (9\text{-}2.41)$$

where

$$a_n = \frac{1}{2\pi i}\oint \frac{f(z)}{(z-a)^{n+1}}\,dz, \qquad C \text{ is any closed path in } R. \qquad (9\text{-}2.42)$$

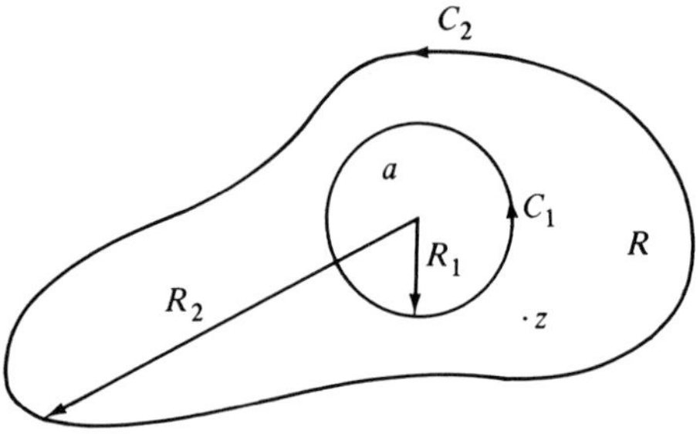

Figure 9-8

We have now obtained series expansions for holomorphic functions and the values of the coefficients of these expansions in terms of boundary values of the holomorphic functions. We shall put these results to good use shortly when we show that the fundamental stresses and displacements of the plane elasticity problem can be expressed in terms of holomorphic functions.

SINGULARITIES

We have defined functions $f(z)$ to be holomorphic if the function possesses a derivative at each point and is single valued. Points at which a function is not analytic are called singular points or singularities. If $f(z)$ is not finite at $z = a$ but if

$$(z-a)^k f(z) \qquad (9\text{-}2.43)$$

is analytic for some integer value of k, then $f(z)$ is said to have a pole of order k at $z = a$.

For example,

$$\frac{1}{z^2+1} = \frac{1}{(z+i)(z-i)} \qquad (9\text{-}2.44)$$

and has poles of order 1 or simple poles at $\pm i$;

$$\frac{1}{z^2} \qquad (9\text{-}2.45)$$

has a double pole at $z = 0$.

Poles associated with the "point" at infinity can also be considered if we substitute $z = 1/t$ into the function $f(z)$ and examine poles at $t = 0$, which corresponds to $z \longrightarrow \infty$.

RESIDUES

It may now be seen that if $f(z)$ has a pole of order n at $z = a$, then $(z - a)^n f(z)$ is analytic within the neighborhood of a and can be expanded in a Taylor series.

Using equation 9-2.35, this function can be written

$$g(z) = (z - a)^n f(z) = \sum_{k=0}^{\infty} A_k (z - a)^k. \tag{9-2.46}$$

Using equations 9-2.35 and 9-2.38, the coefficients become

$$A_k = \frac{1}{k!}\frac{d}{dk}[g(z)] = \frac{1}{k!}\left[\frac{d^k}{dz^k}\{(z - a)^n f(z)\}\right]_{z=a}. \tag{9-2.47}$$

The function $f(z)$ is

$$f(z) = \frac{A_0}{(z - a)^n} + \frac{A_1}{(z - a)^{n-1}} + \cdots + \frac{A_{n-1}}{(z - a)} + A_n + A_{n+1}(z - a) + \cdots. \tag{9-2.48}$$

Using Cauchy's integral theorem, we integrate around a closed contour C, including but not passing through a:

$$\oint_C f(z)\, dz = 2\pi i A_{n-1}. \tag{9-2.48}$$

A_{n-1} is called the residue of $f(z)$ at $z = a$, and equation 9-2.49 may be generalized as follows: If $f(z)$ has a pole of order n at $z = a$ and C_a includes this singularity and no others,

$$\oint_{C_a} f(z)\, dz = 2\pi i \operatorname{Res}\{a\}, \tag{9-2.50}$$

where
$$\operatorname{Res}\{a\} = \frac{1}{(n - 1)!}\left\{\frac{d^{n-1}}{dz^{n-1}}[(z - a)^n f(z)]\right\}_{z=a}.$$

This procedure is called the calculus of residues and is extremely useful in evaluating complex integrals. We shall use this method in the sections that follow.

3 Complex stress formulation

In the general development of plane strain, we obtained stress-displacement relations in the following form related to the Airy stress function:

$$\sigma_{xx} = \lambda\left(\frac{\partial u_x}{\partial x} + \frac{\partial u_y}{\partial y}\right) + 2G\frac{\partial u_x}{\partial x} = \frac{\partial^2 \varphi}{\partial y^2}, \tag{9-3.1}$$

$$\sigma_{yy} = \lambda\left(\frac{\partial u_x}{\partial x} + \frac{\partial u_y}{\partial y}\right) + 2G\frac{\partial u_y}{\partial y} = \frac{\partial^2 \varphi}{\partial x^2}, \tag{9-3.2}$$

$$\sigma_{xy} = G\left(\frac{\partial u_x}{\partial y} + \frac{\partial u_y}{\partial x}\right) = -\frac{\partial^2 \varphi}{\partial x\,\partial y}. \tag{9-3.3}$$

Adding equations 9-3.1 and 9-3.2 gives

$$\frac{\partial u_x}{\partial x} + \frac{\partial u_y}{\partial y} = \frac{1}{2(\lambda + G)}\nabla^2\varphi. \tag{9-3.4}$$

If we let $P = \nabla^2\varphi$, we write equations 9-3.1 and 9-3.2 as

$$2G\frac{\partial u_x}{\partial x} = \left\{\frac{\partial^2\varphi}{\partial y^2} - \frac{\lambda}{2(\lambda + G)}P\right\} = -\frac{\partial^2\varphi}{\partial x^2} + \frac{\lambda + 2G}{2(\lambda + G)}P, \tag{9-3.5}$$

$$2G\frac{\partial u_y}{\partial y} = \left\{\frac{\partial^2\varphi}{\partial x^2} - \frac{\lambda}{2(\lambda + G)}P\right\} = -\frac{\partial^2\varphi}{\partial y^2} + \frac{(\lambda + 2G)P}{2(\lambda + G)}. \tag{9-3.6}$$

P is a harmonic function:

$$\nabla^2\nabla^2\varphi = \nabla^2 P = 0. \tag{9-3.7}$$

We introduce the conjugate harmonic function Q satisfying the Cauchy-Riemann conditions:

$$\frac{\partial P}{\partial x} = \frac{\partial Q}{\partial y}, \qquad \frac{\partial P}{\partial y} = -\frac{\partial Q}{\partial x}. \tag{9-3.8}$$

Let $f(z) = P + iQ$, which is holomorphic in the region S occupied by a simply connected body. Let us now introduce a function

$$\gamma(z) = p + iq = \tfrac{1}{4}\int f(z)\,dz. \tag{9-3.9}$$

One may easily show that $\gamma(z)$ is also holomorphic in this region. Differentiation yields

$$\gamma'(z) = \frac{\partial p}{\partial x}\frac{\partial x}{\partial z} + \frac{\partial p}{\partial y}\frac{\partial y}{\partial z} + i\left(\frac{\partial q}{\partial x}\frac{\partial x}{\partial z} + \frac{\partial q}{\partial y}\frac{\partial y}{\partial z}\right). \tag{9-3.10}$$

Noting that

$$x = \tfrac{1}{2}(z + \bar{z}), \qquad y = -\tfrac{1}{2}i(z - \bar{z}),$$

equation 9-3.10 can be written

$$\gamma'(z) = \frac{1}{2}\left\{\frac{\partial p}{\partial x} - i\frac{\partial p}{\partial y} + i\frac{\partial q}{\partial x} + \frac{\partial q}{\partial y}\right\}. \tag{9-3.11}$$

$\gamma(z)$ is analytic and the Cauchy–Riemann conditions apply to p and q. Equation 9-3.11 can be written

$$\gamma'(z) = \left\{\frac{\partial p}{\partial x} + i\frac{\partial q}{\partial x}\right\} = \frac{1}{4}(P + iQ). \tag{9-3.12}$$

Equating the real and imaginary parts and using the Cauchy–Riemann relations yields

$$\frac{\partial p}{\partial x} = \frac{\partial q}{\partial y} = \frac{1}{4}P, \qquad (9\text{-}3.13)$$

$$\frac{\partial p}{\partial y} = -\frac{\partial q}{\partial x} = -\frac{1}{4}Q. \qquad (9\text{-}3.14)$$

Introducing the two different forms of equations 9-3.13 into equations 9-3.5 and 9-3.6 yields

$$2G\frac{\partial u_x}{\partial x} = -\frac{\partial^2\varphi}{\partial x^2} + \frac{2(\lambda + 2G)}{\lambda + G}\frac{\partial p}{\partial x}, \qquad (9\text{-}3.15)$$

$$2G\frac{\partial u_y}{\partial y} = -\frac{\partial^2\varphi}{\partial y^2} + \frac{2(\lambda + 2G)}{\lambda + G}\frac{\partial q}{\partial y}. \qquad (9\text{-}3.16)$$

Integrating these expressions and using equation 9-3.3 and finally setting rigid-body displacements equal to zero yields

$$2Gu_x = -\frac{\partial\varphi}{\partial x} + \frac{2(\lambda + 2G)}{(\lambda + G)}p, \qquad (9\text{-}3.17)$$

$$2Gu_y = -\frac{\partial\varphi}{\partial y} + \frac{2(\lambda + 2G)}{(\lambda + G)}q. \qquad (9\text{-}3.18)$$

We wish to simplify these expressions by showing that the Airy stress function may be expressed in terms of two complex functions. The following expression is valid:

$$\nabla^2(\varphi - px - qy) = 0. \qquad (9\text{-}3.19)$$

This can be integrated in terms of an arbitrary harmonic function p_1:

$$\varphi = px + qy + p_1. \qquad (9\text{-}3.20)$$

As before, we shall define a function q_1 to be the conjugate function to p_1, and the stress function can now be written

$$\varphi = \text{Re}\,\{\bar{z}\gamma(z) + \chi(z)\}, \qquad (9\text{-}3.21)$$

where

$$\chi(z) = p_1 + iq_1.$$

An alternative method for obtaining equation 9-3.21 is to write the biharmonic equation in terms of the complex variables z and $\bar{z}$:

$$\nabla^2\nabla^2\varphi = \frac{\partial^2}{\partial z\,\partial\bar{z}}\left(\frac{\partial^2}{\partial z\,\partial\bar{z}}\varphi\right) = 0.$$

Integrating yields

$$\frac{\partial^2}{\partial z\,\partial\bar{z}}\varphi = f(z) + g(\bar{z}).$$

If $\nabla^2\varphi$ is real, as the sum of the normal stresses must be, this can be written

$$\frac{\partial^2\varphi}{\partial z\,\partial\bar{z}} = f(z) + \bar{f}(\bar{z}).$$

Integrating again yields, for real φ,

$$\varphi = \tfrac{1}{2}\{\bar{z}\gamma(z) + z\bar{\gamma}(\bar{z}) + \chi(z) + \bar{\chi}(\bar{z})\},$$

where
$$\gamma(z) = 2\int f(z)\,dz.$$

This equation agrees with equation 9-3.21. A similar development using operators is used in anisotropic elasticity theory.*

Taking the derivative of φ with respect to x and y gives

$$\frac{\partial\varphi}{\partial x} + i\frac{\partial\varphi}{\partial y} = \gamma(z) + z\overline{\gamma'(z)} + \overline{\chi'(z)}. \qquad (9\text{-}3.22)$$

Using equations 9-3.17, 9-3.18, and 9-3.22, we introduce the complex displacement

$$2G(u_x + iu_y) = \eta\gamma(z) - z\overline{\gamma'(z)} - \overline{\chi'(z)}, \qquad (9\text{-}3.23)$$

where
$$\eta = \frac{\lambda + 2G}{\lambda + G}, \qquad \eta = \begin{cases} (3 - 4\nu) < 1, & \text{plane strain} \\ \dfrac{3 - \nu}{1 + \nu} < 1, & \text{plane stress} \end{cases}$$

The stresses can now be calculated by use of equation 9-3.21 and the following "so-called fundamental" stress combinations are obtained:

$$(\sigma_{xx} + \sigma_{yy}) = 2\{\gamma'(z) + \overline{\gamma'(z)}\} = 4\,\mathrm{Re}\,\{\Gamma(z)\}, \qquad (9\text{-}3.24)$$

where
$$\Gamma(z) = \gamma'(z),$$

and
$$\begin{aligned}(\sigma_{yy} - \sigma_{xx} + 2i\sigma_{xy}) &= 2\{\bar{z}\gamma''(z) + \chi''(z)\}, \\ &= 2\{\bar{z}\Gamma'(z) + \Psi(z)\},\end{aligned} \qquad (9\text{-}3.25)$$

where $\Psi(z) = \chi''(z)$. We shall use the notation, $\psi(z) = \chi'(z)$ in the displacement equation.

These are all analytic functions of the variables x and y. By equating the real and imaginary parts of equation 9-3.25 and using equation 9-3.24, the three stresses are determined.

Now consider an element ds under the following state of stress. (Figure 9-9).

$$t_x^n = \sigma_{xx}\cos(n, x) + \sigma_{xy}\cos(n, y) = \frac{\partial^2\varphi}{\partial y^2}\cos(n, x) - \frac{\partial^2\varphi}{\partial x\,\partial y}\cos(n, y),$$

$$t_y^n = \sigma_{xy}\cos(n, x) + \sigma_{yy}\cos(n, y) = -\frac{\partial^2\varphi}{\partial x\,\partial y}\cos(n, x) + \frac{\partial^2\varphi}{\partial x^2}\cos(n, y),$$

Note that

$$\cos(n, x) = \frac{dy}{ds},$$

$$\cos(n, y) = -\frac{dx}{ds}.$$

* Milne-Thompson, *op. cit.*

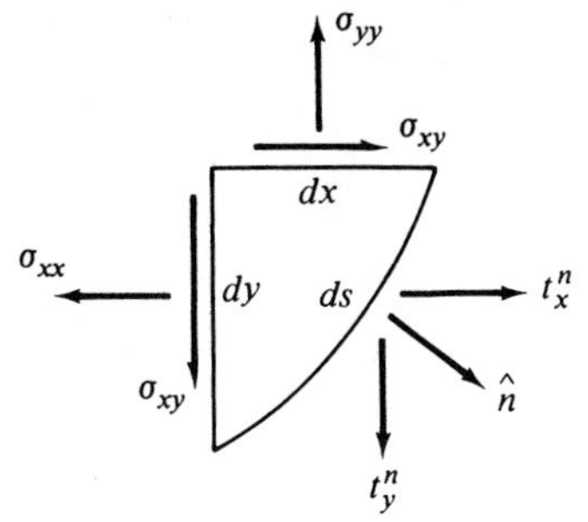

Figure 9-9

Therefore,

$$t_x^n = \frac{d}{ds}\left(\frac{\partial \varphi}{\partial y}\right), \qquad t_y^n = -\frac{d}{ds}\left(\frac{\partial \varphi}{\partial x}\right),$$

$$t_x^n + it_y^n = \frac{d}{ds}\left(\frac{\partial \varphi}{\partial y} - i\frac{\partial \varphi}{\partial x}\right) = -i\frac{d}{ds}\left(\frac{\partial \varphi}{\partial x} + i\frac{\partial \varphi}{\partial y}\right), \qquad (9\text{-}3.26)$$

$$= -\frac{d}{ds}\{\gamma(z) + z\overline{\gamma'(z)} + \overline{\psi(z)}\}.$$

Examination of equations 9-3.23 to 9-3.25 indicates that the functions Γ, γ, and ψ are not completely determined. This indeterminancy will not influence the state of stress in the body. If we note that an imaginary constant may be added to Γ and not affect the stresses produced in equation 9-3.24, we replace this function by the following:

$$\Gamma_1(z) = \Gamma(z) + iC, \qquad C \text{ is a real constant.} \qquad (9\text{-}3.27)$$

In a similar manner, we obtain

$$\gamma_1(z) = \gamma(z) + iCz + A, \qquad (9\text{-}3.28)$$

$$\psi_1(z) = \psi(z) + B, \qquad A \text{ and } B \text{ are complex constants.} \qquad (9\text{-}3.29)$$

Substitution of these results into the displacement equation yields

$$2G(u_x + iu_y)_1 = 2(u_x + iu_y) + (1 + \eta)iCz + \eta A - \bar{B}. \qquad (9\text{-}3.30)$$

It should be noted that these additional terms represent rigid-body motions, and the stresses determine the displacements up to rigid-body motions. If the stresses are given, we choose the rigid-body motions such that

$$\gamma(0) = 0, \qquad \operatorname{Im}\gamma'(0) = 0, \qquad \psi(0) = 0. \qquad (9\text{-}3.31)$$

MULTIPLY CONNECTED REGIONS

Consider a region bounded by simple (nonintersecting) closed contours L_k (Figure 9-10). Note that we assumed γ and ψ were holomorphic

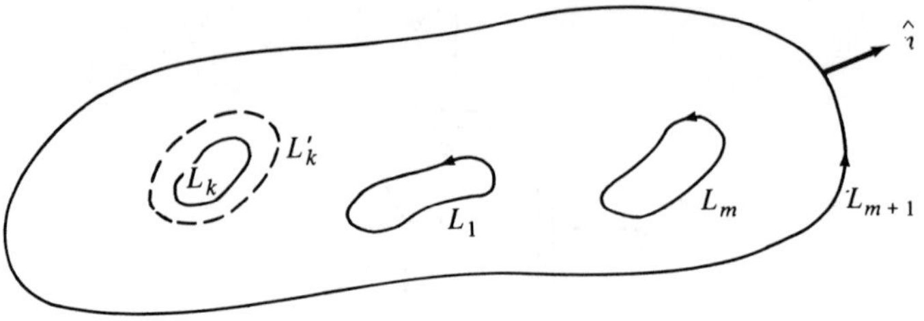

Figure 9-10

(therefore single-valued functions) in a simply connected region. However, in multiply connected regions they need not be single valued.

In the study of multivalue functions it is interesting to consider *cyclic functions.** Let $f(z, \bar{z})$ be a function defined in the region D, and C be a simple closed contour in D. The point z_1 is a nonsingular point of $f(z, \bar{z})$ on C. Let $f(z, \bar{z})$ be equal to $f_1(z, \bar{z})$ for definiteness when we start at z_1 (Figure 9-11).

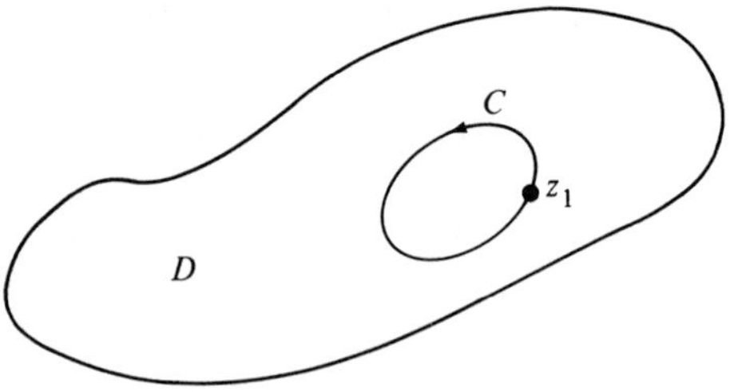

Figure 9-11

Going around the circuit C, let the value of $f(z, \bar{z})$ become $f_2(z, \bar{z})$:

$$[f(z, \bar{z})]_C = f_2(z, \bar{z}) - f_1(z, \bar{z}).$$

↖ cyclic function of $f(z, \bar{z})$ relative to the circuit C

The properties of cyclic functions are

1. The cyclic function of any function f for which $\partial^2 f/\partial z\,\partial \bar{z} = \partial^2 f/\partial \bar{z}\,\partial z$ is the same for C and for any curve C' that passes through z_1 and is reconcilable with C without passing over a singularity of $f(z, \bar{z})$.
2. The cyclic function of the sum of two functions is equal to the sum of the cyclic functions of each separately.
3. If f and g are two functions such that $[g]_C = 0$, then $[fg]_C = g[f]_C$.
4. If $f(z, \bar{z})$ is one valued in D, then $[f(z, \bar{z})]_C = 0$ for any circuit C in D.

* See Milne-Thompson, *op. cit.*

5. The partial derivative of the cyclic function with respect to z or $\bar{z}$ is equal to the cyclic function of the derivative

$$\frac{\partial}{\partial z}[f(z,\bar{z})]_C = \left[\frac{\partial}{\partial z}f(z,\bar{z})\right]_C.$$

Therefore,

$$\left[\frac{\partial f}{\partial z}\right]_C = 0$$

implies that $[f(z)]_C = \alpha$, a constant that may be set to zero:

$$\left[\frac{\partial^2 f}{\partial z^2}\right]_C = 0,$$

$$\left[\frac{\partial f}{\partial z}\right]_C = \alpha,$$

$$[f]_C = \alpha z + \beta.$$

If the complex stresses are required to be single valued,

$$\sigma_{xx} + \sigma_{yy} = 4\,\mathrm{Re}\,[\Gamma(z)] = 2[\Gamma(z) + \overline{\Gamma(z)}], \qquad (9\text{-}3.32)$$

$$[\Gamma(z) + \overline{\Gamma(z)}]_{L_k'} = 0, \qquad (9\text{-}3.33)$$

where L_k' is some anticlockwise contour surrounding L_k. This avoids the need to discuss the properties of the functions at the boundaries.

$$[\Gamma(z)]_{L_k'} = -[\overline{\Gamma(z)}]_{L_k'},$$

$$[\Gamma(z)]_{L_k} = \alpha_k \qquad \Longrightarrow \qquad [\overline{\Gamma(z)}]_{L_k'} = \bar{\alpha}_k,$$

$$\alpha_k + \bar{\alpha}_k = 0,$$

$$\alpha_k = 2\pi i A_k, \qquad \text{where } A_k \text{ is real.}$$

Noting the cyclic properties of the log function, we may write

$$[\Gamma^*(z) + A_k \log(z - z_k)]_{L_k'} = 2\pi i A_k,$$

where $\Gamma^*(z)$ is a holomorphic function.

$$\Gamma(z) = \Gamma^*(z) + \sum_{k=1}^{n} A_k \log(z - z_k) \qquad \text{for all circuits.} \qquad (9\text{-}3.34)$$

$$\gamma(z) = \int \Gamma(z)\,dz, \qquad [\gamma(z)]_{L_k'} = 2\pi i A_k z + 2\pi i B_k,$$

$$\gamma(z) = \gamma^*(z) + z\sum_{k=1}^{m} A_k \log(z - z_k) + \sum_{k=1}^{m} B_k \log(z - z_k). \qquad (9\text{-}3.35)$$

($\gamma^*(z)$: holomorphic function)

In a similar manner, we note

$$\psi(z) = \psi^*(z) + \sum_{k=1}^{m} B_k' \log(z - z_k). \qquad (9\text{-}3.36)$$

The cyclic properties of $u_x + iu_y$ become

$$\begin{aligned} 2G[u_x + iu_y]_{L_k'} &= [\eta\gamma(z) - z\overline{\gamma'(z)} - \overline{\psi(z)}]_{L_k'}, \\ &= 2\pi i[A_k(\eta + 1)z + \eta B_k + \bar{B}_k']. \end{aligned} \tag{9-3.37}$$

If the displacements are not single valued, these correspond to the dislocations discussed earlier. For single valued displacements the following conditions hold:

$$A_k = 0, \qquad \eta B_k + \bar{B}_k' = 0. \tag{9-3.38}$$

We can obtain a physical interpretation of the constants B_k and B_k' when the stresses and displacements are single valued. Consider an inner contour L_k' around a boundary contour L_k (Figure 9-12). To obtain the influ-

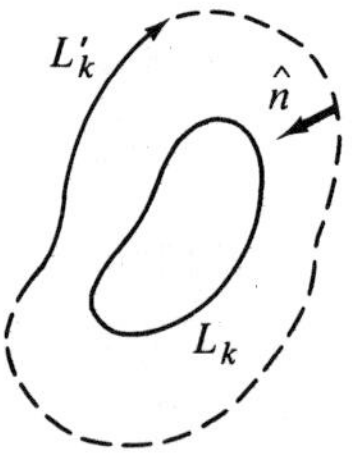

Figure 9-12

ence of the external forces on the region, we calculate the net resultant force, $X + iY$:

$$X + iY = i\{\gamma(z) + z\overline{\gamma'(z)} + \overline{\psi(z)}\}_{L_k} \tag{9-3.39}$$

the plus sign reverses the contour

$$\begin{aligned} X + iY &= i\{[\gamma(z)]_{L_k} + [z\overline{\gamma'(z)}]_{L_k} + [\overline{\psi(z)}]_{L_k}\} \\ &= -2\pi[B_k - \bar{B}_k']. \end{aligned}$$

Because $\bar{B}_k' = -\eta B_k$, we obtain

$$B_k = -\frac{X + iY}{2\pi(1 + \eta)}, \qquad B_k' = \frac{\eta(X - iY)}{2\pi(1 + \eta)}. \tag{9-3.40}$$

The fundamental stress functions for multiply connected regions are

$$\gamma(z) = -\sum_{k=1}^{m} \frac{F_k}{2\pi(\eta + 1)} \log(z - z_k) + \gamma^*(z), \tag{9-3.41}$$

$$\psi(z) = \sum_{k=1}^{m} \frac{\eta \bar{F}_k}{2\pi(\eta + 1)} \log(z - z_k) + \psi^*(z), \tag{9-3.42}$$

where F_k represents the force vector $X + iY$ acting on the contour L_k.

INFINITE REGIONS

Before using these functions, we shall investigate their form in the special domain of the infinite region. To consider the behavior of the functions $\gamma(z)$ and $\psi(z)$ at the point infinity, we take a circle enclosing all the contours L_k and examine a point outside this circle. The log function is written $\log(z - z_k) = \log z + \log[1 - (z_k/z)]$. Expanding the second term in series form yields

$$\begin{aligned}\log(z - z_k) &= \log z - \frac{z_k}{z} - \frac{1}{2}\left(\frac{z_k}{z}\right)^2 + \cdots \\ &= \log z + \text{a holomorphic function for large } z.\end{aligned} \tag{9-3.43}$$

The complex function $\gamma(z)$ becomes

$$\gamma(z) = -\frac{\log z}{2\pi(1+\eta)} \sum F_k + \gamma^{**}(z) \text{ holomorphic not including } \infty. \tag{9-3.44}$$

In a similar manner, we obtain

$$\psi(z) = \frac{\eta \log z}{2\pi(1+\eta)} \sum \bar{F}_k + \psi^{**}(z). \tag{9-3.45}$$

Using a Laurent series for γ and ψ outside the enclosing circle L_k,

$$\gamma^{**}(z) = \sum_{-\infty}^{+\infty} a_n z^n, \tag{9-3.46}$$

$$\psi^{**}(z) = \sum_{-\infty}^{+\infty} a'_n z^n. \tag{9-3.47}$$

If we require that all stresses remain bounded at infinity, we obtain

$$\begin{aligned}\sigma_{xx} + \sigma_{yy} &= 2\{\gamma'(z) + \overline{\gamma'(z)}\} \\ &= -\frac{1}{2\pi(1+\eta)}\left[\frac{F}{z} + \frac{\bar{F}}{\bar{z}}\right] + \sum_{-\infty}^{\infty} n(a_n z^{n-1} + \bar{a}_n \bar{z}^{n-1}),\end{aligned} \tag{9-3.48}$$

$$\sigma_{yy} - \sigma_{xx} + 2i\sigma_{xy} = 2[\bar{z}\gamma''(z) + \psi'(z)], \tag{9-3.49}$$

and

$$\begin{aligned} a_n = \bar{a}_n &= 0, \qquad n \geq 2, \\ a'_n &= 0, \qquad n \geq 2. \end{aligned} \tag{9-3.50}$$

These functions in terms of the state of stress at infinity are

$$\gamma(z) = -\frac{\log z}{2\pi(1+\eta)} \sum F_k + \frac{XX + YY}{4} z + \sum_{n=1}^{\infty} a_{-n} z^{-n}, \tag{9-3.51}$$

$$\psi(z) = \frac{\eta \log z}{2\pi(1+\eta)} \sum \bar{F}_k + \frac{YY - XX + 2iXY}{2} z + \sum_{n-1}^{\infty} a'_{-n} z^{-n}, \tag{9-3.52}$$

where XX, YY, and XY are the Cartesian components of stress at infinity. Examination of the displacements at infinity implies that they will become infinite unless $\sum F_k$, XX, YY, and XY are zero. This is not disturbing physically if one notes that a constant strain over an infinite length will produce infinite displacements.

4 Polar coordinates

By use of the tensor transformation equations we write

$$\begin{aligned}\sigma_{rr} &= \sigma_{xx}\cos^2\theta + \sigma_{yy}\sin^2\theta + 2\sigma_{xy}\sin\theta\cos\theta,\\ \sigma_{\theta\theta} &= \sigma_{xx}\sin^2\theta + \sigma_{yy}\cos^2\theta - 2\sigma_{xy}\sin\theta\cos\theta,\\ \sigma_{r\theta} &= (-\sigma_{xx} + \sigma_{yy})\sin\theta\cos\theta + \sigma_{xy}(\cos^2\theta - \sin^2\theta).\end{aligned} \tag{9-4.1}$$

Using these equations we obtain

$$\sigma_{rr} + \sigma_{\theta\theta} = \sigma_{xx} + \sigma_{yy}, \tag{9-4.2}$$

$$\sigma_{\theta\theta} - \sigma_{rr} + 2i\sigma_{r\theta} = (\sigma_{yy} - \sigma_{xx} + 2i\sigma_{xy})e^{2i\theta}. \tag{9-4.3}$$

In a similar manner, the displacements become

$$u_r + iu_\theta = e^{-i\theta}(u_x + iu_y). \tag{9-4.4}$$

Equations 9-3.23, 9-3.24, and 9-3.25 can be substituted into these equations to give the required expressions in polar coordinates.

5 Interior problem

The fundamental stress combinations in polar coordinates are

$$\sigma_{rr} + \sigma_{\theta\theta} = 2[\Gamma(z) + \overline{\Gamma(z)}], \tag{9-5.1}$$

$$\sigma_{\theta\theta} - \sigma_{rr} + 2i\sigma_{r\theta} = 2e^{2i\theta}[\bar{z}\Gamma'(z) + \Psi(z)], \tag{9-5.2}$$

$$\sigma_{rr} - i\sigma_{r\theta} = \Gamma(z) + \overline{\Gamma(z)} - e^{2i\theta}[\bar{z}\Gamma'(z) + \Psi(z)]. \tag{9-5.3}$$

Consider a disk loaded by $\sigma_{rr}|_{r=R} = f_1(\theta)$ and $\sigma_{r\theta}|_{r=R} = -f_2(\theta)$. Let

$$f = f_1 + if_2 = \sigma_{rr} - i\sigma_{r\theta}|_{r=R}. \tag{9-5.4}$$

If the stress tractions are specified, we can write the boundary condition as

$$t_x^r + it_y^r = -i\frac{d}{ds}[\gamma(z) + z\overline{\gamma'(z)} + \overline{\psi(z)}]|_{r=R}. \tag{9-5.5}$$

Integrating yields

$$i\int (t_x^r + it_y^r)\,ds = [\gamma(z) + z\overline{\gamma'(z)} + \overline{\psi(z)}]|_{r=R} = g. \tag{9-5.6}$$

Along the circle $r = R$, $ds = R\,d\theta$, and g becomes a specified function of θ. When using equation 9-5.6, $\gamma(0)$ or $\psi(0)$ is not specified so that the constant of integration may be absorbed.

Consider the problem where the boundary conditions are given in the form of equation 9-5.4. We know from our examination of holomorphic functions that $\gamma(z)$ and $\psi(z)$ can be expressed as a Taylor's series expansion about the origin:

$$\gamma(z) = \sum_{n=0}^{\infty} a_n z^n, \tag{9-5.7}$$

$$\psi(z) = \sum_{n=0}^{\infty} b_n z^n. \tag{9-5.8}$$

Note that $\gamma(0)$ and $\psi(0)$ and the imaginary part of $\gamma'(0)$ will not be determined in the stress formulation. Differentiation of equations 9-5.7 and 9-5.8 and substitution into equation 9-5.3 yields

$$\sigma_{rr} - i\sigma_{r\theta} = \sum_{n=1}^{\infty} a_n n z^{n-1} + \bar{a}_n n \bar{z}^{n-1} - e^{2i\theta}[\bar{z} a_n n(n-1) z^{n-2} + b_n n z^{n-1}]. \tag{9-5.9}$$

This can be written

$$\sigma_{rr} - i\sigma_{r\theta} = a_1 + \bar{a}_1 + \sum_{k=1}^{\infty} \{-[a_{k+1}(k^2 - 1)r^k + b_{k-1}(k-1)r^{k-2}]e^{ik\theta} + \bar{a}_{k+1}(k+1)r^k e^{-ik\theta}\}. \tag{9-5.10}$$

This is a complex Fourier series expansion for $\sigma_{rr} - i\sigma_{r\theta}$. At the boundary $\sigma_{rr} - i\sigma_{r\theta}|_{r=R} = f$, and f may be expanded in a Fourier series also:

$$f = \sum_{-\infty}^{+\infty} C_k e^{ik\theta}, \tag{9-5.11}$$

where the constants C_k are obtained by use of the orthogonality relation

$$C_k = \frac{1}{2\pi} \int_0^{2\pi} f(\theta) e^{-ik\theta}\, d\theta. \tag{9-5.12}$$

Equating coefficients of the exponentials of θ at the boundary yields

$$a_1 + \bar{a}_1 = C_0 = 2\,\mathrm{Re}\,(a_1). \tag{9-5.13}$$

As predicted, the imaginary part of a_1 or $\gamma'(0)$ is not obtained:

$$\bar{a}_{k+1}(k+1)R^k = C_{-k}, \qquad k \text{ positire} > 0, \tag{9-5.14}$$

$$a_{k+1}(k^2 - 1)R^k + b_{k-1}(k-1)R^{k-2} = C_k, \qquad k \text{ positire} > 0. \tag{9-5.15}$$

Equations 9-5.14 and 9-5.15 represent two complex equations for the complex constants a_k and b_k. They can be written as four real equations by equating the real and imaginary parts of each.

This solution is in essence the same as the Fourier analysis solution using the Michell solution presented in section 8-5.1.

The annulus and the exterior problem may be solved in a similar manner. In the case of the infinite plate, $\gamma(z)$ and $\psi(z)$ are taken in the form of equations 9-3.51 and 9-3.52 In the case of the annulus, equations 9-3.41 and 9-3.42 are used and $\gamma^*(z)$ and $\psi^*(z)$ are expanded in a Laurent series.

The real power of the complex variable approach does not appear in this form of solution, which duplicates the work done previously with Fourier analysis, but only when the Cauchy integral formulas are used.

Consider the case of a disk bounded by the unit circle and stress tractions prescribed on the boundary in the form of equation 9-5.6:

$$[\gamma(z) + z\overline{\gamma'(z)} + \overline{\psi(z)}]|_{z=\zeta} = g, \tag{9-5.16}$$

where

$$\zeta = z\,|_{r=1} = e^{i\theta}.$$

Note that

$$\bar{\zeta} = e^{-i\theta} = \frac{1}{\zeta}.$$

As before, the complex functions are expanded in a Taylor series expansion:

$$\gamma(z) = \sum a_n z^n, \tag{9-5.17}$$

$$\psi(z) = \sum b_n z^n. \tag{9-5.18}$$

We multiply equation 9-5.16 by $(1/2\pi i)[1/(\zeta - z)]\,d\zeta$ and integrate around the boundary contour of the unit circle designated by C:

$$\frac{1}{2\pi i}\oint_C \frac{\gamma(\zeta)}{(\zeta - z)}\,d\zeta + \frac{1}{2\pi i}\oint_C \frac{\zeta\overline{\gamma'(\zeta)}}{\zeta - z}\,d\zeta + \frac{1}{2\pi i}\oint_C \frac{\overline{\psi(\zeta)}}{\zeta - z}\,d\zeta = \frac{1}{2\pi i}\oint_C \frac{g(\zeta)}{\zeta - z}\,d\zeta. \tag{9-5.19}$$

The first term is evaluated by the Cauchy integral formula, noting that $\gamma(z)$ is holomorphic inside and on the unit circle and z is inside C:

$$\frac{1}{2\pi i}\oint_C \frac{\gamma(\zeta)}{\zeta - z}\,d\zeta = \gamma(z). \tag{9-5.20}$$

The next two integrals require further examination:

$$\begin{aligned} \overline{\gamma'(\zeta)} &= \sum \bar{a}_n n\zeta^{-(n-1)} = \bar{a}_1 + \frac{2\bar{a}_2}{\zeta} + \frac{3\bar{a}_3}{\zeta^2} + \cdots, \\ \zeta\overline{\gamma'(\zeta)} &= \bar{a}_1\zeta + 2\bar{a}_2 + 0\left(\frac{1}{\zeta}\right). \end{aligned} \tag{9-5.21}$$

The first two terms are holomorphic inside and on C and can be handled by the Cauchy integral formula. Let us examine the integral

$$\frac{1}{2\pi i}\oint_C \frac{1}{\zeta^n(\zeta - z)}\,d\zeta. \tag{9-5.22}$$

We can evaluate this by use of the calculus of residues. If a function $f(z)$ has a pole of order m at $z = a$,

$$\oint f(z)\,dz = 2\pi i \operatorname{Res}(a), \tag{9-5.23}$$

where

$$\operatorname{Res}(a) = \frac{1}{(m-1)!}\left[\frac{d^{m-1}}{dz^{m-1}}\{(z - a)^m f(z)\}\right]_{z=a}. \tag{9-5.24}$$

The integral 9-5.22 has a pole of order n at $\zeta = 0$ and a pole of order 1 at

$\zeta = z$; therefore,

$$\frac{1}{2\pi i}\oint_C \frac{1}{\zeta^n(\zeta - z)}\,d\zeta = \frac{1}{(n-1)!}\left[\frac{d^{n-1}}{d\zeta^{n-1}}\left\{\zeta^n \frac{1}{\zeta^n(\zeta - z)}\right\}\right]_{\zeta=0} + \left[(\zeta - z)\frac{1}{\zeta^n(\zeta - z)}\right]_{\zeta=z}. \qquad (9\text{-}5.25)$$

Expanding yields

$$\frac{1}{2\pi i}\oint_C \frac{1}{\zeta^n(\zeta - z)}\,d\zeta = \frac{1}{(n-1)!}\left[(n-1)!\frac{(-1)^{n-1}}{(\zeta - z)^n}\right]_{\zeta=0} + \frac{1}{z^n} = 0. \qquad (9\text{-}5.26)$$

We now write the useful formula

$$\frac{1}{2\pi i}\oint_C \frac{1}{\zeta^n(\zeta - z)}\,d\zeta = \begin{cases} 0, & n > 0 \\ 1, & n = 0 \end{cases}. \qquad (9\text{-}5.27)$$

Using equation 9-5.27 for the third term in equation 9-5.19, this equation can be written

$$\gamma(z) + \bar{a}_1 z + 2\bar{a}_2 + \overline{\psi(0)} = \frac{1}{2\pi i}\oint_C \frac{g(\zeta)}{\zeta - z}\,d\zeta. \qquad (9\text{-}5.28)$$

Noting that

$$\gamma(z) = a_0 + a_1 z + a_2 z^2,$$

differentiating equation 9-5.28 with respect to z, and evaluating the equation at $z = 0$ yields

$$a_1 + \bar{a}_1 = \frac{1}{2\pi i}\oint_C \frac{g(\zeta)}{\zeta^2}\,d\zeta. \qquad (9\text{-}5.29)$$

In a similar manner

$$a_2 = \frac{1}{2\pi i}\oint \frac{g(\zeta)}{\zeta^3}\,d\zeta. \qquad (9\text{-}5.30)$$

We have now evaluated $\gamma(z)$ and will return to equation 9-5.16 to determine $\psi(z)$. If we take the complex conjugate of this equation, we obtain

$$\overline{\gamma(\zeta)} + \bar{\zeta}\gamma'(\zeta) + \psi(\zeta) = \bar{g}. \qquad (9\text{-}5.31)$$

Operating on equation 9-5.31 with $(1/2\pi i)\oint [1/(\zeta - z)]\,d\zeta$ yields

$$\frac{1}{2\pi i}\oint_C \frac{\bar{\gamma}(\zeta)}{\zeta - z}\,d\zeta + \frac{1}{2\pi i}\oint_C \frac{\bar{\zeta}\gamma'(\zeta)}{\zeta - z}\,d\zeta + \frac{1}{2\pi i}\oint_C \frac{\psi(\zeta)}{\zeta - z}\,d\zeta = \frac{1}{2\pi i}\oint_C \frac{\bar{g}\,d\zeta}{\zeta - z}. \qquad (9\text{-}5.32)$$

This can now be written

$$\bar{\gamma}(0) + \psi(z) = \frac{1}{2\pi i}\oint_C \frac{\bar{g}\,d\zeta}{\zeta - z} - \frac{1}{2\pi i}\oint \frac{\gamma'(\zeta)}{\zeta(\zeta - z)}\,d\zeta. \qquad (9\text{-}5.33)$$

By use of calculus of residues, the last term that has single poles at $\zeta = 0$ and $\zeta = z$ is equal to

$$\frac{1}{2\pi i}\oint_C \frac{\gamma'(\zeta)}{\zeta(\zeta - z)}\,d\zeta = -\frac{a_1}{z} + \frac{\gamma'(z)}{z}. \qquad (9\text{-}5.34)$$

Therefore,

$$\psi(z) = \frac{1}{2\pi i}\oint_C \frac{\bar{g}\, d\zeta}{\zeta - z} + \frac{a_1}{z} - \frac{\gamma'(z)}{z} - \bar{\gamma}(0). \qquad (9\text{-}5.35)$$

The constant terms in equations 9-5.28 and 9-5.35 do not effect the solution, and the final form becomes

$$\gamma(z) = \frac{1}{2\pi i}\oint_C \frac{g}{(\zeta - z)}\, d\zeta - \bar{a}_1 z, \qquad (9\text{-}5.36)$$

$$\psi(z) = \frac{1}{2\pi i}\oint_C \frac{\bar{g}}{\zeta - z}\, d\zeta - \frac{\gamma'(z)}{z} + \frac{a_1}{z}, \qquad (9\text{-}5.37)$$

where a_1 is evaluated by use of equation 9-5.29.

A simple example of the solution in this form may be seen by examining the case of a disk under uniform external pressure (Figure 9-13). The

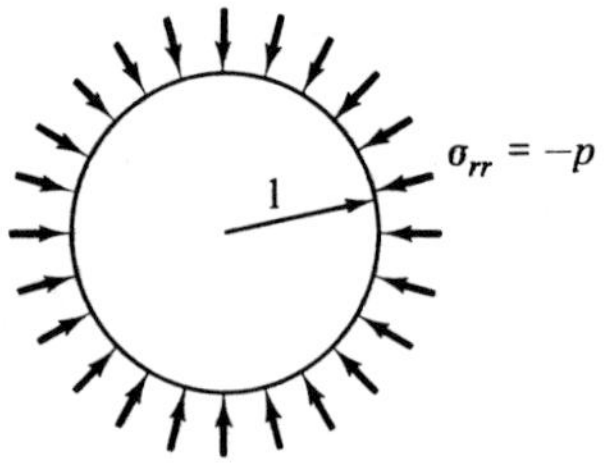

Figure 9-13

stress tractions are

$$t_x^r + i t_y^r = (\sigma_{rr} + i\sigma_{r\theta}) e^{i\theta} = -pe^{i\theta}.$$

Therefore, the function g becomes

$$g = i\int_0^\theta (t_x^r + i t_y^r)\, d\theta = -i\int_0^\theta pe^{i\theta}\, d\theta = -pe^{i\theta} = -p\zeta.$$

Substitution into equation 9-5.36 yields

$$\gamma(z) = -\frac{1}{2\pi i}\oint_C \frac{p\zeta}{\zeta - z}\, d\zeta - \bar{a}_1 z = -pz - \bar{a}_1 z.$$

Equation 9-5.26 gives

$$a_1 + \bar{a}_1 = -\frac{1}{2\pi i}\oint_C \frac{p}{z}\, d\zeta = -p.$$

Substituting these results into equation 9-5.37 gives

$$\psi(z) = -\frac{1}{2\pi i}\oint_C \frac{p}{\zeta(\zeta - z)}\, d\zeta + \frac{p}{z} + \frac{a_1 + \bar{a}_1}{z} = 0.$$

The stresses can now be calculated:

$$\sigma_{rr} + \sigma_{\theta\theta} = 2[-p - \bar{a}_1 - p - a_1] = -2p,$$

$$\sigma_{\theta\theta} - \sigma_{rr} + 2i\sigma_{r\theta} = 0.$$

Therefore,

$$\sigma_{r\theta} = 0, \qquad \sigma_{rr} = \sigma_{\theta\theta} = -p.$$

6 Conformal transformations

The general solution to the interior problem has been presented using the Cauchy integral formulas. The interior problem was bounded by the unit circle, which appears to greatly restrict the practical application of this solution. However, the opposite is actually true if we use conformal transformations. Consider two complex regions S and Σ in the z and ξ planes, respectively (Figure 9-14). Let z and ξ be related by the following equation:

$$z = w(\xi), \tag{9-6.1}$$

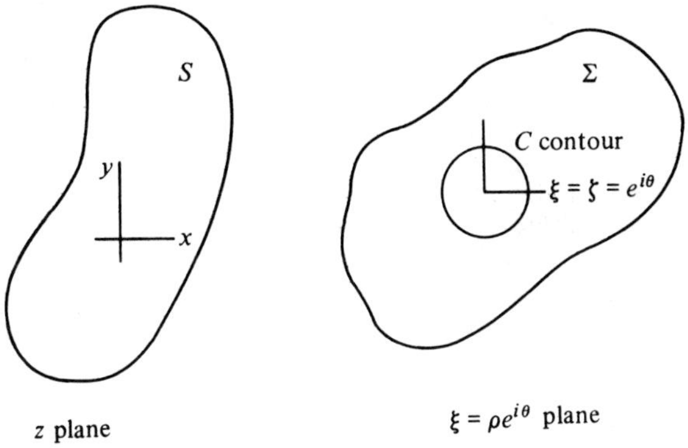

Figure 9-14

where w is a holomorphic function and therefore is single valued. This represents a conformal transformation from ξ to the z plane. Note that the derivatives are related as follows:

$$dz = \frac{dw}{d\xi}\,d\xi. \tag{9-6.2}$$

The regions Σ and S may be infinite or finite. For example, if Σ is finite and S is infinite, there must be a point in the Σ plane to correspond to the point ∞ in the S plane. If this point corresponds to the origin,

$$w(\xi) = \frac{c}{\xi} + \text{holomorphic function}\left(\sum_{n=0}^{\infty} a_n\xi^n\right). \tag{9-6.3}$$

If Σ and S are both infinite, there must be a point corresponding to infinity for both regions:

$$w(\xi) = R\xi + \text{holomorphic function}\left(\sum_{n=0}^{\infty} a_n \xi^{-n}\right). \qquad (9\text{-}6.4)$$

It is beyond the scope of this work to give but a few of the many conformal tranformations that are useful. A few examples are shown in Figure 9-15. Note that the boundary directions are reversed when the exterior is mapped onto the interior and preserved when the exterior is mapped onto the exterior.

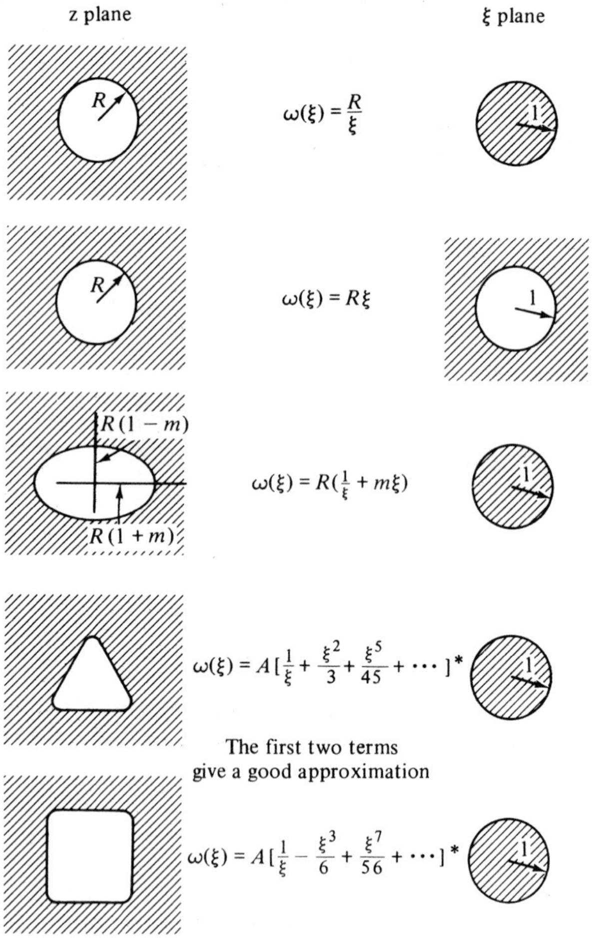

Figure 9-15

* The last two conformal transformations use the Schwarz–Christoffel relations.

Consider the stress functions as functions of ξ instead of z and use the notation

$$\gamma(\xi) = \gamma_1(z) = \gamma_1[w(\xi)], \tag{9-6.5}$$

$$\psi(\xi) = \psi_1(z) = \psi_1[w(\xi)]. \tag{9-6.6}$$

Note that

$$\frac{d\gamma(\xi)}{dz} = \frac{d\gamma}{d\xi}\frac{d\xi}{dz} = \frac{\gamma'(\xi)}{w'(\xi)}. \tag{9-6.7}$$

We write the boundary tractions in the form

$$\begin{aligned} i\int (t_x^n + it_y^n)\,ds &= \gamma_1(z) + z\overline{\gamma_1'(z)} + \overline{\psi_1(z)} \\ &= \gamma(\xi) + \frac{w(\xi)}{\overline{w'(\xi)}}\overline{\gamma'(\xi)} + \overline{\psi(\xi)}. \end{aligned} \tag{9-6.8}$$

After lengthy calculations the stresses become

$$\sigma_{\rho\rho} + \sigma_{\theta\theta} = 2\left\{\frac{\gamma'(\xi)}{w'(\xi)} + \frac{\overline{\gamma'(\xi)}}{\overline{w'(\xi)}}\right\}, \tag{9-6.9}$$

$$\sigma_{\theta\theta} - \sigma_{\rho\rho} + 2i\sigma_{\rho\theta} = \frac{2\xi^2}{\rho^2\overline{w'(\xi)}}\left\{\overline{w(\xi)}\left[\frac{\gamma''(\xi)}{w'(\xi)} - \frac{\gamma'(\xi)w''(\xi)}{[w'(\xi)]^2}\right] + \psi'(\xi)\right\}. \tag{9-6.10}$$

If an external region is mapped into the interior of the unit circle, the equations for the stress functions are obtained by use of equations 9-3.51 and 9-3.52:

$$\gamma_1(z) = -\frac{\log z}{2\pi(1+\eta)}F + \frac{XX + YY}{4}z + \gamma_1^{**}(z), \tag{9-6.11}$$

$$\psi_1(z) = \frac{\eta \log z}{2\pi(1+\eta)}\bar{F} + \frac{YY - XX + 2iXY}{2}z + \psi_1^{**}(z), \tag{9-6.12}$$

where F is the complex resultant force acting on the interior of the hole, and $\psi_1^{**}(z)$ and $\gamma_1^{**}(z)$ are holomorphic functions exterior to the hole. These can be written

$$\gamma(\xi) = -\frac{F}{2\pi(1+\eta)}\log[w(\xi)] + \frac{XX + YY}{4}w(\xi) + \gamma^*(\xi), \tag{9-6.13}$$

$$\psi(\xi) = \frac{\eta\bar{F}}{2\pi(1+\eta)}\log[w(\xi)] + \frac{YY - XX + 2iXY}{2}w(\xi) + \psi^*(\xi). \tag{9-6.14}$$

$\gamma^*(\xi)$ and $\psi^*(\xi)$ are holomorphic inside the unit circle. Note that the conformal transformation takes the form

$$w(\xi) = \frac{c}{\xi} + \text{holomorphic function.} \tag{9-6.15}$$

Therefore, the log $[w(\xi)]$ is written

$$\log[w(\xi)] = -\log\xi + \text{holomorphic function.} \tag{9-6.16}$$

The complex stress functions become

$$\gamma(\xi) = \frac{F}{2\pi(1+\eta)} \log \xi + \frac{XX + YY}{4} \frac{c}{\xi} + \gamma^*(\xi), \tag{9-6.17}$$

$$\psi(\xi) = -\frac{\eta \bar{F}}{2\pi(1+\eta)} \log \xi + \frac{YY - XX + 2iXY}{2} \frac{c}{\xi} + \psi^*(\xi). \tag{9-6.18}$$

We summarize the external and the internal problems where stress tractions are given on the boundary, as follows:

$$\gamma^*(\zeta) + \frac{w(\zeta)}{\overline{w'(\zeta)}} \overline{\gamma'^*(\zeta)} + \overline{\psi^*(\zeta)} = g^0, \qquad \zeta = e^{i\theta},$$

where $g^0 = g = i \int (t_x^n + it_y^n)\, d\zeta,$ interior problem,

$$\begin{aligned} g^0 = g &- \frac{F}{2\pi} \log \zeta - \frac{XX + YY}{4} \frac{c}{\zeta} \\ &- \frac{w(\zeta)}{\overline{w'(\zeta)}} \left[\frac{\bar{F}\zeta}{2\pi(1+\eta)} - \frac{XX + YY}{4} \bar{c}\zeta^2 \right] \\ &- \frac{YY - XX - 2iXY}{2} \bar{c}\zeta, \qquad \text{exterior problem.} \end{aligned} \tag{9-6.19}$$

The approach used earlier for the interior may be used to solve more complex geometries. As an example, let us consider the case of an elliptical hole in an infinite plate. The mapping function needed to map this region onto the interior of the unit circle is

$$w(\xi) = R\left(\frac{1}{\xi} + m\xi\right). \tag{9-6.20}$$

The function required in equation 9-6.19 is

$$\frac{w(\zeta)}{\overline{w'(\zeta)}} = \frac{1}{\zeta}\left(\frac{1 + m\zeta^2}{m - \zeta^2}\right). \tag{9-6.21}$$

The defining equation becomes

$$\gamma^*(\zeta) + \frac{1}{\zeta}\left(\frac{1 + m\zeta^2}{m - \zeta^2}\right)\overline{\gamma'^*(\zeta)} + \overline{\psi^*(\zeta)} = g^0. \tag{9-6.22}$$

Note that $\gamma^*(\zeta)$ and $\psi^*(\zeta)$ are holomorphic inside the unit circle. Operating on equation 9-6.22 with the operator $(1/2\pi i) \oint 1/(\zeta - \xi)\, d\zeta$ yields

$$\gamma^*(\xi) + \frac{1}{2\pi i} \oint \frac{1 + m\zeta^2}{\zeta(m - \zeta^2)} \frac{\overline{\gamma'^*(\zeta)}}{(\zeta - \xi)}\, d\zeta + \overline{\psi^*(0)} = \frac{1}{2\pi i} \oint \frac{g^0\, d\zeta}{(\zeta - \xi)}. \tag{9-6.23}$$

We evaluate the second term by extending the definition of the Cauchy integral formula. Let $f(\alpha)$ be holomorphic outside C. Let C_1 approach ∞ and remember that $f(\alpha)$ is bounded. The second integral approaches $f(\infty)$ (let $\alpha = Re^{i\theta}$ and $R \to \infty$):

$$\frac{1}{2\pi i} \oint_C \frac{f(\alpha)}{\alpha - z}\, d\alpha + f(\infty) = \begin{cases} f(z), & z \text{ outside } C \\ 0, & z \text{ inside } C \end{cases}.$$

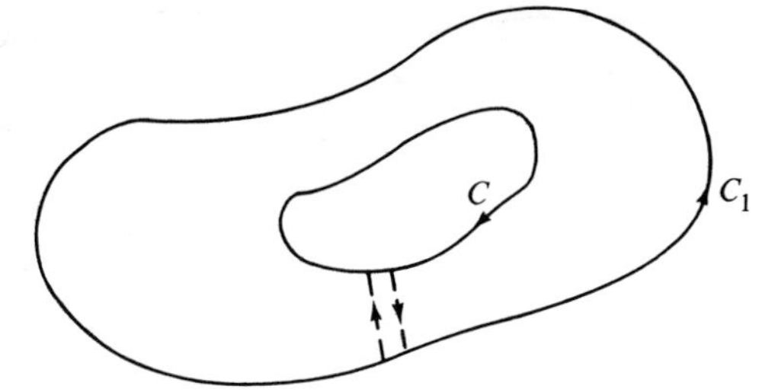

Figure 9-16

Reversing the direction of the integral gives

$$\frac{1}{2\pi i}\oint_C \frac{f(\alpha)}{\alpha - z}\,d\alpha = \begin{cases} -f(z) + f(\infty), & z \text{ outside } C \\ f(\infty), & z \text{ inside } C \end{cases}.$$

This is a generalization of Cauchy's integral formula. Now let a be some finite point of the z plane and let the function $f(z)$ have the form

$$f(z) = f_0(z) + G(z)$$

in the neighborhood of this point, where

$$G(z) = \frac{A_1}{z - a} + \frac{A_2}{(z - a)^2} + \cdots + \frac{A_n}{(z - a)^n}.$$

At this point, $f(z)$ has a pole of order n with principal part $G(z)$. In the neighborhood of $z = \infty$, $f_0(z)$ is holomorphic and

$$G_\infty(z) = A_0 + A_1 z + \cdots + A_n z^n,$$

or a pole of order n at infinity.

Let $f(z)$ be holomorphic inside and on C except at $a_1 \ldots a_n$, where $f(z)$ has poles; then

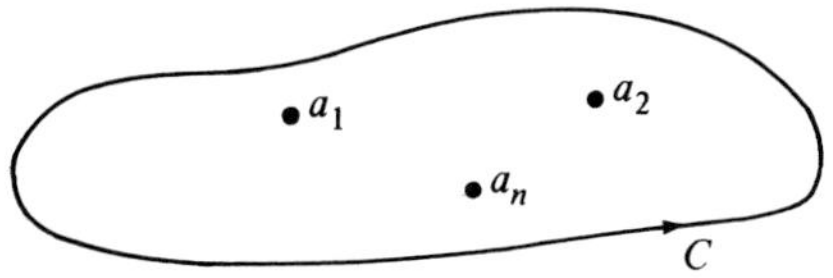

Figure 9-17

$$\frac{1}{2\pi i}\oint \frac{f(\alpha)\,d\alpha}{\alpha - g} = \begin{cases} f(z) - G_1(z) \cdots - G_n(z), & z \text{ outside } C \\ -G_1(z) \cdots - G_n(z), & z \text{ inside } C \end{cases}.$$

Let $f(z)$ be holomorphic outside of C except at $a_1 \ldots a_n$ and ∞, then

$$\frac{1}{2\pi i}\oint_C \frac{f(\alpha)}{\alpha - z}\,d\alpha = \begin{cases} -f(z) + G_1(z) + \cdots + G_n(z) + G_\infty(z), & z \text{ outside } C \\ G_1(z) + \cdots + G_n(z) + G_\infty(z), & z \text{ inside } C \end{cases}.$$

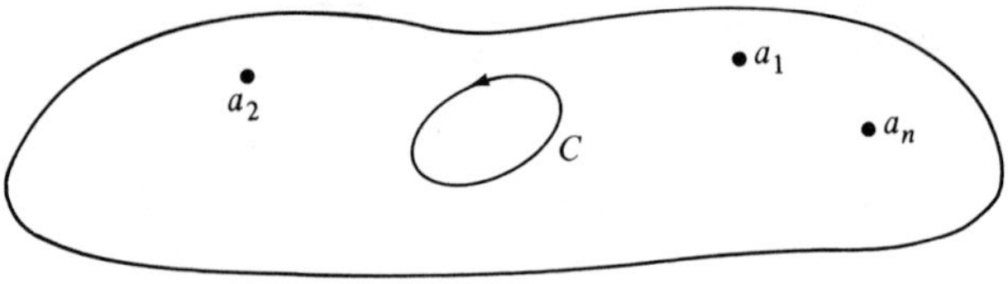

Figure 9-18

Let us now return to the elasticity problem and designate the integrand of the second term by $q(\zeta)$:

$$q(\zeta) = \frac{1 + m\zeta^2}{\zeta(m - \zeta^2)}\overline{\gamma^{*\prime}(\zeta)}. \tag{9-6.24}$$

This function is holomorphic outside the unit circle and

$$\frac{1}{2\pi i}\oint \frac{q(\zeta)}{\zeta - \xi}\,d\zeta = +\,q(\infty) = 0. \tag{9-6.25}$$

The boundary equation becomes

$$\gamma^*(\xi) + \overline{\psi^*(0)} = \frac{1}{2\pi i}\oint \frac{g^0}{(\zeta - \xi)}\,d\zeta. \tag{9-6.26}$$

If we set $\gamma^*(0) = 0$ and equate ξ to zero, we obtain

$$\overline{\psi^*(0)} = \frac{1}{2\pi i}\oint \frac{g^0}{\zeta}\,d\zeta. \tag{9-6.27}$$

The stress function is now written

$$\gamma^*(\xi) = \frac{1}{2\pi i}\oint g^0\left[\frac{1}{\zeta - \xi} - \frac{1}{\zeta}\right]d\zeta = \frac{\xi}{2\pi i}\oint \frac{g^0}{\zeta(\zeta - \xi)}\,d\zeta. \tag{9-6.28}$$

Returning to equation 9-6.22 and taking the complex conjugate of this equation yields

$$\overline{\gamma^*(\zeta)} + \frac{\overline{w(\zeta)}}{w'(\zeta)}\gamma^{*\prime}(\zeta) + \psi^*(\zeta) = \bar{g}^0. \tag{9-6.29}$$

Operating on this equation with $(1/2\pi i)\oint 1/(\zeta - \xi)\,d\zeta$ yields

$$\overline{\gamma^*(0)} + \frac{1}{2\pi i}\oint \frac{\overline{w(\zeta)}}{w'(\zeta)}\frac{\gamma^{*\prime}(\zeta)}{(\zeta - \xi)}\,d\zeta + \psi^*(\xi) = \frac{1}{2\pi i}\oint \frac{\bar{g}^0}{\zeta - \xi}\,d\zeta. \tag{9-6.30}$$

We have taken $\gamma^*(0) = 0$, and the second term can be written

$$-\frac{1}{2\pi i}\oint \frac{\zeta(\xi^2 + m)}{(1 - m\zeta^2)}\frac{\gamma^{*\prime}(\zeta)}{(\zeta - \xi)}\,d\zeta = -\frac{\xi(\xi^2 + m)}{1 - m\xi^2}\gamma^{*\prime}(\xi). \tag{9-6.31}$$

This result can also be obtained by calculus of residues by noting that the numerator is a holomorphic function and there is a simple pole at $\zeta = \xi$.

Because $m < 1$ and $|\zeta| \leq 1$, there are no other poles. The second stress function becomes

$$\psi^*(\xi) = \frac{\xi(\xi^2 + m)}{1 - m\xi^2}\gamma^{*\prime}(\xi) + \frac{1}{2\pi i}\oint \frac{\bar{g}^0}{\zeta - \xi}\,d\zeta. \qquad (9\text{-}6.32)$$

Equations 9-6.28 and 9-6.32 yield the desired solution.

If, for example, we consider the case of an infinite plate subjected to pure tension with an elliptic hole, examination of equation 9-6.19 gives

$$\begin{aligned} g^0 &= -\frac{XX}{4}\frac{R}{\zeta} + \frac{1}{\zeta}\left(\frac{1 + m\zeta^2}{m - \zeta^2}\right)\frac{XX}{4}R\zeta^2 + \frac{XX}{2}R\zeta \\ &= \frac{XX}{4}R\left\{\left(2\zeta - \frac{1}{\zeta}\right) + \zeta\left(\frac{1 + m\zeta^2}{m - \zeta^2}\right)\right\}. \end{aligned} \qquad (9\text{-}6.33)$$

The required integrals are

$$\gamma^*(\xi) = \frac{\xi}{2\pi i}\oint \frac{g^0}{\zeta(\zeta - \xi)}\,d\zeta = \frac{XX}{4}R\xi(2 - m), \qquad (9\text{-}6.34)$$

and

$$\psi^*(\xi) = \frac{1}{2\pi i}\oint \frac{\bar{g}^0}{\zeta - \xi}\,d\zeta + \frac{\xi(\xi^2 + m)}{1 - m\xi^2}\frac{XX}{4}R(2 - m) \qquad (9\text{-}6.35)$$

$$= \frac{XX}{2}R\frac{\xi}{\xi^2 m - 1}(m^2 + 1 - \xi^2 - m). \qquad (9\text{-}6.36)$$

Substitution into equations 9-6.17 and 9-6.18 yields

$$\gamma(\xi) = \frac{XXR\xi}{4}\left\{2 - m + \frac{1}{\xi^2}\right\}, \qquad (9\text{-}6.37)$$

$$\psi(\xi) = \frac{XXR\xi}{2(\xi^2 m - 1)}[m^2 - 1 - (\xi^2 + m)] - \frac{XXR}{2\xi}. \qquad (9\text{-}6.38)$$

The stresses are calculated by use of equations 9-6.9 and 9-6.10. Equation 9-6.9 yields

$$\begin{aligned} \sigma_{\rho\rho} + \sigma_{\theta\theta} &= 2\left\{\frac{\gamma'(\xi)}{w'(\xi)} + \overline{\frac{\gamma'(\xi)}{w'(\xi)}}\right\} \\ &= XX\,\mathrm{Re}\left\{\frac{(2\xi^2 - m\xi^2 - 1)(m\bar{\xi}^2 - 1)}{m^2(\xi\bar{\xi})^2 - m(\xi^2 + \bar{\xi}^2) + 1}\right\}. \end{aligned} \qquad (9\text{-}6.39)$$

The highest stresses occur on the boundary where $\xi = e^{i\theta}$ and $\sigma_{\rho\rho}$ is zero. At the boundary, equation 9-6.39 can be written

$$\sigma_{\theta\theta} = XX\left\{\frac{2m + 1 - 2\cos 2\theta - m^2}{m^2 - 2m\cos 2\theta + 1}\right\}, \qquad (9\text{-}6.40)$$

at

$$\theta = 0, \qquad \sigma_{\theta\theta} = -XX,$$

$$\theta = \frac{\pi}{2}, \qquad \sigma_{\theta\theta} = -XX\left(\frac{m - 3}{m + 1}\right).$$

Note that $m = 0$ corresponds to a circular hole, and the solution is the same as the solution given previously. If m approaches -1, the ellipse is perpen-

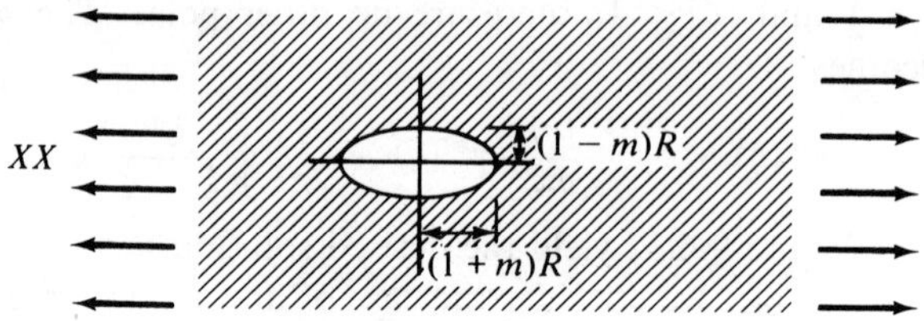

Figure 9-19

dicular to the loading and the stresses at $\theta = \pi/2$ become unbounded as one would expect. Note that these solutions have been computed in the ξ plane. $\sigma_{\theta\theta}$ corresponds to the tangential stresses on the boundary.*

PROBLEMS

9.1. If $3x^2y - y^3$ is the real part of an analytic function of z, determine the imaginary part.

9.2. Prove that xy^2 cannot be the real part of an analytic function.

9.3. Determine if $2xy + i(x^2 - y^2)$ is an analytic function.

9.4. Consider the function $f(z) = u + iv$, where $u = e^x(x \cos y - y \sin y)$ and $v = e^x(y \cos y + x \sin y)$.
(a) Establish that $f(z)$ is an analytic function.
(b) Write $f(z)$ in terms of the single variable z.

9.5. Show that the real and imaginary parts of any twice differentiable function of the form $f(\bar{z})$ satisfy Laplace's equation, but that such a function is nowhere an analytic function of z unless it is a constant.

9.6. Find the cyclic function of the following with respect to a circuit C that encloses the origin.
(a) $z \ln z$, (b) $\bar{z} \ln z$, (c) $z \ln \bar{z}$, (d) $z \ln z + \bar{z} \ln \bar{z}$.

9.7. Find the stresses in the disk shown.

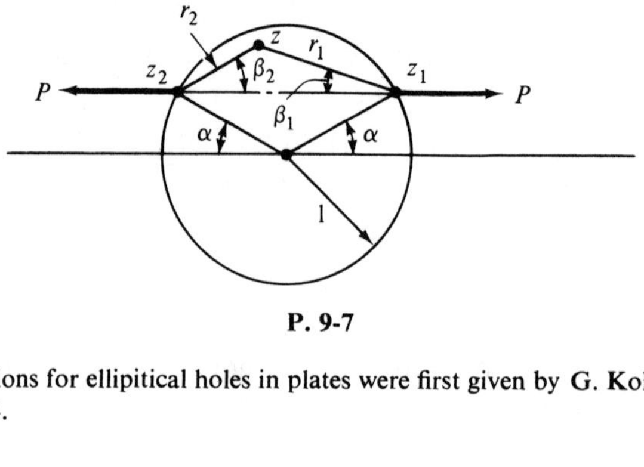

P. 9-7

* Solutions for ellipitical holes in plates were first given by G. Kolosoff, *Z. Math. Physik*, **62**, 1914.

9.8. Solve the problem of an infinite plate in uniaxial tension with a circular hole of radius 1.

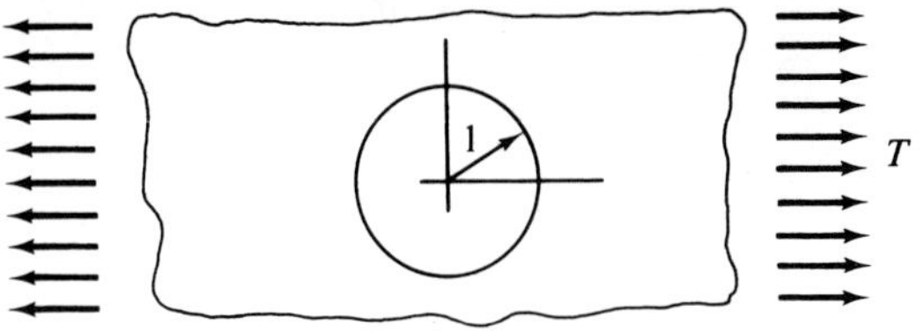

P. 9-8

9.9. Evaluate the integral $(1/2\pi i)\oint_C f(\alpha)/(\alpha - z)\,d\alpha$, where $f(\alpha) = (\alpha^3 + 2\alpha + 3)/[\alpha(\alpha - 2)]$, C is the unit circle, and z is inside the unit circle.
(a) Consider $f(\alpha)$ to be the boundary value of a function holomorphic outside C except at a finite number of points, and use the Cauchy formulas.
(b) Consider $f(\alpha)$ to be the boundary value of a function holomorphic inside C except at a finite number of points, and use Cauchy's formulas.

9.10. Determine the complex stress function for an infinite plane with a triangular hole under the influence of equal biaxial tension T at infinity. Use the first two terms of the mapping function given in Figure 9-15.

9.11. Determine the stress functions for an annulus under external radial stress $\sigma_{rr}|_{r,b} = p \sin 2\theta$ and zero shear. The inner boundary is stress free.

9.12. Determine the stress functions for an infinite plate with a square hole under uniaxial tension at ∞, T, along the x axis.

CHAPTER 10

FINITE-DIFFERENCE AND FINITE-ELEMENT METHODS

1 Introduction

In this chapter we shall present two methods of analyzing elastic bodies that discretize the body to some degree and obtain approximate solutions at particular points. These two methods are called the *finite-difference* and the *finite-element methods* and both require the use of digital computers in the inversion of large matricies.

Finite-difference methods have a long history, including contributions by Newton, Laplace, Gauss, Bessel, and many others. On the other hand, the finite-element method is relatively new, being developed by members of the aircraft industry in the early 1950s. This method, first named by Argyris in 1955,* has been used principally to solve problems in structural mechanics.

* J. H. Argyris, *Energy Theorems and Structural Analysis*, Butterworth & Company (Publishers) Ltd., London, 1960 (Reprinted from *Aircraft Eng.*, 1954–1955).

The method of finite differences replaces the defining, differential equation with equivalent difference equations. The boundary conditions are satisfied at discrete points by specifying either the function or its derivatives, replaced by equivalent difference expressions. The results of this analysis are numerical values of the function at discrete points or nodes throughout the body.

The finite-element method divides the elastic body into a number of standard shapes called elements. The elements are analyzed approximately by elasticity theory and assembled to form the elastic body.

2 Finite-element method

The finite-element method follows the procedures used in structural analysis in which the structure is broken into discrete elements, which are in turn analyzed independently and assembled by matrix methods. This method has been extended to include not only beams but plates, shells and two- and three-dimensional elastic elements. It is used as a basic tool of analysis by those who have access to large computer facilities. Although its principal area of application occurs where many different types of structural elements (beams, plates, membranes, etc.) are used together, we shall restrict ourselves to applications in two-dimensional elasticity. These applications indicate all the essential features of the method. We shall use matrix notation when applicable, and shall assume the reader is familiar with this notation.

3 Displacement functions

The fundamental assumption used in this method of analysis is that a plane elastic body may be satisfactorily represented by an assemblage of discrete elements having simplified deformation fields. In order to analyze the body, we divide it by a system of grid lines, the intersections of which are referred to as nodes or nodal points. For the triangular elements used in this analysis, there are three nodal points for each element. The displacements of each element are compatible at these nodal points to the displacements of neighboring elements.

The first step in the analysis is to select a displacement function that gives the displacement of every point within the element. In plane elasticity, the displacement function is of the form of a vector $[g]$:

$$[g] = \begin{bmatrix} u_x(x, y) \\ u_y(x, y) \end{bmatrix}, \qquad (10\text{-}3.1)$$

where $u_x(x, y)$ is the displacement in the x direction and $u_y(x, y)$ is the displacement in the y direction.

If the element is as shown in Figure 10-1, the nodal points can be designated by i, j, and k.

The displacement vector must satisfy the conditions that it does not produce strain in an element when the nodal displacements are caused by rigid body displacements and that it yields a constant strain field when constant strain would, in fact, occur.

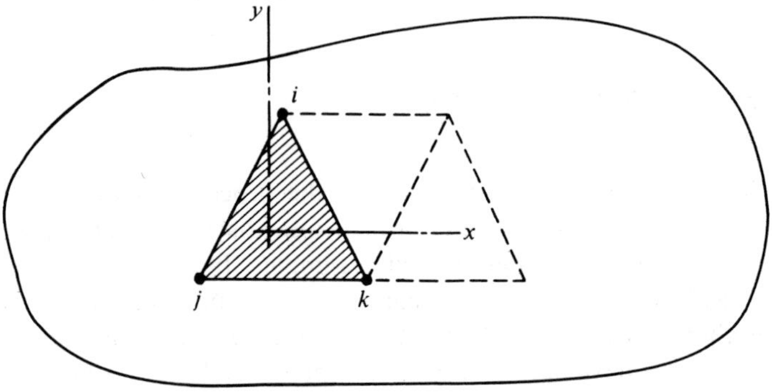

Figure 10-1

These two conditions imply that we must be able to express the displacement function in terms of the nodal displacements. Such an expression takes the following matrix form:

$$[g] = [A][d^e]. \tag{10-3.2}$$

A simple example of this form and the restrictions on the displacement function can be seen by considering the bar in uniaxial tension shown in Figure 10-2.

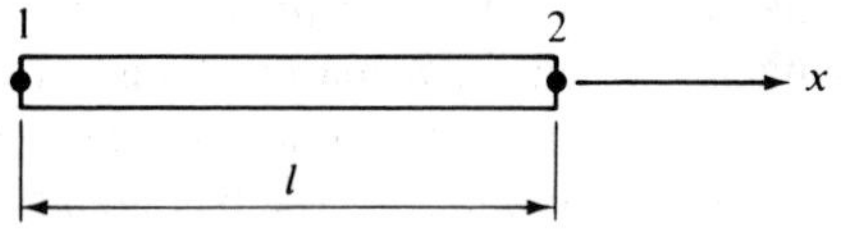

Figure 10-2

In this example the displacement function $[g]$ consists of the displacement u_x only. If u_x is assumed to be a linear function, the necessary conditions are satisfied:

$$u_x = C_1 x + C_2. \tag{10-3.3}$$

If the element has length, l, and the nodal points are numbered 1 and 2, the displacement function may be written in terms of these nodal displacements as

$$u_x = \frac{u_2 - u_1}{l}x + u_1. \qquad (10\text{-}3.4)$$

Note that a rigid-body motion would occur if u_1 equals u_2, and this function would give only a rigid-body displacement in the interior. In a similar manner, this function corresponds to constant strain so that the second condition is also satisfied. This simple function may be written in matrix notation to correspond to equation 10-3.2 as

$$u_x = \left[1 - \frac{x}{l}, \frac{x}{l}\right]\begin{bmatrix} u_1 \\ u_2 \end{bmatrix}. \qquad (10\text{-}3.5)$$

In this case, $[d^e]$ is the matrix giving the nodal displacements of the element.

For plane elasticity, the nodal displacement vector is expressed as a six-component vector involving the u_x and u_y displacements at i, j, and k. This column matrix would take the form

$$[d^e] = \begin{bmatrix} u_x^i \\ u_y^i \\ u_x^j \\ u_y^j \\ u_x^k \\ u_y^k \end{bmatrix}. \qquad (10\text{-}3.6)$$

Equation 10-3.2 is normally formed by assuming a general displacement expression with arbitrary constants and evaluating these constants in terms of the nodal displacements. Inversion of this equation would yield equation 10-3.2. The analysis will be carried out in symbolic notation initially and the matrix entries will be developed later for a triangular element.

4 Stresses and strains

The strains are obtained in terms of the nodal displacements and have the form

$$[\epsilon] = [B][g] = [B][A][d^e]. \qquad (10\text{-}4.1)$$

The general stress–strain relation is written as the following matrix equation:

$$[\epsilon] = [C][\sigma] + [\epsilon_0], \qquad (10\text{-}4.2)$$

where $[C]$ is a matrix relating the elastic stress to the elastic strain and $[\epsilon_0]$ is the initial strain caused by temperature changes, shrinkage, crystal growth, etc. Solving for the stresses yields the following matrix equation:

$$[\sigma] = [D][\epsilon - \epsilon_0], \quad [D] = [C]^{-1}$$

5 Nodal force—displacement relations

We introduce two more generalized vectors. The nodal force vector has the form

$$[P^e] = \begin{bmatrix} [P^i] \\ [P^j] \\ [P^k] \end{bmatrix}, \tag{10-5.1}$$

where $[P^i]$, $[P^j]$, and $[P^k]$ represent the force vector at i, j, and k nodal points, respectively. The nodal forces are equivalent statically to the boundary stresses and distributed surface loads on the element. The second vector is the body force vector $[q]$, which corresponds to the distributed loads acting on a unit volume of material.

We shall satisfy equilibrium conditions by use of the method of virtual work. Let us designate the virtual displacement of the nodal point by $[d^e*]$. The virtual-displacement method uses a fictitious (virtual) displacement from equilibrium and calculates the virtual work done by the real loads when moved through this virtual displacement. If an isolated system, or portion of a system, is in static equilibrium under the forces acting on it, the resultant external force must equal zero. Therefore, a virtual displacement of a free body in equilibrium will cause no work to be done on or by the system. It follows that the real forces acting on a free body in equilibrium must be such that the total virtual work done by these forces during any virtual displacement is zero. This statement is sometimes called the *principle of virtual displacements.*

The displacement function and the strains, in terms of the virtual displacements, become

$$[g^*] = [A][d^{e*}], \tag{10-5.2}$$

$$[\epsilon^*] = [B][A][d^{e*}]. \tag{10-5.3}$$

The work done by the nodal forces as they move a virtual displacement is

$$[d^{e*}]^T[p^e], \tag{10-5.4}$$

where $[\ \ ]^T$ denotes the transpose.
The work done by the internal forces per unit volume can be expressed in a similar manner:

$$[\epsilon^*]^T[\sigma] - [g^*]^T[q], \tag{10-5.5}$$

where $[q]$ is a distributed body force. Note that the minus sign is associated with the second term as it is really an external loading, as compared with the stresses that are truly internal forces. Using equations 10-5.2 and 10-5.3, the term 10-5.5 can be written

$$[d^{e*}]^T\{[A]^T[B]^T[\sigma] - [A]^T[q]\}. \qquad (10\text{-}5.6)$$

Equating the total external and internal work for the element yields

$$[d^{e*}]^T[P^e] = [d^{e*}]^T \int_{\text{Vol}} \{[A]^T[B]^T[\sigma] - [A]^T[q]\}\, dV. \qquad (10\text{-}5.7)$$

This relationship must be valid for any arbitrary virtual displacement; therefore,

$$[P^e] = \int_{\text{Vol}} \{[A]^T[B]^T[\sigma] - [A]^T[q]\}\, dV. \qquad (10\text{-}5.8)$$

Using equation 10-4.3 for the stress vector yields

$$[P^e] = \int_{\text{Vol}} \{[A]^T[B]^T[D][B][A]\, dV\, [d^e],$$

$$- \int_{\text{Vol}} [A]^T[B][D][\epsilon_0]\, dV - \int_{\text{Vol}} [A]^T[q]\, dV. \qquad (10\text{-}5.9)$$

If we introduce a stiffness matrix defined as

$$[K] = \int_{\text{Vol}} [A]^T[B]^T[D][B][A]\, dV, \qquad (10\text{-}5.10)$$

and nodal forces due to distributed loads,

$$[P^e_q] = -\int_{\text{Vol}} [A]^T[q]\, dV, \qquad (10\text{-}5.11)$$

and those due to initial strain,

$$[P^e_{\epsilon_0}] = -\int_{\text{Vol}} [A]^T[B]^T[D][\epsilon_0]\, dV, \qquad (10\text{-}5.12)$$

we write equation 10-5.9 as follows:

$$[P^e] = [K][d^e] + [P^e_{\epsilon_0}] + [P^e_q], \qquad \text{for the element } e. \qquad (10\text{-}5.13)$$

The stiffness matrix is symmetric because the matrix $[D]$ must be symmetric.

It should be noted that the displacement vector of the element is made up of three parts: the displacement of each of the three nodal points. These correspond to the three parts of the nodal force vector. We might then consider the stiffness matrix to be composed of submatrices as follows:

$$[K^e] = \begin{bmatrix} [k_{ii}] & [k_{ij}] & [k_{ik}] \\ [k_{ji}] & [k_{jj}] & [k_{jk}] \\ [k_{ki}[& [k_{kj}] & [k_{kk}] \end{bmatrix}. \qquad (10\text{-}5.14)$$

It the displacement at nodal point k does not influence the force at nodal point i, then the submatrix $[k_{ik}]$ would be zero. Because we have already noted that the stiffness matrix is symmetric, we also may note that the submatrices $[k_{ij}]$ and $[k_{ji}]$ are equal. This is an expression of the Maxwell–Betti reciprocal theorem.

6 Analysis of a structure

A system of elements must satisfy two conditions: (1) displacement compatibility, and (2) equilibrium. The first condition is automatically satisfied by matching nodal point displacements of neighboring elements.

The second condition, that of overall equilibrium, will now be discussed. Consider a structure loaded by external forces $[R]$, which are considered to be acting at the nodal points. If we consider a structure having n nodal points, $[R]$ becomes

$$[R] = \begin{bmatrix} R_1 \\ R_2 \\ \cdot \\ \cdot \\ \cdot \\ R_n \end{bmatrix} \qquad (10\text{-}6.1)$$

The nodal resultant, R_i at the ith nodal point, is composed of the nodal forces acting at that point. In the two-dimensional-plane problem considered here, R_i would have two components, X_i and Y_i, and R_i would be a two-component subcolumn matrix.

Now for equilibrium at a specific nodal point, say the ith one, the external force must be equal to the sum of the force components at that node:

$$R_i = \sum P_i. \qquad (10\text{-}6.2)$$

The summation is taken over all the elements sharing that nodal point. For example, with the triangular elements shown in Figure 10-3 have six different elements sharing the ith nodal point.

Now examining a nodal force $[P_i]^e$ for one element from equation 10-5.13, we obtain

$$[P_i]^e = \sum_{m=1}^{n} [k_{im}][d_m] + [P_{i\epsilon_0}] + [P_{iq}]. \qquad (10\text{-}6.3)$$

We have extended the sum on m to include all n nodal displacements, placing zero submatrices in those places where the nodal displacements do

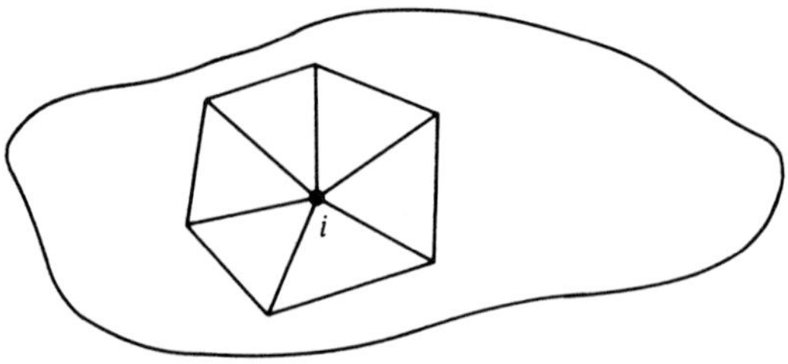

Figure 10-3

not influence the nodal force. Substituting equation 10-6.3 into equation 10-6.2 yields

$$[R_i] = \sum_{m=1}^{n} \{\sum^{e} [k_{im}][d_m]\} + \sum^{e} [P_{i\epsilon_0}]^e + \sum^{e} [P_{iq}]^e. \qquad (10\text{-}6.4)$$

The inner sum in the double sum is the sum over the elements containing the ith nodal point. For example, each of these elements might be expected to contain the submatrix $[k_{ii}]$, so that the sum over these elements would give the sum of these submatrices.

If we order the n equations of the form 10-6.4, we obtain the general structure equation

$$[R] = [K][d] + [P_{\epsilon_0}] + [P_q], \qquad (10\text{-}6.5)$$

where R, d, P_{ϵ_0}, and P_q are all column matrices covering the n nodal points. The imth submatrix of K is

$$[K_{im}] = \sum^{e} [k_{im}], \qquad \text{sum over elements.} \qquad (10\text{-}6.6)$$

In a similar manner, the column sub-matrices $[P_{i\epsilon_0}]$ and $[P_{iq}]$ are

$$[P_{i\epsilon_0}] = \sum^{e} [P_{i\epsilon_0}]^e, \qquad \text{sum over elements,} \qquad (10\text{-}6.7)$$

$$[P_{iq}] = \sum^{e} [P_{iq}]^e, \qquad \text{sum over elements.} \qquad (10\text{-}6.8)$$

Equation 10-6.5 may now be written in the form

$$[K][d] = [R] - [P_{\epsilon_0}] - [P_q]. \qquad (10\text{-}6.9)$$

We know that a one-dimensional body has a single degree of freedom corresponding to a rigid-body displacement, a two-dimensional element has three rigid-body motions, and a three-dimensional body has six rigid-body degrees of freedom. It is not too surprising that the matrix $[K]$ is singular. We must specify not less than one, three, or six rigid-body motions, that is, one, three, or six nodal displacements. These displacements are generally specified as zero by replacing the corresponding row and column in the stiffness matrix with zero, except for a value of unity on the diagonal, and placing a zero in the corresponding row in the column matrices on the right side. The stiffness matrix is thereby made nonsingular. Another alternative is to eliminate the row and column relating to the specified displacement entry.

A simple examination of tensile member may make this procedure clearer. The length of the element will be taken as l and the nodal points will be designated as i and j (Figure 10-4). The displacement function for the

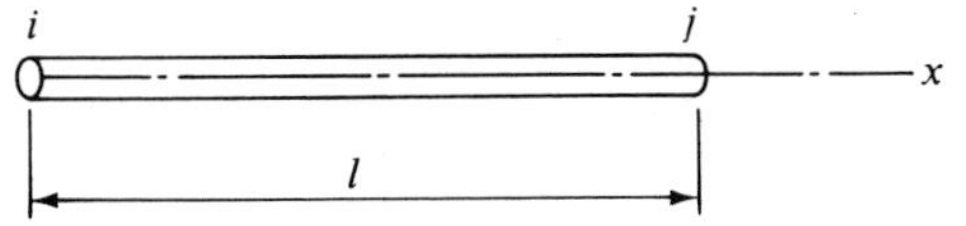

Figure 10-4

axial displacement within the element is chosen as

$$g = a_1 x + a_2. \qquad (10\text{-}6.10)$$

This function satisfies the requirements of an admissible displacement function. We can solve for a_1 and a_2 in terms of the nodal displacements:

$$g = [A][d^e], \qquad (10\text{-}6.11)$$

where

$$[A] = \left[1 - \frac{x}{l}, \frac{x}{l}\right], \qquad [d^e] = \begin{bmatrix} u_x^i \\ u_x^j \end{bmatrix}. \qquad (10\text{-}6.12)$$

The normal strain in the x direction is

$$\epsilon_{xx} = \frac{\partial g}{\partial x} = \frac{1}{l}[-1, 1][d^e]. \qquad (10\text{-}6.13)$$

For an isotropic material the single nonvanishing stress is

$$\sigma_{xx} = \frac{E}{l}[-1, 1][d^e] - E[\epsilon_0]. \qquad (10\text{-}6.14)$$

The nodal forces are

$$[P^e] = \begin{bmatrix} P_x^i \\ P_x^j \end{bmatrix}. \qquad (10\text{-}6.15)$$

The work done by the nodal forces during a virtual displacement becomes

$$[d^{e*}]^T[P^e] = \int_V [\epsilon^*]^T[\sigma]\, dV. \qquad (10\text{-}6.16)$$

Therefore, the nodal forces are written as

$$[P^e] = [K][d^e] - [P_{\epsilon 0}], \qquad (10\text{-}6.17)$$

where

$$\begin{aligned} [K] &= \int \frac{E}{l^2} \begin{bmatrix} -1 \\ 1 \end{bmatrix} [-1, 1]\, dV \\ &= \frac{EA}{l} \begin{bmatrix} 1 & -1 \\ -1 & 1 \end{bmatrix}. \end{aligned} \qquad (10\text{-}6.18)$$

Consider the structure shown in Figure 10-5. The reaction matrix is

$$[R] = \begin{bmatrix} R_1 \\ F \\ 0 \\ P \end{bmatrix} = [\sum_{m=1} \sum_{\text{elements}} [k_{im}][d_m]]. \qquad (10\text{-}6.19)$$

This can be written

$$[R] = \frac{EA}{l} \begin{bmatrix} 1 & -1 & 0 & 0 \\ -1 & 2 & -1 & 0 \\ 0 & -1 & 2 & -1 \\ 0 & 0 & -1 & 1 \end{bmatrix} \begin{bmatrix} d_1 \\ d_2 \\ d_3 \\ d_4 \end{bmatrix}. \qquad (10\text{-}6.20)$$

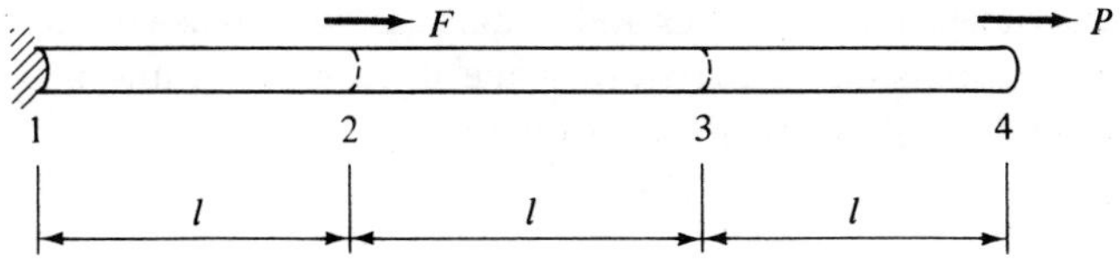

Figure 10-5

The boundary condition is that d_1 is zero. We note that the determinant of $[K]$ is zero. If $[K]$ is replaced by

$$[K_r] = \frac{EA}{l}\begin{bmatrix} 1 & 0 & 0 & 0 \\ 0 & 2 & -1 & 0 \\ 0 & -1 & 2 & -1 \\ 0 & 0 & -1 & 1 \end{bmatrix}. \qquad (10\text{-}6.21)$$

The determinant is 1.

The inverse of $[K]$ is

$$[K]^{-1} = \frac{l}{EA}\begin{bmatrix} 1 & 0 & 0 & 0 \\ 0 & 1 & 1 & 1 \\ 0 & 1 & 2 & 2 \\ 0 & 1 & 2 & 3 \end{bmatrix}. \qquad (10\text{-}6.22)$$

The displacements can now be calculated:

$$d_1 = 0$$

given

$$d_2 = \frac{l}{EA}[F + P],$$

$$d_3 = \frac{l}{EA}[F + 2P], \qquad (10\text{-}6.23)$$

$$d_4 = \frac{l}{EA}[F + 3P].$$

7 Facet stiffness matrix

If we consider only the in-plane forces in a triangular element, we shall have a plane stress elasticity problem. In general, a finite-element analysis would include bending, and this effect would be added to the plane stress considerations. Our attempt here is to familiarize the reader with the finite-element method of analysis and we shall restrict ourselves to the two-dimensional analysis.

The generalized Hooke's law may be written

$$[\sigma] = [D]\,([\epsilon] - [\epsilon_0].) \qquad (10\text{-}7.1)$$

If the three-dimensional stresses and strains are written as column matrices, $[D]$ becomes a six-by-six matrix. For plane stress σ_{zz}, σ_{yz}, and σ_{xz} are taken to be zero and the stress vector is written as

$$\begin{bmatrix} \sigma_{xx} \\ \sigma_{yy} \\ \sigma_{xy} \\ \sigma_{zz} \\ \sigma_{xz} \\ \sigma_{yz} \end{bmatrix} = \begin{bmatrix} D_{11} & D_{12} & D_{13} & D_{14} & D_{15} & D_{16} \\ & D_{22} & D_{23} & D_{24} & D_{25} & D_{26} \\ & & D_{33} & D_{34} & D_{35} & D_{36} \\ & \text{Sym.} & & D_{44} & D_{45} & D_{46} \\ & & & & D_{55} & D_{56} \\ & & & & & D_{66} \end{bmatrix} \begin{bmatrix} \epsilon_{xx} \\ \epsilon_{yy} \\ \gamma_{xy} \\ \epsilon_{zz} \\ \gamma_{xz} \\ \gamma_{yz} \end{bmatrix}. \tag{10-7.2}$$

The fourth equation yields

$$\sigma_{zz} = 0 = D_{41}\epsilon_{xx} + D_{42}\epsilon_{yy} + D_{43}\gamma_{xy} + D_{44}\epsilon_{zz} + D_{45}\gamma_{xz} + D_{46}\gamma_{yz}. \tag{10-7.3}$$

Equation 10-7.1 may be reduced to a form involving only the three plane stresses and strains. We shall use the engineering shear strain, as it is a little easier to work with in this case. Note that it is not necessary to assume isotropy in this development:

$$[\sigma] = \begin{bmatrix} \sigma_{xx} \\ \sigma_{yy} \\ \sigma_{xy} \end{bmatrix}, \qquad [\epsilon] = \begin{bmatrix} \epsilon_{xx} \\ \epsilon_{yy} \\ \gamma_{xy} \end{bmatrix}. \tag{10-7.4}$$

The initial strain vector due to temperature effects is

$$[\epsilon_0] = \begin{bmatrix} \alpha T \\ \alpha T \\ 0 \end{bmatrix}. \tag{10-7.5}$$

The displacement function of each displacement component is assumed to be a linear function of the coordinates:

$$\begin{aligned} u_x &= a_1 x + b_1 y + c_1, \\ u_y &= a_2 x + b_2 y + c_2, \end{aligned} \tag{10-7.6}$$

These displacement functions satisfy the required conditions. Consider the facet shown in Figure 10-6. Let the coordinates of the three nodal points be (x_i, y_i), (x_j, y_j) and (x_k, y_k). The constants in equation 10-7.6 are evaluated in terms of the nodal displacements:

$$\begin{aligned} u_x^i &= a_1 x_i + b_1 y_i + c_1, \\ u_x^j &= a_1 x_j + b_1 y_j + c_1, \\ u_x^k &= a_1 x_k + b_1 y_k + c_1. \end{aligned} \tag{10-7.7}$$

The determinant of the coefficient matrix multiplying the constants is

$$x_i y_j + x_k y_i + x_j y_k - x_k y_j - x_i y_k - x_j y_i. \tag{10-7.8}$$

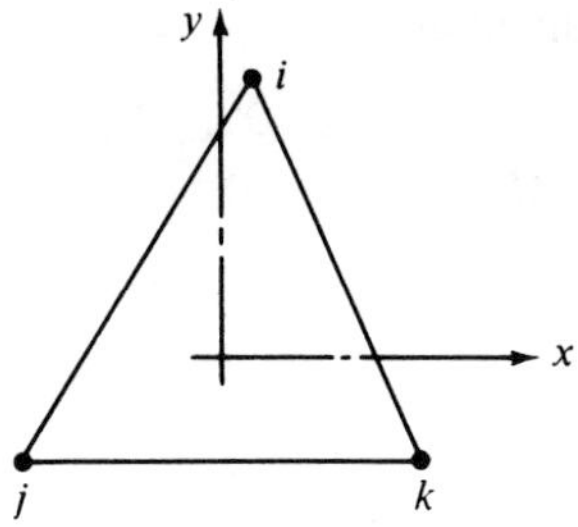

Figure 10-6

If we note from the sketch of the facet that $y_j = y_k$, the coefficient determinant becomes

$$y_i(x_k - x_j) - y_j(x_k - x_j) = (y_i - y_j)(x_k - x_j). \qquad (10\text{-}7.9)$$

Noting that $(y_k - y_j)$ is the altitude of the triangle and $(x_k - x_j)$ is the base, the coefficient determinant can be interpreted as twice the facet area. Using Cramer's rule, the coefficients become

$$a_1 = \frac{1}{2A}\{u_x^i y_{jk} + u_x^j y_{ki} + u_x^k y_{ij}\}, \qquad (10\text{-}7.10)$$

where A is the area and

$$y_{jk} = y_j - y_k. \qquad (10\text{-}7.11)$$

In a similar manner,

$$b_1 = \frac{1}{2A}\{u_x^i x_{kj} + u_x^j x_{ik} + u_x^k x_{ij}\}, \qquad (10\text{-}7.12)$$

$$c_1 = \frac{1}{2A}\{u_x^i r_i + u_x^j r_j + u_x^k r_k\}, \qquad (10\text{-}7.13)$$

where

$$\begin{aligned} x_{jk} &= x_j - x_k, \\ r_i &= x_j y_k - x_k y_j, \\ r_j &= x_k y_i - x_i y_k, \\ r_k &= x_i y_j - x_j y_i. \end{aligned} \qquad (10\text{-}7.14)$$

The general displacement function is written

$$[u_x, u_y] = \frac{1}{2A}[x \quad y \quad 1]\begin{bmatrix} y_{jk} & y_{ki} & y_{ij} \\ x_{kj} & x_{ik} & x_{ji} \\ r_i & r_j & r_k \end{bmatrix}\begin{bmatrix} u_x^i & u_y^i \\ u_x^j & u_y^j \\ u_x^k & u_y^k \end{bmatrix}. \qquad (10\text{-}7.15)$$

We now compute the strain vector, noting the relation between strains and displacements:

$$[\epsilon] = \frac{1}{2A}[M \mid N][d^e], \qquad (10\text{-}7.16)$$

where $[M]$ and $[N]$ are defined as

$$[M] = \begin{bmatrix} y_{jk} & y_{ki} & y_{ij} \\ 0 & 0 & 0 \\ x_{kj} & x_{ki} & x_{ji} \end{bmatrix}, \qquad [N] = \begin{bmatrix} 0 & 0 & 0 \\ x_{kj} & x_{ki} & x_{ji} \\ y_{jk} & y_{ki} & y_{ij} \end{bmatrix}. \tag{10-7.17}$$

Defining the virtual work as has been done before yields

$$[d^{*e}]^T[P^e] = \int_V [\epsilon^*]^T[D][\epsilon]\, dV, \tag{10-7.18}$$

$$[P^e] = \frac{t}{4A}\begin{bmatrix} [M]^T \\ [N]^T \end{bmatrix}[D][[M] \mid [N]][d^e]. \tag{10-7.19}$$

The stiffness matrix becomes

$$[K] = \frac{t}{4A}\begin{bmatrix} [M]^T[D][M] & [M]^T[D][N] \\ [N]^T[D][M] & [N]^T[D][N] \end{bmatrix}. \tag{10-7.20}$$

The nodal force vector on the facet consists of forces in the x and y directions. The actual membrane stress distribution must be modeled by statically equivalent nodal forces. It is apparent that for finite elements this modeling is not unique. The stiffness matrix for the element is complete and can be assembled to form the elastic body.

8 Local and global coordinates

The stiffness matrices have been developed with respect to a local coordinate system attached to the element. We have discussed that after the stiffness matrices for all the elements are known, we may combine these to form a stiffness matrix for the entire structure. We cannot directly combine elements that have different coordinate systems.

Let us consider two coordinate systems designated by primed and unprimed symbols (Figure 10-7). We have examined how the components in one coordinate system are related to those in the other. If we designate the direction cosine between the x_i' and the x_j axes by a_{ij}, the transformation of a vector $\bar{u}$ may be written

$$[u'] = [A][u]. \tag{10-8.1}$$

The transformation matrix $[A]$ has some special properties, which have already been examined in Chapter 1. The most important of these is that its inverse is equal to its transpose:

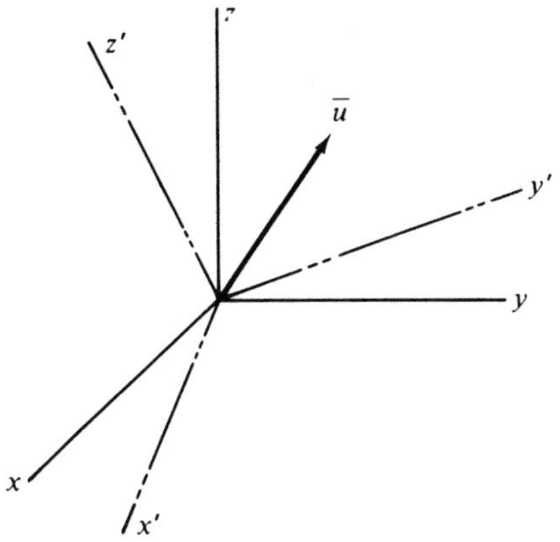

Figure 10-7

$$[A]^{-1} = [A]^T. \tag{10-8.2}$$

A second property is that the determinant of the transformation matrix is ± 1; $+1$ for rotations and -1 for inversions.

Now let us examine the stiffness relationship between a force and displacement in the local coordinates:

$$\begin{bmatrix} P_x \\ P_y \end{bmatrix} = [K] \begin{bmatrix} u_x \\ u_y \end{bmatrix}, \tag{10-8.3}$$

where $\bar{P}$ and $\bar{u}$ are only the force and displacement of one nodal point.

If we establish one overall coordinate system, the global coordinate system, we can relate all vectors to that system, and in that system add corresponding components. We shall designate this system as the x' system.

If $[A]$ is the transformation matrix between a local and the global system,

$$[P]' = [A][P] \quad \text{and} \quad [u]' = [A][u]. \tag{10-8.4}$$

We use these relations with the properties of the transformation matrix to write

$$[A][P] = [A][K][A]^T[A][u], \tag{10-8.5}$$

$$[P]' = [A][K][A][u]'. \tag{10-8.6}$$

Note that $[A][K][A]^T$ is symmetric and is the stiffness matrix in the global coordinate system. We can generalize this when $[P]$ and $[u]$ are six-component column matrices including the displacements at each nodal point of the element:

$$[P] = \begin{bmatrix} \bar{P}^i \\ \bar{P}^j \\ \bar{P}^k \end{bmatrix}, \qquad [d] = \begin{bmatrix} \bar{u}^i \\ \bar{u}^j \\ \bar{u}^k \end{bmatrix}. \tag{10-8.7}$$

The transformation matrix for these vectors becomes

$$[A] = \begin{bmatrix} [A] & 0 & 0 \\ 0 & [A] & 0 \\ 0 & 0 & [A] \end{bmatrix}, \qquad (10\text{-}8.8)$$

where each of the submatrices is the ordinary transformation matrix for a two-component vector.

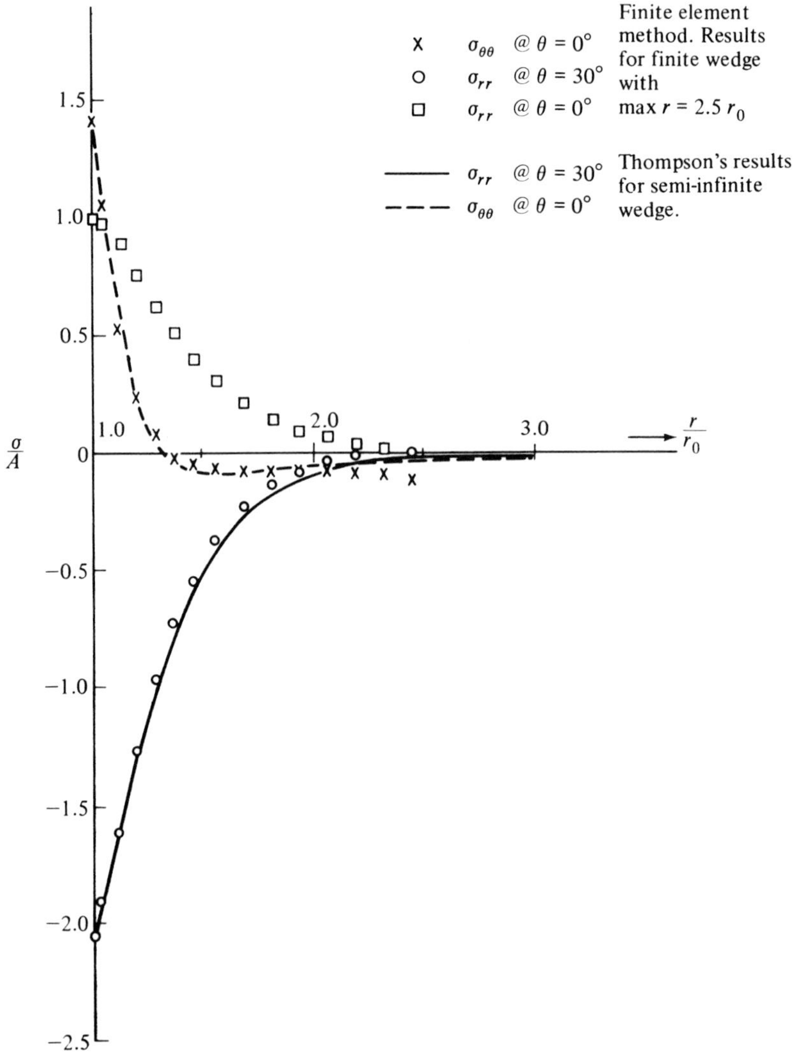

Figure 10-8

9 Finite-element example

The finite-element method depends upon access to a large digital computer. Most computer programs require only the input data of the coordinates of the nodal points and the applied nodal forces. The program will

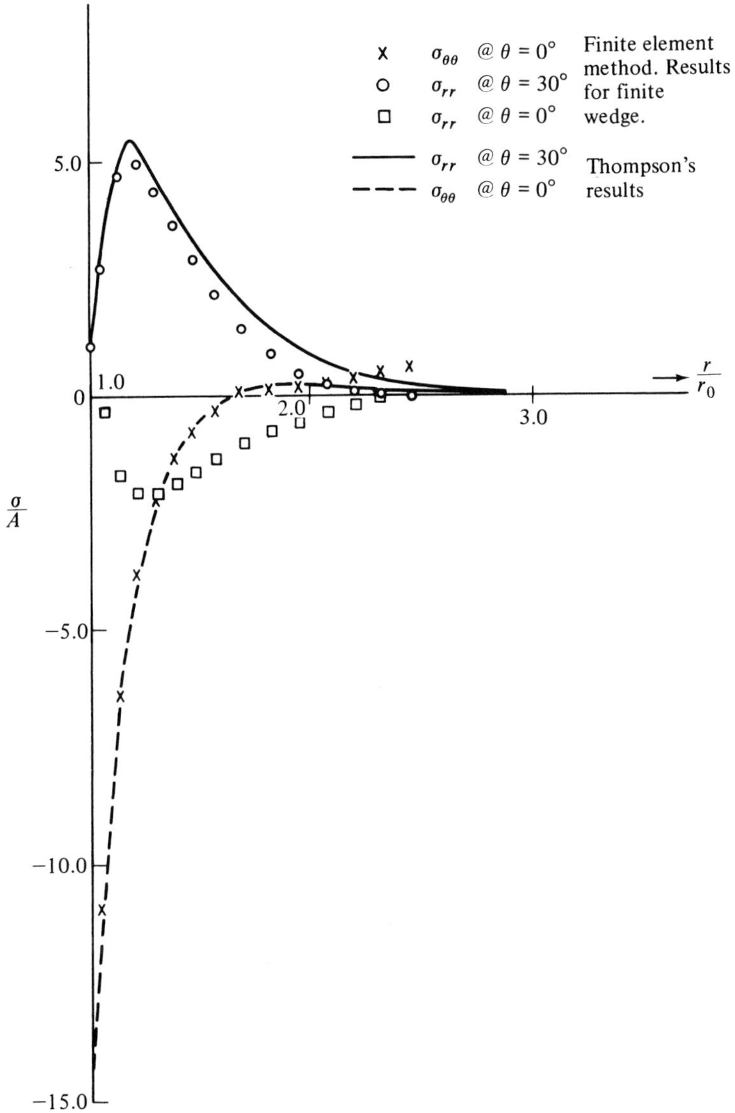

Figure 10-9

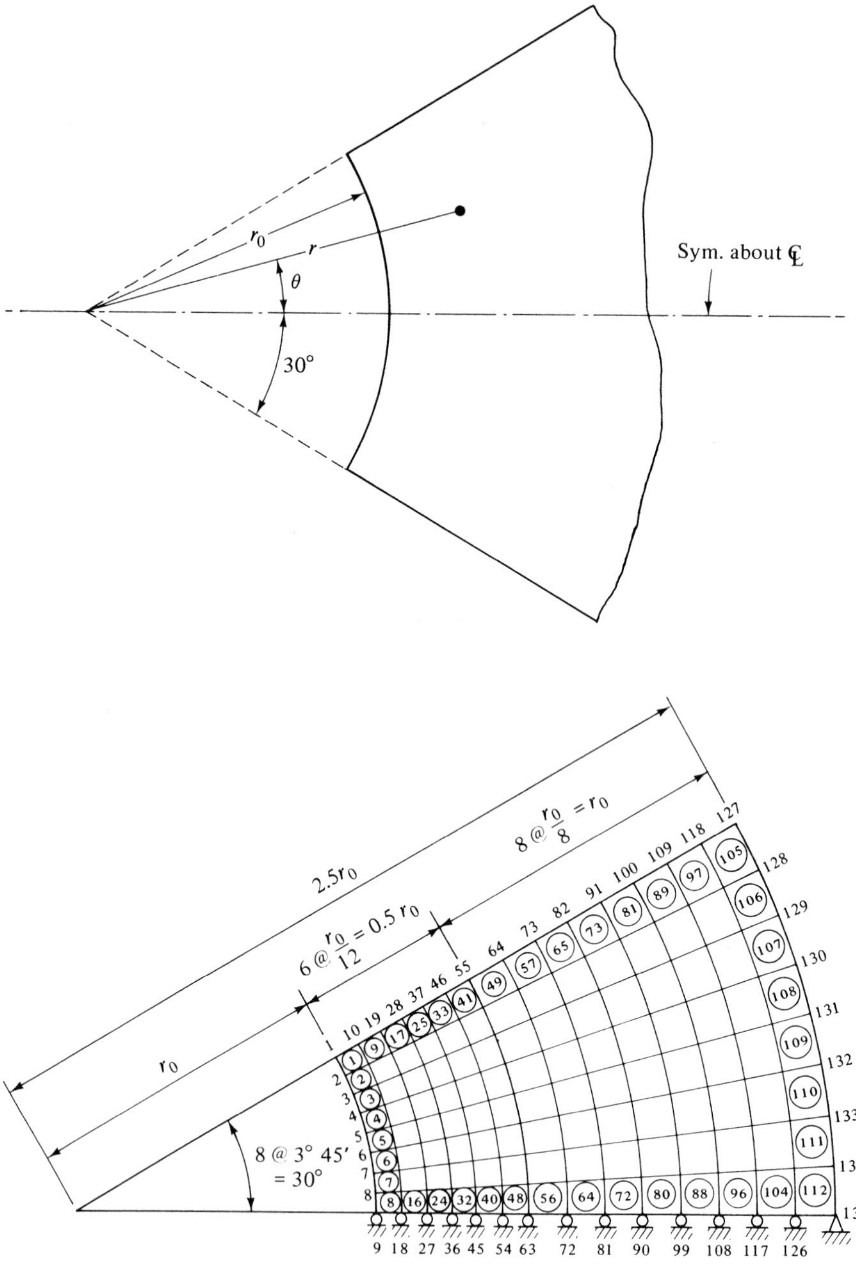

r0
r
θ
30°
Sym. about ℄
2.5r0
8 @ r0/8 = r0
6 @ r0/12 = 0.5 r0
r0
8 @ 3° 45′
= 30°
1 10 19 28 37 46 55 64 73 82 91 100 109 118 127
128 129 130 131 132 133 134 135
2 3 4 5 6 7 8
9 18 27 36 45 54 63 72 81 90 99 108 117 126
1 2 3 4 5 6 7 8 9 17 25 33 41 49 57 65 73 81 89 97 105
106 107 108 109 110 111 112
16 24 32 40 48 56 64 72 80 88 96 104

Figure 10-10

internally compute the necessary stiffness matrices and the transformations between local and global coordinate systems. The stiffness matrices are assembled and inverted to yield the desired answers.

As an example of the accuracy of a finite-element solution, an analysis of the truncated semi-infinite wedge discussed in section 8-7.3 is shown*. The program used developed triangular elements, and 135 nodal points were used in the analysis. Results of this analysis comparing the finite-element method with the eigenfunction expansions shown in section 8-7.3 are presented in Figures 10-8 and 10-9. The nodal point distribution is shown in Figure 10-10.

10 Finite-difference methods

The finite-difference methods, as indicated earlier, are much older and were developed as a basic numerical method of solution of differential equations. It is not unexpected that these methods are applicable to the equations of elasticity. Before discussing details of these applications, we shall review briefly some of the fundamentals of finite differences†

If we consider a smooth function $f(x)$, we can calculate the change of this function over an increment h:

$$\Delta f = f(x + h) - f(x). \qquad (10\text{-}10.1)$$

This difference is called the forward difference of $f(x)$. If we designate discrete values of x by integer subscripts, x_0, x_1, x_2, etc., we can write this forward difference as

$$\Delta f_0 = f(x_1) - f(x_0). \qquad (10\text{-}10.2)$$

The spacing between x_1 and x_0 is h. A backward difference at some point x_k can also be used:

$$\nabla f_k = f(x_k) - f(x_{k-1}). \qquad (10\text{-}10.3)$$

If we consider equal spacings of the variable x, we denote these points as

$$x_k = x_0 + kh. \qquad (10\text{-}10.4)$$

The value of the function at the point x_k can be written as f_k. The first difference is used to approximate the value of the derivative of the function at the point x_k:

$$\left.\frac{df}{dx}\right|_{x=x_k} = \frac{f_{k+1} - f_k}{h}. \qquad (10\text{-}10.5)$$

* This numerical work was done by William A. Bradley of Michigan State University.

† For additional information, see F. B. Hildebrand, *Methods of Applied Mathematics*, Prentice-Hall, Inc., Englewood Cliffs, N.J., 1952.

This uses the forward difference to approximate the derivative, and we could have used either the backward or the central difference to approximate this derivative:

$$\left.\frac{df}{dx}\right|_{x=x_k} = \frac{f_k - f_{k-1}}{h}, \tag{10-10.6}$$

$$\left.\frac{df}{dx}\right|_{x=x_k} = \frac{f_{k+1/2} - f_{k-1/2}}{h}. \tag{10-10.7}$$

The second derivative can be written using the difference representation of the first derivative:

$$\begin{aligned}\left.\frac{d^2f}{dx^2}\right|_{x=x_k} &= \frac{\left.\frac{df}{dx}\right|_{x=x_k} - \left.\frac{df}{dx}\right|_{x=x_{k-1}}}{h}, \\ &= \frac{f_{k+1} - 2f_k + f_{k-1}}{h^2}.\end{aligned} \tag{10-10.8}$$

Note that the second derivative has been formed by using the backward difference of the first derivative, which, in turn, was formed by use of a forward difference.

If we have a smooth function of two variables, x and y, and determine the partial derivatives in terms of evenly spaced values of the two variables, we obtain

$$\frac{\partial f_{i,j}}{\partial x} = \frac{f_{i+1,j} - f_{i,j}}{h}, \tag{10-10.9}$$

$$\frac{\partial f_{i,j}}{\partial y} = \frac{f_{i,j+1} - f_{i,j}}{h}, \tag{10-10.10}$$

where

$$f_{i,j} = f(x_i, y_j),$$

$$x_i = x_0 + ih,$$

$$y_j = y_0 + jh,$$

The second derivatives become

$$\frac{\partial^2 f_{i,j}}{\partial x^2} = \frac{1}{h^2}(f_{i+1,j} - 2f_{i,j} + f_{i-1,j}), \tag{10-10.11}$$

$$\frac{\partial^2 f_{i,j}}{\partial y^2} = \frac{1}{h^2}(f_{i,j+1} - 2f_{i,j} + f_{i,j-1}). \tag{10-10.12}$$

The Laplace equation can now be written in difference form for the point x_i, y_j:

$$\nabla^2 f = \frac{1}{h^2}[f_{i+1,j} + f_{i,j+1} - 4f_{i,j} + f_{i-1,j} + f_{i,j-1}]. \tag{10-10.13}$$

This is sometimes expressed in pictorial form as shown in Figure 10-11.

Boundary value problems lend themselves to solution by finite-difference methods if the region can be subdivided without an excessive number of

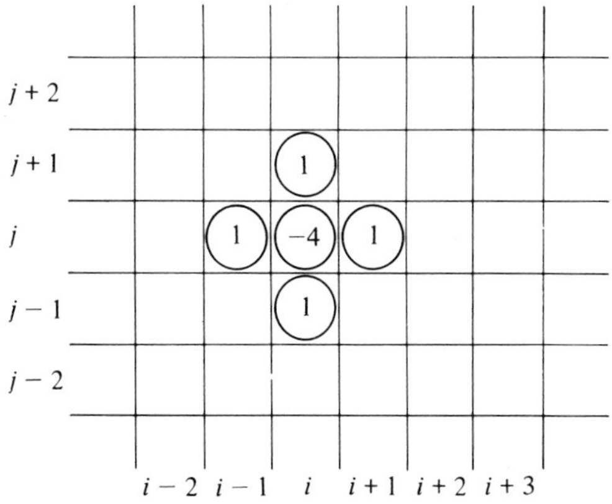

Figure 10-11

nodal points. If two-dimensional elasticity problems are formulated by use of the Airy stress function, this function is governed, in the absense of body forces, by the biharmonic equation:

$$\nabla^2\nabla^2\varphi = 0. \qquad (10\text{-}10.14)$$

This is written in difference form as

$$\begin{aligned}&\varphi_{i-2,j} + \varphi_{i,j-2} + \varphi_{i+2,j} + \varphi_{i,j+2} + 2(\varphi_{i-1,j-1} + \varphi_{i-1,j+1}\\ &+ \varphi_{i+1,j-1} + \varphi_{i+1,j+1}) - 8(\varphi_{i-1,j} + \varphi_{i,j-1} + \varphi_{i+1,j} + \varphi_{i,j+1})\\ &+ 20\varphi_{i,j} = 0.\end{aligned} \qquad (10\text{-}10.15)$$

This finite-difference pattern is shown in Figure 10-12.

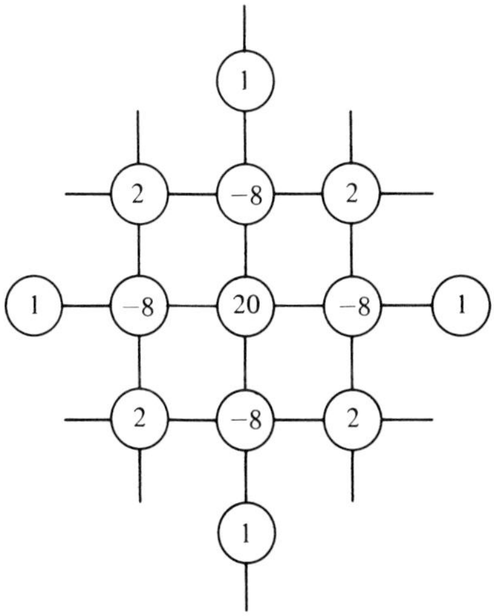

Figure 10-12

Consider a simply connected region as shown in Figure 10-13. One can classify three types of nodal points in Figure 10-13: interior, boundary, and exterior. The interior points satisfy the biharmonic difference equation (10-10.15), and the boundary points and exterior points are used to satisfy the two boundary conditions. If the stress traction vector is given on the

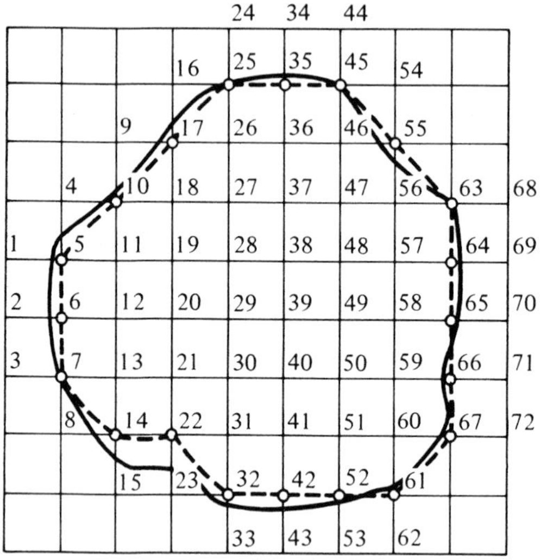

Figure 10-13

boundary, we obtain

$$\bar{t}^n = \hat{n}\cdot\bar{\bar{\sigma}}. \tag{10-10.16}$$

Expanding yields

$$\bar{t}^n = \left[(\hat{n}\cdot\hat{i})\frac{\partial^2\varphi}{\partial y^2} - (\hat{n}\cdot\hat{j})\frac{\partial^2\varphi}{\partial x\,\partial y}\right]\hat{i} - \left[(\hat{n}\cdot\hat{i})\frac{\partial^2\varphi}{\partial x\,\partial y} - (\hat{n}\cdot\hat{j})\frac{\partial^2\varphi}{\partial x^2}\right]\hat{j} \tag{10-10.17}$$

Noting that

$$\hat{n}\cdot\hat{i} = \frac{dy}{ds},$$

and

$$\hat{n}\cdot\hat{j} = -\frac{dx}{ds},$$

we write equation 10-10.17 as

$$\bar{t}^n = \frac{d}{ds}\left(\frac{\partial\varphi}{\partial y}\right)\hat{i} - \frac{d}{ds}\left(\frac{\partial\varphi}{\partial x}\right)\hat{j} \tag{10-10.18}$$

Integrating equation 10-10.18 gives

$$\frac{\partial\varphi}{\partial y} = \int t_x\,ds, \qquad \frac{d\varphi}{dx} = -\int t_y\,ds. \tag{10-10.19}$$

Equation 10-10.19 can be used to establish the value of φ on the boundary. Specification of the stress traction vector is therefore equivalent to specification of the function φ and its normal derivative on the boundary. In the integration called for by equation 10-10.19 and the subsequent integration to determine φ, a linear function of the form $Ax + By + C$ arises. These constants are arbitrarily chosen, but care must be used to maintain consistency in their values around the entire contour.

Equations 10-10.15, 10-10.19, and its integral form a system of banded linear equations for φ at each of the interior, boundary, and neighboring exterior points. This system can be solved by any of the commonly used matrix inversion techniques.

Finite-difference methods can also be applied directly to the reduced Navier equations, equation 6-2.22, giving the two displacements at each nodal point. This procedure is useful when displacements are specified on the boundaries and does not require calculation of higher-order derivatives. Example solutions of a plane elasticity problem* employing finite-difference methods are found in Timoshenko's text.

BIBLIOGRAPHY (Finite-Element Method)

ARGYRIS, J. H., "Matrix analysis of three-dimensional elastic media: Small and large displacements," *J. AIAA*, **3**, 45–51, 1965.

CLOUGH, R. W., "The finite element in plane stress analysis," *Proc. 2nd ASCE Conf. on Electronic Computation*, Pittsburgh, September 1960.

HOLAND, I., and K. BELL, "Finite element methods in stress analysis," Tapir (The Technical University of Norway), 1970.

JOHNSON, M. W., and R. W. MCLAY, "Convergence of the finite element method in the theory of elasticity," *J. Appl. Mech. Trans. ASME*, 274–278, June 1968.

NEWMARK, N. M., "Numerical methods of analysis in bars, plates and elastic bodies, in *Numerical Methods of Analysis in Engineering*, edited by L. E. Grinter, The Macmillan Company, New York, 1949.

ODEN, J. T., "A general theory of finite elements: 1. Topological considerations, II. Application," *Int. J. Num. Meth. Eng.*, **1**, 205–221, 247–260, 1969.

PIAN, T. H. H., "Derivation of element stiffness matrices," *J. ATAA*, **2**, 576–577, 1964.

* Finite-difference solutions in elasticity date back to C. Runge, *Z. Moth. Physik*, **56**, 225, 1908. R. V. Southwell employed it extensively, and much of his work is summarized in his text, *Relaxation Methods in Theoretical Physics*, Oxford University, Press, Inc., New York, 1946.

PIAN, T. H. H., and P. TONG, "Basis of finite element methods for solid continua," *Int. J. Num. Meth. Eng.*, **1**, 3–28, 1969.

PRZEMIENIECKI, J. S., *Theory of Matrix Structural Analysis*, McGraw-Hill New York, 1968.

ZIENKIEWICZ, O. C., *The Finite Element Method*, McGraw-Hill Book Company, New York, 1967.

"On General Purpose Finite Element Computer Programs," Papers presented at winter annual meeting of ASME, November 1970.

CHAPTER 11

ENERGY THEOREMS AND VARIATIONAL TECHNIQUES

1 Introduction

We have considered methods of solution of elasticity problems that involve sums of continuous functions, each term of which satisfies the defining differential equation, while the sum satisfies the boundary conditions. In these methods we approximated the boundary conditions in a least-squares sense. An alternative procedure would be to select functions that satisfy some, if not all, of the boundary conditions and satisfy, in their sum, the differential equations of elasticity and the remaining boundary conditions. This type of solution results from the use of a variational technique associated with one of the energy theorems of linear elasticity.

In 1877, the third Lord Rayleigh published his Rayleigh principle for vibrations of elastic systems and thus introduced variational techniques into elasticity. Surprisingly, the work of Rayleigh went unnoticed on the European continent and in 1908 his method was independently discovered and expanded upon by Ritz. These investigations were extended and formalized by R. Courant, B. Galerkin, S. P. Timoshenko, L.V. Kantorovich, E. Trefftz, E. Reissner, and others.

It is necessary to examine the energy-balance equation and some of the fundamentals of the calculus of variations to fully understand this approach. The reader who wishes to examine these in more detail than is presented here should read the excellent treatment in I.S. Sokolnikoff's *Mathematical Theory of Elasticity* (McGraw-Hill Book Company, New York 1956).

2 Calculus of variations

The basic problem involved in the mathematical field of claculus of variations is to minimize or maximize a certain definite integral containing a particular function. This maximization or minimization is accomplished by proper selection of this function. In this process we must establish what conditions the trial functions must satisfy to be admissible. To gain a better understanding, let us consider an integral involving a particular function and its derivative. Let us assume that u is a function of x and the integral in question is

$$I(u) = \int_{x_1}^{x_2} F(u(x), u'(x), x)\, dx. \qquad (11\text{-}2.1)$$

It is apparent that different choices of the function u will produce different values of $I(u)$. A quantity such as $I(u)$, which takes particular values depending upon the choice of the function u, is called a *functional.* Simple examples of functionals are

$$I(u) = \int_a^b u(x)\, dx, \qquad (11\text{-}2.2a)$$

$$I(u) = \int_a^b \frac{u(x)}{[1 - u'(x)]^2}\, dx. \qquad (11\text{-}2.2b)$$

In each of these cases a choice of u will produce a particular numerical value for the functional $I(u)$.

Suppose now that the problem in question is to choose the function u such that the functional $I(u)$ will assume a stationary value, for example, a minimum, and that in addition u will satisfy certain prescribed boundary conditions:

$$u(x_1) = u_1, \qquad u(x_2) = u_2. \qquad (11\text{-}2.3)$$

Suppose that we designate u to be the correct function which makes $I(u)$ a minimum and satisfies the boundary condition in equation 11-2.3. We may represent other functions by $u(x) + \epsilon\eta(x)$, where ϵ is any constant and $\eta(x)$ satisfies homogeneous conditions at x_1 and x_2. Then $\eta(x)$ will be a continuously differentiable function that vanishes at the end points. Equation 11-2.1 may be written in terms of this trial function as

$$I(\epsilon) = \int_{x_1}^{x_2} F[x, u + \epsilon\eta, u' + \epsilon\eta']\, dx. \tag{11-2.4}$$

We have assumed that u is the correct function which minimizes I and this can occur only if I(ϵ) takes its minimum value when $\epsilon = 0$:

$$\frac{dI(\epsilon)}{d\epsilon} = 0, \qquad \text{when } \epsilon = 0. \tag{11-2.5}$$

Differentiation of equation 11-2.4 with respect to ϵ yields

$$\frac{dI}{d\epsilon} = \int_{x_1}^{x_2} \left[\frac{\partial F}{\partial u}\eta + \frac{\partial F}{\partial u'}\eta'\right] dx. \tag{11-2.6}$$

Integrating the last term by parts and equating equation 11-2.6 to zero when $\epsilon = 0$ yields

$$\int_{x_1}^{x_2} \left\{\frac{\partial F}{\partial u} - \frac{d}{dx}\left(\frac{\partial F}{\partial u'}\right)\right\}\eta\, dx + \left[\frac{\partial F}{\partial u'}\eta\right]_{x_1}^{x_2} = 0. \tag{11-2.7}$$

Since $\eta(x)$ is an arbitrary function vanishing at the end point, but not everywhere zero, the following Euler equation must be satisfied during the minimization process:

$$\frac{d}{dx}\left(\frac{\partial F}{\partial u'}\right) - \frac{\partial F}{\partial u} = 0. \tag{11-2.8}$$

Notice that if $\eta(x)$ is not zero at the end points [$u(x)$ not prescribed at these points], the Euler equation is still satisfied if

$$\frac{\partial F}{\partial u'} = 0 \qquad \text{at } x = x_1 \text{ and } x_2. \tag{11-2.9}$$

These are known as the natural boundary conditions.

A variational notation can be introduced if the variation of u is denoted by

$$\delta u = \epsilon\eta(x). \tag{11-2.10}$$

Note that this variation occurs for a fixed value of x so that in this process the variation of x, (δx), is zero. An analogy to differentiation can be written as

$$\partial F = \frac{\partial F}{\partial u}\delta u + \frac{\partial F}{\partial u'}\delta u'. \tag{11-2.11}$$

If F is a function of two variables $u(x, y)$ and $v(x, y)$ and their derivatives, each of which is a function of the spatial variables x and y, then the variation of F becomes

$$\begin{aligned}\delta F = \frac{\partial F}{\partial u}\delta u + \frac{\partial F}{\partial v}\delta v + \frac{\partial F}{\partial u_x}\delta u_x \\ + \frac{\partial F}{\partial u_y}\delta u_y + \frac{\partial F}{\partial v_x}\delta v_x + \frac{\partial F}{\partial v_y}\delta v_y,\end{aligned} \tag{11-2.12}$$

where the notation u_x is used to denote the partial derivative of u with respect to x.

If we consider the functional shown in equation 11-2.1, then the variation of this functional becomes

$$\begin{aligned}\delta I &= \int_{x_1}^{x_2} \delta F\, dx \\ &= \int_{x_1}^{x_2} \left[\frac{\partial F}{\partial u} - \frac{d}{dx}\left(\frac{\partial F}{\partial u'}\right)\right] \delta u\, dx + \left[\frac{\partial F}{\partial u'}\delta u\right]_{x_1}^{x_2}.\end{aligned} \tag{11-2.13}$$

We have used other properties of the variational notation in obtaining equation 11-2.13. Some of these are

$$\begin{aligned}\frac{d}{dx}(\delta u) &= \delta\left(\frac{du}{dx}\right), \\ \delta(uv) &= u\delta v + v\delta u, \\ \delta\left(\frac{u}{v}\right) &= \frac{v\delta u - u\delta v}{v^2}.\end{aligned} \tag{11-2.14}$$

We may note then, that the variation is analogous to differentiation and satisfies similar laws. The variational notation can be used to express the following necessary condition that an integral be stationary. This condition on the integral, I, defined by

$$I = \int_{x_1}^{x_2} F(x, u, u')\, dx, \tag{11-2.15}$$

is that is first variation vanish:

$$\delta I = \delta \int_{x_1}^{x_2} F(x, u, u') = 0. \tag{11-2.16}$$

We now write the general two-dimensional case in which F is the function defined as in equation 11-2.12 and the integration is carried out over a simple region S:

$$I = \iint_S F(x, y, u, v, u_x, u_y, v_x, v_y)\, dx\, dy. \tag{11-2.17}$$

If this integral is to assume a stationary value, the condition becomes

$$\delta I = 0. \tag{11-2.18}$$

Integration by parts yields

$$\delta I = \iint_S \left\{ \left[\frac{\partial F}{\partial u} - \frac{\partial}{\partial x}\left(\frac{\partial F}{\partial u_x}\right) - \frac{\partial}{\partial y}\left(\frac{\partial F}{\partial u_y}\right) \right] \delta u + \left[\frac{\partial F}{\partial v} - \frac{\partial}{\partial x}\left(\frac{\partial F}{\partial v_x}\right) - \frac{\partial}{\partial y}\left(\frac{\partial F}{\partial v_y}\right) \right] \delta v \right\} dx\, dy = 0. \tag{11-2.19}$$

For greater detail, the reader is referred to the texts by Kantorovich and Krylov* or Hildebrand,† or Washizu.‡

3 Strain energy methods

Let us consider the elastic body shown in Figure 11-1. A portion of the development in Sect. 4-5 will be reproduced here for continuity. The surface tractions are $t_j^n = \sigma_{ij} n_i$. If we designate the body forces by X_i, we

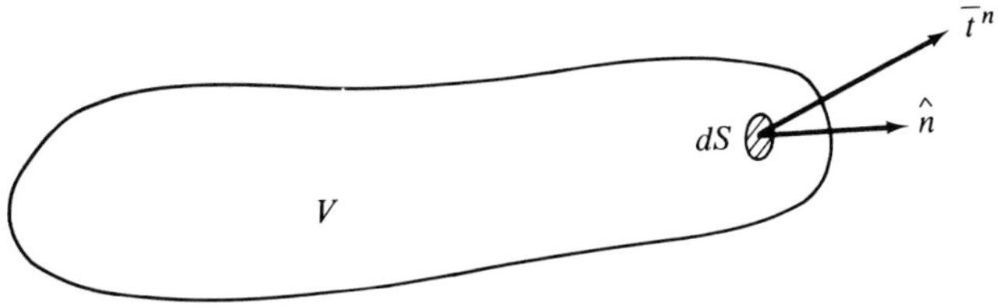

Figure 11-1

can denote the power (work rate) of the external forces on the volume as

$$P = \int_V X_i u_i \, dV + \int_S n_i \sigma_{ij} \dot{u}_j \, dS. \tag{11-3.1}$$

Using the divergence theorem, the last integral over the surface is

$$\int_S n_i \sigma_{ij} \dot{u}_j \, dS = \int_V \frac{\partial}{\partial x_i} (\sigma_{ij} \dot{u}_j) \, dV. \tag{11-3.2}$$

Expanding yields

$$\int_V \frac{\partial}{\partial x_i} (\sigma_{ij} \dot{u}_j) \, dV = \int_V \left[\frac{\partial \sigma_{ij}}{\partial x_i} \dot{u}_j + \sigma_{ij} \frac{\partial \dot{u}_j}{\partial x_i} \right] dV. \tag{11-3.3}$$

* L. V. Kantorovich and V. I. Krylov, *Approximate Methods of Higher Analysis*, Walters-Noordhoff Publishing, Groningen, 1958.

† F. B. Hildebrand, *Methods of Applied Mathematics*, Prentice-Hall, Inc., Englewood Cliffs, N.J., 1952.

‡ K. Washizu, *Variational Methods in Elasticity and Plasticity*, Pergamon Press, Inc., Oxford, 1968.

The last term can be further expanded to give

$$\sigma_{ij}\frac{\partial \dot{u}_j}{\partial x_i} = \frac{\sigma_{ij}}{2}\left(\frac{\partial \dot{u}_j}{\partial x_i} + \frac{\partial \dot{u}_i}{\partial x_j}\right) + \frac{\sigma_{ij}}{2}\left(\frac{\partial \dot{u}_j}{\partial x_i} - \frac{\partial \dot{u}_i}{\partial x_j}\right). \tag{11-3.4}$$

By introduction of the definitions of the strain tensor and the rotation tensor thus can be written

$$\sigma_{ij}\frac{\partial \dot{u}_j}{\partial x_i} = \sigma_{ij}\dot{\epsilon}_{ij} + \sigma_{ij}\dot{\omega}_{ij}. \tag{11-3.5}$$

Because $\omega_{ij} = -\omega_{ji}$ and a sum on i and j is implied, the last term is zero. The external work rate now becomes

$$P = \int_V \left\{\left[\frac{\partial \sigma_{ij}}{\partial x_i} + X_j\right]\dot{u}_j + \sigma_{ij}\dot{\epsilon}_{ij}\right] dV. \tag{11-3.6}$$

Using the Cauchy equation of motion,

$$\frac{\partial \sigma_{ij}}{\partial x_i} + X_j = \rho \ddot{u}_j, \tag{11-3.7}$$

and noting that

$$\rho \ddot{u}_j \dot{u}_j = \frac{\rho}{2}\frac{d}{dt}(\dot{u}_j)^2, \tag{11-3.8}$$

and that

$$\dot{K} = \frac{d}{dt}\left[\frac{1}{2}\rho \dot{u}_j \dot{u}_j\right] \tag{11-3.9}$$

is the time rate of change of the kinetic energy, the external work rate term becomes

$$P = \dot{K} + \int_V \sigma_{ij}\dot{\epsilon}_{ij}\, dV. \tag{11-3.10}$$

The first law of thermodynamics (see equation 4-5.4) can be written

$$\dot{K} + \int_V \dot{U}\, dV = P + \dot{Q}, \tag{11-3.11}$$

where U is the internal energy per unit volume and Q corresponds to the heat energy term. Substituting for the external work rate, this equation becomes

$$\int_V \dot{U}\, dV = \int_V \sigma_{ij}\dot{\epsilon}_{ij}\, dV + \dot{Q}. \tag{11-3.12}$$

Let δU be the increment of U per unit volume in δt so that equation 11-3.12 becomes

$$\int_V \delta U\, dV = \int_V \sigma_{ij}\delta\epsilon_{ij}\, dV + \delta Q. \tag{11-3.13}$$

Consider the reversible adiabatic case (no heat transfer). The increment of internal energy becomes

$$\delta U = \sigma_{ij}\delta\epsilon_{ij}. \tag{11-3.14}$$

δU is the differential of a function $U[\epsilon_{ij}, T(\epsilon_{ij})]$. For the adiabatic case this can be written $V_a = U[\epsilon_{ij}, T(\epsilon_{ij})]$ and expressed as a function of the strains:

$$\delta V_a = \sigma_{ij}\delta\epsilon_{ij}, \tag{11-3.15}$$

$$\frac{\partial V_a}{\partial \epsilon_{ij}}\delta\epsilon_{ij} = \sigma_{ij}\delta\epsilon_{ij},$$

$$\sigma_{ij} = \frac{\partial V_a}{\partial \epsilon_{ij}}. \tag{11-3.16}$$

Therefore, the adiabatic reversible case assures existence of a strain energy function V_a. By a slightly different argument, a different strain energy function can be shown to exist in the case of reversible isothermal cycles.

For an isotropic linearly elastic solid, Hooke's law yields

$$\sigma_{ij} = \lambda\epsilon_{kk}\delta_{ij} + 2G\epsilon_{ij}. \tag{11-3.17}$$

Therefore, the incremental change in the strain energy function becomes

$$\delta V = [\lambda\epsilon_{kk}\delta_{ij} + 2G\epsilon_{ij}]\delta\epsilon_{ij}, \tag{11-3.18}$$

or

$$V = \tfrac{1}{2}[\lambda\epsilon_{kk}^2 + 2G\epsilon_{ij}\epsilon_{ij}]. \tag{11-3.19}$$

For the case of plane stress, in terms of engineering constants, the strain energy per unit volume is

$$V = \frac{E}{2(1+\nu)}\left\{\frac{1}{(1-\nu)}(\epsilon_{xx}^2 + \epsilon_{yy}^2 + 2\nu\epsilon_{xx}\epsilon_{yy}) + 2\epsilon_{xy}^2\right\}. \tag{11-3.20}$$

The total strain energy in a body can be written in terms of the displacements as

$$V^T = \frac{E}{2(1+\nu)}\int_{\text{body}}\left\{\frac{1}{1-\nu}\left[\left(\frac{\partial u_x}{\partial x}\right)^2 + \left(\frac{\partial u_y}{\partial y}\right)^2 + 2\nu\left(\frac{\partial u_x}{\partial x}\right)\left(\frac{\partial u_y}{\partial y}\right)\right] + \frac{1}{2}\left(\frac{\partial u_x}{\partial y} + \frac{\partial u_y}{\partial x}\right)^2\right\} dV. \tag{11-3.21}$$

4 Theorem of stationary potential energy

We would like to show that when a body is in a state of equilibrium the total potential energy of the system is a stationary value and, in fact, a minimum in certain cases. The increment in external work W_{ext} is

$$\delta W_{\text{ext}} = \delta K + \delta V. \tag{11-4.1}$$

For the case of a body in static equilibrium, $\delta K = 0$, and this equation can be written

$$\delta(V - W_{\text{ext}}) = 0. \tag{11-4.2}$$

Let $\pi = (V - W)$ and note that it equals the internal energy plus the negative of the work done by external forces, and is therefore equal to the total potential energy of the body. We now state

$$\delta\pi = 0. \tag{11-4.3}$$

Therefore, for a body in equilibrium π is a stationary value.

Let us consider the set of displacements satisfying the equilibrium conditions and designated by u_i, and an arbitrary increment or variation of these displacements satisfying the displacement conditions on the boundary. We shall designate the variations by δu_i (Figure 11-2). Consider the displace-

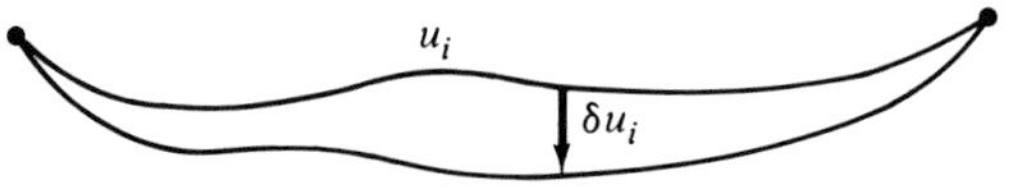

Figure 11-2

ments $u_i + \delta u_i$; the total potential energy plus its variation is

$$\begin{aligned}
\pi + \delta\pi &= V^T + \delta V^T - W - \delta W \\
&= \tfrac{1}{2}\int_{\text{Vol}} \{\lambda(\epsilon_{kk} + \delta\epsilon_{kk})^2 + 2G(\epsilon_{ij} + \delta\epsilon_{ij})(\epsilon_{ij} + \delta\epsilon_{ij})\}\, dV \\
&\quad - \int_{\text{Vol}} X_i(u_i + \delta u_i)\, dV - \int_S n_i\sigma_{ij}(u_j + \delta u_j)\, dS \\
&= \tfrac{1}{2}\int_{\text{Vol}} (\lambda\epsilon_{kk}^2 + 2G\epsilon_{ij}\epsilon_{ij})\, dV - \int_{\text{Vol}} X_i u_i\, dV - \int_S n_i\sigma_{ij}u_j\, ds \\
&\quad + \int_{\text{Vol}} (\lambda\epsilon_{kk}\delta\epsilon_{kk} + 2G\epsilon_{ij}\delta\epsilon_{ij})\, dV - \int_{\text{Vol}} X_i\delta u_i\, dV \\
&\quad - \int_S n_i\sigma_{ij}\,\delta u_j\, dS + \underbrace{\tfrac{1}{2}\int_{\text{Vol}} [\lambda(\delta\epsilon_{kk})^2 + 2G\delta\epsilon_{ij}\delta\epsilon_{ij}]\, dV.}_{R}
\end{aligned} \tag{11-4.4}$$

$$\begin{aligned}
\delta\pi = \int_{\text{Vol}} &\left[\lambda\epsilon_{kk}\frac{\partial u_j}{\partial x_j} + G\epsilon_{ij}\left(\frac{\partial\delta u_j}{\partial x_i} + \frac{\partial\delta u_i}{\partial x_j}\right)\right] dV \\
&- \int_{\text{Vol}} X_i\delta u_i\, dV - \int_S n_i\sigma_{ij}\,\delta u_j\, dS + R.
\end{aligned} \tag{11-4.5}$$

Integration by parts of the first term yields

$$\begin{aligned}
\delta\pi = &\int_S n_i[\lambda\epsilon_{kk}\delta_{ij} + 2G\epsilon_{ij}]\delta u_j\, dS - \int_{\text{Vol}} \frac{\partial}{\partial x_i}[\lambda\epsilon_{kk}\delta_{ij} + 2G\epsilon_{ij}]\delta u_j\, dV \\
&- \int_{\text{Vol}} X_i\delta u_i\, dV - \int_S n_i\sigma_{ij}\delta u_j\, dS + R.
\end{aligned} \tag{11-4.6}$$

If one notes that the body is in equilibrium,

$$\delta\pi = R \geq 0. \tag{11-4.7}$$

R represents the higher-order term and is positive. Therefore, the change is positive and the potential energy is a minimum.

5 Rayleigh–Ritz method*

We note that when the displacements correspond to the correct ones for equilibrium, they must be such as to make the potential energy a minimum. Suppose that we assume the displacements in the form of a series of functions which satisfy the kinematic or displacement boundary conditions, and are such that the series will converge to the correct functions:

$$u_i = \sum_n C_n^{(i)} \varphi_n^{(i)}. \qquad (11\text{-}5.1)$$

For the case of a two-dimensional model the potential energy can be written

$$\pi = V - W, \qquad (11\text{-}5.2)$$

where

$$V = \frac{E}{2(1+\nu)} \int_{\text{Vol}} \left\{ \frac{1}{1-\nu} \left[\left(\frac{\partial u_x}{\partial x} \right)^2 + \left(\frac{\partial u_y}{\partial y} \right)^2 + 2\nu \left(\frac{\partial u_x}{\partial x} \right) \left(\frac{\partial u_y}{\partial y} \right) \right] + \frac{1}{2} \left(\frac{\partial u_x}{\partial y} + \frac{\partial u_y}{\partial x} \right)^2 \right\} dV, \qquad (11\text{-}5.3)$$

and

$$W = \int_{\text{Vol}} X_i u_i \, dV + \int_s t_j^n u_j \, ds. \qquad (11\text{-}5.4)$$

For notational purposes, let us assume that

$$u_x = \sum A_n \varphi_n, \qquad (11\text{-}5.5)$$

$$u_y = \sum B_n \psi_n. \qquad (11\text{-}5.6)$$

Substitution of equations 11-5.5 and 11-5.6 into equation 11-5.2 yields

$$\begin{aligned} \pi = \frac{E}{2(1+\nu)} \int_{\text{Vol}} & \left\{ \frac{1}{1-\nu} \left[\sum_{m,n} A_m A_n \frac{\partial \varphi_n}{\partial x} \frac{\partial \varphi_m}{\partial x} + B_m B_n \frac{\partial \psi_m}{\partial y} \frac{\partial \psi_n}{\partial y} \right. \right. \\ & \left. + 2\nu A_m B_n \frac{\partial \varphi_m}{\partial x} \frac{\partial \psi_n}{\partial y} \right] + \frac{1}{2} \sum_{m,n} \left[A_m A_n \frac{\partial \varphi_m}{\partial y} \frac{\partial \varphi_n}{\partial y} \right. \\ & \left. \left. + 2 A_m B_n \frac{\partial \varphi_m}{\partial y} \frac{\partial \psi_n}{\partial x} + B_m B_n \frac{\partial \psi_m}{\partial x} \frac{\partial \psi_n}{\partial x} \right] \right\} dV \\ & - \int_{\text{Vol}} \left[X_x \sum_n A_n \varphi_n + X_y \sum_n B_n \psi_n \right] dV \\ & - \int_s \left[t_x^n \sum_n A_n \varphi_n + t_y^n \sum_n B_n \psi_n \right] ds. \end{aligned} \qquad (11\text{-}5.7)$$

* W. Ritz, *J. Reine Angew, Math.*, **135**, 1–61, 1908; or *Gesammelte Werke*, Gauthier-Villars, Paris, 192–250, 1911.

Minimizing the potential energy with respect to the coefficients yields

$$\frac{\partial \pi}{\partial A_j} = 0 = \frac{E}{(1+\nu)}\int_{\text{Vol}}\left[\frac{1}{(1-\nu)}\left\{\sum_m A_m \frac{\partial \varphi_m}{\partial x}\frac{\partial \varphi_j}{\partial x} + \nu B_m \frac{\partial \varphi_j}{\partial x}\frac{\partial \psi_m}{\partial y}\right\}\right.$$
$$\left. + \frac{1}{2}\sum_m A_m \frac{\partial \varphi_m}{\partial y}\frac{\partial \varphi_j}{\partial y} + B_m \frac{\partial \varphi_j}{\partial y}\frac{\partial \psi_m}{\partial x}\right] dV \tag{11-5.8}$$
$$- \int_{\text{Vol}} X_x \varphi_j \, dV - \int_s t_x^n \varphi_j \, ds,$$

$$\frac{\partial \pi}{\partial B_j} = 0 = \frac{E}{(1+\nu)}\int_{\text{Vol}}\left[\frac{1}{1-\nu}\left\{\sum_m B_m \frac{\partial \psi_m}{\partial y}\frac{\partial \psi_j}{\partial y} + \nu A_m \frac{\partial \varphi_m}{\partial x}\frac{\partial \psi_j}{\partial y}\right\}\right.$$
$$\left. + \frac{1}{2}\sum_m \left\{A_m \frac{\partial \varphi_m}{\partial y}\frac{\partial \psi_j}{\partial x} + B_m \frac{\partial \psi_j}{\partial x}\frac{\partial \psi_m}{\partial x}\right\}\right] dV \tag{11-5.9}$$
$$- \int_{\text{Vol}} X_y \psi_j \, dV - \int_s t_y^n \psi_j \, ds.$$

This represents a double system of equations for the unknowns A_n and B_n.

The solution shown has been simplified from what actually is involved if one attempts to make the necessary calculations. For numerical calculations using a digital computer, it is easy to assume φ and ψ in terms of polynomial series expansions. As an example, let us consider a rectangular plate loaded at the ends (Figure 11-3).* The symmetry of the problem would indicate that $u_x(x, y) = -u_x(-x, y)$ or u_x is odd in x and even in y, whereas u_y is odd in y and even in x. We assume the solution in the form

$$u_x = \sum_{m,n} A_{mn}\left(\frac{x}{l}\right)^m \left(\frac{y}{h}\right)^n, \qquad m \text{ odd}, \ n \text{ even} \tag{11-5.10}$$

$$u_y = \sum_{ij} B_{ij}\left(\frac{x}{l}\right)^i \left(\frac{y}{h}\right)^j, \qquad i \text{ even}, \ j \text{ odd} \tag{11-5.11}$$

Equation 11-5.8 can now be written

$$\frac{E}{1+\nu}\int_{-l}^{+i}\int_{-h}^{+h}\left\{\frac{1}{1-\nu}\sum_{rs}\frac{mr}{l^2}\left(\frac{x}{l}\right)^{m+r-2}\left(\frac{y}{h}\right)^{n+s} A_{rs}\right.$$
$$+ \frac{1}{2}\sum_{rs}\frac{ns}{h^2}\left(\frac{x}{l}\right)^{m+r}\left(\frac{y}{h}\right)^{n+s-2} A_{rs} + \frac{\nu}{1-\nu}\sum_{pq}\frac{mq}{hl}\left(\frac{x}{l}\right)^{m+p-1}\left(\frac{y}{h}\right)^{n+q-1} B_{pq}$$
$$\left. + \sum_{pq}\frac{np}{2hl}\left(\frac{x}{l}\right)^{m+p-1}\left(\frac{y}{h}\right)^{n+q-1} B_{pq}\right\} dx\, dy - 2\int_{-h}^{+h} S\left(1 - \frac{y^2}{h^2}\right)\left(\frac{y}{h}\right)^n = 0.$$
$$\tag{11-5.12}$$

This is an equation for each mn term, and can be integrated and written as

* See S. P. Timoshenko, *Phil. Mag.*, **47**, 1095, 1924; C. E. Inglis, *Proc. Roy. Soc.* (London), **A 103**, 1923; and G. Pickett, *J. Appl. Mech.*, **11**, 176, 1944.

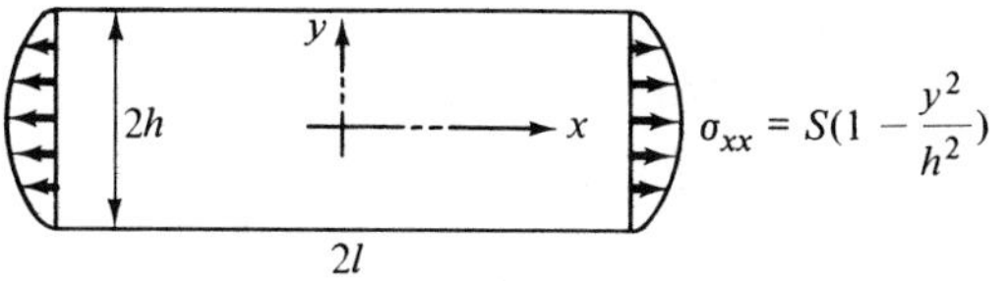

Figure 11-3

$$\frac{E}{1+\nu}\left\{\sum_{rs}\left[\frac{h}{(1-\nu)l}\frac{mr}{(m+r-1)(n+s-1)} + \frac{1}{2}\frac{l}{h}\frac{ns}{(m+r+1)(n+s-1)}\right]A_{rs} + \sum_{pq}\left[\frac{\nu}{1-\nu}\frac{mq}{(m+p)(n+q)} + \frac{np}{2(m+p)(n+q)}\right]B_{pq}\right\} = s\left[\frac{h}{n+1} - \frac{h}{n+3}\right]. \quad (11\text{-}5.13)$$

In a similar manner, equation 11-5.9 can be written

$$\frac{E}{1+\nu}\left\{\sum_{rs}\left[\frac{\nu}{1-\nu}\frac{rj}{(r+i)(s+j)} + \frac{si}{2(r+i)(s+j)}\right]A_{rs} + \sum_{pq}\left[\frac{l}{(1-\nu)h}\frac{jq}{(i+p+1)(j+q-1)} + \frac{1}{2}\frac{h}{l}\frac{ip}{(i+p-1)(j+q+1)}\right]B_{pq}\right\} = 0. \quad (11\text{-}5.14)$$

Let us consider a few terms in the series for displacements:

$$u_x = A_{10}\left(\frac{x}{l}\right) + A_{30}\left(\frac{x}{l}\right)^3 + A_{12}\left(\frac{x}{l}\right)\left(\frac{y}{h}\right)^2, \quad (11\text{-}5.15)$$

$$u_y = B_{01}\left(\frac{y}{h}\right) + B_{03}\left(\frac{y}{h}\right)^3 + B_{21}\left(\frac{x}{l}\right)^2\left(\frac{y}{h}\right). \quad (11\text{-}5.16)$$

Consider only the first term in each series. The two equations from minimization of the potential energy are

$$\begin{aligned}\frac{h}{l}\frac{1}{(1-\nu)}A_{10} + \frac{\nu}{1-\nu}B_{01} &= \frac{1+\nu}{E}\frac{Sh}{4},\\ \frac{\nu}{1-\nu}A_{10} + \frac{1}{(1-\nu)}\frac{l}{h}B_{01} &= 0,\end{aligned} \quad (11\text{-}5.17)$$

$$\begin{aligned}A_{10} &= \frac{S}{4El},\\ B_{01} &= \frac{\nu S}{4El}\left(\frac{h}{l}\right).\end{aligned} \quad (11\text{-}5.18)$$

This is the first approximation for the displacement functions. We can add terms to either displacement function and improve the accuracy of the solution. We can also check the accuracy of the solution by checking the boundary conditions on the stresses and by substituting the results into the Navier equations to see how well these are satisfied.

6 Energy method for problems involving multiply connected domains

In the discussion of problems dealing with the annulus or the external problem in the complex variable analysis or the general polar coordinates analysis, we noted that the displacements will include log terms if the body is multiply connected (equations 9-3.41 and 9-3.42). These log terms and the negative powers of r or z represent the effects on the displacement functions due to the presence of the holes. If we are to solve problems for bodies having holes in them by use of the energy method, it would appear necessary that the displacement trial functions should have terms of this form in them. These terms create problems in integration and require special consideration. Few solutions have been attempted for these cases using energy techniques, and the reader is referred to articles by Schlack* and Pickett† for additional comments. This is not to imply that many problems of this type have not been solved by other techniques, such as complex variable methods, but energy solutions are difficult.

BIBLIOGRAPHY

ARGYRIS, J. H., and S. KELSEY, *Energy Theorems and Structural Analysis*, Butterworth, & Company (Publishers) Ltd., London, 1960.

BIEZENO, C., and R. GRAMMEL, *Technische Dynamik*, Springer-Verlag, Berlin, 1939.

COLLATZ, L. *Eigenwertaufgaben mit technischen Aswendungen*, Akademische Verlagsgesellschaft, Leipzig, 1949.

FRIEDRICHS, K., "Ein Verfahren der Variationsrechnung das Minimum eines Integrals als das Maximum eines anderen Ausdruckes darzustellen," in *Nachrichten der Academie der Wissenschaften in Gottingen*, pp. 13–20, 1929.

HELLINGER, E., "Der allgemeine Ansatz der Mechanik der Kontinua," in *Encyclopadie der Mathematischen Wissenschaften*, Vol. 4, Part 4, pp. 602–94, 1914.

HOFF, N. J., *The Analysis of Structures*, John Wiley & Sons, Inc., New York, 1956.

HU, H. C., "On some variational principles in the theory of elasticity and plasticity," *Sci. Sinica* (Peking), **4**, No. 1, pp. 33–54, March 1955.

* A. L. Schlack, "Elastic stability of pierced square plates," *Experimental Mechanics*, June 1964.

† C. J. Rao and G. Pickett, *J. Aero. Soc. India*, 13(3) 83–88, 1961.

IAI, T., "A method of solution of elastic problems by the theorem of maximum energy" (in Japanese), *Journal of the Society of Aeronautical Science of Nippon*, **10**(96), 276–98, April 1943.

KOHN, W., "A note on Weinstein's variational method," *Phys. Rev.*, **71**(12), 902–904, June 1947.

MIKHLIN, S. G., *Variational Methods in Mathematical Physics*, Pergamon Press, Inc., Elmsford, N.Y., 1964.

NAGHDI, P. M., "On a variational theorem in elasticity and its application to shell theory," *J. Appl. Mech.*, **31**(4), 647–653, December 1964.

PRAGER, W. and J. L. SYNGE, "Approximations in elasticity based on the concept of function space," *Quart. Appl. Math.*, **5**(3), 241–269, October 1947.

REISSNER, E., "Note on the method of complementary energy," *J. Math. Phys.*, **27**, 159–160, 1948.

——— "On a variational theorem in elasticity," *J. Math. Phys.*, **29**(2), 90–95, July 1950.

——— "On variational principles in elasticity," *Proceedings of Symposia in Applied Mathematics*, Vol. 8, McGraw-Hill Book Company, 1958, New York, pp. 1–6.

SOUTHWELL, R. V., "Some extensions of Rayleigh's principle," *Quart. J. Mech. Appl. Math.*, **6**, pt. 3, 257–272, October 1953.

SYNGE, J. L., *The Hypercircle in Mathematical Physics*, Cambridge University Press, New York, 1957.

TEMPLE, G., and W. G. BICKLEY, *Rayleigh's Principle and Its Applications to Engineering*, Oxford University Press, Inc., New York, 1933.

WASHIZU, K., "Bounds for solutions of boundary value problems in elasticity," *J. Math. Phys.*, **32** (2–3), 119–128, July–October 1953.

——— *Variational Methods in Elasticity and Plasticity*, Pergamon Press, Inc., Elmsford, N. Y., 1968.

WEINSTEIN, A., "Etude des spectres des équations aux derivées partielles de la théorie des plaques élastiques," *Memorial des Sciences Mathematiques*, **88**, Paris, 1937.

WESTERGAARD, H. M., "On the method of complementary energy and its application to structures stressed beyond the proportional limit, to buckling and vibrations, and to suspension bridges," *Proc. Am. Soc. Civil Engs*, **67**(2), 199–227, February 1941.

PART III

THREE-DIMENSIONAL ELASTICITY

CHAPTER

12

SAINT-VENANT TORSION AND BENDING THEORY

1 Torsion of circular cylinder

We can frequently estimate the distribution of stress by examination of the displacements. Consider, for example, a circular cylinder of length l with one of its bases fixed in the xy plane while the other is acted upon by a moment T along the z axis (Figure 12-1).

Under the action of the moment T the cylinder is twisted, and the generators of the cylinder form helical curves. After examination of the symmetry involved, it is reasonable to assume that sections of the cylinder in planes normal to the z axis will remain plane after deformation. The amount of the rotation will depend upon the distance from the base, $z = 0$; therefore, we can write the twist at any section β if we assume a linear dependence:

$$\beta = \alpha z, \qquad (12\text{-}1.1)$$

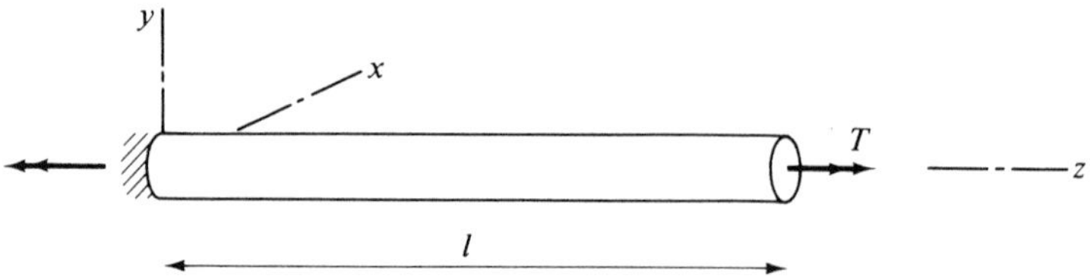

Figure 12-1

where α is the twist per unit length. This process of reasoning was first proposed by Coulomb.*

If cross sections remain plane after deformation, one might assume that the deformation in the z direction is zero.

$$u_z = 0. \tag{12-1.2}$$

A plane normal to the z axis is shown in Figure 12-2 and a point having

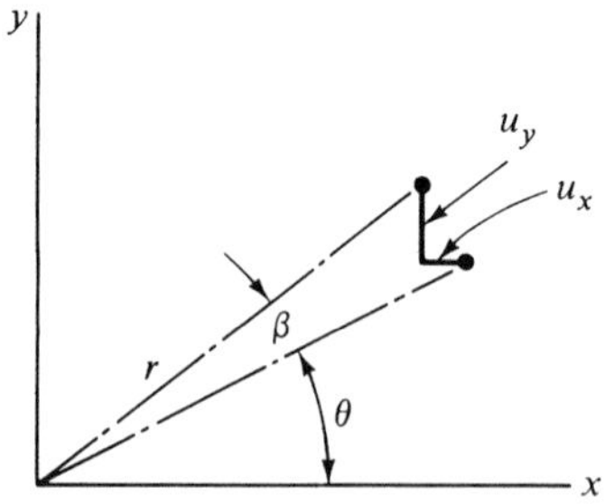

Figure 12-2

initial coordinates r and θ is considered. We can examine geometrically the displacements u_x and u_y in terms of the twist at the section, $z =$ constant,

$$u_x = r\cos(\beta + \theta) - r\cos\theta, \tag{12-1.3}$$

$$u_y = r\sin(\beta + \theta) - r\sin\theta. \tag{12-1.4}$$

Using the following identities for small values of β,

$$\cos(\beta + \theta) = \cos\beta\cos\theta - \sin\beta\sin\theta = \cos\theta - \beta\sin\theta, \tag{12-1.5}$$

$$\sin(\beta + \theta) = \sin\beta\cos\theta + \cos\beta\sin\theta = \beta\cos\theta + \sin\theta, \tag{12-1.6}$$

we obtain

$$u_x = -yz\alpha, \tag{12-1.7}$$

$$u_y = x\beta = xz\alpha, \tag{12-1.8}$$

$$u_z = 0. \tag{12-1.9}$$

* C. A. Coulomb, *Historie de l'Académie*, Paris, 1787.

If these displacements are to constitute a solution to the problem in elasticity, they must satisfy the Navier equations for the displacements and the proper boundary conditions. The Navier equations for elastostatics are

$$G\nabla^2\bar{u} + (\lambda + G)\bar{\nabla}(\bar{\nabla}\cdot\bar{u}) = 0. \tag{12-1.10}$$

Noting $\bar{\nabla}\cdot\bar{u} = 0$ and expanding yields

$$\begin{aligned} G\nabla^2(-yz\alpha) &= 0, \\ G\nabla^2(xz\alpha) &= 0. \end{aligned} \tag{12-1.11}$$

The Navier equations are satisfied by these displacements. The boundary condition on the lateral surface is that the stress traction $\bar{t}^{(n)}$ be zero. On an

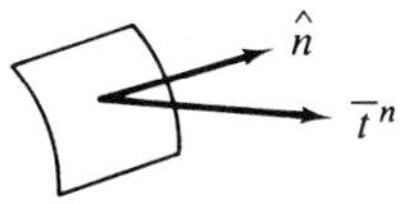

Figure 12-3

element of surface S_n, this condition takes the form

$$\bar{t}_i^n = n_i\sigma_{ij}, \tag{12-1.12}$$

or in dyadic notation

$$\bar{t}^n = \hat{n}\cdot\bar{\bar{\sigma}} = 0. \tag{12-1.13}$$

To obtain the proper components of the stress tensor to use in this condition, we shall first calculate the strains from the displacements:

$$\begin{aligned} \epsilon_{xx} &= \frac{\partial u_x}{\partial x} = 0, \\ \epsilon_{yy} &= \frac{\partial u_y}{\partial y} = 0, \\ \epsilon_{xy} &= \frac{1}{2}\left(\frac{\partial u_x}{\partial y} + \frac{\partial u_y}{\partial x}\right) = \frac{1}{2}(-z\alpha + z\alpha) = 0, \\ \epsilon_{zz} &= \frac{\partial u_z}{\partial z} = 0, \\ \epsilon_{xz} &= \frac{1}{2}\left(\frac{\partial u_x}{\partial z} + \frac{\partial u_z}{\partial x}\right) = -\frac{1}{2}y\alpha, \\ \epsilon_{yz} &= \frac{1}{2}\left(\frac{\partial u_y}{\partial z} + \frac{\partial u_z}{\partial y}\right) = \frac{1}{2}x\alpha. \end{aligned} \tag{12-1.14}$$

The stresses obtained from Hooke's Law for an isotropic material are

$$\begin{aligned} \sigma_{xx} &= \sigma_{yy} = \sigma_{zz} = \sigma_{xz} = 0, \\ \sigma_{xz} &= -Gy\alpha, \\ \sigma_{yz} &= Gx\alpha. \end{aligned} \tag{12-1.15}$$

For the problem under consideration the dyadic form of the stress tensor becomes

$$\bar{\bar{\sigma}} = \hat{i}_{(x)}\sigma_{xz}\hat{i}_{(z)} + \hat{i}_{(y)}\sigma_{yz}\hat{i}_{(z)} + \hat{i}_{(z)}\sigma_{zx}\hat{i}_{(x)} + \hat{i}_{(z)}\sigma_{zy}\hat{i}_{(y)}. \qquad (12\text{-}1.16)$$

In the case of a circular cylinder the normal vector $\hat{n}$ is perpendicular to the unit vector $\hat{i}_{(z)}$ and $\hat{n}\cdot\hat{i}_{(z)} = 0$. The boundary condition becomes

$$\begin{aligned} \hat{n}\cdot\bar{\bar{\sigma}} &= [\hat{n}\cdot\hat{i}_{(x)}\sigma_{xz} + \hat{n}\cdot\hat{i}_{(y)}\sigma_{yz}]\hat{i}_{(z)} = 0, \\ \sigma_{xz}&\cos(n, x) + \sigma_{yz}\cos(n, y) = 0. \end{aligned} \qquad (12\text{-}1.17)$$

For the circular boundary in question, we consider the radius of the cylinder to be a and obtain by geometrical considerations

$$\begin{aligned} \cos(n, x) &= \frac{x}{a}, \\ \cos(n, y) &= \frac{y}{a}. \end{aligned} \qquad (12\text{-}1.18)$$

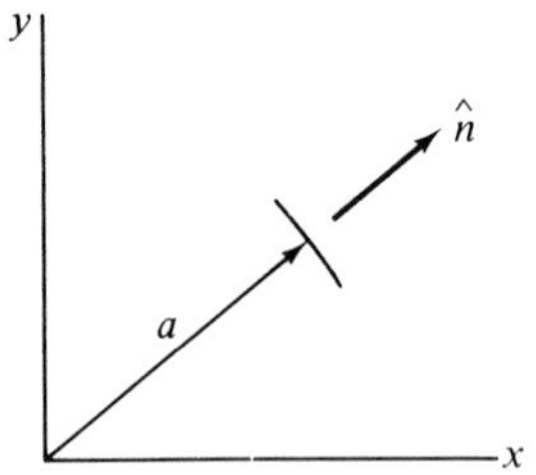

Figure 12-4

Substituting equations 12-1.15 and 12-1.18 into 12-1.17 yields

$$-G\alpha\frac{yx}{a} + G\alpha\frac{xy}{a} = 0. \qquad (12\text{-}1.19)$$

We can now note that this boundary condition is satisfied.

The constant α is not yet determined and will be determined in terms of the applied loads to satisfy the boundary conditions at the ends of the cylinder. The boundary conditions at the ends $z = 0$ and $z = l$ are not clearly stated, for all that is specified is that the moment about the z axis is equal to T and that there are no other forces or moments. We shall not attempt to specify an exact distribution of the stresses on these faces, but shall specify integral conditions and use the Saint-Venant approximation. This assumes that for sufficiently long cylinders the local effects of the various possible distributions of stress will decay quickly as one proceeds from the ends toward the center (see section 14). The boundary conditions can then be written

$$F_x = \int_A \sigma_{xz}\, dA = 0, \qquad (12\text{-}1.20)$$

$$F_y = \int_A \sigma_{yz}\, dA = 0, \qquad (12\text{-}1.21)$$

$$T = \int_A \bar{r} \times \bar{t}^z\, dA = \int_A (x\sigma_{zy} - y\sigma_{zx})\, dA. \qquad (12\text{-}1.22)$$

If the origin is taken to coincide with the centroid, the integrals of the first moment of the area are zero, and the first two conditions are satisfied:

$$\int_A \sigma_{xz}\, dA = -G\alpha \int_A y\, dA = 0, \qquad (12\text{-}1.23)$$

$$\int_A \sigma_{yz}\, dA = G\alpha \int_A x\, dA = 0. \qquad (12\text{-}1.24)$$

The final boundary condition becomes

$$\begin{aligned} T &= \int_A (x\sigma_{zy} - y\sigma_{zx})\, dA = G\alpha \int_A (x^2 + y^2)\, dA, \\ &= G\alpha \int_A r^2\, dA = G\alpha J, \end{aligned} \qquad (12\text{-}1.25)$$

$$\alpha = \frac{T}{GJ}, \qquad \text{where } J = \int_A r^2\, dA. \qquad (12\text{-}1.26)$$

This determines the constant α and yields the elementary solution for the torsion of circular cylinders.

Examining the stress vector on an element of the face normal to the z axis yields

$$\bar{t}^z\, dA = \sigma_{xz}\, dA\hat{i}_{(x)} + \sigma_{yz}\, dA\hat{i}_{(y)}. \qquad (12\text{-}1.27)$$

Using equations 12-1.15 this becomes

$$\bar{t}^z\, dA = G\alpha\, dA(-y\hat{i}_{(x)} + x\hat{i}_{(y)}). \qquad (12\text{-}1.28)$$

A radius vector to the differential area is

$$\bar{r} = x\hat{i}_{(x)} + y\hat{i}_{(y)}. \qquad (12\text{-}1.29)$$

Because it can be noted that $\bar{r} \cdot \bar{t}^z\, dA = 0$, the stress vector is normal to the radius vector. The magnitude of the stress vector is

$$\begin{aligned} (\bar{t}^z\, dA) &= G\alpha\sqrt{x^2 + y^2}\, dA = G\alpha r\, dA, \\ \sigma_{z\theta} &= G\alpha r = \frac{Tr}{J}. \end{aligned} \qquad (12\text{-}1.30)$$

2 Noncircular section

After Coulomb's development, it was natural to attempt to extend this line of reasoning to noncircular cylinders. Navier,* having formulated

* Navier, *Résumé des leçons sur l'application de la mécanique*, 3rd ed., Paris, 1864, edited by B. de Saint-Venant.

the basic equations of elasticity, attempted to apply the reasoning of plane sections remaining plane to noncircular sections. He had no success in this venture, being unaware of the previous work by Augustin Cauchy, who in his investigation of narrow rectangular bars showed that the cross sections of a twisted bar do not, in general, remain plane, but warp during torsion.

Barrie de Saint-Venant, a student of Navier's, proposed a natural modification of Coulomb's assumptions.* Feeling that direct solutions of the Navier equations were too difficult, he assumed that the warping could be described by a function which did not vary with z.

He assumed the following set of displacements:

$$u_x = -\alpha z y, \tag{12-2.1}$$

$$u_y = \alpha z x, \tag{12-2.2}$$

$$u_z = \alpha \varphi(x, y), \tag{12-2.3}$$

where $\varphi(x, y)$ is the warping function describing the z displacement of each section independent of position along the twisted axis. The divergence of the displacement vector taken in this manner is zero:

$$\bar{\nabla} \cdot \bar{u} = 0. \tag{12-2.4}$$

The Navier equation becomes

$$\nabla^2 \varphi(x, y) = 0. \tag{12-2.5}$$

This is the defining equation for φ.

φ is independent of z and the equation reduces to

$$\frac{\partial^2 \varphi}{\partial x^2} + \frac{\partial^2 \varphi}{\partial y^2} = 0. \tag{12-2.6}$$

To examine the boundary conditions of the lateral edges

$$\bar{t}^{\nu} = \hat{\nu} \cdot \bar{\bar{\sigma}} = 0, \tag{12-2.7}$$

we shall calculate the strains and stresses from the specified displacements:

$$\epsilon_{xx} = \epsilon_{yy} = \epsilon_{zz} = \epsilon_{xy} = 0, \tag{12-2.8}$$

$$\epsilon_{xz} = \frac{1}{2}\left(-\alpha y + \alpha \frac{\partial \varphi}{\partial x}\right), \tag{12-2.9}$$

$$\epsilon_{yz} = \frac{1}{2}\left(\alpha x + \alpha \frac{\partial \varphi}{\partial y}\right), \tag{12-2.10}$$

$$\sigma_{xx} = \sigma_{yy} = \sigma_{zz} = \sigma_{xz} = 0, \tag{12-2.11}$$

$$\sigma_{xz} = \alpha G\left(\frac{\partial \varphi}{\partial x} - y\right), \tag{12-2.12}$$

$$\sigma_{yz} = \alpha G\left(\frac{\partial \varphi}{\partial y} + x\right). \tag{12-2.13}$$

* B. de Saint-Venant, "Mémorie sur la torsion des prismes," *Mem. Acad. Sci. des Savants Etrangers*, **14**, 1855.

The boundary condition becomes

$$\left\{\alpha G\left(\frac{\partial\varphi}{\partial x} - y\right)\hat{\imath}_{(x)}\cdot\hat{n} + \alpha G\left(\frac{\partial\varphi}{\partial y} + x\right)\hat{\imath}_{(y)}\cdot\hat{n}\right\}\hat{\imath}_{(z)} = 0, \qquad (12\text{-}2.14)$$

$$\left(\frac{\partial\varphi}{\partial x} - y\right)\cos(n, x) + \left(\frac{\partial\varphi}{\partial y} + x\right)\cos(n, y) = 0. \qquad (12\text{-}2.15)$$

Noting that

$$\frac{\partial\varphi}{\partial x}\cos(n, x) + \frac{\partial\varphi}{\partial y}\cos(n, y) = \frac{\partial\varphi}{\partial n},$$

the boundary condition now becomes

$$\frac{\partial\varphi}{\partial n} = y\cos(n, x) - x\cos(n, y). \qquad (12\text{-}2.16)$$

The warping function is then specified by a Neumann-type problem:

$$\nabla^2\varphi = 0,$$
$$\frac{\partial\varphi}{\partial n} = \text{given function on the boundary}. \qquad (12\text{-}2.17)$$

The uniqueness of this particular type of problem is discussed separately and will be considered in section 3.

The constant α has not yet been specified, but will be chosen to satisfy the end conditions. The conditions at $z = l$ are as before:

$$\iint_A \sigma_{xz}\,dx\,dy = 0. \qquad (12\text{-}2.18)$$

Using equation 12-2.17, this may be written

$$\iint_A \left(\frac{\partial\varphi}{\partial x} - y\right)dx\,dy = \iint_A \left\{\frac{\partial}{\partial x}\left[x\left(\frac{\partial\varphi}{\partial x} - y\right)\right] + \frac{\partial}{\partial y}\left[x\frac{\partial\varphi}{\partial y} + x\right)\right]\right\}dx\,dy. \qquad (12\text{-}2.19)$$

The advantage of this relation is that it is now in the form

$$\iint_A \left(\frac{\partial Q}{\partial x} + \frac{\partial P}{\partial y}\right)dx\,dy. \qquad (12\text{-}2.20)$$

By use of Green's theorem this may be written as a contour integral:

$$\iint_A \left\{\frac{\partial Q}{\partial x} + \frac{\partial P}{\partial y}\right\}dx\,dy = \oint_C [Q\,dy - P\,dx]. \qquad (12\text{-}2.21)$$

However, we can write dx and dy in terms of ds as follows:

$$dx = -ds\cos(n, y),$$
$$dy = ds\cos(n, x). \qquad (12\text{-}2.22)$$

Using Green's theorem and the forms of dx and dy given, the boundary condition becomes

$$\iint_A \sigma_{xz}\, dA = G\alpha \oint_C x\left[\frac{\partial\varphi}{\partial n} - y\cos(n, x) + x\cos(n, y)\right] ds. \qquad (12\text{-}2.23)$$

The boundary condition of φ required that the integrand of equation 12-2.23 be zero everywhere on the contour C. In a similar manner, we can show that

$$\iint_A \sigma_{yz}\, dA = 0.$$

The final condition of summation of moments allows us to evaluate the constant α:

$$\begin{aligned} T &= \iint_A (\bar{r} \times \bar{t}^z)\, dA, \\ &= \iint_A (x\sigma_{zy} - y\sigma_{zx})\, dA, \qquad (12\text{-}2.24) \\ &= G\alpha \iint_A \left[x\frac{\partial\varphi}{\partial y} - y\frac{\partial\varphi}{\partial x} + x^2 + y^2\right] dA. \end{aligned}$$

Analogous to the problem of circular cylinders, we let the torsional rigidity constant be

$$J = \iint_A \left[x\frac{\partial\varphi}{dy} - y\frac{\partial\varphi}{\partial x} + x^2 + y^2\right] dA. \qquad (12\text{-}2.25)$$

3 Uniqueness of Saint-Venant torsion problem

Consider the cross section described in Figure 12-5. The normal derivative is related to the divergence as follows:

$$\frac{\partial\varphi}{\partial n} = \hat{n}\cdot\bar{\nabla}\varphi. \qquad (12\text{-}3.1)$$

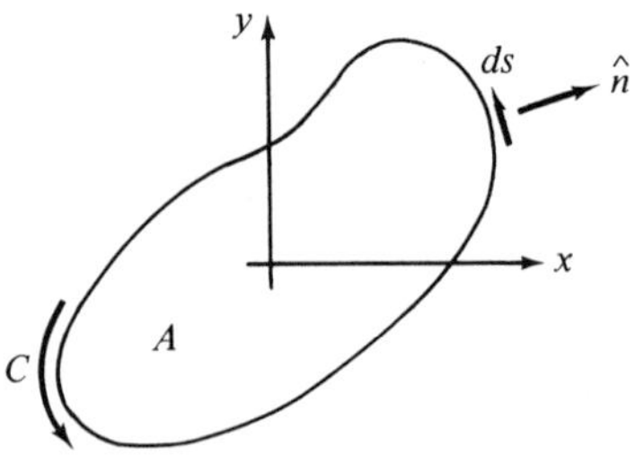

Figure 12-5

Let us consider the contour integral

$$\oint_c \varphi \frac{\partial \varphi}{\partial n} ds = \oint_c \hat{n} \cdot (\varphi \vec{\nabla} \varphi)\, ds. \tag{12-3.2}$$

The divergence theorem yields

$$\begin{aligned} \oint_c \varphi \frac{\partial \varphi}{\partial n} ds &= \iint_A \vec{\nabla} \cdot (\varphi \vec{\nabla} \varphi)\, dA, \\ &= \iint_A \{\vec{\nabla}\varphi \cdot \vec{\nabla}\varphi + \underbrace{\varphi \nabla^2 \varphi}_{=0}\}\, dA, \\ &= \iint_A (\vec{\nabla}\varphi \cdot \vec{\nabla}\varphi)\, dA. \end{aligned} \tag{12-3.3}$$

After this mathematical exercise, we can consider the uniqueness of our solution by examining two solutions, φ_1 and φ_2, which are harmonic and satisfy the same condition:

$$\frac{\partial \varphi}{\partial n} \text{ given on } C. \tag{12-3.4}$$

Considering the difference solution $(\varphi_1 - \varphi_2)$ and noting that $(\partial/\partial n)(\varphi_1 - \varphi_2) = 0$ everywhere on C yields

$$\iint_A [\vec{\nabla}(\varphi_1 - \varphi_2) \cdot \vec{\nabla}(\varphi_1 - \varphi_2)]\, dA = 0. \tag{12-3.5}$$

Since the integrand is everywhere positive,

$$\vec{\nabla}(\varphi_1 - \varphi_2) = 0, \tag{12-3.6}$$

$$\varphi_1 - \varphi_2 = \text{constant}. \tag{12-3.7}$$

Note that if we had specified φ on the boundary, we could now conclude that $\varphi_1 = \varphi_2$, as this must be true on C.

These two problems are well known and are called the Dirichlet and Neumann problems:

$$\nabla^2 \varphi = 0, \qquad \begin{array}{ll} \text{Dirichlet} & \varphi \text{ given}, \\ \text{Neumann} & \dfrac{\partial \varphi}{\partial n} \text{ given}. \end{array}$$

Note that in the Neumann problem φ is determined only up to a constant. In the torsion problem the constant contributes nothing to the stresses and corresponds to a rigid-body displacement in u_z.

The uniqueness property is not the only thing of interest in this solution, as one might also question the effect of a shift of the origin. Remember the displacements are based on the rotation about the origin.* The old coordi-

* I. S. Sokolnikoff, *Mathematical Theory of Elasticity*, McGraw-Hill Book Company, New York, 1956.

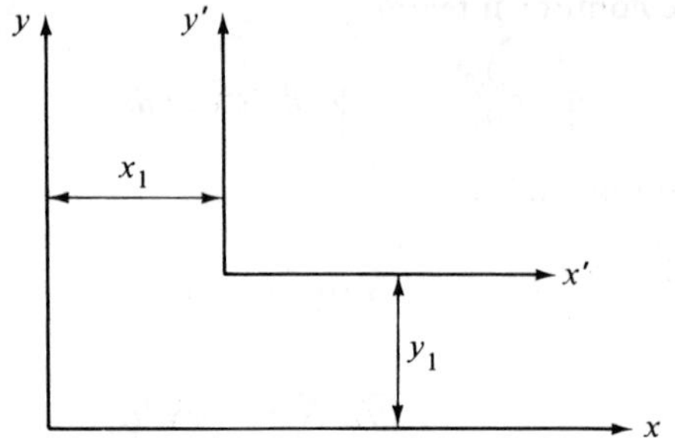

Figure 12-6

nates are related to the new ones by

$$x = x' + x_1, \qquad y = y' + y_1. \tag{12-3.8}$$

The displacements with respect to the new axes are

$$\begin{aligned} u'_x &= -\alpha z(y - y_1), \\ u'_y &= \alpha z(x - x_1), \\ u'_z &= \alpha\varphi_1(x, y). \end{aligned} \tag{12-3.9}$$

In general, φ_1 does not equal φ. Calculating the stresses yields

$$\begin{aligned} \sigma'_{zy} &= G\alpha\left(\frac{\partial\varphi_1}{\partial y} + x - x_1\right), \\ \sigma'_{zx} &= G\alpha\left(\frac{\partial\varphi_1}{\partial x} - y + y_1\right). \end{aligned} \tag{12-3.10}$$

Substituting the displacement equations into Navier's equation yields

$$\nabla^2\varphi_1 = 0. \tag{12-3.11}$$

The boundary condition is

$$\left(\frac{\partial\varphi_1}{\partial x} - y + y_1\right)\cos(n, x) + \left(\frac{\partial\varphi_1}{\partial y} + x - x_1\right)\cos(n, y) = 0. \tag{12-3.12}$$

We can write this as

$$\begin{aligned} \frac{\partial}{\partial x}(\varphi_1 + y_1 x - x_1 y)\cos(n, x) + \frac{\partial}{\partial y}(\varphi_1 - x_1 y + y_1 x)\cos(n, y) \\ = y\cos(n, x) - x\cos(n, y), \end{aligned}$$

or

$$\frac{\partial}{\partial n}[\varphi_1 + y_1 x - x_1 y] = y\cos(n, x) - x\cos(n, y). \tag{12-3.13}$$

We note that $[\varphi_1 + y_1 x - x_1 y]$ satisfies the same condition as the original function φ and is also harmonic. Using the uniqueness property, these two functions may differ only by a constant:

$$\varphi_1 = \varphi - y_1 x + x_1 y + C. \tag{12-3.14}$$

Using equations 12-3.10, we see that σ'_{zx} and σ'_{zy} are the same as about the old origin. The two solutions differ only by rigid-body displacements.

4 Semi-inverse approach—problems of Saint-Venant

Let us examine some harmonic functions and see what torsion problem they solve. This is indeed an inverse approach to the problem. The entire problem may be restated as

$$
\begin{aligned}
\nabla^2\varphi &= 0,\\
\left(\frac{\partial\varphi}{\partial x} - y\right)\cos(n, x) + \left(\frac{\partial\varphi}{\partial y} + x\right)\cos(n, y) &= 0,\\
\cos(n, y) &= -\frac{dx}{ds},\\
\cos(n, x) &= \frac{dy}{ds},
\end{aligned}
\tag{12-4.1}
$$

$$
\begin{aligned}
\sigma_{xz} &= G\alpha\left(\frac{\partial\varphi}{\partial x} - y\right),\\
\sigma_{yz} &= G\alpha\left(\frac{\partial\varphi}{\partial y} + x\right).
\end{aligned}
\tag{12-4.2}
$$

Special cases are as follows:

I. $\varphi = C$.

Note that C will not be uniquely determined since φ is determined only up to a constant.

$$
\begin{aligned}
\nabla^2\varphi &= 0,\\
-y\cos(n, x) + x\cos(n, y) &= 0, \qquad \text{on boundary.}\\
-y\frac{dy}{ds} - x\frac{dx}{ds} &= 0.
\end{aligned}
$$

Therefore,

$$
\frac{d}{ds}\left(\frac{x^2 + y^2}{2}\right) = 0,
$$

$$
x^2 + y^2 = R^2, \qquad \text{on boundary.}
$$

This is the Coulomb solution.

II. $\varphi = Axy,$

$\nabla^2\varphi = 0.$

Boundary condition:

$$(Ay - y)\cos(n, x) + (Ax + x)\cos(n, y) = 0,$$

$$(A - 1)y\frac{dy}{ds} - (A + 1)x\frac{dx}{ds} = 0,$$

$$x^2 + \frac{1 - A}{1 + A}y^2 = C, \qquad \text{on boundary.}$$

The equation for an ellipse with origin at the center is

$$x^2 + \frac{a^2}{b^2}y^2 = a^2.$$

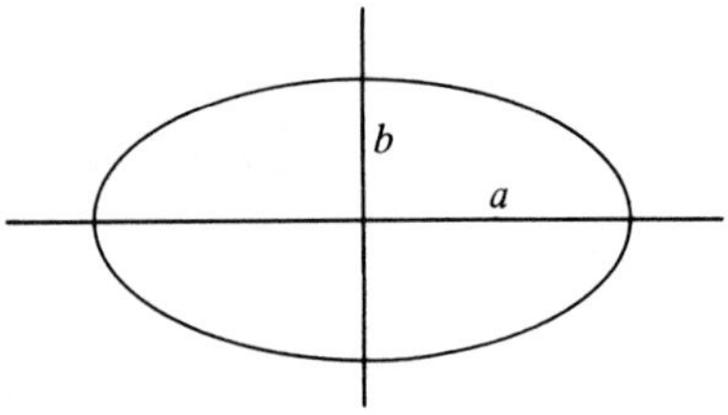

Figure 12-7

Comparing, we see

$$\frac{1 - A}{1 + A} = \frac{a^2}{b^2}, \qquad A = \frac{b^2 - a^2}{b^2 + a^2},$$

$$\varphi = \frac{b^2 - a^2}{b^2 + a^2}xy.$$

This is the warping function for an elliptical cross section,

$$\sigma_{xz} = -\frac{2a^2 y}{b^2 + a^2}G\alpha, \qquad \sigma_{yz} = \frac{2b^2 x}{b^2 + a^2}G\alpha.$$

The torsional rigidity constant can now be calculated:

$$J = \iint_A \left[x\frac{\partial\varphi}{\partial y} - y\frac{\partial\varphi}{\partial x} + x^2 + y^2\right] dA,$$

$$= (1 + A)\iint_A x^2\, dA + (1 - A)\iint_A y^2\, dA,$$

$$= (1 + A)I_y + (1 - A)I_x,$$

where I_x and I_y are the flexural moments of inertia:

$$I_y = \frac{\pi a^3 b}{4}, \qquad I_x = \frac{\pi a b^3}{4},$$

$$J = \frac{\pi a^3 b^3}{a^2 + b^2}.$$

We may relate α to T by the relation

$$T = G\alpha J,$$

$$\sigma_{xz} = -\frac{2Ty}{\pi ab^3},$$

$$\sigma_{yz} = \frac{2Tx}{\pi a^3 b}.$$

Let us see how J compares to the polar moment of inertia I_p:

$$I_p = I_x + I_y = \frac{\pi}{4} ab(a^2 + b^2).$$

Noting that the area is $A = \pi ab$, we can write

$$J = \frac{A^4}{4\pi^2 I_p}.$$

Twenty-five years after publishing his memoir on torsion, Saint-Venant noted this relation and observed that all solid sections except a few elongated ones fit this formula to within 10 per cent. Replacing $4\pi^2$ by 40 brought the accuracy to within 8 per cent. The maximum stress in all cases examined by Saint-Venant occurred on the boundary at the point nearest to the centroid.*

5 Torsion of rectangular bars using Fourier analysis

Let us consider a more direct approach to the problem and use as an example the torsion of rectangular cross sections. Consider the rectangular section shown in Figure 12-8. The boundary conditions are

$$x = \pm a, \quad \frac{\partial \varphi}{\partial x} = y, \qquad y = \pm b, \quad \frac{\partial \varphi}{\partial y} = -x. \tag{12-5.1}$$

The boundary conditions take a simpler form if the following substitution is made:

$$\varphi = xy - \varphi_1, \tag{12-5.2}$$

$$\nabla^2 \varphi_1 = 0, \tag{12-5.3}$$

where the boundary conditions on φ_1 become

$$x = \pm a, \quad \frac{\partial \varphi_1}{\partial x} = 0, \qquad y = \pm b, \quad \frac{\partial \varphi_1}{\partial y} = 2x. \tag{12-5.4}$$

* Filon indicated that there were cases in which the maximum stress did not occur at the boundary point nearest the centroid but at other boundary points. See L. N. G. Filon, *Trans. Roy. Soc.* (London), **A193**, 1900.

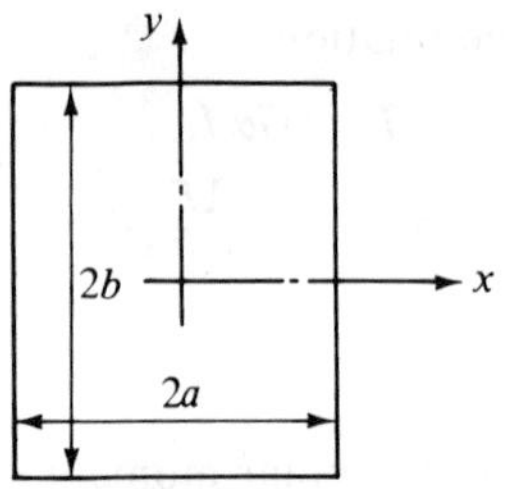

Figure 12-8

Separation of variables yields

$$\varphi_1 = \sum_{n=0}^{\infty} A_n \sin k_n x \operatorname{sh} k_n y + B_n \sin k_n x \operatorname{ch} k_n y + C_n \cos k_n x \operatorname{sh} k_n y + D_n \cos k_n x \operatorname{ch} k_n y. \tag{12-5.5}$$

From the boundary conditions, one notes that φ_1 should be odd in both x and y:

$$\varphi_1 = \sum_{n=0}^{\infty} A_n \sin k_n x \operatorname{sh} k_n y. \tag{12-5.6}$$

Satisfying the first boundary condition term by term yields

$$\cos k_n a = 0 \Longrightarrow k_n = \frac{2n+1}{2a}\pi. \tag{12-5.7}$$

The second condition yields

$$\frac{\partial \varphi_1}{\partial y} = \sum A_n k_n \operatorname{ch} k_n b \sin k_n x = 2x. \tag{12-5.8}$$

Let

$$\bar{A}_n = A_n k_n \operatorname{ch} k_n b:$$

$$\sum \bar{A}_n \sin k_n x = 2x. \tag{12-5.9}$$

Taking the finite Fourier transform gives

$$\bar{A}_n = \frac{1}{a}\int_{-a}^{+a} 2x \sin k_n x \, dx = \frac{16(-1)^n a}{\pi^2(2n+1)^2}, \tag{12-5.10}$$

$$A_n = \frac{32(-1)^n a^2}{\pi^3(2n+1)^3 \operatorname{ch} k_n b}. \tag{12-5.11}$$

Calculating J yields

$$J = \iint_A \left(x^2 + y^2 + x\frac{\partial \varphi_1}{\partial y} - y\frac{\partial \varphi_1}{\partial x}\right) dx\, dy = 16a^3 b\left[\frac{1}{3} - \frac{64a}{\pi^5 b}\sum_{n=0}^{\infty}\frac{\tanh k_n b}{(2n+1)^5}\right]. \tag{12-5.12}$$

The maximum shear stress occurs on the long sides at their midpoint and is equal to

$$\tau_{\max} = \frac{2Ta}{J}\left[1 - \frac{8}{\pi^2}\sum_n \frac{1}{(2n+1)^2 \operatorname{ch} k_n b}\right]. \qquad (12\text{-}5.13)$$

If $b \gg a$, we may use just the first terms:

$$J = \frac{16}{3}a^3 b, \qquad \tau_{\max} = \frac{2Ta}{J} = \frac{3}{8}\frac{T}{a^2 b}, \qquad (12\text{-}5.14)$$

or for the member shown in Figure 12-9

$$J = \frac{1}{3}lt^3, \qquad \tau_{\max} = \frac{3T}{lt^2}, \qquad \alpha = \frac{3T}{lt^3 G}. \qquad (12\text{-}5.15)$$

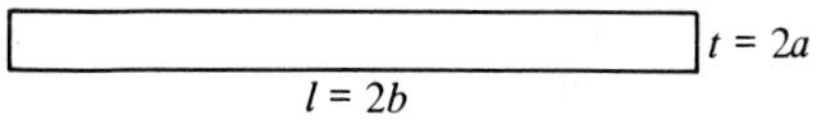

Figure 12-9

We may summarize these results for different b/a ratios and compare these values to Saint-Venant's approximation formula discussed in section 4. The approximation formula may be written

$$J = \frac{A^4}{40 I_p} = \frac{4.8a^3 b}{[1 + (a^2/b^2)]}. \qquad (12\text{-}5.16)$$

The results of this comparison are shown in Table 12-1.

Table 12-1

b/a	K_1	K_1 approx.	Error (%)	K_2	K_3
1.0	2.250	2.4	6.7	1.350	0.600
1.2	2.656	2.83	6.6	1.518	0.571
1.5	3.136	3.31	5.4	1.696	0.541
2.0	3.664	3.84	4.9	1.860	0.508
2.5	3.984	4.14	4.0	1.936	0.484
3.0	4.208	4.32	2.6	1.970	0.468
4.0	4.496	4.500	0.1	1.994	0.443
5.0	4.656	4.610	1.0	1.998	0.430
10.0	4.992	4.750	4.8	2.000	0.401
∞	5.328	4.8	9.9	2.000	0.37

$J = K_1 a^3 b$, $\qquad K_1$ approx. $= 4.8/[1 + (a^2/b^2)]$,

$\tau_{\max} = K_2(Ta/J) = K_3(T/a^2 b)$.

6 Prandtl stress function

An alternative approach to the problem of torsion was proposed by L. Prandtl in 1903.* He based his assumptions on the stress distribution, instead of the displacement distribution as Saint-Venant had done. In this manner, the Prandtl solution is also a semi-inverse solution. This solution has the advantage that it greatly simplifies the boundary conditions. We shall assume that the following stresses are zero:

$$\sigma_{xx} = \sigma_{yy} = \sigma_{zz} = \sigma_{xy} = 0. \tag{12-6.1}$$

The equilibrium equations are now

$$\begin{aligned} \frac{\partial \sigma_{xz}}{\partial z} &= 0, \\ \frac{\partial \sigma_{yz}}{\partial z} &= 0, \\ \frac{\partial \sigma_{xz}}{\partial x} + \frac{\partial \sigma_{yz}}{\partial y} &= 0. \end{aligned} \tag{12-6.2}$$

We can solve these by relating σ_{xz} and σ_{yz} to a stress function $\psi(x, y)$ as follows:

$$\sigma_{xz} = \frac{\partial \psi}{\partial y}, \tag{12-6.3}$$

$$\sigma_{yz} = -\frac{\partial \psi}{\partial x}. \tag{12-6.4}$$

Note that $\psi(x, y)$ does not depend upon z and the first two equations are automatically satisfied. The Beltrami–Michell compatibility equations (see problem 5.2) can be written in this case as

$$\nabla^2 \sigma_{ij} + \frac{1}{1+\nu} \nabla_i \nabla_j \sigma_{kk} = 0. \tag{12-6.5}$$

These reduce to the following two equations:

$$\nabla^2 \sigma_{yz} = 0, \tag{12-6.6}$$

$$\nabla^2 \sigma_{xz} = 0, \tag{12-6.7}$$

$$\frac{\partial}{\partial x}(\nabla^2 \psi) = 0, \qquad \frac{\partial}{\partial y}(\nabla^2 \psi) = 0 \Longrightarrow \nabla^2 \psi = \text{constant}. \tag{12-6.8}$$

The boundary condition on the lateral surfaces parallel to the z axis is given in equation 12-6.9.

$$\sigma_{zx} \cos(n, x) + \sigma_{zy} \cos(n, y) = 0, \tag{12-6.9}$$

* L. Prandtl, *Physik. Z.*, **4**, 1903.

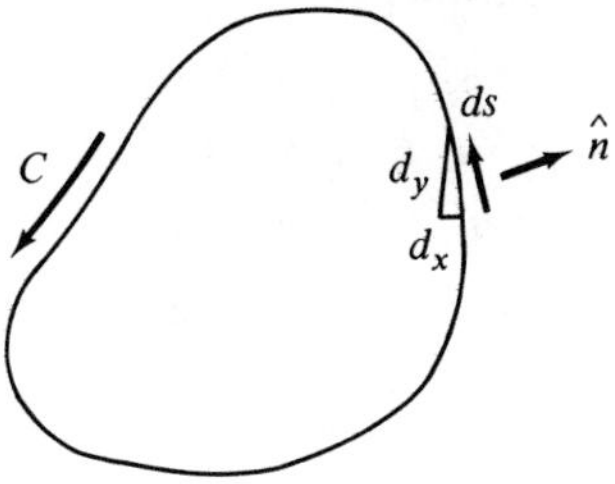

Figure 12-10

$$\cos(n, x) = \frac{dy}{ds}, \qquad \cos(n, y) = -\frac{dx}{ds}. \tag{12-6.10}$$

This boundary condition becomes

$$\begin{aligned} \frac{\partial \psi}{\partial y}\frac{dy}{ds} + \frac{\partial \psi}{\partial x}\frac{dx}{\partial s} &= 0, \\ \frac{d}{ds}\psi &= 0, \end{aligned} \tag{12-6.11}$$

$$\psi = \text{constant on the boundary.} \tag{12-6.12}$$

We can determine the constant in Poisson's equation that defines ψ by relating the two definitions for stress from the stress function and warping function:

$$\sigma_{xz} = \frac{\partial \psi}{\partial y} = G\alpha\left(\frac{\partial \varphi}{\partial x} - y\right), \tag{12-6.13}$$

$$\sigma_{yz} = -\frac{\partial \psi}{\partial x} = G\alpha\left(\frac{\partial \varphi}{\partial y} + x\right). \tag{12-6.14}$$

Differentiating the first with respect to y and the second with respect to x and subtracting yields

$$\nabla^2 \psi = -2G\alpha. \tag{12-6.15}$$

The constant in equation 12-6.8 has the value $-2G\alpha$. The displacements can be calculated from the strain displacement relations and the stress–strain relations:

$$\begin{gathered} \frac{\partial u_x}{\partial x} = 0, \qquad \frac{\partial u_y}{\partial y} = 0, \qquad \frac{\partial u_z}{\partial z} = 0, \\ \frac{\partial u_x}{\partial y} + \frac{\partial u_y}{\partial x} = 0, \\ \frac{\partial u_x}{\partial z} + \frac{\partial u_z}{\partial x} = \frac{\sigma_{xz}}{G}, \\ \frac{\partial u_y}{\partial z} + \frac{\partial u_z}{\partial y} = \frac{\sigma_{yz}}{G}. \end{gathered} \tag{12-6.16}$$

Integrating the first three equations yields

$$\begin{aligned} u_x &= u_x(y, z), \\ u_y &= u_y(x, z), \\ u_z &= u_z(x, y). \end{aligned} \tag{12-6.17}$$

Substitution into the first shear equation gives

$$\frac{\partial u_x}{\partial y} + \frac{\partial u_y}{\partial x} = 0. \tag{12-6.18}$$

Because u_x is only a function of y and z and u_y a function of x and z, $\partial u_x/\partial y$ must be the negative of the function of z that equals $\partial u_y/\partial x$.

$$\begin{aligned} \frac{\partial u_x}{\partial y} &= -\frac{\partial u_y}{\partial x} = f(z), \\ u_x &= yf(z) + g(z), \\ u_y &= -xf(z) + r(z). \end{aligned} \tag{12-6.19}$$

u_z, σ_{xz}, and σ_{yz} are independent of z, yielding

$$\begin{aligned} u_x &= C_1 yz + C_2 z + A_1 y + A_2, \\ u_y &= -C_1 xz + C_3 z - A_1 x + A_3. \end{aligned} \tag{12-6.20}$$

The constants A_1, A_2, and A_3 represent rigid-body motions and are neglected:

$$\begin{aligned} u_x &= C_1 yz + C_2 z, \\ u_y &= -C_1 xz + C_3 z. \end{aligned} \tag{12-6.21}$$

Equation 12-6.16 is now used to obtain the following expressions:

$$\begin{aligned} \frac{\partial u_z}{\partial x} &= \frac{\sigma_{xz}}{G} - C_1 y - C_2, \\ \frac{\partial u_z}{\partial y} &= \frac{\sigma_{yz}}{G} + C_1 x - C_3. \end{aligned} \tag{12-6.22}$$

If we relate these expressions to those obtained in terms of the warping function, we note that C_2 and C_3 are also related to rigid-body motions and the expressions 12-6.21 and 12-6.22 can be written

$$\begin{aligned} u_x &= -\alpha yz, \\ u_y &= \alpha xz, \\ \frac{\partial u_z}{\partial x} &= \frac{1}{G}\frac{\partial \psi}{\partial y} + \alpha y, \\ \frac{\partial u_z}{\partial y} &= -\frac{1}{G}\frac{\partial \psi}{\partial x} - \alpha x. \end{aligned} \tag{12-6.23}$$

Substitution of ψ and integrating yields u_z. α can be determined from the remaining boundary conditions on the ends. Consider first the condition

$$\iint_A \sigma_{xz}\, dA = 0.$$

Substitution for σ_{xz} yields

$$\iint_A \frac{\partial \psi}{\partial y}\, dA = \iint_A \vec{\nabla}\cdot[\hat{i}_{(y)}\psi]\, dA, \qquad (12\text{-}6.24)$$
$$= \oint \hat{i}_{(y)}\cdot\hat{n}\psi\, ds = \oint \psi \cos(n, y)\, ds.$$

Because ψ is specified to be a constant on the boundary, this condition is satisfied. Although it is easily shown for simply connected regions, we shall have to say more about multiply connected domains.

In a similar manner,

$$\iint_A \sigma_{yz}\, dA = 0. \qquad (12\text{-}6.25)$$

The final boundary condition is

$$T = \iint_A [x\sigma_{yz} - y\sigma_{xz}]\, dA, \qquad (12\text{-}6.26)$$
$$= -\iint_A \left[x\frac{\partial \psi}{\partial x} + y\frac{\partial \psi}{\partial y}\right] dA,$$

$$= -\iint_A \left[\frac{\partial}{\partial x}(x\psi) + \frac{\partial}{\partial y}(y\psi)\right] dA + \iint_A 2\psi\, dA. \qquad (12\text{-}6.27)$$

Using Green's theorem, this becomes

$$T = -\oint_C \{x\psi \cos(n, x) + y\psi \cos(n, y)\}\, ds + \iint_A 2\psi\, dA. \qquad (12\text{-}6.28)$$

Using the condition that ψ vanishes on the boundary for simply connected regions yields the following expression for T:

$$T = 2\iint_A \psi\, dA. \qquad (12\text{-}6.29)$$

Note that ψ is known and the relation between α and ψ is established so that we can now determine α knowing T.

Consider the multiply connected domain shown in Figure 12-11. Let $A = A_0 + \sum_{i=1} A_i$ be the total area inclosed by C_0. If we examine the integral corresponding to zero lateral force on the cross section, we obtain

$$\iint_{A_0} \frac{\partial \psi}{\partial x}\, dA = \iint_A \frac{\partial \psi}{\partial x}\, dA - \sum_{i=1} \iint_{A_i} \frac{\partial \psi}{\partial x}\, dA \qquad (12\text{-}6.30)$$
$$= \oint_{C_0} \psi \cos(n, x)\, ds - \sum_{i=1} \oint_{C_i} \psi \cos(n_i, x)\, ds.$$

Since ψ equals a constant on each on boundary, we see that this integral

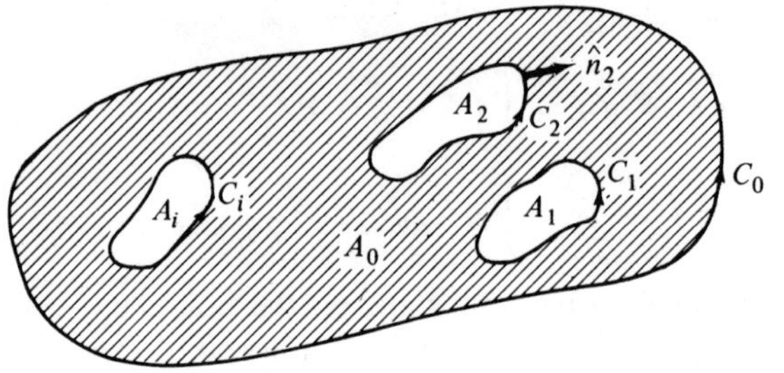

Figure 12-11

is zero even if the constant is not chosen to be zero:

$$\oint \cos(n, x)\, ds = \oint dy = 0. \qquad (12\text{-}6.31)$$

Let the value of ψ on the boundary C_i be given as K_i for reference purposes. Equation 12-6.28 gives

$$\begin{aligned} T = &-\oint_{C_0} \{x\psi \cos(n, x) + y\psi \cos(n, y)\}\, ds \\ &+ \sum_{i=1} \oint_{C_i} \{x\psi \cos(n, x) + y\psi \cos(n, y)\}\, ds + \iint_{A_0} 2\psi\, dA. \end{aligned} \qquad (12\text{-}6.32)$$

Noting that

$$\oint [x \cos(n, x) + y \cos(n, y)\, ds] = \oint x\, dy - y\, dx,$$

and by Green's theorem

$$\oint x\, dy - y\, dx = \iint_A \left(\frac{\partial x}{\partial x} + \frac{\partial y}{\partial y}\right) dx\, dy = 2A,$$

we write equation 12-6.32 as

$$T = \iint_{A_0} 2\psi\, dA + 2 \sum_{i=1} K_i A_i - 2K_0 A_0. \qquad (12\text{-}6.33)$$

K_0 is set equal to zero, and equation 12-6.33 becomes

$$T = \iint_{A_0} 2\psi\, dA + 2 \sum_{i=1} K_i A_i. \qquad (12\text{-}6.34)$$

To understand the necessity of leaving K_i unspecified, we shall examine the requirement that the displacement u_z be single valued. When the warping function is used, φ is single valued and so also are all the stresses and displacements.

For a multiply connected region, we must examine φ so as we make a circuit around one of the inner contours to ensure that it is single valued:

$$\oint_{C_i} \left(\frac{\partial \varphi}{\partial x} dx + \frac{\partial \varphi}{\partial y} dy\right) = 0. \tag{12-6.35}$$

Noting that

$$\begin{aligned} \sigma_{yz} &= \alpha G\left(\frac{\partial \varphi}{\partial y} + x\right) = -\frac{\partial \psi}{\partial x}, \\ \sigma_{xz} &= \alpha G\left(\frac{\partial \varphi}{\partial x} - y\right) = \frac{\partial \varphi}{\partial y}, \end{aligned} \tag{12-6.36}$$

and substituting these relations into 12-6.35 yields

$$\oint \left[\frac{\partial \psi}{\partial y} dx - \frac{\partial \psi}{\partial x} dy - G\alpha(x\, dy - y\, dx)\right] = 0. \tag{12-6.37}$$

Using Green's theorem yields

$$\oint x\, dy - y\, dx = 2A_i, \tag{12-6.38}$$

where A_i is the area inside the contour C_i. Noting that

$$\begin{aligned} dx &= -ds \cos(n, y), \\ dy &= ds \cos(n, x), \end{aligned} \tag{12-6.39}$$

equation 12-6.37 becomes

$$\begin{aligned} \oint_{C_i} \left[\frac{\partial \psi}{\partial y} \cos(n, y) + \frac{\partial \psi}{\partial x} \cos(n, x)\right] ds &= -2G\alpha A_i, \\ \oint_{C_i} \frac{\partial \psi}{\partial n} ds &= -2G\alpha A_i. \end{aligned} \tag{12-6.40}$$

Another form of this equation is obtained in terms of stresses:

$$\frac{\partial \psi}{\partial n} = \sigma_{xy} \cos(n, y) - \sigma_{yz} \cos(n, x) = -\tau_{si}, \tag{12-6.41}$$

where τ_{si} is the shear stress tangential to C_i. Therefore, the required condition becomes

$$\oint_{C_i} \tau_{si}\, ds = 2G\alpha A_i. \tag{12-6.42}$$

This form of the equation is used in shear flow equations for thin-walled sections. For details see any advanced strength of materials book. It is obvious that equation 12-6.40 serves as another condition on ψ which must be satisfied to ensure that the displacements are single valued. There are not a great number of solutions of torsion problems of multiply connected domains, but we shall examine by one simple example how these conditions are applied.

We compare the two formulations of the torsion problem in Table 12-2.

These two solutions are related directly by use of the theory of complex variables. We define the complex conjugate function λ of φ by use of the Cauchy–Riemann equations as follows:

$$\frac{\partial\varphi}{\partial x} = \frac{\partial\lambda}{\partial y}, \qquad \frac{\partial\varphi}{\partial y} = -\frac{\partial\lambda}{\partial x}. \tag{12-6.43}$$

Table 12-2

Warping Function		Stress Function
$\nabla^2\varphi = 0$		$\nabla^2\psi = -2G\alpha$
$\frac{\partial\varphi}{\partial n} = y\cos(n, x) - x\cos(y, n)$	on C	$\psi = K_i$ on C_i, where K_0 may be taken as zero.
$\sigma_{xz} = G\alpha\left(\frac{\partial\varphi}{\partial x} - y\right)$		$\sigma_{xz} = \frac{\partial\psi}{\partial y}, \quad \sigma_{yz} = -\frac{\partial\psi}{\partial x}$
$\sigma_{yz} = G\alpha\left(\frac{\partial\varphi}{\partial y} + x\right)$	$J = \frac{T}{G\alpha}$	$u_x = -\alpha yz$
$u_x = -\alpha yz$		$u_y = +\alpha xz$
$u_y = \alpha xz$		$\frac{\partial u_z}{\partial x} = \frac{1}{G}\frac{\partial\psi}{\partial y} + \alpha y$
$u_z = \alpha\varphi$		$\frac{\partial u_z}{\partial y} = -\frac{1}{G}\frac{\partial\psi}{\partial x} - \alpha x$
$T = \alpha G \iint_A \left[x^2 + y^2 + x\frac{\partial\varphi}{\partial y} - y\frac{\partial\varphi}{\partial x}\right] dA$		$T = 2\iint_A \psi\, dx\, dy + 2\sum_{i=1} A_i K_i$

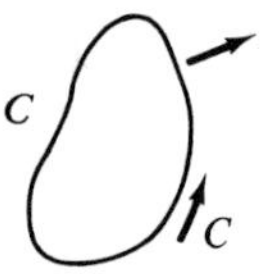

Note that this requires both φ and λ to be harmonic functions. The stress equations relate the warping function to the stress function as follows:

$$\frac{\partial\psi}{\partial y} = G\alpha\left[\frac{\partial\varphi}{\partial x} - y\right], \qquad \frac{\partial\psi}{\partial x} = -G\alpha\left[\frac{\partial\varphi}{\partial y} + x\right]. \tag{12-6.44}$$

Replacing φ by its conjugate function yields

$$\frac{\partial\psi}{\partial y} = G\alpha\left[\frac{\partial\lambda}{\partial y} - y\right], \qquad \frac{\partial\psi}{\partial x} = G\alpha\left[\frac{\partial\lambda}{\partial x} - x\right]. \tag{12-6.45}$$

Integrating yields

$$\psi = G\alpha\left\{\lambda - \frac{x^2 + y^2}{2} + C\right\}. \tag{12-6.46}$$

where C is a constant of integration that can be arbitrarily chosen. The function λ satisfies Laplace's equation:

$$\nabla^2\lambda = 0. \tag{12-6.47}$$

7 Solution of a hollow cylinder

To understand how the boundary conditions are applied in a multiply connected domain, consider a hollow cylinder with inner radius a and outer radius b. Equations 12-6.46 and 12-6.47 yield

$$\lambda = C_1 \ln r + C_2, \tag{12-7.1}$$

$$\psi = G\alpha\left[C_1 \ln r + C_2 - \frac{r^2}{2}\right]. \tag{12-7.2}$$

The boundary condition on the inner edge is (equation 12-6.40)

$$\oint \frac{\partial \psi}{\partial r} r \Big|_{r=a} d\theta = 2\pi G\alpha[C_1 - a^2] = -2G\alpha\underbrace{\pi a^2}_{\text{area}}, \tag{12-7.3}$$

$$C_1 - a^2 = -a^2, \qquad \Longrightarrow C_1 = 0. \tag{12-7.4}$$

The value of ψ on the outer boundary is chosen to be zero, yielding

$$\psi\,|_{r=b} = 0 = G\alpha\left[C_2 - \frac{b^2}{2}\right] \qquad \Longrightarrow C_2 = \frac{b^2}{2}. \tag{12-7.5}$$

The Prandlt stress function becomes

$$\psi = \frac{G\alpha}{2}(b^2 - r^2). \tag{12-7.6}$$

The torsional rigidity constant J is

$$J = 2\pi \int_a^b \frac{1}{2}(b^2 - r^2)r\,dr + (b^2 - a^2)\pi a^2, \tag{12-7.7}$$

$$= \frac{\pi}{2}(b^4 - a^4). \tag{12-7.8}$$

Although this yields the expected elementary solution, it does illustrate the use of the boundary conditions for multiconnected regions.

8 Solution by use of orthogonal series

The Laplace equation may be solved by separation of variables in a number of different coordinate systems, for example, rectangular, parabolic, polar, and elliptic.* The advantage of this is that all solutions of the partial differential equation can be built up out of linear combinations of the members of the family of separated solutions. We shall examine a few cases in particular for rectangular and polar coordinates. In rectangular coordinates

* P. M. Morse and H. Feshbach, *Methods of Theoretical Physics, Part I*, McGraw-Hill Book Company, New York, 1953.

the separation constant can be taken as real or complex, yielding the following families of separated solutions:

$$\begin{aligned}
\lambda^1 &= A_m \sin \alpha_m x \begin{Bmatrix} \operatorname{sh} \alpha_m y \\ \operatorname{ch} \alpha_m y \end{Bmatrix}, \\
\lambda^2 &= B_m \cos \alpha_m x \begin{Bmatrix} \operatorname{sh} \alpha_m y \\ \operatorname{ch} \alpha_m y \end{Bmatrix}, \\
\lambda^3 &= C_m \sin \beta_m y \begin{Bmatrix} \operatorname{sh} \beta_m x \\ \operatorname{ch} \beta_m x \end{Bmatrix}, \\
\lambda^4 &= D_m \cos \beta_m y \begin{Bmatrix} \operatorname{sh} \beta_m x \\ \operatorname{ch} \beta_m x \end{Bmatrix}.
\end{aligned} \tag{12-8.1}$$

As an example of a solution in this coordinate system, consider a direct solution of the torsion of a rectangular bar. This problem was discussed

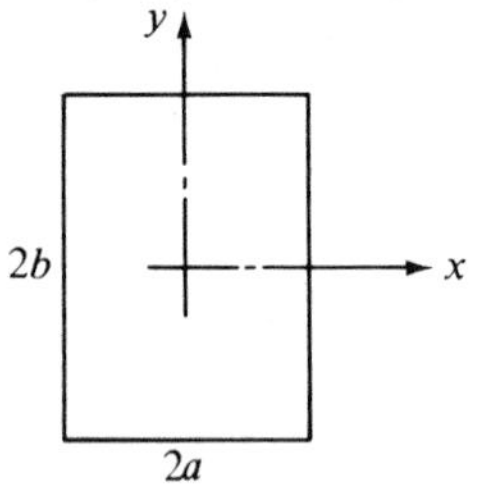

Figure 12-12

in section 5. On the boundaries ψ equals zero, and we shall take the solution in the form

$$\psi = G\alpha\left[\lambda - \frac{x^2 + y^2}{2}\right], \tag{12-8.2}$$

where λ satisfies the equation

$$\nabla^2 \lambda = 0. \tag{12-8.3}$$

The boundary conditions on λ become

$$\lambda\,|_{x=\pm a} = \frac{a^2 + y^2}{2}, \qquad \lambda\,|_{y=\pm b} = \frac{x^2 + b^2}{2}. \tag{12-8.4}$$

We can simplify the solution by considering λ to be the sum of two harmonic functions:

$$\lambda = \lambda_1 + \frac{y^2 - x^2}{2} + a^2, \qquad \nabla^2 \lambda_1 = 0. \tag{12-8.5}$$

The stress function now becomes

$$\psi = G\alpha[\lambda_1 - x^2 + a^2], \tag{12-8.6}$$

and the boundary conditions on λ_1 are

$$\begin{aligned} \lambda_1 |_{x=\pm a} &= 0, \\ \lambda_1 |_{y=\pm b} &= x^2 - a^2. \end{aligned} \tag{12-8.7}$$

Since the boundary conditions imply even functions in both x and y, the solution for λ_1 is

$$\lambda_1 = \sum A_m \cos \beta_m x \operatorname{ch} \beta_m y. \tag{12-8.8}$$

The boundary condition at $x = \pm a$ will be satisfied by selecting β to satisfy the condition

$$\cos \beta_m a = 0, \qquad \beta_m = \frac{2m+1}{2a}\pi. \tag{12-8.9}$$

To satisfy the boundary conditions on $y = \pm b$, we select the coefficients A_m so that the series sums to the required function on the boundary:

$$\sum_{m=0} A_m \operatorname{ch} \beta_m b \cos \beta_m x = x^2 - a^2. \tag{12-8.10}$$

A_m will be obtained by use of the orthogonality property of the trigonometric series

$$\int_{-a}^{+a} \cos \beta_m x \cos \beta_n x \, dx = \begin{cases} 0, & m \neq n \\ a, & m = n \end{cases}, \tag{12-8.11}$$

$$A_m = -\frac{4(-1)^m}{\beta_m^3 a \operatorname{ch} \beta_m b}, \tag{12-8.12}$$

$$\lambda_1 = -\sum_{m=0}^{\infty} \frac{4(-1)^m}{\beta_m^3 a \operatorname{ch} \beta_m b} \operatorname{ch} \beta_m y \cos \beta_m x. \tag{12-8.13}$$

The stress function ψ becomes

$$\psi = G\alpha\left[a^2 - x^2 - \sum_{m=0}^{\infty} \frac{4(-1)^m}{\beta_m^3 a \operatorname{ch} \beta_m b} \cos \beta_m x \operatorname{ch} \beta_m y\right]. \tag{12-8.14}$$

The stresses become

$$\sigma_{xz} = -G\alpha \sum \frac{(-1)^m 4}{\beta_m^2 a \operatorname{ch} \beta_m b} \operatorname{sh} \beta_m y \cos \beta_m x, \tag{12-8.15}$$

$$\sigma_{yz} = -G\alpha\left[\sum \frac{4(-1)^m}{\beta_m^2 a \operatorname{ch} \beta_m b} \operatorname{ch} \beta_m y \sin \beta_m x - 2x\right]. \tag{12-8.16}$$

The torsional rigidity constant is

$$T = 2 \iint \psi \, dA, \tag{12-8.17}$$

$$T = G\alpha\, 2 \iint_A \left[a^2 - x^2 - \sum \frac{4(-1)^m}{\beta_m^3 a \operatorname{ch} \beta_m b} \cos \beta_m x \operatorname{ch} \beta_m y\right] dA, \tag{12-8.18}$$

$$J = 16a^3 b\left\{\frac{1}{3} - \frac{64}{\pi^5}\frac{a}{b} \sum \frac{\tanh \beta_m b}{(2m+1)^5}\right\}, \tag{12-8.19}$$

where the summation identity $\sum_{n=0}^{\infty} [1/(2n+1)^5] = \pi^4/96$ has been used. As would be expected, this agrees with the expression obtained by use of the Saint-Venant warping function approach.

9 Polar coordinates

The Laplace equation in polar coordinates is

$$\left(\frac{\partial^2}{\partial r^2} + \frac{1}{r}\frac{\partial}{\partial r} + \frac{1}{r^2}\frac{\partial^2}{\partial \theta^2}\right)\lambda = 0. \tag{12-9.1}$$

This equation may be solved by separation of variables approach or by using its equidimensional characteristics. The solution takes the form

$$\lambda = A_\mu r^\mu \sin \mu\theta + B_\gamma r^\gamma \cos \gamma\theta, \tag{12-9.2}$$

where μ and γ may assume both positive and negative values. If μ or γ is imaginary, we obtain solutions in the form of a product of r raised to an imaginary power and a hyperbolic function of θ. This form is not as useful as that shown in equation 12-9.2 and will not be discussed. For zero eigenvalues we obtain a multiplicity of roots and the solution takes the form

$$\lambda = A_0 + B_0 \ln r + (A_1 + B_1 \ln r)\theta. \tag{12-9.3}$$

Let us consider as an example the torsional properties of a bar with the cross section of a sector* (Figure 12-13). The stress function takes the form

$$\psi = G\alpha\left[\lambda - \frac{r^2}{2} + b\right], \tag{12-9.4}$$

where

$$\lambda = Ar^\gamma \cos \gamma\theta + Br^\mu \sin \mu\theta. \tag{12-9.5}$$

We shall seek a solution that satisfies the boundary conditions exactly on

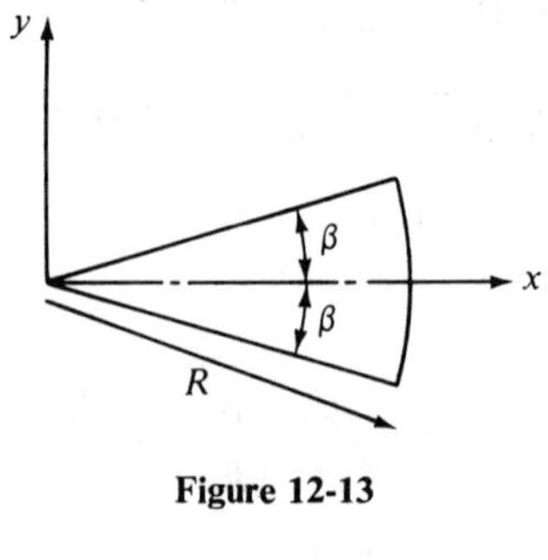

Figure 12-13

* See B. de Saint-Venant, *Compt. Rend.*, **87**, 849, 893, 1878; and A. G. Greenhill, *Messenger of Math.*, **10**, 83, 1880.

the radial sides and yields a series at $r = R, -\beta \leq \theta \leq \beta$. The problem is symmetric about the x-axis and we use only the cosine terms. To remove the $r^2/2$ term on the boundaries, we choose

$$\lambda = Ar^2 \cos 2\theta, \tag{12-9.6}$$

where

$$A = \frac{1}{2\cos 2\beta}. \tag{12-9.7}$$

To satisfy the boundary condition at $r = R$, add terms of the form

$$\lambda = \sum_n C_n r^{\gamma_n} \cos \gamma_n \theta, \qquad \gamma_n = \frac{2n+1}{2\beta}\pi. \tag{12-9.8}$$

Note that these terms are all zero on the radial edge. Taking the constant b to be zero yields the following for the stress function:

$$\psi = G\alpha\left\{\frac{r^2}{2}\left[\frac{\cos 2\theta}{\cos 2\beta} - 1\right] + \sum_{n=0}^{\infty} C_n r^{\gamma_n} \cos \gamma_n \theta\right\}. \tag{12-9.9}$$

At $r = R, \psi = 0$; therefore,

$$\frac{R^2}{2}\left(1 - \frac{\cos 2\theta}{\cos 2\beta}\right) = \sum_{n=0}^{\infty} C_n R^{\gamma_n} \cos \gamma_n \theta, \tag{12-9.10}$$

$$C_n = \frac{1}{R^{\gamma_n}\beta}\frac{R^2}{2}\int_{-\beta}^{\beta}\left(1 - \frac{\cos 2\theta}{\cos 2\beta}\right)\cos \gamma_n \theta \, d\theta. \tag{12-9.11}$$

Carrying out the required integration yields

$$\psi = G\alpha\left[\frac{r^2}{2}\left(\frac{\cos 2\theta}{\cos 2\beta} - 1\right) + \sum_{n=0}^{\infty} \frac{4R^2}{\beta}\left(\frac{r}{R}\right)^{\gamma_n} \frac{(-1)^{n+1}}{\gamma_n(\gamma_n^2 - 4)} \cos \gamma_n \theta\right]. \tag{12-9.12}$$

Although the first term of the solution becomes unbounded when $\beta = 45°$, so also does the term $n = 0$ in the series ($\gamma_n = 2$) and the sum remains finite. The validity of this solution would, however, be limited to values of β less than $\pi/2$. If larger than this, β would involve a reentry corner and one would have to correct the solution with the necessary singularity.

10 Nonorthogonal functions on boundary

Numerical methods for solving boundary value problems have been discussed in Section 7-10. Laplace's equation is particularly well suited for solution by use of these methods as it can be shown that the maximum error will occur on the boundary. The approximation of the numerical solution is concentrated on the boundary and the error may be easily calculated.

To examine these methods, we take as an example a cylindrical shaft with a flat "face" on one side (Figure 12-14). We assume the stress function

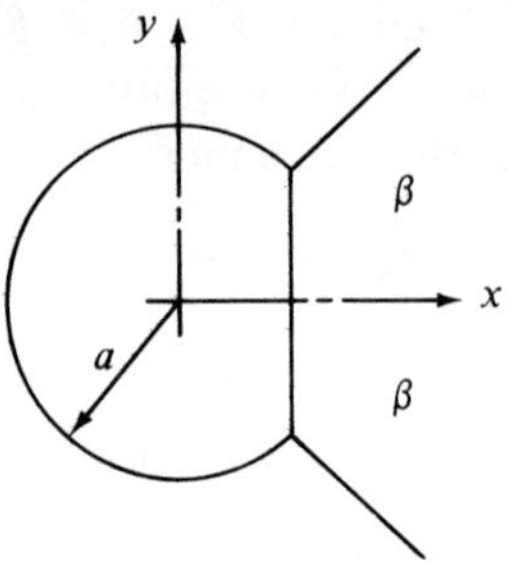

Figure 12-14

in the form

$$\psi = G\alpha\left\{-\frac{r^2}{2} + \sum_n C_n r^{\gamma_n} \cos \gamma_n \theta\right\}. \tag{12-10.1}$$

Only the cosine terms are used because of the symmetry about the x axis. γ_n is selected so that the functions will be single valued:

$$\gamma_n = n. \tag{12-10.2}$$

The boundary conditions become

$$0 \leq \theta \leq \beta, \qquad \psi = 0 \quad \text{at} \quad x = a \cos \beta = r \cos \theta.$$

Therefore,

$$r = \frac{a \cos \beta}{\cos \theta}. \tag{12-10.3}$$

This condition becomes

$$0 \leq \theta \leq \beta, \qquad \frac{a^2 \cos^2 \beta}{2 \cos^2 \theta} = \sum C_n \left[\frac{a \cos \beta}{\cos \theta}\right]^n \cos n\theta. \tag{12-10.4}$$

On the remaining part of the boundary

$$\beta \leq \theta \leq \pi, \qquad r = a, \qquad \psi = 0, \tag{12-10.5}$$

$$\frac{a^2}{2} = \sum_n C_n a^n \cos n\theta. \tag{12-10.6}$$

Introducing the constant defined as $A_n = 2C_n a^{n-2}$, these conditions can be written

$$0 \leq \theta \leq \beta, \qquad \sum_n A_n \left[\frac{\cos \beta}{\cos \theta}\right]^{n-2} \cos n\theta = 1, \tag{12-10.7}$$

$$\beta \leq \theta \leq \pi, \qquad \sum_n A_n \cos n\theta = 1. \tag{12-10.8}$$

We now have the solution in a form that term by term satisfies the differential equation, and on the boundary has the form

$$\sum A_n \varphi_n(\theta) = 1, \tag{12-10.9}$$

where

$$\varphi_n(\theta) = \begin{cases} \left[\dfrac{\cos\beta}{\cos\theta}\right]^{(n-2)} \cos n\theta, & 0 \leq \theta \leq \beta \\ \cos n\theta, & \beta \leq \theta \leq \pi \end{cases} \tag{12-10.10}$$

We compute the stresses in polar coordinates by using the transformation equations

$$\begin{aligned} \sigma_{rz} &= \sigma_{xz}\cos\theta + \sigma_{yz}\sin\theta, \\ \sigma_{\theta z} &= -\sigma_{xz}\sin\theta + \sigma_{yz}\cos\theta, \\ \sigma_{xz} &= \frac{\partial\psi}{\partial y} = \frac{\partial\psi}{\partial r}\sin\theta + \frac{\partial\psi}{\partial\theta}\frac{\cos\theta}{r}, \\ \sigma_{yz} &= -\frac{\partial\psi}{\partial x} = -\frac{\partial\psi}{\partial r}\cos\theta + \frac{\partial\psi}{\partial\theta}\frac{\sin\theta}{r}, \end{aligned} \tag{12-10.11}$$

$$\begin{aligned} \sigma_{rz} &= \frac{1}{r}\frac{\partial\psi}{\partial\theta}, \\ \sigma_{\theta z} &= -\frac{\partial\psi}{\partial r}. \end{aligned} \tag{12-10.12}$$

11 Complex variable solution of torsion problems

We can obtain a different form of boundary condition for simply connected regions in terms of the harmonic function λ. Noting that $\psi = 0$ and

$$\psi = G\alpha[\lambda - \tfrac{1}{2}(x^2 + y^2)], \tag{12-11.1}$$

we can conclude that on the boundary

$$\lambda = \frac{x^2 + y^2}{2} = \frac{r^2}{2}. \tag{12-11.2}$$

λ and φ are related by the Cauchy–Riemann conditions, and we define an analytic function of the complex variable z as

$$F(z) = \varphi + i\lambda. \tag{12-11.3}$$

This function is holomorphic inside the region enclosed by the boundary of the cross section (see Chapter 9 for details).

Consider a cross section in the z plane (Figure 12-15), which may be obtained from a mapping from the interior of the unit circle in the ξ plane of the form

$$z = w(\xi).$$

In the z plane, λ on the boundary must satisfy the condition

$$\lambda = \tfrac{1}{2}(x^2 + y^2) = \tfrac{1}{2}z\bar{z}. \tag{12-11.4}$$

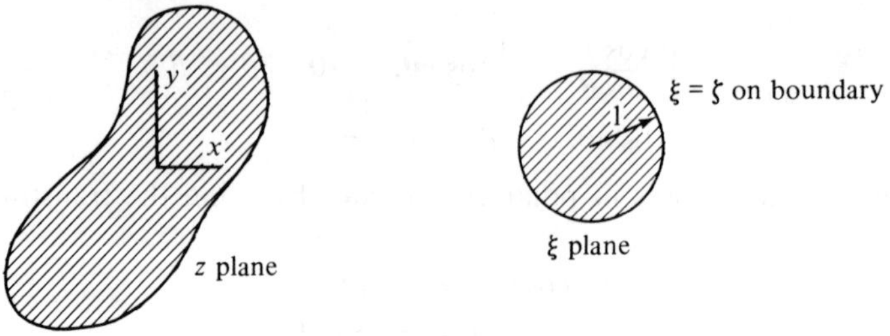

Figure 12-15

In the ξ plane we define the problem as

$$f(\xi) = F(z) = F(w[\xi]), \qquad (12\text{-}11.5)$$

with the boundary condition as

$$-\tfrac{1}{2}i(F - \bar{F}) = \tfrac{1}{2}z\bar{z}, \qquad (12\text{-}11.6)$$

or

$$-i(f - \bar{f}) = w(\xi)\overline{w(\xi)}, \qquad (12\text{-}11.7)$$

on the boundary $\xi = \zeta$(unit circle). If we operate on both sides of the boundary equation with $1/2\pi i \oint_C 1/(\zeta - \xi)\, d\zeta$, we obtain

$$-\frac{i}{2\pi i}\oint_c \frac{f}{\zeta - \xi}d\zeta + \frac{i}{2\pi i}\oint_c \frac{\bar{f}}{\zeta - \xi}d\zeta = \frac{1}{2\pi i}\oint_c \frac{w(\zeta)\overline{w(\zeta)}}{\zeta - \xi}d\zeta. \qquad (12\text{-}11.8)$$

Using Cauchy's integral formula yields

$$f(\xi) - \overline{f(0)} = \frac{i}{2\pi i}\oint \frac{w(\zeta)\overline{w(\zeta)}}{\zeta - \xi}d\zeta. \qquad (12\text{-}11.9)$$

At $\xi = 0$,

$$f(0) - \overline{f(0)} = 2i\mathrm{Im}[f(0)]. \qquad (12\text{-}11.10)$$

Note that the real part of the constant $f(0)$ is not determined, but this is to be expected, as φ is determined only up to an arbitrary constant. The torsional rigidity constant J may be calculated by use of the expression

$$J = \iint_A \left(x^2 + y^2 + x\frac{\partial\varphi}{\partial y} - y\frac{\partial\varphi}{\partial x}\right) dA. \qquad (12\text{-}11.11)$$

We may write this as

$$J = \iint_A \left[\frac{\partial}{\partial y}(x^2 y) + \frac{\partial}{\partial x}(y^2 x) + \frac{\partial}{\partial y}(x\varphi) - \frac{\partial}{\partial x}(y\varphi)\right] dA. \qquad (12\text{-}11.12)$$

By use of Green's theorem, this becomes

$$J = -\oint \{xy(x\,dx - y\,dy) + \varphi(x\,dx + y\,dy)\}. \qquad (12\text{-}11.13)$$

Noting that $x\,dx + y\,dy = d(r^2/2)$ and $r^2 = z\bar{z} = w(\zeta)\overline{w(\zeta)}$ on the contour C, and that $\varphi = \frac{1}{2}[f(\zeta) + \bar{f}(\zeta)]$, the last term in the expression for J may be written

$$-\oint \varphi(x\,dx + y\,dy) = -\tfrac{1}{4}\oint_C [f(\zeta) + \overline{f(\zeta)}]d[w(\zeta)\overline{w(\zeta)}]. \qquad (12\text{-}11.14)$$

If we note that

$$x = \frac{z+\bar{z}}{2}, \qquad y = \frac{z-\bar{z}}{2i}, \qquad (12\text{-}11.15)$$

the first term in J can be written

$$-\oint xy(x\,dx - y\,dy) = -\frac{1}{8i}\oint (z^2 - \bar{z}^2)(z\,dz + \bar{z}\,d\bar{z}). \qquad (12\text{-}11.16)$$

However,

$$\oint z^2\,dz = 0, \qquad \oint \bar{z}^3\,d\bar{z} = 0,$$

and

$$\oint z^3\bar{z}\,d\bar{z} = -\oint \bar{z}^2 z\,dz,$$

so that J becomes

$$J = -\frac{i}{4}\oint_c [\overline{w(\zeta)}]^2 w(\zeta)\,d[w(\zeta)] - \frac{1}{4}\oint_c [f(\zeta) + \overline{f(\zeta)}]\,d[w(\zeta)\overline{w)\zeta)}]. \qquad (12\text{-}11.17)$$

These integrals can be evaluated by use of calculus of residues when $w(\zeta)$ is a rational function. The stresses are calculated by forming the expression

$$\sigma_{zx} - i\sigma_{zy} = G\alpha\left\{\frac{\partial\varphi}{\partial x} - i\frac{\partial\varphi}{\partial y} - y - ix\right\},$$
$$= G\alpha\left\{\frac{\partial}{\partial x}(\varphi + i\lambda) - i(x - iy)\right\}, \qquad (12\text{-}11.18)$$
$$= G\alpha\left\{\frac{\partial F}{\partial z} - i\bar{z}\right\}. \qquad (12\text{-}11.19)$$

This is written

$$\sigma_{zx} - i\sigma_{zy} = G\alpha\left\{\frac{f'(\xi)}{w'(\xi)} - i\overline{w(\xi)}\right\}. \qquad (12\text{-}11.20)$$

Although mathematically this solution presents no difficulties, lack of simple mapping functions of the interior of the cross section onto the interior of the unit circle makes the method impractical. The Laplace equation may be solved numerically easier than seeking approximate conformal mappings. Sample solutions for certain cross sections by complex variables may be found in I.S. Sokolnikoff's *Mathematical Theory of Elasticity* (McGraw-Hill Book Company, New York, 1956).

12 Membrane analogy for torsion

Figure 12-16 shows a membrane subjected to uniform pressure stretching it under uniform tension F per unit length. Let w designate the deformation of the membrane in the z direction. The equilibrium conditions can be examined by use of Figure 12-17. The angle β is related to the deflection w

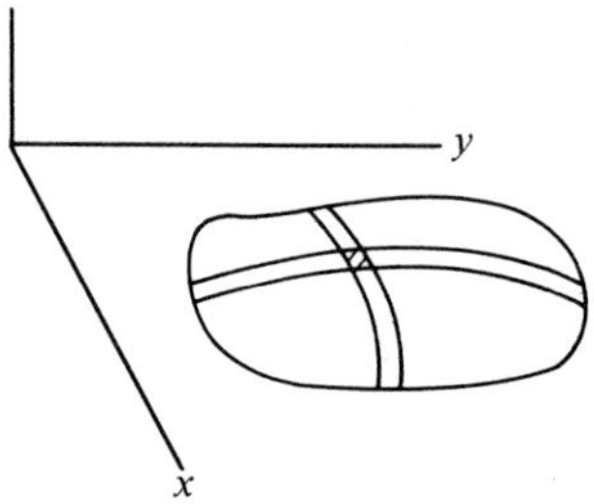

Figure 12-16

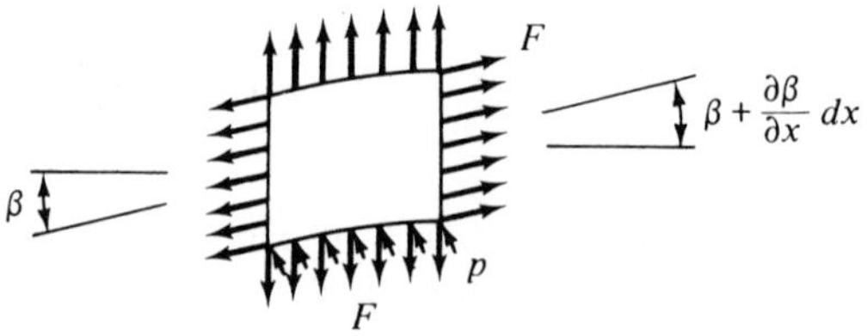

Figure 12-17

by $\beta = \partial w/\partial x$. Note that

$$\sin\left(\beta + \frac{\partial \beta}{\partial x} dx\right) = \beta + \frac{\partial \beta}{\partial x} dx$$

if only small slope and deflections are considered. The equation of equilibrium becomes

$$F\left\{\frac{\partial^2 w}{\partial x^2} + \frac{\partial^2 w}{\partial y^2}\right\} + p = 0, \qquad (12\text{-}12.1)$$

or

$$\nabla^2 w = -\frac{p}{F}. \qquad (12\text{-}12.2)$$

The boundary conditions are that w is a constant on the contour and, as the constant produces only rigid translation of the membrane, it can be set to zero.

Comparing the membrane equation to that of the Prandtl stress function, one may note a mathematical analogy.* The stress function ψ is related to the deflection of the membrane w and we can now compare cross sections by forming membranes across openings and subjecting them to pressure. The volume under the membrane gives a measure of the relative stiffness, and the slope gives a measure of the stress. One can also observe that at a reentrant corner, the slope becomes very high (approaching infinite), thereby corresponding to a very high stress concentration at that point,

The membrane analogy also provides an insight into methods of solution of open thin-wall sections such as T and I sections so commonly used in structures (see advanced strength books). If the volumetric differences between a membrane stretched across two long thin sections joined and not joined are small (see Figure 12-18), the torsional rigidity is thought of as the sum of the individual rigidities.

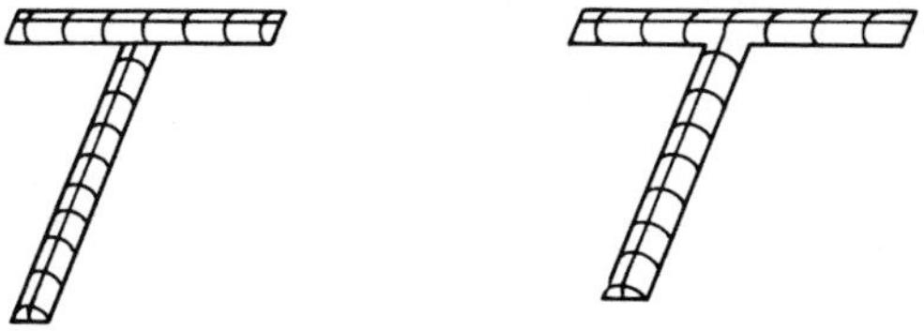

Figure 12-18

We have already noted for the problem of long thin sections that $J = \frac{1}{3}lt^3$, and J for a thin open section becomes

$$J = \frac{1}{3}\sum_i t_i^3 l_i, \qquad (12\text{-}12.3)$$

and the stresses may be found for any individual part of the cross section independently.

SPECIAL NOTE

If the ends of a torsional member are constrained, preventing warping, the torsional stiffness of the member will be greatly increased. Timoshenko† has approximated this increase for certain sections by considering the induced bending of parts of the section near the ends. Before torsional constants are used, this effect should be examined. This is particularly true for open thin-wall sections.

* This was introduced by L. Prandtl, *Physik. Z.*, **4**, 1903. Also see A. A. Griffith and G. I. Taylor, *Tech. Rept. Adv. Comm. Aeron.*, **3**, 910, 938, 1917–1918.

† S. P. Timoshenko and J. N. Goodier, *Theory of Elasticity*, 3rd ed., McGraw-Hill Book Company, New York, 1970.

13 Torsion of circular shafts of variable diameter

It is easier to use cylindrical coordinates in the discussion of circular shafts of variable diameter. The equilibrium equations become

$$\frac{1}{r}\frac{\partial}{\partial r}(r\sigma_{nr}) + \frac{1}{r}\frac{\partial \sigma_{r\theta}}{\partial \theta} + \frac{\partial}{\partial z}\sigma_{rz} - \frac{\sigma_{\theta\theta}}{r} + F_r = 0,$$

$$\frac{1}{r^2}\frac{\partial}{\partial r}(r^2\sigma_{\theta r}) + \frac{1}{r}\frac{\partial \sigma_{\theta\theta}}{\partial \theta} + \frac{\partial}{\partial z}\sigma_{\theta z} + F_\theta = 0, \qquad (12\text{-}13.1)$$

$$\frac{1}{r}\frac{\partial}{\partial r}(r\sigma_{zr}) + \frac{1}{r}\frac{\partial \sigma_{z\theta}}{\partial \theta} + \frac{\partial}{\partial z}\sigma_{zz} + F_z = 0.$$

We assume that the problem exhibits axial symmetry, and therefore the stresses, strains, and displacements are independent of θ.

Guided by the work on uniform cylinders, we assume the only nonzero displacement is u_θ; that is, $u_r = u_z = 0$ and u_θ, as has been mentioned, is independent of θ. The stresses in this case become

$$\sigma_{rr} = \sigma_{\theta\theta} = \sigma_{zz} = \sigma_{rz} = 0,$$

$$\sigma_{r\theta} = Gr\frac{\partial}{\partial r}\left(\frac{u_\theta}{r}\right), \qquad \sigma_{z\theta} = Gr\frac{\partial}{\partial z}\left(\frac{u_\theta}{r}\right). \qquad (12\text{-}13.2)$$

Substitution into the Navier equation yields

$$G\nabla^2\bar{u} + (\lambda + G)\vec{\nabla}\vec{\nabla}\cdot\bar{u} = 0. \qquad (12\text{-}13.3)$$

The equations reduce to the Laplacian operator on the displacement vector as

$$\bar{\nabla}\cdot\bar{u} = 0,$$

$$\nabla^2(u_\theta\hat{i}_\theta) = 0, \qquad (12\text{-}13.4)$$

$$\left(\frac{\partial^2}{\partial r^2} + \frac{1}{r}\frac{\partial}{\partial r} - \frac{1}{r^2}\right)u_\theta + \frac{\partial^2 u_\theta}{\partial z^2} = 0.$$

This can be written as

$$\frac{1}{r^2}\frac{\partial}{\partial r}\left[r^3\frac{\partial}{\partial r}\left(\frac{u_\theta}{r}\right)\right] + \frac{\partial}{\partial z}\left[r\frac{\partial}{\partial z}\left(\frac{u_\theta}{r}\right)\right] = 0. \qquad (12\text{-}13.5)$$

We note that written in this form this is just the second equilibrium equation (equation 12-13.2). Equation 12-13.5 becomes

$$\frac{\partial}{\partial r}\left[r^3\frac{\partial}{\partial r}\left(\frac{u_\theta}{r}\right)\right] + \frac{\partial}{\partial z}\left[r^3\frac{\partial}{\partial z}\left(\frac{u_\theta}{r}\right)\right] = 0. \qquad (12\text{-}13.6)$$

This suggests the introduction of a function ψ in the following manner:

$$\frac{\partial\psi}{\partial z} = -r^3\frac{\partial}{\partial r}\left(\frac{u_\theta}{r}\right) = -\frac{r^2\sigma_{r\theta}}{G},$$
$$\frac{\partial\psi}{\partial r} = r^3\frac{\partial}{\partial z}\left(\frac{u_\theta}{r}\right) = \frac{r^2\sigma_{z\theta}}{G}. \qquad (12\text{-}13.7)$$

ψ is usually called a stress function*. If the two expressions are to be compatible, ψ must satisfy the following equation:

$$\frac{\partial^2\psi}{\partial r^2} - \frac{3}{r}\frac{\partial\psi}{\partial r} + \frac{\partial^2\psi}{\partial z^2} = 0. \qquad (12\text{-}13.8)$$

Along the boundary of the body the stresses are zero and the following condition must hold:

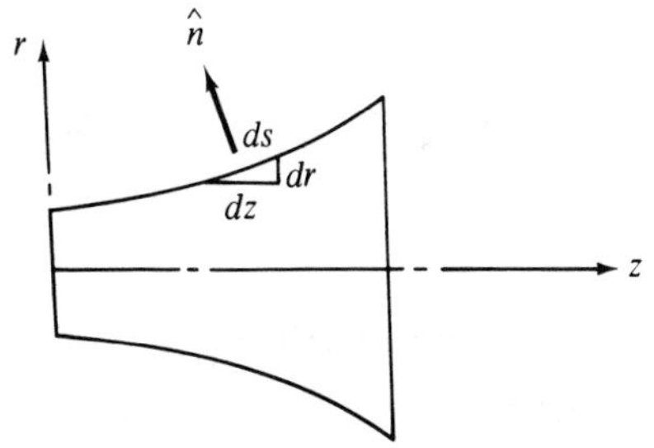

Figure 12-19

$$\sigma_{r\theta}\cos(n, r) + \sigma_{z\theta}\cos(n, z) = 0,$$
$$\cos(n, r) = \frac{dz}{ds}, \qquad (12\text{-}13.9)$$
$$\cos(n, z) = -\frac{dr}{ds}.$$

Therefore, the boundary condition becomes

$$\frac{G}{r^2}\left\{\frac{\partial\psi}{\partial r}\frac{dr}{ds} + \frac{\partial\psi}{\partial z}\frac{dz}{ds}\right\} = 0,$$

or

$$\left.\begin{aligned}\frac{\partial\psi}{\partial s} &= 0,\\ \psi &= \text{constant},\end{aligned}\right\} \text{on boundary.}^\dagger \qquad (12\text{-}13.10)$$

The particular case of a conical shaft with cone angle α (Figure 12-20) has been obtained by Föppl‡ by noting that the ratio $z/(r^2 + z^2)^{1/2}$ is a constant

* J. H. Michell, *Proc. London Math. Soc.* **31**, 141, 1899.

† Solutions in cylindrical coordinates are given by H. Reissner and B. J. Wennagel, *J. Appl. Mech.*, **17**, 275–282, 1950.

‡ A. Föppl, *Sitzber. Bayer. Akad. Wiss., Math. Naturw. K.*, **35**, 249, 504, 1905.

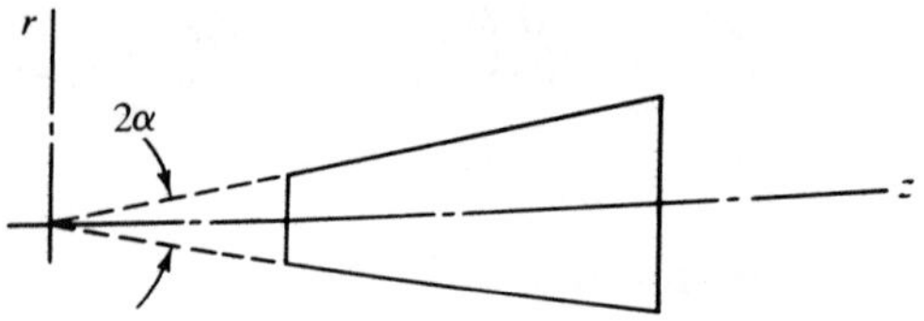

Figure 12-20

on the boundary. The solution for ψ becomes

$$\psi = C\left\{\frac{z}{(r^2+z^2)^{1/2}} - \frac{1}{3}\left[\frac{z}{(r^2+z^2)^{1/2}}\right]^3\right\}, \qquad (12\text{-}13.11)$$

where

$$C = -\frac{T}{2\pi(\frac{2}{3} - \cos\alpha + \frac{1}{3}\cos^3\alpha)}. \qquad (12\text{-}13.12)$$

14 Saint-Venant approximation for circular cylinders

F. Purser* investigated the Saint-Venant approximation for circular cross sections by assuming a solution of equation 12-13.8 in the form

$$\psi = R(r)e^{-\alpha z}.$$

Consider a semi-infinite circular cylinder with self-equilibrating tractions applied to the end. We assume that the solution decays to zero at infinity, thus motivating the form of the solution shown. Substitution of this form into equation 12-13.8 yields the following equation for $R(r)$:

$$\left(\frac{d^2}{dr^2} - \frac{3}{r}\frac{d}{dr} + \alpha^2\right)R = 0. \qquad (12\text{-}14.1)$$

This can be written as

$$\left(\frac{d^2}{dr^2} + \frac{1}{r}\frac{d}{dr} + \alpha^2 - \frac{4}{r^2}\right)\left(\frac{R}{r^2}\right) = 0. \qquad (12\text{-}14.2)$$

Equation 12-14.2 is now in the form comparable to Bessel's equation:

$$x^2\frac{d^2y}{dx^2} + x\frac{dy}{dx} + (x^2 - n^2)y = 0. \qquad (12\text{-}14.3)$$

The solution of 12-14.2 is

$$R = r^2 J_2(\alpha r). \qquad (12\text{-}14.4)$$

* F. Purser, "Some applications of Bessel's functions to physics," *Proc. Roy. Irish Acad.*, Dublin, **26**, 54, 1906.

The general form of a Bessel function is

$$J_n(x) = \sum_{k=0}^{\infty} \frac{(-1)^k}{k!\Gamma(n+k+1)}\left(\frac{x}{2}\right)^{n+2k},$$
$$= \frac{x^n}{2^n n!}\left\{1 - \frac{x^2}{2^2 \cdot 1!(n+1)} + \frac{x^4}{2^4 \cdot 2!(n+1)(n+2)} + \cdots\right\}. \qquad (12\text{-}14.5)$$

These functions satisfy the following recurrence formulas:

$$J_{n-1}(x) + J_{n+1}(x) = \frac{2n}{x} J_n(x),$$
$$J_{n-1}(x) - J_{n+1}(x) = 2J_n'(x),$$
$$nJ_n(x) + xJ_n'(x) = xJ_{n-1}(x), \qquad (12\text{-}14.6)$$
$$nJ_n(x) - xJ_n'(x) = xJ_{n+1}(x).$$

The behavior of these functions is seen in Figure 12-21.

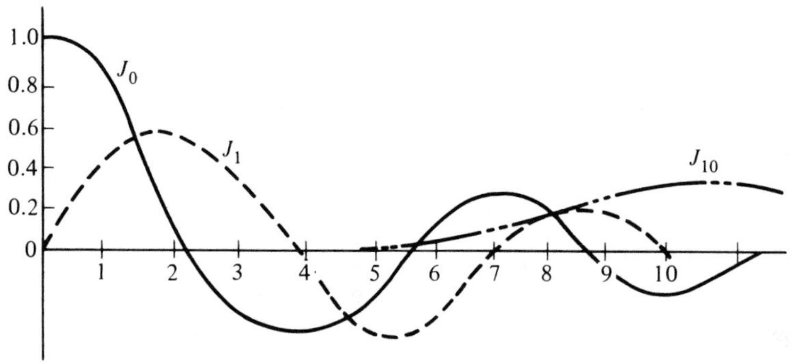

Figure 12-21

For large values of x, the Bessel functions are approximated by

$$J_n(x) = \sqrt{\frac{2}{\pi x}}\left[\cos\left(x - \frac{1}{2}n\pi - \frac{1}{4}\pi\right)\right]. \qquad (12\text{-}14.7)$$

Note that the roots of J_n for large values are about π distance apart.

These functions also exhibit an orthogonality condition, which will be used later but can be noted now:

$$\int_0^1 rJ_n(\alpha_k r)J_n(\alpha_j r)\,dr = 0, \qquad k \neq j,$$
$$= \tfrac{1}{2}[J_n'(\alpha_j)]^2, \qquad k = j, \qquad (12\text{-}14.8)$$

if α_k are the roots of $J_n(x) = 0$;

$$= \frac{1}{2\alpha_j^2}\left[\frac{a^2}{b^2} + \alpha_j^2 - n^2\right]\left[J_n(\alpha_j)\right]^2, \qquad k = j,$$

if α_j are the roots of $aJ_n(x) + bxJ_n'(x) = 0$.

The solution for ψ satisfying the condition ψ equals a constant at the outer radius $r = 1$ is

$$\psi = \sum_n A_n r^2 e^{-\alpha_n z} J_2(\alpha_n r), \tag{12-14.9}$$

where α_n are the roots of $J_2(\alpha_n) = 0$. Noting that

$$J_2(\alpha r) = -J_1'(\alpha r) + \frac{1}{\alpha r} J_1(\alpha r), \tag{12-14.10}$$

the transcendental equation for α_n becomes

$$J_1'(\alpha_n) = \frac{1}{\alpha_n} J_1(\alpha_n). \tag{12-14.11}$$

The torque across the end $z = 0$ is

$$\begin{aligned} T &= \iint_A \sigma_{z\theta} r \, dA = 2\pi G \int_0^1 \frac{1}{r^2} \frac{\partial \psi}{\partial r} r^2 \, dr \\ &= 2\pi G \sum A_n \int_0^1 [2rJ_2(\alpha_n r) + r^2 J_2'(\alpha_n r)] \, dr. \end{aligned} \tag{12-14.12}$$

Integrating by parts yields

$$T = 2\pi G \sum_n A_n \{r^2 J_2(\alpha_n r)\big|_0^1\} = 0. \tag{12-14.13}$$

This verifies that the solution does indeed correspond to a self-equilibrating stress distribution across the terminal end. We can see that the lowest root of J_2 gives a measure of the exponential decay parameter. The first three roots are given in Table 12-3, and one may note that, as Saint-Venant had speculated for this geometry, the effects of the end distribution of stresses decay quite rapidly.

Table 12-3. Roots of J_2

n	α_n
1	5.1356
2	8.4172
3	11.6194

One can observe that the minimum decay length reduces the local effect to less than 1 per cent of its maximum value in one radius away from the end. However, this analysis is for a particular geometry and should not be interpreted as a general verification of the Saint-Venant approximation for torsional members.

15 Flexure of beams by transverse end loads

Bending of beams is the subject of intense interest in elementary strength of materials courses. Using Euler–Bernoulli theory, one extends the exact solution of a long bar subjected to end moments to analyze bars bent by various transverse loads. We have investigated the stress distribution in rectangular bars by the methods of plane stress when the uniform thickness of the bar was assumed small enough that variations of stress across it could be neglected. These approximate methods give satisfactory solutions for long bars when shear effects and local effects in the regions of loading are neglected. For bars where shear effects are important, we have no approximate means for determining these effects, except for thin-walled sections where equilibrium considerations may be used to obtain approximations. In this section we shall review Saint-Venant's semi-inverse method for determining the distribution of these shear stresses for solid sections.* We shall develop this solution taking full advantage of the elementary solution.

Let us consider a cantilever beam of uniform cross section, fixed at one end, $z = 0$, and loaded by transverse forces at the other end, $z = l$. Although at first this may seem like a restricted problem, if one remembers that the shear distribution across the cross section is our desired result, we can see that this may indeed be the most direct method to obtain this result. We assume the z axis coincides with the central line of the beam passing through the centroid of the cross section (Figure 12-22).

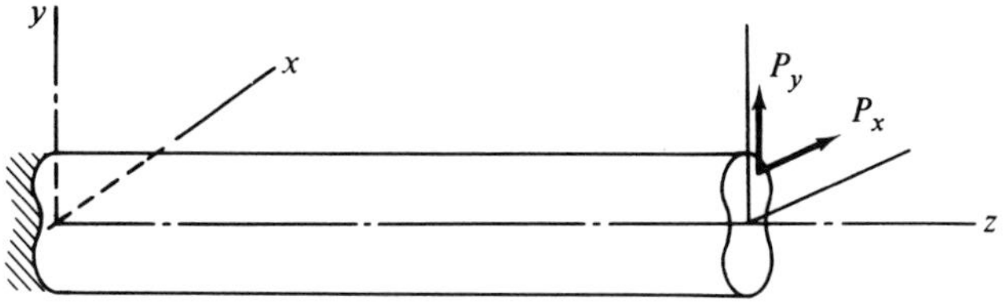

Figure 12-22

The boundary conditions are that the lateral surfaces are free of stresses, and at the end, $z = l$, the transverse loads P_x and P_y are applied at a point (x_0, y_0, l). We shall not meet this last condition exactly, but will use the Saint-Venant approximation by replacing these loads with statically equivalent loads.

* B. de Saint-Venant, *J. Mathemat.* (Liouville), Ser. 2, **1**, 1856.

We assume the following stresses are zero:

$$\sigma_{xx} = \sigma_{xy} = \sigma_{yy} = 0. \tag{12-15.1}$$

The other normal stress is assumed to vary as the bending moment and, guided by the elementary solution, they are assumed to vary linearly across the cross section:

$$\sigma_{zz} = (Bx + Cy)(l - z). \tag{12-15.2}$$

The constants are determined from the conditions on the end $z = l$:

$$\begin{gathered}\int_A \sigma_{zx}\, dA = P_x,\\ \int_A \sigma_{zy}\, dA = P_y,\\ \int_A (x\sigma_{zy} - y\sigma_{zx})\, dA = x_0 P_y - y_0 P_x.\end{gathered} \tag{12-15.3}$$

The equations of equilibrium become in the absence of body forces

$$\begin{gathered}\frac{\partial \sigma_{xz}}{\partial z} = 0,\\ \frac{\partial \sigma_{yz}}{\partial z} = 0,\\ \frac{\partial \sigma_{xz}}{\partial x} + \frac{\partial \sigma_{yz}}{\partial y} - Bx - Cy = 0.\end{gathered} \tag{12-15.4}$$

These can be rewritten by noting that σ_{xz} and σ_{yz} are independent of z:

$$\frac{\partial}{\partial x}\left[\sigma_{xz} - \frac{1}{2}Bx^2\right] + \frac{\partial}{\partial y}\left[\sigma_{yz} - \frac{1}{2}Cy^2\right] = 0. \tag{12-15.5}$$

This implies a function F such that

$$\frac{\partial F}{\partial y} = \sigma_{xz} - \frac{1}{2}Bx^2, \qquad -\frac{\partial F}{\partial x} = \sigma_{yz} - \frac{1}{2}Cy^2. \tag{12-15.6}$$

F is termed a stress function, as the stresses are written

$$\begin{gathered}\sigma_{xz} = \frac{\partial F}{\partial y} + \frac{1}{2}Bx^2,\\ \sigma_{yz} = -\frac{\partial F}{\partial x} + \frac{1}{2}Cy^2.\end{gathered} \tag{12-15.7}$$

Substitution of equations 12-15.7 and 12-15.2 into the Beltrami–Michell compatibility equations yields

$$\begin{gathered}\frac{\partial}{\partial y}(\nabla^2 F) + \frac{\nu B}{(1 + \nu)} = 0,\\ -\frac{\partial}{\partial x}(\nabla^2 F) + \frac{\nu C}{(1 + \nu)} = 0.\end{gathered} \tag{12-15.8}$$

Integration gives

$$\nabla^2 F = \frac{\nu}{1+\nu}(Cx - By) - 2G\alpha. \tag{12-15.9}$$

The constant of integration has been chosen to be $-2G\alpha$, which suggests an analogy to the torsion problem that will now be justified. The rotation of an element about the z axis is

$$\omega_z = \frac{1}{2}\left(\frac{\partial u_y}{\partial x} - \frac{\partial u_x}{\partial y}\right).$$

The rotation per unit length is

$$\begin{aligned}\frac{\partial \omega_z}{\partial z} &= \frac{1}{2}\left(\frac{\partial^2 u_y}{\partial x\,\partial z} - \frac{\partial^2 u_x}{\partial y\,\partial z}\right) = \frac{1}{2G}\left(\frac{\partial \sigma_{yz}}{\partial x} - \frac{\partial \sigma_{xz}}{\partial y}\right)\\ &= -\frac{1}{2G}\nabla^2 F.\end{aligned} \tag{12-15.10}$$

Substituting from equation 12-15.9 yields

$$\frac{\partial \omega_z}{\partial z} = \alpha - \frac{1}{2G}\frac{\nu}{(1+\nu)}(Cx - By). \tag{12-15.11}$$

The constant α therefore represents the rate of twist of an element at the origin and also the average twist of the cross section, and is correctly related to the torsion problem.*

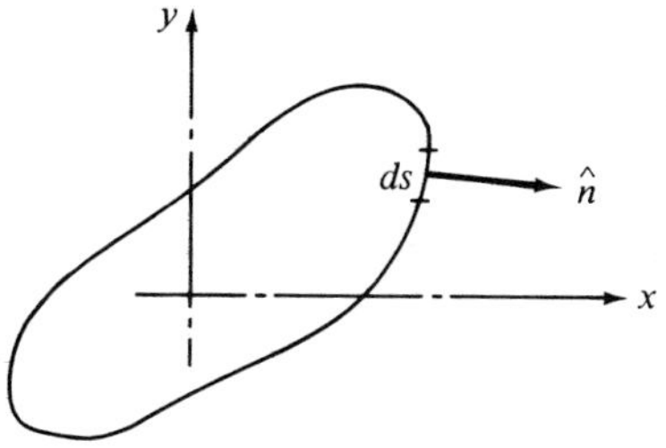

Figure 12-23

The boundary condition on the lateral surface yields

$$\begin{aligned}\sigma_{xz}\cos(n, x) + \sigma_{yz}\cos(n, y) &= 0,\\ \left(\frac{\partial F}{\partial y} + \frac{1}{2}Bx^2\right)\cos(n, x) - \left(\frac{\partial F}{\partial x} - \frac{1}{2}Cy^2\right)\cos(n, y) &= 0.\end{aligned} \tag{12-15.12}$$

Using the relations $\cos(n, x) = dy/ds$ and $\cos(n, y) = -dx/ds$ yields

$$\left(\frac{\partial F}{\partial y}\frac{dy}{ds} + \frac{\partial F}{\partial x}\frac{dx}{ds}\right) + \frac{1}{2}[Bx^2\cos(n, x) + Cy^2\cos(n, y)] = 0. \tag{12-15.13}$$

* See J. N. Goodier, *J. Aeron. Sci.*, **11**, 173, 1944.

This can be written

$$\frac{\partial F}{\partial s} = -\frac{1}{2}\{Bx^2 \cos(n, x) + Cy^2 \cos(n, y)\}. \qquad (12\text{-}15.14)$$

We shall separate F into a torsional part and a flexural part by the following substitution:

$$F = \psi + \beta, \qquad (12\text{-}15.15)$$

where

$$\begin{aligned} \nabla^2\psi &= -2G\alpha, \\ \frac{\partial \psi}{\partial s} &= 0 \qquad \text{on the contour } C, \end{aligned} \qquad (12\text{-}15.16)$$

$$\begin{aligned} \nabla^2\beta &= \frac{\nu}{1+\nu}[Cx - By], \\ \frac{\partial \beta}{\partial s} &= -\frac{1}{2}\{Bx^2 \cos(n, x) + Cy^2 \cos(n, y)\} \\ &\qquad\qquad \text{on the contour } C. \end{aligned} \qquad (12\text{-}15.17)$$

We have previously discussed the solution of the problem related by 12-15.16 and shall now consider 12-15.17. The solution for β is written as

$$\beta = \Gamma + \frac{1}{6}\frac{\nu}{1+\nu}(Cx^3 - By^3), \qquad (12\text{-}15.18)$$

where Γ is a harmonic function. The boundary condition of the end becomes

$$\begin{aligned} \int_A \sigma_{xz}\, dA &= P_x, \\ \iint_A \left\{\frac{\partial}{\partial y}[\beta + \psi] + \frac{1}{2}Bx^2\right\} dA &= P_x. \end{aligned} \qquad (12\text{-}15.19)$$

We can note from the torsion problem that

$$\iint_A \frac{\partial \psi}{\partial y}\, dA = 0, \qquad (12\text{-}15.20)$$

and write equation 12-15.19 in the form

$$\iint_A \left\{\frac{\partial}{\partial x}\left[x\frac{\partial \beta}{\partial y}\right] - \frac{\partial}{\partial y}\left[x\frac{\partial \beta}{\partial x}\right]\right\} dA + \iint_A \frac{1}{2}Bx^2\, dA = P_x. \qquad (12\text{-}15.21)$$

The first integral can be written by use of Green's theorem and the boundary condition 12-15.17 as

$$\iint_A \left\{\frac{\partial}{\partial x}\left[x\frac{\partial \beta}{\partial y}\right] - \frac{\partial}{\partial y}\left[x\frac{\partial \beta}{\partial x}\right]\right\} dA = -\iint_A \left[\frac{3}{2}Bx^2 + Cxy\right] dA. \qquad (12\text{-}15.22)$$

Equation 12-15.21 now becomes

$$BI_{yy} + CI_{xy} = -P_x, \qquad (12\text{-}15.23)$$

where I_{yy} and I_{xy} are the moments of inertia defined as

$$I_{yy} = \iint_A x^2\,dA, \qquad I_{xy} = \iint_A xy\,dA.$$

In a similar manner, the boundary condition on σ_{yz} yields

$$\iint_A \sigma_{yz}\,dA = P_y,$$

$$BI_{xy} + CI_{xx} = -P_y, \tag{12-15.24}$$

where

$$I_{xx} = \iint_A y^2\,dA.$$

The constants B and C are evaluated from equations 12-15.23 and 12-15.24 as follows:

$$B = -\left\{\frac{P_x I_{xx} - P_y I_{xy}}{I_{xx}I_{yy} - I_{xy}^2}\right\},$$
$$C = -\left\{\frac{P_y I_{yy} - P_x I_{xy}}{I_{xx}I_{yy} - I_{xy}^2}\right\}. \tag{12-15.25}$$

The final boundary condition becomes

$$\iint_A (x\sigma_{zy} - y\sigma_{zx})\,dA = x_0 P_y - y_0 P_x, \tag{12-15.26}$$

$$-\iint_A \left[x\frac{\partial\psi}{\partial x} + y\frac{\partial\psi}{\partial y}\right] dA + \iint_A \left\{-\left[x\frac{\partial\beta}{\partial x} + y\frac{\partial\beta}{\partial y}\right] + \frac{1}{2}Cxy^2 - \frac{1}{2}Bx^2y\right\} dA = x_0 P_y - y_0 P_x. \tag{12-15.27}$$

Equation 12-6.29 gives

$$2\iint_A \psi\,dA + \iint_A \left\{-\left[x\frac{\partial\beta}{\partial x} + y\frac{\partial\beta}{\partial y}\right] + \frac{1}{2}Cxy^2 - \frac{1}{2}Bx^2y\right\} dA = x_0 P_y - y_0 P_x. \tag{12-15.28}$$

Since $2\iint \psi\,dA$ is equal to $JG\alpha$, we see that this equation determines the final constant α. Note that only if

$$x_0 P_y - y_0 P_x = \iint_A \left\{-\left[x\frac{\partial\beta}{\partial x} + y\frac{\partial\beta}{\partial y}\right] + \frac{1}{2}Cxy^2 - \frac{1}{2}Bx^2y\right\} dA \tag{12-15.29}$$

does α equal zero or is there no induced torsion. Examination of the boundary condition 12-15.17 and equation 12-15.18 shows that if the x axis is an axis of symmetry, β will be an odd function of y; and, in a similar manner, if y is an axis of symmetry, β will be an odd function of x.

We obtain the value of x_0 and y_0 for no induced torsion by considering independently the cases when either P_x or P_y is zero. If $P_x = 0$, β can be written from equation 12-15.18 as

$$\beta = P_y\left[\Gamma_1 - \frac{1}{6}\frac{\nu}{1+\nu}\left(\frac{I_{yy}}{\Delta}x^3 + \frac{I_{xy}}{\Delta}y^3\right)\right]. \tag{12-15.30}$$

$$x_0 = \iint_A \left\{-\left[x\frac{\partial\Gamma_1}{\partial x} + y\frac{\partial\Gamma_1}{\partial y}\right] + \frac{\nu}{2(1+\nu)}\left[\frac{I_{yy}}{\Delta}x^3 + \frac{I_{xy}}{\Delta}y^3\right] - \frac{1}{2}\left[\frac{I_{yy}}{\Delta}xy^2 + \frac{I_{xy}}{\Delta}x^2y\right]\right\} dA, \tag{12-15.31}$$

where $\Delta = I_{xx}I_{yy} - I_{xy}^2$ and P_y has been factored from Γ to form Γ_1. In a similar manner,

$$y_0 = \iint_A \left\{-\left[x\frac{\partial\Gamma_2}{\partial x} + y\frac{\partial\Gamma_2}{\partial y}\right] - \frac{\nu}{2(1+\nu)}\left[\frac{I_{xy}x^3}{\Delta} + \frac{I_{xx}y^3}{\Delta}\right] + \frac{1}{2}\left[\frac{I_{xy}xy^2}{\Delta} + \frac{I_{xx}x^2y}{\Delta}\right]\right\} dA, \tag{12-15.32}$$

where in this case P_x has been factored from Γ to form Γ_2. If the x axis is an axis of symmetry, $I_{xy} = 0$ and $y_0 = 0$, or if the y axis is an axis of symmetry, $x_0 = 0$. In general, the *center of flexure*, or that point through which the load must be placed so that no torsion will be introduced, will not coincide with the centroid nor, indeed, need it lie in the cross section at all.

16 Flexure of a circular beam under a load P_y through the centroid

Consider the cylinder shown in Figure 12-24. The example chosen simplifies the algebra as P_y passes through the origin and P_x has been taken to be zero. The torsional part of the problem is zero and will therefore not be considered. Equation 12-15.17 becomes

$$\nabla^2\beta = -\frac{\nu}{1+\nu}\frac{P_y}{I_{xx}}r\cos\theta. \tag{12-16.1}$$

$$\frac{1}{a}\frac{\partial\beta}{\partial\theta} = \frac{1}{2}\frac{P_y}{I_{xx}}a^2\sin^3\theta \qquad \text{on the contour} \quad r = a. \tag{12-16.2}$$

β is taken as

$$\beta = \frac{P_y}{I_{xx}}\left[\Gamma - \frac{1}{6}\frac{\nu}{1+\nu}r^3\cos^3\theta\right]. \tag{12-16.3}$$

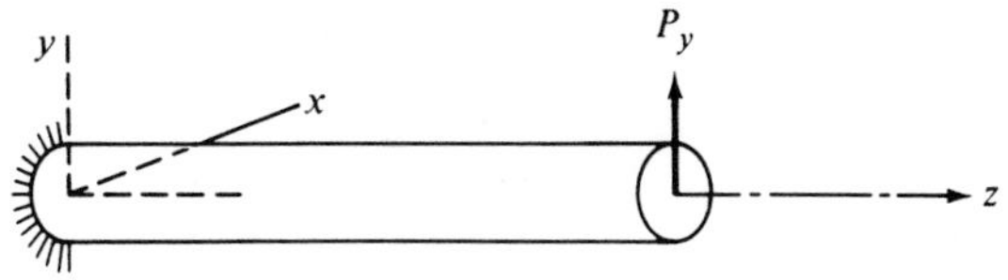

Figure 12-24

We write this in a more workable form by using the relations

$$\cos^3\theta = \tfrac{1}{4}\cos 3\theta + \tfrac{3}{4}\cos\theta, \qquad (12\text{-}16.4)$$
$$\sin^3\theta = -\tfrac{1}{4}\sin 3\theta + \tfrac{3}{4}\sin\theta.$$

β can now be written

$$\beta = \frac{P_y}{I_{xx}}\left[\Gamma - \frac{1}{24}\frac{\nu}{1+\nu}r^3(\cos 3\theta + 3\cos\theta)\right]. \qquad (12\text{-}16.5)$$

The boundary condition becomes

$$\frac{\partial\beta}{\partial\theta} = \frac{1}{8}\frac{P_y}{I_{xx}}a^3(-\sin 3\theta + 3\sin\theta). \qquad (12\text{-}16.6)$$

We take the solution for Γ in the form

$$\Gamma = \sum_n A_n r^n \cos n\theta. \qquad (12\text{-}16.7)$$

In particular, we consider the two terms

$$\Gamma = A_1 \cos\theta r + A_3 r^3 \cos 3\theta. \qquad (12\text{-}16.8)$$

β becomes

$$\beta = \frac{P_y}{I_{xx}}\left[\left\{A_1 r - \frac{\nu r^3}{8(1+\nu)}\right\}\cos\theta + \left\{A_3 - \frac{\nu}{24(1+\nu)}\right\}r^3\cos 3\theta\right]. \qquad (12\text{-}16.9)$$

The boundary condition yields

$$A_1 = -\frac{a^2}{8}\frac{3+2\nu}{1+\nu}, \qquad A_3 = \frac{1}{24}\left[1 + \frac{\nu}{1+\nu}\right]. \qquad (12\text{-}16.10)$$

β becomes

$$\beta = \frac{P_y}{I_{xx}}\left\{\left[\frac{(a^2 - r^2)\nu}{(1+\nu)} - 3a^2\right]\frac{r\cos\theta}{8} + \frac{r^3\cos 3\theta}{24}\right\}. \qquad (12\text{-}16.11)$$

Using the relation

$$\cos 3\theta = 4\cos^3\theta - 3\cos\theta,$$

we write β in the form

$$\beta = \frac{P_y}{I_{xx}}\left\{-\frac{3+2\nu}{1+\nu}\frac{a^2x}{8} - \frac{1+2\nu}{1+\nu}\frac{xy^2}{8} + \frac{1-2\nu}{1+\nu}\frac{x^3}{24}\right\}. \qquad (12\text{-}16.12)$$

The stresses become·

$$\sigma_{xz} = \frac{\partial \beta}{\partial y} = -\frac{P_y}{4I_{xx}}\frac{1+2\nu}{1+\nu}xy,$$

$$\sigma_{yz} = -\frac{\partial \beta}{\partial x} - \frac{1}{2}\frac{P_y}{I_{xx}}y^2 = \frac{P_y}{I_{xx}}\frac{3+2\nu}{8(1+\nu)}\left[a^2 - y^2 - \frac{1-2\nu}{3+2\nu}x^2\right], \quad (12\text{-}16.13)$$

$$\sigma_{zz} = -\frac{P_y}{I_{xx}}y(l-z).$$

Now that the stresses have been obtained, the displacements can be obtained by integration.

PROBLEMS

12.1 Show that the warping function $\varphi = A(y^3 - 3x^2y)$ is the correct solution for the equilateral triangular bar shown under pure torsion. Find J and the maximum shearing stress. The equations of the boundaries are

$$x - a = 0,$$
$$x + 2a - \sqrt{3}\,y = 0,$$
$$x + 2a + \sqrt{3}\,y = 0.$$

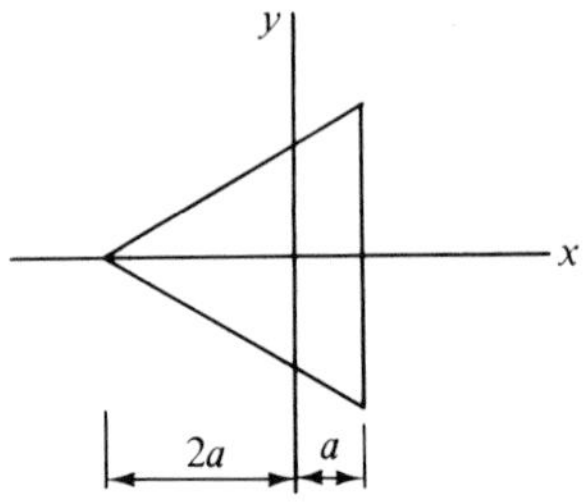

Problem 12-1

12.2. Consider the section that is symmetric with respect to the x and y axes and has a torsional stress function in the form

$$\varphi = m\left[x^2 - c^2\left(1 - \frac{y^2}{b^2}\right)\right],$$

which is zero along the boundaries. Pick m so that the differential equation is satisfied and find the torsional stiffness J and the maximum shearing stress. Show that the solution corresponds to that for a circular cross section when $b = c$.

12.3. Solve the torsion problem of the modified sector.

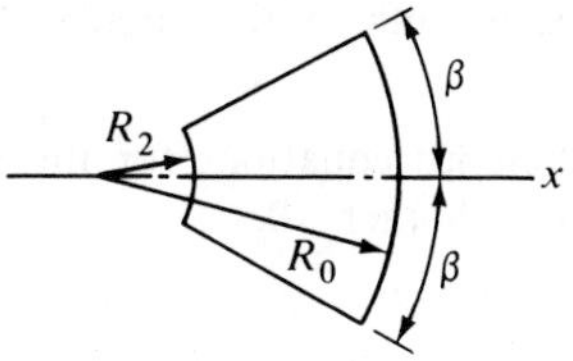

Problem 12-3

12.4. Set up the form of the stress function for the following cross section.

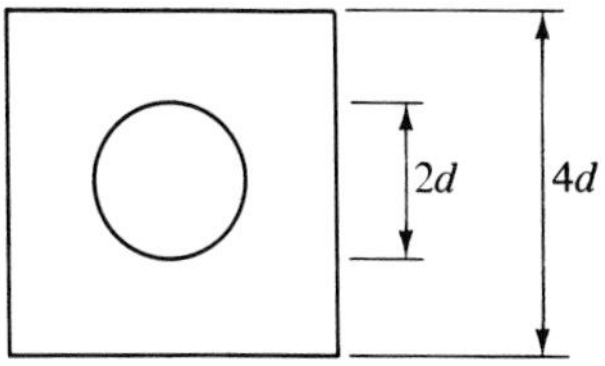

Problem 12-4

12.5. Set up the form of the stress function for the following cross section.

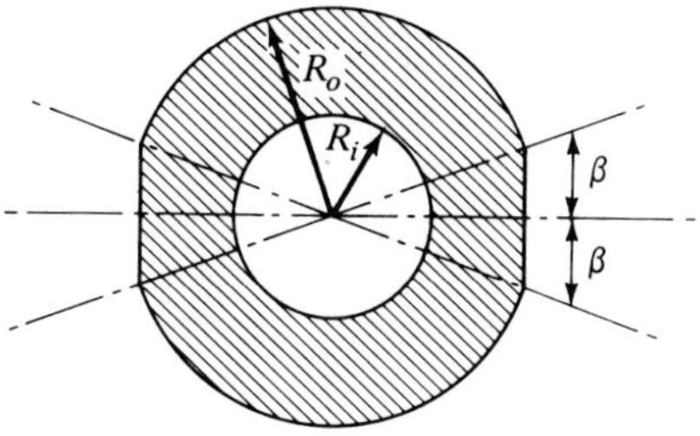

Problem 12-5

12.6. Verify equation 12-15.24.

12.7. Determine the shear distribution in a rectangular beam loaded as shown.

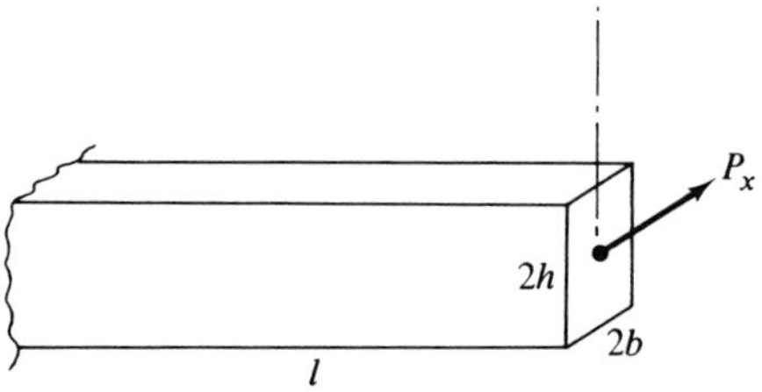

Problem 12-7

12.8. Determine the displacements in the circular beam discussed in section **12-16.**

12.9. Set up the finite difference equations for the torsion problem using the methods indicated in Chapter 10.

12.10. Set up a variational approach to the torsional problem similar to that used in Chapter 11, based on minimum potential energy methods.

CHAPTER 13

NAVIER EQUATION AND THE GALERKIN VECTOR

1 Solution of the Navier equation in three-dimensional elastostatics

Almost all solutions of three-dimensional problems in elastostatics involve the use of the Navier equation. These methods involve the introduction of certain stress or displacement functions that decouple or otherwise reduce the complexity of the defining differential equations. Review articles discussing the history of the development of these functions and their relative merits may be found in the survey papers by Marguerre,* Truesdell,† and Sternberg.‡

* K. Marguerre, "Ansötze zur Lösung der Grundgleichungen der Elostytätstheorie," *Z. Angew. Math. Mech.*, **35**, 1955.

† C. Truesdell, "Invariant and complete stress functions for general continua," *Archive Rat'l Mech. Anal.*, **4**, 1959–1960.

‡ E. Sternberg, "On some recent developments in the linear theory of elasticity," Goodier and Hoff (eds.), *Structural Mechanics*, Pergamon Press, Inc., Elmsford, New York, 1960.

Before considering these approaches, let us examine a few special cases where the Navier equations decouple and may be easily solved.* These examples come from physical situations where one may assume all but one of displacement components are zero and the remaining one is independent of some of the coordinate variables. We have already examined one such case. Axisymmetry was assumed in problems of plane strain (section 8-2). Another problem of this type arises when one considers a hollow spherical body under internal and external pressure (Figure 13-1). It is reasonable to

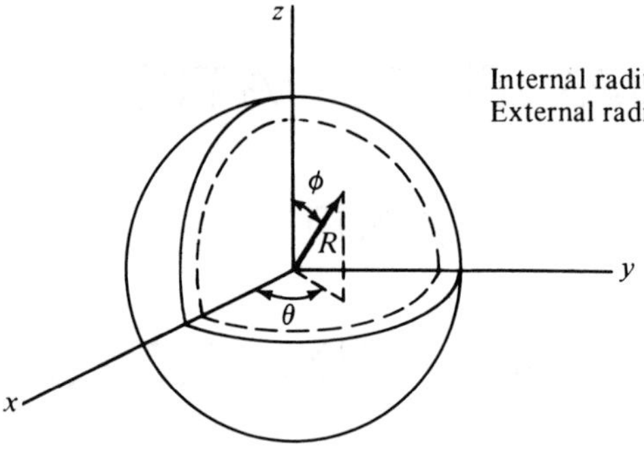

Figure 13-1

assume that elements of the body move only radially; therefore, the displacement components u_θ and u_φ are zero, and u_R is a function of R only. The displacement vector for this case becomes

$$\bar{u} = u_R \hat{i}_R. \tag{13-1.1}$$

The Navier equation can be written in the forms

$$G\nabla^2\bar{u} + (\lambda + G)\vec{\nabla}(\vec{\nabla}\cdot\bar{u}) + \rho\bar{f} = 0, \tag{13-1.2}$$

or

$$G\left[\nabla^2\bar{u} + \frac{1}{1-2\nu}\vec{\nabla}(\vec{\nabla}\cdot\bar{u})\right] + \bar{F} = 0. \tag{13-1.3}$$

Using the vector identity

$$\nabla^2\bar{u} = \vec{\nabla}(\vec{\nabla}\cdot\bar{u}) - \vec{\nabla}\times(\vec{\nabla}\times\bar{u}), \tag{13-1.4}$$

equation 13-1.1 becomes

$$(\lambda + 2G)\vec{\nabla}(\vec{\nabla}\cdot\bar{u}) - G\vec{\nabla}\times(\vec{\nabla}\times\bar{u}) + \rho\bar{f} = 0. \tag{13-1.5}$$

The curl of the displacement vector 13-1.1 is zero and the body forces are neglected, so that equation 13-1.5 becomes

$$\vec{\nabla}(\vec{\nabla}\cdot\bar{u}) = 0, \tag{13-1.6}$$

* Lamé examined simple potential functions as possible solutions of Navier's equation. See "Leçons sur la théorie mathématique de l'élasticité des corps solides," 1852.

or
$$\frac{1}{R^2}\frac{d}{dR}[R^2u_R] = \text{constant}. \qquad (13\text{-}1.7)$$

This leads to the solution
$$u_R = AR + \frac{B}{R^2}. \qquad (13\text{-}1.8)$$

The shear stresses $\sigma_{R\theta}$, $\sigma_{R\varphi}$, and $\sigma_{\theta\varphi}$ are all zero, and the boundary conditions at the inner and outer radii are
$$\begin{aligned} \sigma_{RR} &= -p_1 \quad \text{at} \quad R = a, \\ \sigma_{RR} &= -p_2 \quad \text{at} \quad R = b. \end{aligned} \qquad (13\text{-}1.9)$$

The constants A and B become
$$A = \frac{p_1a^3 - p_2b^3}{b^3 - a^3}\frac{(1-2\nu)}{E}, \qquad B = \frac{a^3b^3(p_1 - p_2)}{b^3 - a^3}\frac{1+\nu}{2E}. \qquad (13\text{-}1.10)$$

These constants lead to the following state of stress when $p_2 = 0$:
$$\sigma_{RR} = \frac{p_1a^3}{b^3 - a^3}\left(1 - \frac{b^3}{R^3}\right), \qquad \sigma_{\theta\theta} = \sigma_{\varphi\varphi} = \frac{p_1a^3}{b^3 - a^3}\left(1 + \frac{b^3}{R^3}\right). \qquad (13\text{-}1.11)$$

Although this solution is easy to obtain and is of importance to a limited audience, it does not lead to a method for solving many three-dimensional problems. Indeed, if we were to limit ourselves to cases of this type, we should be able to solve only a half-dozen or so problems.

2 Stress functions

A logical approach to seek solutions to three-dimensional problems would be to try a similar procedure as the one used in two-dimensional cases.* This is to introduce some function or functions that satisfy the equilibrium equations and to use the compatibility equations to define these functions. We do not wish to limit ourselves to Cartesian coordinate systems; therefore, it will be necessary to examine the equations of elasticity in symbolic notation. The equation of equilibrium is
$$\vec{\nabla}\cdot\bar{\bar{\sigma}} + \bar{F} = 0. \qquad (13\text{-}2.1)$$

We can develop a general symbolic form of the Beltrami–Michell compatibility equations by examining the Cartesian tensor identity (see section 1-5):
$$\begin{aligned} e_{npj}e_{lmi} = {} & \delta_{nl}\delta_{pm}\delta_{ij} + \delta_{nm}\delta_{pi}\delta_{jl} + \delta_{ni}\delta_{pl}\delta_{jm} \\ & - \delta_{nl}\delta_{pi}\delta_{jm} - \delta_{ij}\delta_{nm}\delta_{pl} - \delta_{ni}\delta_{pm}\delta_{lj}. \end{aligned} \qquad (13\text{-}2.2)$$

* See A. I. Luré, *Three-Dimensional Problems of the Theory of Elasticity*, John Wiley & Sons, Inc. (Interscience Division), 1964, for additional details.

Since

$$\vec{\nabla} \times \bar{\bar{A}} \times \overleftarrow{\nabla} \Longrightarrow e_{npj} e_{lmi} \frac{\partial^2 A_{jm}}{\partial x_p \, \partial x_i}, \tag{13-2.3}$$

we write in symbolic notation

$$\begin{aligned} \vec{\nabla} \times \bar{\bar{A}} \times \overleftarrow{\nabla} = -\{\mathbf{I}\nabla^2 a + (\vec{\nabla}\cdot\bar{\bar{A}})\overleftarrow{\nabla} + \vec{\nabla}(\bar{\bar{A}}\cdot\overleftarrow{\nabla}) \\ - \mathbf{I}(\vec{\nabla}\cdot\bar{\bar{A}}\cdot\overleftarrow{\nabla}) - \vec{\nabla}\overleftarrow{\nabla} a - \vec{\nabla}\cdot\vec{\nabla}\bar{\bar{A}}\}, \end{aligned} \tag{13-2.4}$$

where a is the first invariant of the tensor. If the tensor is of the form that $\bar{\bar{A}} = \bar{\bar{\mathbf{I}}}\varphi$, we can write this identity as

$$\vec{\nabla} \times \bar{\bar{\mathbf{I}}}\varphi \times \overleftarrow{\nabla} = -\bar{\bar{\mathbf{I}}}\nabla^2\varphi + \vec{\nabla}\overleftarrow{\nabla}\varphi. \tag{13-2.5}$$

The equation of compatibility is

$$\vec{\nabla} \times \bar{\bar{\epsilon}} \times \overleftarrow{\nabla} = 0. \tag{13-2.6}$$

The strain tensor is symmetric so that $\vec{\nabla}\cdot\bar{\bar{\epsilon}} = \bar{\bar{\epsilon}}\cdot\overleftarrow{\nabla}$ and we can use the stress–strain relation in the form

$$\bar{\bar{\epsilon}} = \frac{1+\nu}{E}\left[\bar{\bar{\sigma}} - \frac{\nu}{1+\nu} S\bar{\bar{\mathbf{I}}}\right], \tag{13-2.7}$$

where S is the first invariant of the stress 13-2.7 tensor. The compatibility equation becomes

$$\begin{aligned} \vec{\nabla}\cdot\vec{\nabla}\bar{\bar{\sigma}} + \vec{\nabla}\overleftarrow{\nabla} S - \bar{\bar{\mathbf{I}}}\nabla^2 S + \bar{\bar{\mathbf{I}}}\frac{\nu}{(1+\nu)}\nabla^2 S - \frac{\nu}{(1+\nu)}\vec{\nabla}\overleftarrow{\nabla} S \\ - (\vec{\nabla}\cdot\bar{\bar{\sigma}})\overleftarrow{\nabla} - \vec{\nabla}(\vec{\nabla}\cdot\bar{\bar{\sigma}}) + \bar{\bar{\mathbf{I}}}\vec{\nabla}\cdot(\vec{\nabla}\cdot\bar{\bar{\sigma}}) = 0. \end{aligned} \tag{13-2.8}$$

Equation 13-2.8 can be written

$$\vec{\nabla}\cdot\vec{\nabla}\bar{\bar{\sigma}} + \frac{1}{1+\nu}[\vec{\nabla}\overleftarrow{\nabla} S - \bar{\bar{\mathbf{I}}}\nabla^2 S] - (\vec{\nabla}\cdot\bar{\bar{\sigma}})\overleftarrow{\nabla} - \vec{\nabla}(\vec{\nabla}\cdot\bar{\bar{\sigma}}) + \bar{\bar{\mathbf{I}}}\vec{\nabla}\cdot(\vec{\nabla}\cdot\bar{\bar{\sigma}}) = 0. \tag{13-2.9}$$

From the equilibrium equation

$$\vec{\nabla}\cdot\bar{\bar{\sigma}} = -\bar{F},$$

we obtain

$$\vec{\nabla}\cdot\vec{\nabla}\bar{\bar{\sigma}} + \frac{1}{(1+\nu)}(\vec{\nabla}\overleftarrow{\nabla} S - \bar{\bar{\mathbf{I}}}\nabla^2 S) + \bar{F}\overleftarrow{\nabla} + \vec{\nabla}\bar{F} - \bar{\bar{\mathbf{I}}}\vec{\nabla}\cdot\bar{F} = 0. \tag{13-2.10}$$

Taking the trace of this tensor equation yields

$$\begin{aligned} \nabla^2 S - \frac{2}{1+\nu}\nabla^2 S - \vec{\nabla}\cdot\bar{F} = 0, \\ \nabla^2 S + \frac{1+\nu}{1-\nu}\vec{\nabla}\cdot\bar{F} = 0. \end{aligned} \tag{13-2.11}$$

Using 13-2.11, we now write the Beltrami–Michell compatibility equation as

$$\nabla^2\bar{\bar{\sigma}} + \frac{1}{(1+\nu)}\vec{\nabla}\overleftarrow{\nabla} S + \bar{F}\overleftarrow{\nabla} + \vec{\nabla}\bar{F} + \frac{\nu}{(1-\nu)}\bar{\bar{\mathbf{I}}}\vec{\nabla}\cdot\bar{F} = 0. \tag{13-2.12}$$

This equation and the equilibrium equation allow the use of dyadic notation for orthogonal coordinates. If we consider cases in which body forces are zero, we may satisfy the equilibrium equation (13-2.1) by relating the stress tensor to a new tensor by the relation

$$\bar{\bar{\sigma}} = \vec{\nabla} \times \bar{P}. \tag{13-2.13}$$

Since $\bar{\bar{\sigma}}$ is a symmetric tensor, we shall relate $\bar{P}$ to a new symmetric tensor $\bar{\bar{\Phi}}$ such that the stress tensor becomes

$$\bar{\bar{\sigma}} = -\vec{\nabla} \times \bar{\bar{\Phi}} \times \overleftarrow{\nabla}. \tag{13-2.14}$$

Note that although this equation ensures that the equilibrium equation is satisfied, we have not shown that the equilibrium equation ensures the existence of the tensor $\bar{\bar{\Phi}}$. This proof can be constructed by similar arguments as used with vector potentials. If we consider $\bar{\bar{\Phi}}$ to be a diagonal tensor, the stresses in Cartesian coordinates become

$$\begin{aligned} \sigma_{xx} &= \frac{\partial^2 \Phi_{yy}}{\partial z^2} + \frac{\partial^2 \Phi_{zz}}{\partial y^2}, & \sigma_{xy} &= -\frac{\partial^2 \Phi_{zz}}{\partial x\, \partial y}, \\ \sigma_{yy} &= \frac{\partial^2 \Phi_{zz}}{\partial x^2} + \frac{\partial^2 \Phi_{xx}}{\partial z^2}, & \sigma_{yz} &= -\frac{\partial^2 \Phi_{xx}}{\partial y\, \partial z}, \\ \sigma_{zz} &= \frac{\partial^2 \Phi_{xx}}{\partial y^2} + \frac{\partial^2 \Phi_{yy}}{\partial x^2}, & \sigma_{zx} &= -\frac{\partial^2 \Phi_{yy}}{\partial z\, \partial x}. \end{aligned} \tag{13-2.15}$$

These three functions are the Maxwell stress functions and in the special two-dimensional case when Φ_{xx} and Φ_{yy} are zero, Φ_{zz} is equivalent to the Airy stress function.

If the diagonal elements of $\bar{\bar{\Phi}}$ are taken to be zero, the three remaining off-diagonal elements become the Morera* stress functions:†

$$\begin{aligned} \sigma_{xx} &= -2\frac{\partial^2 \Phi_{yz}}{\partial y\, \partial z}, & \sigma_{xy} &= \frac{\partial}{\partial z}\left[\frac{\partial \Phi_{yz}}{\partial x} + \frac{\partial \Phi_{zx}}{\partial y} - \frac{\partial \Phi_{xy}}{\partial z}\right], \\ \sigma_{yy} &= -2\frac{\partial^2 \Phi_{zx}}{\partial z\, \partial x}, & \sigma_{yz} &= \frac{\partial}{\partial x}\left[\frac{\partial \Phi_{zx}}{\partial y} + \frac{\partial \Phi_{xy}}{\partial z} - \frac{\partial \Phi_{yz}}{\partial x}\right], \\ \sigma_{zz} &= -2\frac{\partial^2 \Phi_{xy}}{\partial x\, \partial y}, & \sigma_{zx} &= \frac{\partial}{\partial y}\left[\frac{\partial \Phi_{xy}}{\partial z} + \frac{\partial \Phi_{yz}}{\partial x} - \frac{\partial \Phi_{zx}}{\partial y}\right]. \end{aligned} \tag{13-2.16}$$

If these functions are to be solutions of the elasticity equations, they must satisfy the compatibility equations. If we consider the case of zero body forces, equation 13-2.10 yields

$$\nabla^2 \bar{\bar{\sigma}} + \frac{1}{(1+\nu)}[\vec{\nabla}\vec{\nabla} S - \bar{\bar{I}}\nabla^2 S] = 0. \tag{13-2.17}$$

* J. Maxwell, *Trans. Roy. Soc. Edinburgh*, **26**, p. 27, 1870.

† See C. Truesdell and R. Toupin, "Principles of Classical Mechanics and Field Theory," *Handbuch der Physik*, Vol. III/1, p. 587, for a discussion of reduction of A.

Using 13-2.14 and 13-2.5, this becomes

$$\vec{\nabla} \times \left[\nabla^2 \bar{\bar{\Phi}} - \frac{1}{1+\nu} \bar{\bar{\mathbf{I}}} S \right] \times \overleftarrow{\nabla} = 0. \qquad (13\text{-}2.18)$$

Noting that the tensor in the bracket is symmetric, integration yields

$$\nabla^2 \bar{\bar{\Phi}} - \frac{1}{1+\nu} \bar{\bar{\mathbf{I}}} S = \vec{\nabla} \bar{c} + \bar{c} \overleftarrow{\nabla}. \qquad (13\text{-}2.19)$$

Examining equations 13-2.4 and 13-2.14, S becomes

$$S = \nabla^2 \varphi - \vec{\nabla} \cdot (\vec{\nabla} \cdot \bar{\bar{\Phi}}), \qquad (13\text{-}2.20)$$

where φ is the first invariant of $\bar{\bar{\Phi}}$.

Substituting 13-2.20 into 13-2.19 and contracting or taking the trace of both sides yields

$$\nabla^2 \varphi - \frac{3}{1+\nu} [\nabla^2 \varphi - \vec{\nabla} \cdot (\vec{\nabla} \cdot \bar{\bar{\Phi}})] = 2 \vec{\nabla} \cdot \bar{c}. \qquad (13\text{-}2.21)$$

This may be written as

$$\vec{\nabla} \cdot (2\bar{c}) = \vec{\nabla} \cdot \left[\frac{3}{(1+\nu)} \vec{\nabla} \cdot \bar{\bar{\Phi}} - \frac{2-\nu}{1+\nu} \vec{\nabla} \varphi \right]. \qquad (13\text{-}2.22)$$

The vector $\bar{c}$ is obtained by integration and becomes

$$\bar{c} = \frac{1}{2} \left[\frac{3}{1+\nu} \vec{\nabla} \cdot \bar{\bar{\Phi}} - \frac{2-\nu}{1+\nu} \vec{\nabla} \varphi \right], \qquad (13\text{-}2.23)$$

where the vector of integration, the curl of another vector, has been absorbed into the $\vec{\nabla} \cdot \bar{\bar{\Phi}}$ term. Substitution of 13-2.20 and 13-2.23 into 13-2.19 yields the defining tensorial equation for $\bar{\bar{\Phi}}$:

$$\nabla^2 \bar{\bar{\Phi}} - \frac{1}{1+\nu} \bar{\bar{\mathbf{I}}} [\nabla^2 \varphi - \vec{\nabla} \cdot (\vec{\nabla} \cdot \bar{\bar{\Phi}})]$$

$$= \frac{3}{2(1+\nu)} [\vec{\nabla}(\vec{\nabla} \cdot \bar{\bar{\Phi}}) + (\vec{\nabla} \cdot \bar{\bar{\Phi}}) \overleftarrow{\nabla}] - \frac{2-\nu}{1+\nu} \vec{\nabla} \vec{\nabla} \varphi. \qquad (13\text{-}2.24)$$

The stress tensor and the displacement vector may be written

$$\bar{\bar{\sigma}} = \frac{1-2\nu}{2(1+\nu)} [\vec{\nabla}(\vec{\nabla} \varphi - \vec{\nabla} \cdot \bar{\bar{\Phi}}) + (\vec{\nabla} \varphi - \vec{\nabla} \cdot \bar{\bar{\Phi}}) \overleftarrow{\nabla}]$$

$$+ \frac{\nu}{1+\nu} \bar{\bar{\mathbf{I}}} [\nabla^2 \varphi - \vec{\nabla} \cdot (\vec{\nabla} \cdot \bar{\bar{\Phi}})], \qquad (13\text{-}2.25)$$

$$\bar{u} = \frac{1}{2G} \left(\frac{1-2\nu}{1+\nu} \right) [\vec{\nabla} \varphi - \vec{\nabla} \cdot \bar{\bar{\Phi}}]. \qquad (13\text{-}2.26)$$

It is important to note that only the divergence and the first invariant of $\bar{\bar{\Phi}}$ are uniquely determined.

3 The Galerkin vector

In 1906, A. E. H. Love* introduced a strain function for considering problems of solids of revolution under the influence of axisymmetric loadings. This function is defined by a biharmonic equation in the case of zero body forces. We shall not present Love's development as it can be shown that this is a special case of the Galerkin vector treatment.

B. Galerkin† extended Love's idea and developed, in 1930, three strain functions for the general problem of elasticity. In Cartesian coordinates, these three functions decouple the Navier equation and produce three fourth-order equations for the functions. These three functions may be interpreted as components of a vector, and the term Galerkin vector has been applied to this vector. H. M. Westergaard‡ presented a comprehensive and therefore often-quoted development of this vector. We shall develop this vector from the previous stress functions first.

In section 2 we noted that the displacement can be related to the gradient of a scalar and the divergence of a symmetric tensor. These quantities are not independent of one another and if we take the divergence of equation 13-2.24 we obtain

$$\begin{aligned}\nabla^2\vec{\nabla}\cdot\bar{\bar{\Phi}} &- \frac{1}{1+\nu}[\nabla^2\vec{\nabla}\varphi - \vec{\nabla}(\vec{\nabla}\cdot(\vec{\nabla}\cdot\bar{\bar{\Phi}}))] \\ &= \frac{3}{2(1+\nu)}[\nabla^2(\vec{\nabla}\cdot\bar{\bar{\Phi}}) + \vec{\nabla}(\vec{\nabla}\cdot(\vec{\nabla}\cdot\bar{\bar{\Phi}}))] - \frac{2-\nu}{1+\nu}\nabla^2\vec{\nabla}\varphi. \qquad (13\text{-}3.1)\end{aligned}$$

If we let $\bar{q} = \vec{\nabla}\cdot\bar{\bar{\Phi}}$, we can write 13-3.1, after combining terms, as

$$\nabla^2\bar{q} + \frac{1}{(1-2\nu)}\vec{\nabla}(\vec{\nabla}\cdot\bar{q}) = \frac{2(1-\nu)}{1-2\nu}\nabla^2(\vec{\nabla}\varphi). \qquad (13\text{-}3.2)$$

Boussinesq suggested the introduction of a vector $\bar{G}$ related to $\bar{q}$ as follows:

$$\bar{q} = \nabla^2\bar{G}. \qquad (13\text{-}3.3)$$

Equation 13-3.2 in this case becomes

$$\nabla^2\nabla^2\bar{G} + \frac{\nabla^2\vec{\nabla}}{1-2\nu}[\vec{\nabla}\cdot\bar{G} - 2(1-\nu)\varphi] = 0. \qquad (13\text{-}3.4)$$

* Details of Love's development may be found in his book, *A Mathematical Theory of Elasticity*, and examples of solution will be presented in later sections.

† B. G. Galerkin, *Comptes rendus hebdomadaires des séances de l'académie des sciences*, Paris, **190**, 1930.

‡ H. M. Westergaard, *Theory of Elasticity and Plasticity*, Harvard University Press, Cambridge, Mass., 1952.

He noted that if φ was selected as

$$\varphi = \frac{1}{2(1-\nu)}\vec{\nabla}\cdot\bar{G}, \tag{13-3.5}$$

then $\bar{G}$ was a biharmonic vector. The displacement vector would become

$$\bar{u} = -\frac{1}{4G}\frac{(1-2\nu)}{(1-\nu^2)}[2(1-\nu)\nabla^2\bar{G} - \vec{\nabla}(\vec{\nabla}\cdot\bar{G})]. \tag{13-3.6}$$

Galerkin independently examined Navier's equation in the form

$$G\left[\nabla^2\bar{u} + \frac{1}{1-2\nu}\vec{\nabla}(\vec{\nabla}\cdot\bar{u})\right] + \bar{F} = 0. \tag{13-3.7}$$

He suggested introducing a new vector, which for logical reasons we will also denote by $\bar{G}$, in the following manner:

$$2G\bar{u} = (c\nabla^2\bar{G} - \vec{\nabla}\vec{\nabla}\cdot\bar{G}). \tag{13-3.8}$$

Substitution of 13-3.8 into the Navier equation will yield a biharmonic form if c is chosen to be equal to $2(1-\nu)$. The Navier equation becomes

$$\nabla^2\nabla^2\bar{G} = -\frac{\bar{F}}{(1-\nu)}, \tag{13-3.9}$$

and the displacement is related to $\bar{G}$ as follows:

$$\bar{u} = \frac{1}{2G}[2(1-\nu)\nabla^2\bar{G} - \vec{\nabla}(\vec{\nabla}\cdot\bar{G})]. \tag{13-3.10}$$

Equations 13-3.6 and 13-3.10 differ only by a constant. Galerkin showed the generality of this solution and illustrated applications; therefore, we will refer to $\bar{G}$ as the Galerkin vector and will use 13-3.9 to define it and 13-3.10 to relate it to displacements.

4 Equivalent Galerkin vectors*

If two Galerkin vectors $\bar{G}_1$ and $\bar{G}_2$ can be found such that they define the same displacement field, they are called equivalent Galerkin vectors. If equivalent vectors exist, their difference must yield a vector $\bar{G}_0$ that gives rise to no displacements. The equation that $\bar{G}_0$ satisfies may be obtained from 13-3.10, and is

$$2(1-\nu)\nabla^2\bar{G}_0 - \vec{\nabla}(\vec{\nabla}\cdot\bar{G}_0) = 0. \tag{13-4.1}$$

We would like to relate $\bar{G}_0$ to a biharmonic function if possible. We shall attempt to relate $\bar{G}_0$ to a new vector in a similar manner as was done in equation 13-3.8:

* See Westergaard, *Theory of Elasticity and Plasticity*.

$$\bar{G}_0 = c\nabla^2\bar{f} - \vec{\nabla}\vec{\nabla}\cdot\bar{f}. \quad (13\text{-}4.2)$$

Substitution of 13-4.2 into 13-4.1 yields

$$c = -(1 - 2\nu) \qquad \text{if } \bar{f} \text{ is biharmonic.}$$

If $\bar{G}_1$ has the general form

$$\bar{G}_1 = G_{1x}\hat{\imath} + G_{1y}\hat{\jmath} + G_{1z}\hat{k}, \quad (13\text{-}4.3)$$

we can choose a $\bar{G}_0$ such that the G_{1z} term is eliminated. Pick

$$\bar{G}_0 = -(1 - 2\nu)\nabla^2 f_x\hat{\imath} - \frac{\partial^2 f_x}{\partial x^2}\hat{\imath} - \frac{\partial^2 f_x}{\partial x\,\partial y}\hat{\jmath} - \frac{\partial^2 f_x}{\partial x\,\partial z}\hat{k}. \quad (13\text{-}4.4)$$

Both G_{1z} and f_x are biharmonic in the absence of body forces, so that it is possible to choose f_x such that the following relation is true:

$$\frac{\partial^2 f_x}{\partial x\,\partial z} = G_{1z}. \quad (13\text{-}4.5)$$

Obviously there are many other $\bar{G}_0$'s that might accomplish the same thing, but the important result is that one can eliminate one component of $\bar{G}$ by a proper choice of $\bar{G}_0$. We can therefore conclude that if the body forces are zero, any Galerkin vector may be replaced by an equivalent Galerkin vector having no component in the z direction. The z direction can be chosen to coincide with any direction. If the body force F_z is zero, G_z may be eliminated even if the other two components of the body force are not zero.

5 Mathematical notes on the Galerkin vector

Equation 13-3.9 reduces to a biharmonic vector equation in the absence of body forces:

$$\nabla^2\nabla^2\bar{G} = 0. \quad (13\text{-}5.1)$$

In Cartesian coordinates $\bar{G}$ becomes

$$\bar{G} = G_x\hat{\imath}_{(x)} + G_y\hat{\imath}_{(y)} + G_z\hat{\imath}_{(z)}, \quad (13\text{-}5.2)$$

and because the unit vectors are independent of the coordinates, each of the functions G_x, G_y, and G_z are biharmonic functions.

If other coordinate systems are used, the unit vectors become functions of the coordinates, and the components of the Galerkin vector are not biharmonic. In general, the biharmonic equation does not even decouple. For example, in cylindrical coordinates

$$\bar{G} = G_r\hat{\imath}_{(r)} + G_\theta\hat{\imath}_{(\theta)} + G_z\hat{\imath}_{(z)}. \quad (13\text{-}5.3)$$

We note that

$$\nabla^2[G_r\hat{i}_{(r)}] = \left[\left(\nabla^2 - \frac{1}{r^2}\right)G_r\right]\hat{i}_{(r)} + \frac{2}{r^2}\left[\frac{\partial G_r}{\partial \theta}\hat{i}_{(\theta)}\right],$$

$$\nabla^2[G_\theta\hat{i}_{(\theta)}] = \left[\left(\nabla^2 - \frac{1}{r^2}\right)G_\theta\right]\hat{i}_{(\theta)} - \frac{2}{r^2}\left[\frac{\partial G_\theta}{\partial \theta}\hat{i}_{(r)}\right], \tag{13-5.4}$$

$$\nabla^2[G_z\hat{i}_{(z)}] = [\nabla^2 G_z]\hat{i}_{(z)}.$$

Using 13-5.4, we write the biharmonic as

$$\nabla^2\nabla^2[G_r\hat{i}_{(r)}] = \left[\left(\nabla^2 - \frac{1}{r^2}\right)^2 G_r\right]\hat{i}_{(r)} + 2\left[\frac{1}{r^2}\frac{\partial}{\partial \theta}\left(\nabla^2 - \frac{1}{r^2}\right)G_r\right]\hat{i}_{(\theta)}$$
$$+ 2\left[\left(\nabla^2 - \frac{1}{r^2}\right)\left(\frac{1}{r^2}\frac{\partial G_r}{\partial \theta}\right)\right]\hat{i}_{(\theta)} - \frac{4}{r^4}\frac{\partial^2 G_r}{\partial \theta^2}\hat{i}_{(r)},$$

$$\nabla^2\nabla^2[G_\theta\hat{i}_{(\theta)}] = \left[\left(\nabla^2 - \frac{1}{r^2}\right)^2 G_\theta\right]\hat{i}_{(\theta)} - 2\left[\frac{1}{r^2}\frac{\partial}{\partial \theta}\left(\nabla^2 - \frac{1}{r^2}\right)G_\theta\right]\hat{i}_{(r)} \tag{13-5.5}$$
$$- 2\left[\left(\nabla^2 - \frac{1}{r^2}\right)\frac{1}{r^2}\frac{\partial G_\theta}{\partial \theta}\right]\hat{i}_{(r)} - \frac{4}{r^4}\frac{\partial^2 G_\theta}{\partial \theta^2}\hat{i}_{(\theta)},$$

$$\nabla^2\nabla^2[G_z\hat{i}_{(z)}] = [\nabla^2\nabla^2 G_z]\hat{i}_{(z)}.$$

In this case only the G_z component is a biharmonic function.

As we have noted, the components of the vector are biharmonic in Cartesian coordinates; it is therefore beneficial to make some comments about harmonic and biharmonic functions.

Consider the gradient of the product of two functions U and V:

$$\vec{\nabla}(UV) = (\vec{\nabla}U)V + U(\vec{\nabla}V). \tag{13-5.6}$$

The Laplacian of this product may be formed by taking the divergence of this gradient:

$$\vec{\nabla}\cdot\vec{\nabla}(UV) = (\nabla^2 U)V + 2(\vec{\nabla}U)\cdot(\vec{\nabla}V) + U(\nabla^2 V). \tag{13-5.7}$$

We now use 13-5.7 to note

$$\nabla^2(x\varphi) = x\nabla^2\varphi + 2\frac{\partial\varphi}{\partial x}. \tag{13-5.8}$$

If φ is a harmonic function, $x\varphi$ is biharmonic, as

$$\nabla^2\left(2\frac{\partial\varphi}{\partial x}\right) = 2\frac{\partial}{\partial x}(\nabla^2\varphi) = 0. \tag{13-5.9}$$

We also note that $\bar{R}\cdot\vec{\nabla}\varphi$ is harmonic, if φ is harmonic and $\bar{R}$ is the position vector. We can see this by examining

$$\vec{\nabla}[(\bar{R}\cdot\vec{\nabla})\varphi] = \bar{\bar{\mathbf{I}}}\cdot\vec{\nabla}\varphi + (\bar{R}\cdot\vec{\nabla})\vec{\nabla}\varphi$$
$$= \vec{\nabla}\varphi + (\bar{R}\cdot\vec{\nabla})\vec{\nabla}\varphi, \tag{13-5.10}$$

$$\nabla^2[(\bar{R}\cdot\vec{\nabla})\varphi] = \nabla^2\varphi + (\bar{\bar{\mathbf{I}}}\cdot\vec{\nabla})\cdot\vec{\nabla}\varphi + (\bar{R}\cdot\vec{\nabla})\nabla^2\varphi,$$
$$= 2\nabla^2\varphi + (\bar{R}\cdot\vec{\nabla})\nabla^2\varphi. \tag{13-5.11}$$

The function $\frac{1}{2}R^2\varphi$ is therefore biharmonic, as

$$\nabla^2(\tfrac{1}{2}R^2\varphi) = 3\varphi + 2\bar{R}\cdot\vec{\nabla}\varphi + \tfrac{1}{2}R^2(\nabla^2\varphi). \tag{13-5.12}$$

These equations indicate that

$$f = \varphi_0 + x\varphi_1 + y\varphi_2 + z\varphi_3 + \tfrac{1}{2}R^2\varphi_4 \tag{13-5.13}$$

is biharmonic, where φ_i is harmonic. Not all the last terms are independent, and three of the last four terms can be chosen arbitrarily.

6 Love's strain function

We shall consider in detail one special case of the Galerkin vector. Let us take $\bar{G}$ such that only the z component is nonzero. We expect this to be a valid choice only for a limited class of problems, as we have shown that without body forces only one of the components of the Galerkin vector can be taken to be zero. Two special cases will be considered: first, the infinite half-space and, second, bodies of revolution that are loaded in an axi-symmetric manner.

The Galerkin vector becomes

$$\bar{G} = Z\hat{i}_{(z)}, \tag{13-6.1}$$

where Z is defined by the equation

$$\nabla^2\nabla^2 Z = -\frac{F_z}{(1-\nu)}. \tag{13-6.2}$$

We can express the components of the displacement vector and the stress tensor in either Cartesian or cylindrical coordinates in terms of Z. These are

$$
\begin{aligned}
2Gu_x &= -\frac{\partial^2 Z}{\partial x\,\partial z}, & 2Gu_r &= -\frac{\partial^2 Z}{\partial r\,\partial z},\\
2Gu_y &= -\frac{\partial^2 Z}{\partial y\,\partial z}, & 2Gu_\theta &= -\frac{1}{r}\frac{\partial^2 Z}{\partial\theta\,\partial z},\\
2Gu_z &= \left[2(1-\nu)\nabla^2 - \frac{\partial^2}{\partial z^2}\right]Z, & &\\
\sigma_{xx} &= \frac{\partial}{\partial z}\left[\nu\nabla^2 - \frac{\partial^2}{\partial x^2}\right]Z, & \sigma_{rr} &= \frac{\partial}{\partial z}\left[\nu\nabla^2 - \frac{\partial^2}{\partial r^2}\right]Z,\\
\sigma_{yy} &= \frac{\partial}{\partial z}\left[\nu\nabla^2 - \frac{\partial^2}{\partial y^2}\right]Z, & \sigma_{\theta\theta} &= \frac{\partial}{\partial z}\left[\nu\nabla^2 - \frac{1}{r}\frac{\partial}{\partial r} - \frac{1}{r^2}\frac{\partial^2}{\partial\theta^2}\right]Z,\\
\sigma_{zz} &= \frac{\partial}{\partial z}\left[(2-\nu)\nabla^2 - \frac{\partial^2}{\partial z^2}\right]Z, & &\\
\sigma_{xy} &= -\frac{\partial^3 Z}{\partial x\,\partial y\,\partial z}, & \sigma_{r\theta} &= -\frac{\partial^3}{\partial r\,\partial\theta\,\partial z}\left(\frac{Z}{r}\right),
\end{aligned}
\tag{13-6.3}
$$

$$\sigma_{yz} = \frac{\partial}{\partial y}\left[(1-\nu)\nabla^2 - \frac{\partial^2}{\partial z^2}\right]Z, \qquad \sigma_{\theta z} = \frac{1}{r}\frac{\partial}{\partial \theta}\left[(1-\nu)\nabla^2 - \frac{\partial^2}{\partial z^2}\right]Z,$$

$$\sigma_{zx} = \frac{\partial}{\partial x}\left[(1-\nu)\nabla^2 - \frac{\partial^2}{\partial z^2}\right]Z, \qquad \sigma_{zr} = \frac{\partial}{\partial r}\left[(1-\nu)\nabla^2 - \frac{\partial^2}{\partial z^2}\right]Z,$$

$$\nabla^2 = \frac{\partial^2}{\partial x^2} + \frac{\partial^2}{\partial y^2} + \frac{\partial^2}{\partial z^2}, \qquad \nabla^2 = \frac{\partial^2}{\partial r^2} + \frac{1}{r}\frac{\partial}{\partial r} + \frac{1}{r^2}\frac{\partial^2}{\partial \theta^2} + \frac{\partial^2}{\partial z^2}.$$

When Z is a function of r and z only, u_θ, $\sigma_{r\theta}$, and $\sigma_{\theta z}$ are zero and the function is applicable for solution of axisymmetric problems of solids of revolution. This form of solution produces a strain function identical to the one suggested earlier by Love.* This particular form of the Galerkin vector is called the Love strain function.

We shall consider the restricted case first, that is, when Z is the Love strain function and is therefore independent of θ. The Laplacian operator in this case becomes

$$\nabla^2 = \frac{\partial^2}{\partial r^2} + \frac{1}{r}\frac{\partial}{\partial r} + \frac{\partial^2}{\partial z^2}. \qquad (13\text{-}6.4)$$

We can find the harmonic functions that satisfy this operator by considering the separation of variables solution of the equation

$$\begin{gathered}\nabla^2\varphi = 0,\\ \varphi = [C_1 J_0(\alpha r) + C_2 Y_0(\alpha r)]e^{\pm\alpha z},\end{gathered} \qquad (13\text{-}6.5)$$

where J_0 and Y_0 are the Bessel functions of order zero of the first and second kind, respectively. Using the properties of biharmonic functions, we can consider a form of solution of 13-6.2 for zero body force as

$$\begin{aligned} Z = \Big\{&AJ_0(\alpha r) + BY_0(\alpha r) + r\frac{\partial}{\partial r}[CJ_0(\alpha r) + DY_0(\alpha r)] \\ &+ E\alpha z J_0(\alpha r) + F\alpha z Y_0(\alpha r)\Big\} e^{\pm\alpha z}. \end{aligned} \qquad (13\text{-}6.6)$$

The separation constant α can be real or imaginary. If α is imaginary, we obtain trigonometric behavior in the z direction and modified Bessel functions in the r direction.

If the separation constant is zero, the solution ψ to the harmonic equation is

$$\psi = A + Bz + C\ln r + Dz\ln r. \qquad (13\text{-}6.7)$$

Biharmonic functions may be constructed for the zero separation constant by multiplying ψ by $[r^2 + z^2]$.

Other polynomial solutions may be constructed, but will not have the same facility to satisfy boundary conditions as do the trigonometric and Bes-

* A. E. H. Love, *A Treatise on the Mathematical Theory of Elasticity*, Dover Publications, Inc., New York, 1944, 4th ed. (1926).

sel series. We shall examine some of these polynomials, as they can be superimposed to satisfy simple boundary conditions.

If ψ is a harmonic function, then $R^2\psi$ is biharmonic:

$$R^2\psi = (r^2 + z^2)\psi. \tag{13-6.8}$$

Examine solutions of the form

$$Z = [A + (r^2 + z^2)B]\psi. \tag{13-6.9}$$

We can also construct harmonic polynomials by trial-and-error examination of polynomials of homogeneous order. For example, consider

$$\begin{aligned}\psi_0 &= 1,\\ \psi_1 &= a_1 z + b_1 r,\\ \psi_2 &= a_2 z^2 + b_2 r^2 + c_2 rz,\\ \psi_3 &= a_3 z^3 + b_3 r^3 + c_3 r^2 z + d_3 rz^2.\end{aligned} \tag{13-6.10}$$

The coefficients are now selected to satisfy the harmonic equation:

$$\frac{\partial^2\psi_n}{\partial r^2} + \frac{1}{r}\frac{\partial\psi_n}{\partial r} + \frac{\partial^2\psi_n}{\partial z^2} = 0. \tag{13-6.11}$$

This yields the following functions:

$$\begin{aligned}\psi_0 &= 1,\\ \psi_1 &= z,\\ \psi_2 &= \left(z^2 - \frac{r^2}{2}\right),\\ \psi_3 &= (z^3 - \tfrac{3}{2}zr^2).\end{aligned} \tag{13-6.12}$$

This procedure can be continued, developing higher-order polynomial solutions. An alternative method of obtaining these polynomials is by detailed examination of the Legendre polynomials, which arise if the harmonic equation is written in spherical coordinates. Let $R^2 = r^2 + z^2$ and $\tan\varphi = r/z$. These are the spherical coordinates, as may be seen in Figure 13-2. For axisymmetric problems, the Laplace equation becomes

$$\nabla^2\psi = \frac{\partial^2\psi}{\partial R^2} + \frac{2}{R}\frac{\partial\psi}{\partial R} + \frac{1}{R^2}\frac{\partial^2\psi}{\partial\varphi^2} + \frac{\cot\varphi}{R^2}\frac{\partial\psi}{\partial\varphi} = 0. \tag{13-6.13}$$

We seek solutions of the form

$$\psi_n = R^n P_n(\cos\varphi). \tag{13-6.14}$$

Substitution into the Laplace equation yields

$$\left[n(n+1)P_n + \frac{d^2P_n}{d\varphi^2} + \cot\varphi\frac{dP_n}{d\varphi}\right]R^{n-2} = 0. \tag{13-6.15}$$

The form of this equation may be altered if we introduce $\zeta = \cos\varphi$.

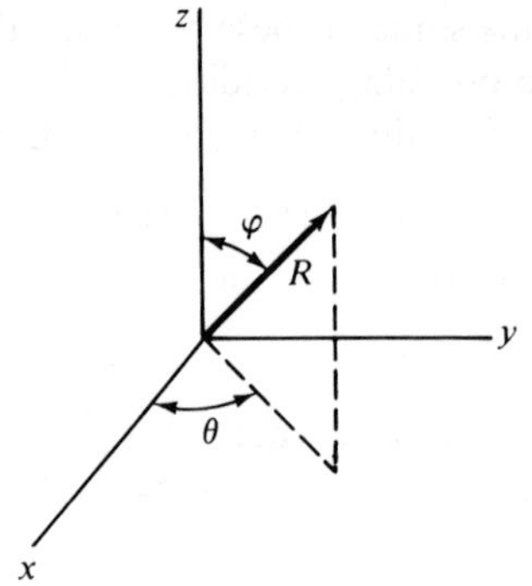

Figure 13-2

Equation 13-6.15 becomes

$$(1-\zeta^2)\frac{d^2P_n}{d\zeta^2}-2\zeta\frac{dP_n}{d\zeta}+n(n+1)P_n=0. \qquad (13\text{-}6.16)$$

This equation is known as Legendre's equation and has a solution of the form

$$P_n(\zeta)=\frac{1}{2^n(n!)}\left[\frac{d^n}{d\zeta^n}(\zeta^2-1)^n\right]. \qquad (13\text{-}6.17)$$

Evaluating the first few values yields

$$\begin{aligned} P_0 &= 1,\\ P_1 &= \zeta,\\ P_2 &= \frac{3\zeta^2-1}{2},\\ P_3 &= \frac{5\zeta^3-3\zeta}{2}. \end{aligned} \qquad (13\text{-}6.18)$$

The solution of the harmonic equation becomes $\psi_n = R^nP_n(\zeta)$, ζ may be replaced by $\cos\varphi = z/R$, and the solutions of the harmonic equation become

$$\begin{aligned} \psi_0 &= 1,\\ \psi_1 &= z,\\ \psi_2 &= z^2-\frac{r^2}{2},\\ \psi_3 &= z^3-\tfrac{3}{2}zr^2,\\ \psi_n &= \frac{R^n}{2^n(n!)}\frac{d^n}{d\zeta^n}[(\zeta^2-1)^n]. \end{aligned} \qquad (13\text{-}6.19)$$

These solutions agree with those generated by trial and error. They are formed by use of the polynomials P_n, which are called the Legendre polynomials.

Further consideration of these functions will be given later when problems in spherical coordinates are considered.

Solutions of the biharmonic equation can now be taken in the form

$$Z_n = \psi_n + (r^2 + z^2)\psi_{n-2}. \qquad (13\text{-}6.20)$$

An alternative form of solutions of the harmonic equation is also obtained in the form

$$\psi_{-(n+1)} = R^{-(n+1)}P_n(\cos\varphi). \qquad (13\text{-}6.21)$$

This leads to the same differential equation (13-6.15) for P_n, so that other harmonic functions are

$$\begin{aligned} \psi_{-1} &= (r^2 + z^2)^{-1/2}, \qquad \psi_{-2} = z(r^2 + z^2)^{-3/2}, \\ \psi_{-3} &= \left(z^2 - \frac{r^2}{2}\right)(r^2 + z^2)^{-5/2}, \qquad \psi_{-n} = \psi_{n-1}(r^2 + z^2)^{-[(2n-1)/2]}. \end{aligned} \qquad (13\text{-}6.22)$$

Before examining the general solutions of this class of problems, let us see the forms of solution one obtains from the polynomials given in equation 13-6.20:

$$Z_2 = A_2\left(z^2 - \frac{r^2}{2}\right) + B_2(r^2 + z^2). \qquad (13\text{-}6.23)$$

This leads to the following stresses and displacements:

$$\begin{aligned} 2Gu_z &= -2A_2 + (10 - 12\nu)B, \\ u_r &= \sigma_{zz} = \sigma_{rr} = \sigma_{\theta\theta} = \sigma_{rz} = 0. \end{aligned} \qquad (13\text{-}6.24)$$

This solution corresponds to a rigid-body motion in the z direction and will be used to adjust the displacement.

Consider homogeneous polynomials of order 3:

$$Z_3 = A_3(z^3 - \tfrac{3}{2}zr^2) + B_3(r^2 + z^2)z. \qquad (13\text{-}6.25)$$

The displacements and stresses become

$$\begin{aligned} 2Gu_z &= -6A_3 z + (14 - 20\nu)B_3 z, \\ 2Gu_r &= 3A_3 r + 2B_3 r, \\ \sigma_{zz} &= -6A_3 + (14 - 10\nu)B_3, \\ \sigma_{rr} &= \sigma_{\theta\theta} = 3A_3 + (10\nu - 2)B_3, \\ \sigma_{rz} &= 0. \end{aligned} \qquad (13\text{-}6.26)$$

This solution is used to solve problems of uniform tension or compression in the z direction if σ_{rr} is set equal to zero. If σ_{zz} is set equal to zero, the solution of uniform radial tension or compression is obtained. Although these solutions are elementary in nature, they are used in superposition with more complex solutions to obtain agreement up to the Saint-Venant conditions on certain boundaries.

The fourth-order polynomial is another useful solution:

$$Z_4 = A_4\left(z^4 - 3z^2r^2 + \frac{3}{8}r^4\right) + B_4(r^2 + z^2)\left(z^2 - \frac{r^2}{2}\right). \qquad (13\text{-}6.27)$$

The stresses and displacements for this case become

$$\begin{aligned} 2Gu_z &= -A_4[12z^2 - 6r^2] + B_4[(16 - 28\nu)z^2 - (13 - 14\nu)r^2], \\ 2Gu_r &= 12A_4zr + 2B_4zr, \\ \sigma_{zz} &= -24A_4z + B_4(32 - 28\nu)z, \qquad (13\text{-}6.28) \\ \sigma_{rr} &= \sigma_{\theta\theta} = 12A_4z + B_4(28\nu - 2)z, \\ \sigma_{rz} &= 12A_4r - B_4(16 - 14\nu)r. \end{aligned}$$

If σ_{zz} is set to zero on the boundaries, the bending solution of a thick plate is obtained. If σ_{rr} is set equal to zero, uniform shear exists on the curved boundaries held in equilibrium by σ_{zz} at the ends.

Solutions that are singular at the origin have been omitted up to this point and are set equal to zero as a regularity condition if the origin is part of the elastic body. In problems of hollow tubes or disks with holes, these terms are required to satisfy certain boundary conditions in the Saint-Venant sense or to provide the required constant term in the Fourier analysis.

Equation 13-6.7 gives two logrithmic terms that are singular at the origin. The $\ln r$ term gives rise to no stresses or displacements and is not of interest by itself. Consider the second term:

$$Z = Dz \ln r \qquad (13\text{-}6.29)$$

The stresses and displacements corresponding to this term are

$$\begin{aligned} 2Gu_r &= -\frac{D}{r}, \qquad u_z = 0, \\ \sigma_{rr} &= \frac{D}{r^2}, \qquad \sigma_{\theta\theta} = -\frac{D}{r^2}, \qquad \sigma_{rz} = \sigma_{zz} = 0. \end{aligned} \qquad (13\text{-}6.30)$$

This term in combination with Z_3 (equation 13-6.25) can be used to place uniform radial stresses on the inner radius and outer radius of a body.

The terms $r^2 \ln r$ and $z^2 \ln r$ are biharmonic functions and stresses, and displacements corresponding to them are

$$Z = Ar^2 \ln r, \qquad (13\text{-}6.31)$$

$$\begin{aligned} 2Gu_z &= 2(1 - \nu)[4 \ln r + 3]A, \\ u_r &= 0, \\ \sigma_{rz} &= \frac{4(1 - \nu)}{r}A, \qquad (13\text{-}6.32) \\ \sigma_{rr} &= \sigma_{\theta\theta} = \sigma_{zz} = 0. \end{aligned}$$

This term places counterbalancing shears on the inner and outer radii of a body.

The other biharmonic solution yields the following stresses and displacements:

$$Z = Bz^2 \ln r, \tag{13-6.33}$$

$$\begin{aligned} 2Gu_z &= 2B(1 - 2\nu) \ln r, \\ 2Gu_r &= -\frac{2Bz}{r}, \end{aligned} \tag{13-6.34}$$

$$\sigma_{rr} = \frac{2Bz}{r^2}, \quad \sigma_{\theta\theta} = -\frac{2Bz}{r^2}, \quad \sigma_{zz} = 0, \quad \sigma_{rz} = -\frac{2B}{r}.$$

This solution when combined with equation 13-6.31 is used to produce a linear distribution of σ_{rr} on an inner and outer radii. This combined solution will in general be used in combination with equation 13-6.27.

7 Long solid cylinders axisymmetrically loaded

As an example of use of the Love strain function, consider long cylinders with some prescribed axisymmetric loading along their lateral surfaces. The boundary conditions in this case are

$$\text{solution regular at } r = 0,$$

$$\sigma_{rr} = f(z) \quad \text{and} \quad \sigma_{rz} = g(z) \qquad \text{at } r = R. \tag{13-7.1}$$

We assume that the boundary conditions at $z = \pm l$ are satisfied in the Saint-Venant sense. The solution uses the imaginary separation constant, as it is desirable to use Fourier series in the z direction. The coefficients of the modified Bessel functions of the second kind are chosen as zero to satisfy the first boundary condition. The solution is separated into Z even or odd in z. The even Z function corresponds to an odd σ_{rr} in z and an even σ_{rz} in z:

$$Z = \sum [AI_0(\alpha r) + B(\alpha r)I_1(\alpha r)] \cos \alpha z, \tag{13-7.2}$$

where $I_n(\alpha r)$ is the modified Bessel function and is defined as

$$I_n(x) = e^{-1/2n\pi i} J_n(ix). \tag{13-7.3}$$

The following recurrence relations hold for the modified Bessel functions:

$$\begin{aligned} I_{n-1}(x) - I_{n+1}(x) &= \frac{2n}{x} I_n(x), \\ I'_n(x) &= I_{n-1}(x) - \frac{n}{x} I_n(x), \\ I_{n-1}(x) + I_{n+1}(x) &= 2I'_n(x), \\ I'_n(x) &= I_{n+1}(x) + \frac{n}{x} I_n(x), \end{aligned} \tag{13-7.4}$$

where

$$I'_n(x) = \frac{d}{dx} I_n(x).$$

This form of solution is motivated by the fact that one desires a Fourier series representation of the two boundary stresses. Noting that

$$\frac{d}{dr} I_0(\alpha r) = \alpha I_1(\alpha r),$$
$$\frac{d}{dr} I_1(\alpha r) = \alpha\left[I_0(\alpha r) - \frac{I_1(\alpha r)}{\alpha r}\right], \tag{13-7.5}$$

the Laplacian of the strain function yields

$$\nabla^2 Z = \sum 2\alpha^2 B I_0(\alpha r)\cos\alpha z. \tag{13-7.6}$$

The stresses become

$$\sigma_{rr} = \sum \alpha^3\left\{A\left[I_0(\alpha r) - \frac{I_1(\alpha r)}{\alpha r}\right] + B[(1-2\nu)I_0(\alpha r) + \alpha r I_1(\alpha r)]\right\}\sin\alpha z, \tag{13-7.7}$$

$$\sigma_{rz} = \sum \alpha^3\{A I_1(\alpha r) + B[\alpha r I_0(\alpha r) + 2(1-\nu)I_1(\alpha r)]\}\cos\alpha z. \tag{13-7.8}$$

The second boundary condition yields

$$\sum \alpha^3\left\{A\left[I_0(\alpha R) - \frac{I_1(\alpha R)}{\alpha R}\right] + B[(1-2\nu)I_0(\alpha R) + \alpha R I_1(\alpha R)]\right\}\sin\alpha z = f(z),$$
$$\sum \alpha^3\{A I_1(\alpha R) + B[\alpha R I_0(\alpha R) + 2(1-\nu)I_1(\alpha R)]\}\cos\alpha z = g(z). \tag{13-7.9}$$

These are in the form of Fourier series expansions for the boundary conditions. If we choose α correctly, we can use the orthogonality properties of the trigonometric functions to evaluate the constants A and B.

Consider, for example, the case where $g(z)$ is zero and the ends are stress free. Noting that σ_{zz} also involves the sin αz term, we choose $\alpha_n = n\pi/l$ so that we have orthogonal functions and σ_{zz} is zero at $z = \pm l$. We can relate A_n and B_n by the second equation of 13-7.9, noting that $g(z)$ is zero:

$$A_n = -B_n\left[2(1-\nu) + \frac{\alpha_n R I_0(\alpha_n R)}{I_1(\alpha_n R)}\right]. \tag{13-7.10}$$

The first equation of 13-7.9 becomes

$$\sum_{n=1}^{\infty} B_n\{2(1-\nu)I_1^2(\alpha_n R) + \alpha_n^2 R^2[I_1^2(\alpha_n R) - I_0^2(\alpha_n R)]\}\frac{\alpha_n^3 \sin\alpha_n z}{\alpha_n R I_1(\alpha_n R)} = f(g). \tag{13-7.11}$$

By Fourier analysis we obtain

$$B_n = \frac{\alpha_n R I_1(\alpha_n R)}{\alpha_n^3[2(1-\nu)I_1^2(\alpha_n R) + \alpha_n^2 R^2(I_1^2(\alpha_n R) - I_0^2(\alpha_n R))]}\frac{2}{l}\int_0^l f(g)\sin\frac{n\pi z}{l}\,dz. \tag{13-7.12}$$

Note that σ_{rz} is not zero at $z = \pm l$ but $\int_A \sigma_{rz}\,dA = 0$. Also note that any cylinder undergoing axisymmetric loading can have this loading separated into an even and odd function and can now be solved in this manner if the Saint-Venant approximation is used on the ends.

As in the case of any Fourier analysis, if zero is an eigenvalue, the solution corresponding to this eigenvalue will be needed for a complete expansion. If σ_{rz} were not zero in the example shown, a constant term would be needed to complete the Fourier series given in equation 13-7.8. Equation 13-6.27 provides this term if the constants are related such that σ_{rr} is zero:

$$A_4 = \frac{28\nu - 2}{12} B_4. \qquad (13\text{-}7.13)$$

σ_{zz} is not zero on the ends in this case and this stress is required for equilibrium.

If the odd problem is considered, the cosine series will arise in the σ_{rr} expansion. Equation 13-6.25 supplies the required constant term if the terms are related so that σ_{zz} is zero:

$$A_3 = \frac{14 - 10\nu}{6} B_3. \qquad (13\text{-}7.14)$$

The addition of the constant terms corresponding to zero eigenvalues presents no difficulties, and they may be evaluated in the normal manner by use of the orthogonality relationship. They correspond to the average value of the functions along the length of the cylinder.

8 Infinite cylinder or hole in an elastic body

The analysis of long solid cylinders by use of Fourier series suggests the application of Fourier transforms* for infinite cylinders or the exterior problem of a hole in a large (theoretically infinite) elastic body. The first of these problems yields a method by which closed solutions can sometimes be obtained for very long cylinders under axisymmetric loads, and the second is of interest in investigations of the stress distributions in long tunnels.

In the absence of body forces, the Love strain function is defined by the binarmonic equation

$$\nabla^2\nabla^2 Z = 0, \qquad \begin{matrix} -\infty < z < +\infty, \\ r > R \quad \text{or} \quad r < R. \end{matrix} \qquad (13\text{-}8.1)$$

If $r < R$, we are considering the case of a solid cylinder, and if $r > R$, we are considering the external problem of a hole in an infinite body.

* See Section 7-8.

We define the Fourier transform of Z to be

$$\mathcal{Z}(r, \beta) = \frac{1}{\sqrt{2\pi}} \int_{-\infty}^{+\infty} Z(r, z)e^{i\beta z}\, dz. \tag{13-8.2}$$

We consider the problem where at the boundary $r = R$, $\sigma_{rz} = 0$, and σ_{rr} is specified as $p(z)$. We can give any value to σ_{rz} but will choose zero to simplify the algebra for presentation of the method. We use the properties of Fourier transforms discussed in section 7-8.2 to write the transformed problem as

$$\left(\frac{d^2}{dr^2} + \frac{1}{r}\frac{d}{dr} - \beta^2\right)^2 \mathcal{Z} = 0. \tag{13-8.3}$$

$$\Sigma_{rr} = \mathfrak{F}[p(z)] = P(\beta), \qquad r = R, \tag{13-8.4}$$

$$\Sigma_{rz} = 0, \qquad r = R \tag{13-8.5}$$

$$\Sigma_{rr} = -i\beta\left[\nu\left(\frac{d^2}{dr^2} + \frac{1}{r}\frac{d}{dr} - \beta^2\right) - \frac{d^2}{dr^2}\right]\mathcal{Z}, \tag{13-8.6}$$

$$\Sigma_{rz} = \frac{d}{dr}\left[(1 - \nu)\left(\frac{d^2}{dr^2} + \frac{1}{r}\frac{d}{dr} - \beta^2\right) + \beta^2\right]\mathcal{Z}. \tag{13-8.7}$$

We take the solution of equation 13-8.3 as

$$\mathcal{Z} = A\zeta_0(\beta r) + B\beta r\zeta_1(\beta r), \tag{13-8.8}$$

where $\zeta_n = I_n$ for interior problem, K_n for exterior problem.*
This form of solution is obtained by imposing a regularity condition upon $\mathcal{Z}$ and noting that $K_n(\beta r)$ becomes unbounded as $r \longrightarrow 0$ and $I_n(\beta r)$ becomes unbounded as $r \longrightarrow \infty$.

The modified Bessel functions have the following behavior as the argument approaches zero:

$$I_n(x) \sim \frac{[(1/2)x]^n}{\Gamma(n + 1)},$$

$$K_0(x) \sim -\ln x,$$

$$K_n(x) \sim \frac{(1/2)\Gamma(n)}{[(1/2)x]^n}.$$

The functions take the following form as the argument approaches infinity:

$$I_n(x) \sim \frac{e^x}{\sqrt{2\pi x}}\left\{1 - 0\left(\frac{1}{x}\right)\right\},$$

$$K_n(x) \sim \sqrt{\frac{\pi}{2x}}\, e^{-x}\left\{1 + 0\left(\frac{1}{x}\right)\right\}.$$

The recurrence relations given in section 7 are valid for ζ_n as well as I_n. We obtain directly from analogy to equation 13-7.10 the conditions upon

* The function K_n is the modified Bessel function of the second kind.

A and B:

$$A = -B\left[2(1-\nu) + \frac{\beta R\zeta_0(\beta R)}{\zeta_1(\beta R)}\right], \tag{13-8.9}$$

$$B = \frac{\beta R\zeta_1(\beta R)P(\beta)}{\beta^3\{2(1-\nu)\zeta_1^2(\beta R) + \beta^2R^2[\zeta_1^2(\beta R) - \zeta_0^2(\beta R)]\}}. \tag{13-8.10}$$

Substitution of these results into the transformed stress expressions Σ_{rr}, Σ_{rz}, $\Sigma_{\theta\theta}$, and Σ_{zz} gives the solution to the transformed problem. Σ_{rr} and Σ_{rz} are defined by Equations 13-8.6 and 13-8.7, respectively, and $\Sigma_{\theta\theta}$ and Σ_{zz} are as follows:

$$\Sigma_{\theta\theta} = -i\beta\left[\nu\left(\frac{d^2}{dr^2} + \frac{1}{r}\frac{d}{dr} - \beta^2\right) - \frac{1}{r}\frac{d}{dr}\right]\mathrm{Z}, \tag{13-8.11}$$

$$\Sigma_{zz} = -\beta\left[(2-\nu)\left(\frac{d^2}{dr^2} + \frac{1}{r}\frac{d}{dr} - \beta^2\right) + \beta^2\right]\mathrm{Z}. \tag{13-8.12}$$

The nonzero displacements are defined in a transformed form by

$$2GU_r(r, \beta) = -i\beta\frac{d\mathrm{Z}}{dr}, \tag{13-8.13}$$

$$2GU_z(r, \beta) = \left[2(1-\nu)\left(\frac{d^2}{dr^2} + \frac{1}{r}\frac{d}{dr} - \beta^2\right) + \beta^2\right]\mathrm{Z}. \tag{13-8.14}$$

The stresses and displacements are obtained by taking the inverse Fourier transform of the expressions. For example,

$$\sigma_{\theta\theta} = \frac{1}{\sqrt{2\pi}}\int_{-\infty}^{+\infty} \Sigma_{\theta\theta}e^{-i\beta z}\,d\beta. \tag{13-8.15}$$

It is obvious that the evaluation of these infinite integrals will not, in general, be a simple task and will require numerical integration. Examples are found in Sneddon's *Fourier Transforms*,* in the discussion of C. J. Tranter and J. W. Cragg's paper, and the external problem is discussed by Brewer and Little.† With large computers, this method remains a powerful tool for consideration of this class of problem.

9 Thick axisymmetrically loaded plate

We have considered the case of a cylindrical body which has a ratio of length to radius such that $l/R \gg 1$. The advantage of this restriction is that one can use a Saint-Venant approximation on the boundaries, $z = \pm l$. At

* I. N. Sneddon, *Fourier Transforms*, McGraw-Hill Book Company, 1951, pp. 504–510.

† W. V. Brewer and R. W. Little, "Interaction of an axisymmetrically loaded plate with the elastic space," *Int. J. Solid Structures*, **6**, 287–299, 1970.

the other end of the scale are problems where $l/R \ll 1$, such that the boundaries $r = R$ can be satisfied only approximately. At first glance this may appear to be too strong an approximation, but the problem was considered by Thomson and Tait in 1867,* and it was shown that local effects are negligible at twice the thickness of the plate from the edge.

In this case, it is desirable that the boundary function at $z = \pm l$ exhibit orthogonal properties. We are therefore motivated to consider solutions that have trigonometric-like behavior in the r direction and hyperbolic behavior in the z direction.

Consider the thick plate shown in Figure 13-3. The boundary condi-

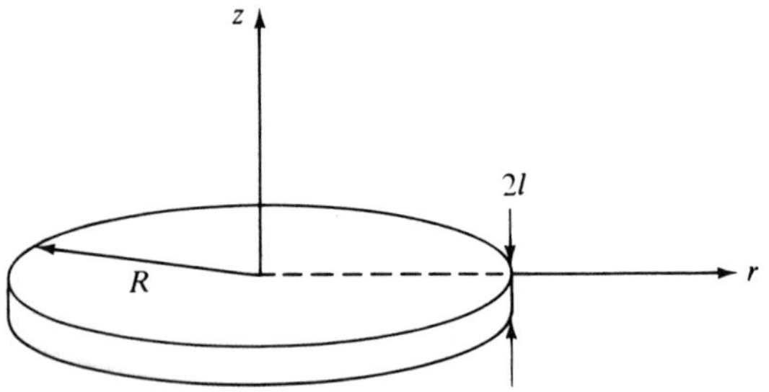

Figure 13-3

tions are as follows:

$$\begin{aligned} z &= +l, & \sigma_{zz} &= f_1(r), & \sigma_{zr} &= g_1(r), \\ z &= -l, & \sigma_{zz} &= f_2(r), & \sigma_{zr} &= g_2(r). \end{aligned} \tag{13-9.1}$$

Along the boundary $r = R$, satisfy equilibrium by using the Saint-Venant approximation. This problem is divided into two Love functions, one even in z and one odd in z. Noting that σ_{zz} is out of phase (evenness or oddness) with Z and σ_{zr} is in phase, the boundary conditions for the odd problem in Z become

$$z = \pm l, \qquad \sigma_{zz} = \tfrac{1}{2}(f_1 + f_2), \qquad \sigma_{rz} = \pm\tfrac{1}{2}(g_1 - g_2), \tag{13-9.2}$$

and for the even problem the boundary conditions become

$$z = \pm l, \qquad \sigma_{zz} = \pm\tfrac{1}{2}(f_1 - f_2), \qquad \sigma_{rz} = \tfrac{1}{2}(g_1 + g_2). \tag{13-9.3}$$

The final solution is, of course, the sum of the even solution and the odd solution. One might note that for the odd problem the equilibrium condition at $r = R$ is

$$\int_{-l}^{+l} \sigma_{rz}\, dz = 0, \qquad \int_{-l}^{+l} \sigma_{rr}\, dz = 0, \qquad \int_{-l}^{+l} z\,\sigma_{rr}\, dz = 0. \tag{13-9.4}$$

* Thomson and Tait, *Treatise on Natural Philosophy*, Oxford, 1867.

The last two boundary conditions are not truly equilibrium conditions but zero loading conditions. In a similar manner, the conditions at $r = R$ for the even problem are

$$2\pi R \int_{-l}^{+l} \sigma_{rz}\, dz = -2\pi \int_{0}^{R} (f_1 - f_2) r\, dr,$$
$$\int_{-l}^{+l} \sigma_{rr}\, dz = 0, \qquad \int_{-l}^{+l} \sigma_{rr} z\, dz = 0. \tag{13-9.5}$$

Some of these conditions can be met by the addition of polynomial solutions to the series solutions.

Let us consider the odd part for the following example:

$$\nabla^2\nabla^2 Z = 0, \tag{13-9.6}$$

$$z = \pm l, \qquad \sigma_{zz} = F(r), \qquad \sigma_{rz} = \pm G(r). \tag{13-9.7}$$

We take as a solution of 13-9.6 that

$$Z = \sum [A \operatorname{sh} \alpha z + B\alpha z \operatorname{ch} \alpha z] J_0(\alpha r). \tag{13-9.8}$$

The term $e^{\alpha z} J_0(\alpha r)$ is a harmonic function, and z times a harmonic function is biharmonic; therefore, 13-9.8 is indeed a solution to 13-9.6.

The boundary stresses are obtained by use of equations 13-6.3 and 13-6.4 and become

$$\sigma_{zz} = \sum \alpha^3\{B[(1 - 2\nu) \operatorname{ch} \alpha z - \alpha z \operatorname{sh} \alpha z] - A \operatorname{ch} \alpha z\} J_0(\alpha z),$$
$$\sigma_{rz} = \sum \alpha^3\{B[2\nu \operatorname{sh} \alpha z + \alpha z \operatorname{ch} \alpha z] + A \operatorname{sh} \alpha z\} J_1(\alpha r),$$
$$\sigma_{rr} = \sum \alpha^3\Big\{B\Big[J_0(\alpha r)\{(1 + 2\nu) \operatorname{ch} \alpha z + \alpha z \operatorname{sh} \alpha z\} \tag{13-9.9}$$
$$- \frac{J_1(\alpha r)}{\alpha r}\{\operatorname{ch} \alpha z + \alpha z \operatorname{sh} \alpha z\}\Big] + A\Big[\operatorname{ch} \alpha z \Big\{J_0(\alpha r) - \frac{J_1(\alpha r)}{\alpha r}\Big\}\Big]\Big\}.$$

The following integral formulas are applicable to Bessel functions;

$$\int x J_n(\alpha x) J_n(\beta x)\, dx = \frac{\beta x J_n(\alpha x) J_{n-1}(\beta x) - \alpha x J_{n-1}(\alpha x) J_n(\beta x)}{\alpha^2 - \beta^2}, \qquad \alpha \neq \beta, \tag{13-9.10}$$

$$\int x J_n^2(\alpha x)\, dx = \frac{x^2}{2}\{J_n^2(\alpha x) - J_{n-1}(\alpha x) J_{n+1}(\alpha x)\}. \tag{13-9.11}$$

Note that we could also have expressed these in terms of J_{n+1} by use of the recurrence relation

$$J_{n-1}(\alpha x) = \frac{2n}{\alpha x} J_n(\alpha x) - J_{n+1}(\alpha x). \tag{13-9.12}$$

We select our eigenvalues αR in equation 13-9.8 as either the zeros of J_0 or J_1, and have orthogonal functions for representation of σ_{zz} and σ_{rz} on the boundary. Each of these two choices leads to two different types of expansions of Bessel functions. An expansion generally termed a Fourier–Bessel expansion takes the form

$$f(x) = \sum_{m=1}^{\infty} a_m J_\nu(\alpha_m x), \tag{13-9.13}$$

where α_m are the roots of $J_\nu(\alpha) = 0$ and

$$a_m = \frac{2}{J^2_{\nu+1}(\alpha_m)} \int_0^1 tf(t) J_\nu(\alpha_m t)\, dt. \tag{13-9.14}$$

The second expansion is called a Dini's series of Bessel functions:

$$f(x) = \sum_{m=1}^{\infty} b_m J_\nu(\lambda_m x), \tag{13-9.15}$$

where λ_m are the roots of $\lambda J'_\nu(\lambda) + HJ_\nu(\lambda) = 0$, and

$$b_m = \frac{2\lambda_m^2}{(\lambda_m^2 - \nu^2)J_\nu^2(\lambda_m) + \lambda_m^2 J_\nu'^2(\lambda_m)} \int_0^1 tf(t) J_\nu(\lambda_m t)\, dt. \tag{13-9.16}$$

If $H \longrightarrow \infty$, the Fourier–Bessel series becomes a special case of the Dini's expansions.

Dini noticed in 1880 (*Serie di Fourier*, Pisa) that if $H + \nu = 0$, an initial term corresponding to a zero eigenvalue must be inserted in the series. He gave this term incorrectly, and it was corrected by Bridgeman.* The initial term is written as b_0 and is defined as

$$b_0 = 2(\nu + 1)x^\nu \int_0^1 t^{\nu+1} f(t)\, dt. \tag{13-9.17}$$

Further details of these expansions are found in Watson's treatise on the theory of Bessel functions.† In particular, if $H + \nu$ is negative, additional consideration must be given to Dini's expansion. Since this case will not arise in our work, no discussion of it is given here.

For the odd problem we select α to be the roots of $J_1(\alpha R) = 0$, so that the shear stress σ_{rz} will be zero on the boundary $r = R$.

The boundary conditions 13-9.7 become

$$\begin{aligned} \sigma_{zz}\big|_{z=\pm l} &= \sum \alpha^3 \{B[(1 - 2\nu)\operatorname{ch} \alpha l - \alpha l \operatorname{sh} \alpha l] - A \operatorname{ch} \alpha l\} J_0(\alpha r), \\ \sigma_{rz}\big|_{z=\pm l} &= \sum \alpha^3 \{B[2\nu \operatorname{sh} \alpha l + \alpha l \operatorname{ch} \alpha l] + A \operatorname{sh} \alpha l\} J_1(\alpha r). \end{aligned} \tag{13-9.18}$$

The expansions in equation 13-9.18 are a Fourier–Bessel expansion for σ_{rz} with $\nu = 1$ and a Dini's expansion for σ_{zz} with $H + \nu = 0$. The expansion for σ_{zz} falls into that special case where an initial constant term is needed. The physical significance of this constant is the measure of the total load applied at the top or bottom surfaces. The necessity of this constant is obvious in this case, as zero is a root of $J_1(\alpha R) = 0$ and $J_0(0) = 1$.

* Bridgeman, *Phil. Mag.*, **6**, XVI, 1908.

† G. N. Watson, *Theory of Bessel Functions*, Cambridge University Press, New York, 1966.

The zero term is added by use of equation 13-6.25. The relation between the constants in equation 13-6.25 is determined to satisfy the boundary condition on the radial stress at the edge $r = R$. We denote these constants as A_0 and B_0 to distinguish them from the terms in the series expansion. The normal stress σ_{zz}^0 corresponding to this additional term becomes

$$\sigma_{zz}^0 = -6A_0 + (14 - 10\nu)B_0 = 2\int_0^R rF(r)\,dr.$$

For $n \neq 0$, we identify the following constants:

$$C_n = \alpha_n^3\{B_n[(1 - 2\nu)\operatorname{ch}\alpha_n l - \alpha_n l \operatorname{sh}\alpha_n l] - A_n \operatorname{ch}\alpha_n l\}, \qquad (13\text{-}9.19)$$

$$D_n = \alpha_n^3\{B_n[2\nu \operatorname{sh}\alpha_n l + \alpha_n l \operatorname{ch}\alpha_n l] + A_n \operatorname{sh}\alpha_n l\}. \qquad (13\text{-}9.20)$$

The series become

$$\sigma_{zz}^0 + \sum_{n=1}^{\infty} C_n J_0(\alpha_n r) = F(r), \qquad (13\text{-}9.21)$$

$$\sum_{n=1}^{\infty} D_n J_1(\alpha_n r) = G(r). \qquad (13\text{-}9.22)$$

Using the orthogonality properties, the constants C_n and D_n are obtained by the following equations:

$$C_n = \frac{2}{R^2 J_0^2(\alpha_n R)} \int_0^R F(r) J_0(\alpha_n r) r\,dr, \qquad (13\text{-}9.23)$$

$$D_n = \frac{2}{R^2 J_0^2(\alpha_n R)} \int_0^R G(r) J_1(\alpha_n r) r\,dr. \qquad (13\text{-}9.24)$$

The boundary conditions at $r = R$ must still be examined. σ_{rz} was chosen equal to zero and the integral of $z\sigma_{rr}$ is zero as it is an odd function. In general, the integral of σ_{rr} across this edge will not be zero and this average radial stress must be removed.* This is the final condition that will be used to select the constants A_0 and B_0. These constants enter into the radial stress term as follows:

$$\sigma_{rr}^0 = 3A_0 + (10\nu - 2)B_0. \qquad (13\text{-}9.25)$$

The two equations for their selection are

$$-6A_0 + (14 - 10\nu)B_0 = 2\int_0^R rF(r)\,dr,$$

$$\int_{-l}^{+l} \sigma_{rr}\,dz\,\Big|_{r=R} = 0 \Longrightarrow [3A_0 - (10\nu - 2)B_0]2l \qquad (13\text{-}9.26)$$

$$+ \sum \int_{-l}^{+l} \alpha_n^3 J_0(\alpha_n R)[B_n\{(1 + 2\nu)\operatorname{ch}\alpha z + \alpha_n z \operatorname{sh}\alpha_n z\}$$

$$+ A_n \operatorname{ch}\alpha_n z]\,dz = 0.$$

* This assumes that the edge is stress free. If the edge is built in, this average stress will not be zero and a condition on displacements must be considered. $\int z\sigma_{rr}\,dz$ would not necessarily be zero and would correspond to a "moment" in plate theory. A zero moment condition may or may not be imposed, depending upon edge conditions.

The balance of the solution is now a matter of algebra and, of course, numerically summing the series expressions for the stresses and displacements.

10 Axisymmetrically loaded plate with a hole

Consider the thick annulus shown in Figure 13-4. We can develop a method to analyze a body of this geometry, assuming that we can approximate the boundary conditions on the radial edges in the Saint-Venant sense.

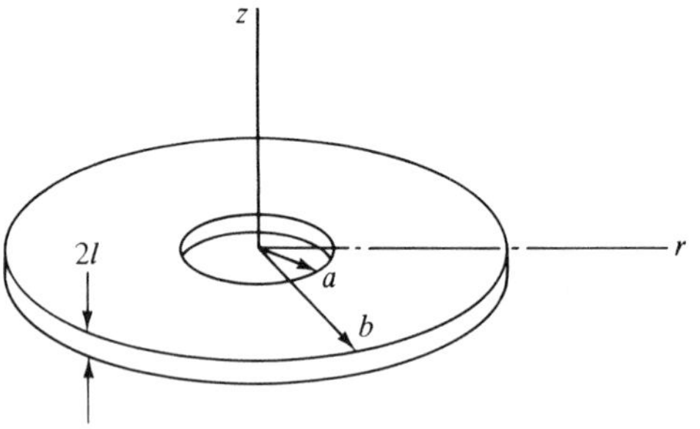

Figure 13-4

This solution is a poorer approximation than that discussed in section 9, because these boundary approximations produce errors over a greater percentage of the body, and the nature of the possible singularity associated with the hole has not been fully examined. On the boundaries z equals constant orthogonal functions are used. The orthogonality property is more complex for this case and involves Bessel functions of both the first and second-kind.

Let $C_n(x)$ be defined to be a cylinder function of order n:

$$C_n(x) = J_n(x) + \beta_n Y_n(x), \tag{13-10.1}$$

where β_n is a real constant and J_n and Y_n are the Bessel functions of the first and second kind. The following relationship holds:

$$\begin{aligned} &\int_a^b xC_n(\lambda_j x)C_n(\lambda_k x)\,dx = 0, \qquad j \neq k, \\ &= \left[\frac{x^2}{2}\left\{\left(1 - \frac{n^2}{\lambda_j^2 x^2}\right)C_n^2(\lambda_j x) + \frac{1}{\lambda_j^2}C_n'^2(\lambda_j x)\right\}\right]_a^b, \qquad j = k, \end{aligned} \tag{13-10.2}$$

if λ_j is a real zero of

$$h_1 \lambda C_{n+1}(\lambda b) - h_2 C_n(\lambda b) = 0, \tag{13-10.3}$$

and if k_1 and k_2 exist (both not zero) so that, for all λ_j,

$$k_1\lambda_j C_{n+1}(\lambda_j a) - k_2 C_n(\lambda_j a) = 0. \qquad (13\text{-}10.4)$$

The Bessel functions of the second kind satisfy the same recurrence relations as those of the first kind, and we can write these as

$$\begin{aligned} C_{n-1}(x) + C_{n+1}(x) &= \frac{2n}{x}C_n(x), \\ C_{n-1}(x) - C_{n+1}(x) &= 2C_n'(x), \\ C_n'(x) &= C_{n-1}(x) - \frac{n}{x}C_n(x), \\ C_n'(x) &= -C_{n+1}(x) + \frac{n}{x}C_n(x). \end{aligned} \qquad (13\text{-}10.5)$$

Let us examine a general solution of the thick annulus to see how these properties are used. The solution of the biharmonic equation is written

$$Z = \sum [A \operatorname{sh} \alpha z + B\alpha z \operatorname{ch} \alpha z]C_0(\alpha r), \qquad (13\text{-}10.6)$$

if we consider the odd problem in z. The boundary stresses are

$$\begin{aligned} \sigma_{zz} &= \sum \alpha^3\{B[(1-2\nu)\operatorname{ch}\alpha z - \alpha z \operatorname{sh}\alpha z] - A\operatorname{ch}\alpha z\}C_0(\alpha r), \\ \sigma_{rz} &= \sum \alpha^3\{B[2\nu \operatorname{sh}\alpha z + \alpha z\operatorname{ch}\alpha z] + A\operatorname{sh}\alpha z\}C_1(\alpha r), \\ \sigma_{rr} &= \sum \alpha^3\Big\{B\Big[C_0(\alpha r)\{(1+2\nu)\operatorname{ch}\alpha z + \alpha z\operatorname{sh}\alpha z\} \\ &\quad - \frac{C_1(\alpha r)}{\alpha r}\{\operatorname{ch}\alpha z + \alpha z \operatorname{sh}\alpha z\}\Big] \\ &\quad + A\Big[\operatorname{ch}\alpha z\Big(C_0(\alpha r) - \frac{C_1(\alpha r)}{\alpha r}\Big)\Big]\Big\}. \end{aligned} \qquad (13\text{-}10.7)$$

We choose α and β such that C_1 is zero at $r = a$ and $r = b$, and thereby set the shear stress zero at those two boundaries. This yields the following two equations:

$$\begin{aligned} J_1(\alpha a) + \beta Y_1(\alpha a) &= 0, \\ J_1(\alpha b) + \beta Y_1(\alpha b) &= 0. \end{aligned} \qquad (13\text{-}10.8)$$

The solution of these homogeneous equations requires that

$$\beta = -\frac{J_1(\alpha a)}{Y_1(\alpha a)} = -\frac{J_1(\alpha b)}{Y_1(\alpha b)}, \qquad (13\text{-}10.9)$$

and that α be the roots of

$$J_1(\alpha a)Y_1(\alpha b) - J_1(\alpha b)Y_1(\alpha a) = 0. \qquad (13\text{-}10.10)$$

We now prescribe σ_{zz} and σ_{rz} on the boundaries $z = \pm l$. Note that in this example σ_{zz} is an even function of z and σ_{rz} is an odd function of z.

The orthogonality condition, equation 13-10.2, becomes

$$\int_a^b rC_n(\alpha_j r)C_n(\alpha_k r)\,dr = 0, \qquad j \neq k, \qquad (13\text{-}10.11)$$

and the normalization constants are

$$\left.\begin{aligned}\int_a^b rC_0^2(\alpha_j r)\,dr &= \\ \int_a^b rC_1^2(\alpha_j r)\,dr &= \end{aligned}\right\} \left[\frac{r^2}{2}C_0^2(\alpha_j r)\right]_a^b. \tag{13-10.12}$$

We use these conditions and proceed with the solution as was done in section 9. It can be seen that zero eigenvalues do not arise in this case and the series are complete without additional eigenfunctions. The solutions discussed in section 6 are, in general, needed to adjust the average radial stress on bending moment on the inner and outer edges. These solutions will include the singular part, as is shown, for example, in equation 13-6.33. To examine the transcendental equation for zero eigenvalues, the following properties of Bessel functions are useful:

$$x \longrightarrow 0 \qquad \begin{aligned} J_0(x) &\longrightarrow 1, \\ Y_0(x) &\longrightarrow \ln x, \\ J_1(x) &\longrightarrow \frac{x}{2}, \\ Y_1(x) &\longrightarrow \frac{1}{2x}. \end{aligned} \tag{13-10.13}$$

The boundary conditions for the odd problem, equation 13-10.6, are

$$\begin{aligned} \sigma_{zz} &= f(r), & z &= \pm l, \\ \sigma_{rz} &= \pm g(r), & z &= \pm l, \\ \sigma_{rr} &= \sigma_{rz} = 0, & r &= a, b. \end{aligned} \tag{13-10.14}$$

Defining two constants D_n and E_n as

$$\begin{aligned} D_n &= \alpha_n^3\{B_n[(1-2\nu)\operatorname{ch}\alpha_n l - \alpha_n l \operatorname{sh}\alpha_n l] - A_n \operatorname{ch}\alpha_n l\}, \\ E_n &= \alpha_n^3\{B_n[2\nu \operatorname{sh}\alpha_n l + \alpha_n l \operatorname{ch}\alpha_n l] + A_n \operatorname{sh}\alpha_n l\}, \end{aligned} \tag{13-10.15}$$

we write the values of these in terms of the boundary stresses:

$$\begin{aligned} D_n &= \frac{2}{b^2C_0^2(\alpha_n b) - a^2C_0^2(\alpha_n a)}\int_a^b rC_0(\alpha_n r)f(r)\,dr, \\ E_n &= \frac{2}{b^2C_0^2(\alpha_n b) - a^2C_0^2(\alpha_n a)}\int_a^b rC_1(\alpha_n r)g(r)\,dr. \end{aligned} \tag{13-10.16}$$

The edge boundary conditions remain to be satisfied, because the average value of σ_{rr} will not be zero. Equations 13-6.25 and 13-6.29 yield a solution comparable to the Lamé solution (see section 8-2.1), and the Lamé solution is used to remove this average stress.

The complete stress function becomes

$$Z = A_0[\tfrac{5}{3}(2 - \nu)z^3 - \tfrac{5}{2}(1 - \nu)zr^2] + B_0 z \ln r + \sum_{n=1}^{\infty} [A_n \operatorname{sh} \alpha_n z + B_n \alpha_n z \operatorname{ch} \alpha_n z] C_0(\alpha_n r). \qquad (13\text{-}10.17)$$

The expressions for the stresses σ_{zz} and σ_{rz} remain the same as given in 13-10.7, but the expression for σ_{rr} becomes

$$\sigma_{rr} = 5(1 + \nu)A_0 + \frac{B_0}{r^2} + \sum_{n=1}^{\infty} \alpha_n^3 \Big\{ B_n \Big[C_0(\alpha_n r)\{(1 + 2\nu) \operatorname{ch} \alpha_n z + \alpha_n z \operatorname{sh} \alpha_n z\} - \frac{C_1(\alpha_n r)}{\alpha_n r}\{\operatorname{ch} \alpha_n z + \alpha_n z \operatorname{sh} \alpha_n z\} \Big] + A_n \Big[\operatorname{ch} \alpha_n z \Big(C_0(\alpha_n r) - \frac{C_1(\alpha_n r)}{\alpha_n r} \Big) \Big] \Big\}. \qquad (13\text{-}10.18)$$

The boundary condition on the edge yields

$$\int \sigma_{rr}\, dz = 0, \qquad 10(1 + \nu) l A_0 + \frac{2l}{r^2} B_0 + \sum_{n=1}^{\infty} \frac{E_n}{\alpha_n} C_0(\alpha_n r) \Big|_a^b = 0. \qquad (13\text{-}10.19)$$

The constants E_n have been determined by Equation 13-10.16, and A_0 and B_0 may now be determined by equation 13-10.19, completing the solution.

11 Hankel transform methods

The combination of operational methods and the Love strain function presents a powerful approach to solutions of axisymmetric problems for the half-space. We now consider an elastic body occupying the region $0 \leq r < \infty$, $0 \leq z < \infty$. The Love strain function will be defined by the biharmonic equation, and a stress or displacement distribution will be prescribed on the surface $z = 0$.

The particular operational method employed is that of the Hankel transform. We shall digress from this problem to present some of the properties of the Hankel transform. The Hankel transform of order n of a function $f(r)$ is defined as

$$\bar{f}^n(\xi) = \mathcal{H}_n\{f(r); \xi\} = \int_0^{\infty} r f(r) J_n(\xi r)\, dr. \qquad (13\text{-}11.1)$$

The transform becomes a function of the transformed variable ξ when this integral exists. The Hankel inversion theorem yields the inverse of the

Hankel transform:

$$f(r) = \int_0^\infty \xi \bar{f}^n(\xi) J_n(\xi r)\, d\xi. \qquad (13\text{-}11.2)$$

We note that there is a symmetric relation between the function and its transform.

Sneddon* lists some of the elementary properties of the Hankel transform as follows:

$$\begin{aligned} \mathcal{H}_n[f(ax); \xi] &= \int_0^\infty x f(ax) J_n(x\xi)\, dx, \qquad a \neq 0, \\ &= \frac{1}{a^2}\mathcal{H}_n\left[f(x); \frac{\xi}{a}\right] = \frac{1}{a^2}\bar{f}^n\left(\frac{\xi}{a}\right). \end{aligned} \qquad (13\text{-}11.3)$$

From the recurrence relation

$$J_{n-1}(\xi x) - \frac{2n}{\xi x} J_n(\xi x) + J_{n+1}(\xi x) = 0,$$

we obtain, by multiplying through by $xf(x)$ and integrating,

$$\mathcal{H}_n[x^{-1}f(x); \xi] = \frac{\xi}{2n}\{\bar{f}^{n-1}(\xi) + \bar{f}^{n+1}(\xi)\}. \qquad (13\text{-}11.4)$$

The transform of the derivative of a function can now be calculated:

$$\bar{f}'^n(\xi) = \int_0^\infty r \frac{df}{dr} J_n(\xi r)\, dr. \qquad (13\text{-}11.5)$$

Integrating by parts yields

$$\begin{aligned} \bar{f}'^n(\xi) &= [rfJ_n(\xi r)]_0^\infty - \int_0^\infty f \frac{d}{dr}[rJ_n(\xi r)]\, dr \\ &= [rfJ_n(\xi r)]_0^\infty - \int_0^\infty fJ_n(\xi r)\, dr - \int_0^\infty rfJ'_n(\xi r)\, dr. \end{aligned} \qquad (13\text{-}11.6)$$

The derivative of $J'_n(\xi r)$ is given by the relation

$$J'_n(\xi r) = \xi J_{n-1}(\xi r) - \frac{n}{r} J_n(\xi r) = -\xi J_{n+1}(\xi r) + \frac{n}{r} J_n(\xi r). \qquad (13\text{-}11.7)$$

We write equation 13-11.6 as

$$\begin{aligned} \bar{f}'^n(\xi) &= [rfJ_n(\xi r)]_0^\infty - (1 - n)\int_0^\infty fJ_n(\xi r)\, dr - \xi \int_0^\infty rfJ_{n-1}(\xi r)\, dr \\ &= [rfJ_n(\xi r)]_0^\infty - (1 - n)\mathcal{H}_n[r^{-1}f; \xi] - \xi \bar{f}^{n-1}(\xi). \end{aligned} \qquad (13\text{-}11.8)$$

Using equation 13-11.4 yields the following equation, when the restrictions $\lim_{r\to 0} r^{n+1}f(r) = 0$ and $\lim_{r\to\infty} r^{1/2}f(r) = 0$ are applied:

* I. N. Sneddon, *Fourier Transforms*, McGraw-Hill Book Company, New York, 1951.

$$\bar{f}'^{n}(\xi) = -\xi\left[\frac{n+1}{2n}\bar{f}^{n-1}(\xi) - \frac{n-1}{2n}\bar{f}^{n+1}(\xi)\right], \qquad n \neq 0. \tag{13-11.9}$$

For the special case of $n = 1$, we obtain

$$\bar{f}'^{1}(\xi) = -\xi\bar{f}^{0}(\xi). \tag{13-11.10}$$

We can obtain a more useful differential formula by considering the following Hankel transform:

$$\mathcal{H}_\nu\left[r^{\nu-1}\frac{\partial}{\partial r}\{r^{1-\nu}f(r)\};\xi\right] = \int_0^\infty r^\nu J_\nu(\xi r)\frac{\partial}{\partial r}[r^{1-\nu}f(r)]\,dr. \tag{13-11.11}$$

Integration by parts yields

$$\mathcal{H}_\nu\left[r^{\nu-1}\frac{\partial}{\partial r}\{r^{1-\nu}f(r)\};\xi\right] = [rJ_\nu(\xi r)f(r)]_0^\infty - \int_0^\infty r^{1-\nu}f(r)\frac{\partial}{\partial r}[r^\nu J_\nu(\xi r)]\,dr. \tag{13-11.12}$$

The first term vanishes if the restrictions $\lim_{r\to 0} r^{\nu+1}f(r) = 0$ and $\lim_{r\to\infty} r^{1/2}f(r) = 0$ hold.

Equation 13-11.12 becomes

$$\mathcal{H}_\nu\left[r^{\nu-1}\frac{\partial}{\partial r}\{r^{1-\nu}f(r)\};\xi\right] = -\xi\mathcal{H}_{\nu-1}[f(r);\xi], \tag{13-11.13}$$

where use has been made of the relation

$$\frac{\partial}{\partial r}[r^\nu J_\nu(\xi r)] = \xi r^\nu J_{\nu-1}(\xi r). \tag{13-11.14}$$

Note that if we let $\nu = 1$, we obtain the same relation obtained in equation 13-11.10. One final observation should be made before we put this formula to good use:

$$\begin{aligned} J_{-1}(\xi x) &= -J_1(\xi x), \\ \mathcal{H}_{\nu-1}[f(r);\xi] &= -\mathcal{H}_{\nu+1}[f(r);\xi], \qquad \text{when } \nu = 0. \end{aligned} \tag{13-11.15}$$

The Laplace operator involves derivatives of r in the form

$$\frac{\partial^2}{\partial r^2} + \frac{1}{r}\frac{\partial}{\partial r} = \frac{1}{r}\frac{\partial}{\partial r}\left(r\frac{\partial}{\partial r}\right). \tag{13-11.16}$$

Equation 13-11.13 yields

$$\mathcal{H}_0\left[\frac{1}{r}\frac{\partial}{\partial r}\left\{r\left[\frac{\partial f}{\partial r}\right]\right\};\xi\right] = \xi\mathcal{H}_1\left[\frac{\partial f}{\partial r};\xi\right]. \tag{13-11.17}$$

Equation 13-11.10 yields

$$\mathcal{H}_0\left[\frac{1}{r}\frac{\partial}{\partial r}\left\{r\frac{\partial f}{\partial r}\right\};\xi\right] = -\xi^2\mathcal{H}_0[f;\xi]. \tag{13-11.18}$$

12 Axisymmetric problem of a half-space

We now return to the problem of the half-space under some axisymmetric loading at $z = 0$. The problem is mathematically stated as

$$\nabla^2\nabla^2 Z = 0. \qquad (13\text{-}12.1)$$

Taking the Hankel transform of order zero yields the following transformed problem:

$$\left(\frac{d^2}{dz^2} - \xi^2\right)^2 \bar{Z}^0 = 0, \qquad (13\text{-}12.2)$$

where $\bar{Z}^0$ is the transform of Z. The solution of 13-12.2 is

$$\bar{Z}^0 = (A + B\xi z)e^{-\xi z}, \qquad (13\text{-}12.3)$$

where the coefficients of the positive exponential are taken as zero, so that the stresses and displacements remain bounded at infinity. The boundary stresses and displacements can be expressed in terms of $\bar{Z}^0$ by taking the proper transform of equations 13-6.3. The boundary functions are u_r, u_z, σ_{rz}, and σ_{zz}. The boundary functions u_z and σ_{zz} can be transformed into functions related to $\bar{Z}^0$ by use of the Hankel transform of order zero. u_r and σ_{rz} will involve the transform of order 1:

$$\begin{aligned} \mathcal{H}_0\{2Gu_z;\xi\} &= \mathcal{H}_0\left\{\left[2(1-\nu)\nabla^2 - \frac{\partial^2}{\partial z^2}\right]Z;\xi\right\}, \\ 2G\bar{u}_z^0 &= -\xi^2 2(1-\nu)\bar{Z}^0 + (1-2\nu)\frac{d^2\bar{Z}^0}{dz^2}, \end{aligned} \qquad (13\text{-}12.4)$$

$$\begin{aligned} \mathcal{H}_1\{2Gu_r;\xi\} &= \mathcal{H}_1\left\{-\frac{\partial}{\partial r}\left(\frac{\partial Z}{\partial z}\right);\xi\right\} = \xi\mathcal{H}_0\left\{\frac{\partial Z}{\partial z};\xi\right\}, \\ 2G\bar{u}_r^1 &= \xi\frac{d\bar{Z}^0}{dz}, \end{aligned} \qquad (13\text{-}12.5)$$

$$\begin{aligned} \mathcal{H}_0\{\sigma_{zz};\xi\} &= \mathcal{H}_0\left\{\frac{\partial}{\partial z}\left[(2-\nu)\nabla^2 - \frac{\partial^2}{\partial z^2}\right]Z;\xi\right\}, \\ \bar{\sigma}_{zz}^0 &= -\xi^2(2-\nu)\frac{d\bar{Z}^0}{dz} + (1-\nu)\frac{d^3\bar{Z}^0}{dz^3}, \end{aligned} \qquad (13\text{-}12.6)$$

$$\begin{aligned} \mathcal{H}_1\{\sigma_{rz};\xi\} &= \mathcal{H}_1\left\{\frac{\partial}{\partial r}\left[(1-\nu)\nabla^2 - \frac{\partial^2}{\partial z^2}\right]Z;\xi\right\} \\ &= -\xi\mathcal{H}_0\left\{\left[(1-\nu)\nabla^2 - \frac{\partial^2}{\partial z^2}\right]Z;\xi\right\}, \\ \bar{\sigma}_{rz}^1 &= \xi^3(1-\nu)\bar{Z}^0 + \nu\xi\frac{d^2\bar{Z}^0}{dz^2}. \end{aligned} \qquad (13\text{-}12.7)$$

The superscript on the stress or displacement component denotes the order of the Hankel transform used.

The boundary conditions specified are now expressed in terms of the transformed Love function. Consider, for example, the case where the stresses σ_{zz} and σ_{rz} are to be specified on the boundary $z = 0$. At $z = 0$,

$$\sigma_{zz} = f(r), \qquad \sigma_{rz} = g(r). \tag{13-12.8}$$

The boundary conditions of the transformed problem are

$$\begin{aligned} \left[-\xi^2(2-\nu)\frac{d\bar{Z}^0}{dz} + (1-\nu)\frac{d^3\bar{Z}^0}{dz^3}\right]_{z=0} &= \bar{f}^0(\xi) \\ &= \int_0^\infty rf(r)J_0(\xi r)\,dr, \\ \left[\xi^3(1-\nu)\bar{Z}^0 + \nu\xi\frac{d^2\bar{Z}^0}{dz^2}\right]_{z=0} &= \bar{g}^1(\xi) \\ &= \int_0^\infty rg(r)J_1(\xi r)\,dr. \end{aligned} \tag{13-12.9}$$

Applying the first of the boundary equations, 13-12.9, to 13-12.3 yields

$$\begin{aligned} \xi^3(2-\nu)(A-B) - \xi^3(1-\nu)(A-3B) &= \bar{f}^0(\xi), \\ A + (1-2\nu)B &= \frac{1}{\xi^3}\bar{f}^0. \end{aligned} \tag{13-12.10}$$

The second boundary condition yields

$$\begin{aligned} \xi^3(1-\nu)A + \nu\xi^3(A-2B) &= \bar{g}^1(\xi), \\ A - 2\nu B &= \frac{1}{\xi^3}\bar{g}^1(\xi). \end{aligned} \tag{13-12.11}$$

Solving for A and B yields

$$\begin{aligned} A &= \frac{1}{\xi^3}[2\nu\bar{f}^0 + (1-2\nu)\bar{g}^1], \\ B &= \frac{1}{\xi^3}[\bar{f}^0 - \bar{g}^1]. \end{aligned} \tag{13-12.12}$$

The transformed Love function becomes

$$\bar{Z}^0 = \frac{1}{\xi^3}\{2\nu\bar{f}^0 + (1-2\nu)\bar{g}^1 + (\bar{f}^0 - \bar{g}^1)\xi z\}e^{-\xi z}. \tag{13-12.13}$$

This function cannot be used directly in the inverse transform to compute the Love function due to the presence of the ξ^{-3} factor. We can, however, compute the stresses and displacements by taking the inverse transforms of equations 13-12.4 to 13-12.7 after substitution of equation 13-12.13. The two remaining stresses σ_{rr} and $\sigma_{\theta\theta}$ require more consideration.

If the expressions for σ_{rr} and $\sigma_{\theta\theta}$ in equation 13-6.3 are added, we obtain

$$\sigma_{rr} + \sigma_{\theta\theta} = \frac{\partial}{\partial z}\left\{\frac{\partial^2 Z}{\partial z^2} - (1-2\nu)\nabla^2 Z\right\}. \tag{13-12.14}$$

This expression lends itself well to the Hankel transform of order zero, whereas the individual expressions for the stresses do not. Taking this transform yields

$$\overline{(\sigma_{rr} + \sigma_{\theta\theta})}^0 = 2\nu \frac{d^3\bar{Z}^0}{dz^3} + (1 - 2\nu)\xi^2 \frac{d\bar{Z}^0}{dz}. \qquad (13\text{-}12.15)$$

The equation of equilibrium in cylindrical coordinates is

$$\frac{\partial}{\partial r}(r^2\sigma_{rr}) = r(\sigma_{rr} + \sigma_{\theta\theta}) - r^2 \frac{\partial \sigma_{rz}}{\partial z}. \qquad (13\text{-}12.16)$$

Equation 13-12.7 and equation 13-12.15 can be inverted and substituted into 13-12.16, yielding

$$\frac{\partial}{\partial r}(r^2\sigma_{rr}) = r \int_0^\infty \left[2\nu \frac{d^3\bar{Z}^0}{dz^3} + (1 - 2\nu)\xi^2 \frac{d\bar{Z}^0}{dz}\right] \xi J_0(\xi r)\, d\xi$$
$$- r^2 \int_0^\infty \left[\xi^3(1 - \nu) \frac{d\bar{Z}^0}{dz} + \nu\xi \frac{d^3\bar{Z}^0}{dz^3}\right] \xi J_1(\xi r)\, d\xi. \qquad (13\text{-}12.17)$$

Integrating with respect to r yields

$$\sigma_{rr} = -\frac{1}{r} \int_0^\infty \xi^2 \frac{d\bar{Z}^0}{dz} J_1(\xi r)\, d\xi + \int_0^\infty \left[\xi^2(1 - \nu) \frac{d\bar{Z}^0}{dz} + \nu \frac{d^3\bar{Z}^0}{dz^3}\right] \xi J_0(\xi r)\, d\xi. \qquad (13\text{-}12.18)$$

The following relations were used in this integration:

$$\int_0^r rJ_0(\xi r)\, dr = \frac{r}{\xi} J_1(\xi r),$$

$$\int_0^r r^2 J_1(\xi r)\, dr = \frac{2r}{\xi^2} J_1(\xi r) - \frac{r^2}{\xi} J_0(\xi r).$$

$\sigma_{\theta\theta}$ can now be calculated by obtaining the difference between the inverse transform of 13-12.15 and equation 13-12.18:

$$\sigma_{\theta\theta} = \frac{1}{r} \int_0^\infty \xi^2 \frac{d\bar{Z}^0}{dz} J_1(\xi r)\, d\xi + \nu \int_0^\infty \left[\frac{d^3\bar{Z}^0}{dz^3} - \xi^2 \frac{d\bar{Z}^0}{dz}\right] \xi J_0(\xi r)\, d\xi. \qquad (13\text{-}12.19)$$

In general the calculation of these inverses is not simple and requires numerical integration. A special case will be given below, leading to the solution of the Boussinesq problem of a concentrated load on the half-space.

13 Boussinesq problem

Boussinesq considered the problem of a concentrated force on the half-space in his paper on potentials in 1885.* We shall consider this problem

* J. Boussinesq, *Applications des potentiels à l'étude de l'équilibre et du mouvement des solides élastiques*, Gauthier-Villars, Paris, 1885.

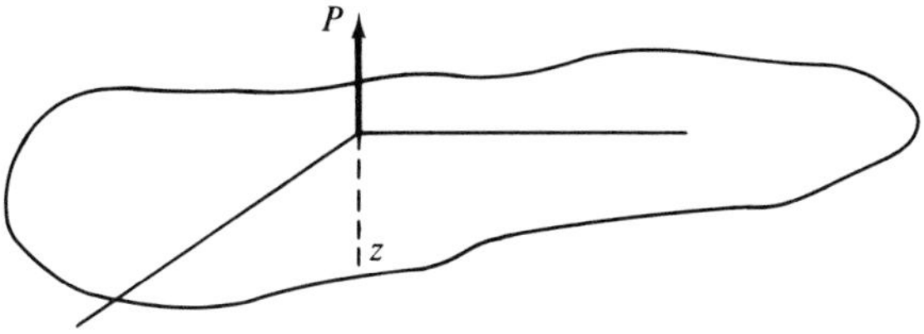

Figure 13-5

as an example of the use of the Hankel transforms discussed in section 11. The boundary conditions on the surface of the half-space are

$$0 < r < a, \qquad z = 0, \qquad \sigma_{zz} = \lim_{a \to 0} \frac{P}{\pi a^2}, \qquad \sigma_{rz} = 0. \tag{13-13.1}$$

Taking the Hankel transform of order zero of $\sigma_{zz}|_{z=0}$ yields

$$\bar{f}^0 = \lim_{a \to 0} \int_0^a \frac{P}{\pi a^2} r J_0(\xi r)\, dr = \lim_{a \to 0} \frac{P}{\pi a^2}\left[\frac{a}{\xi} J_1(\xi a)\right] = \frac{P}{2\pi}. \tag{13-13.2}$$

Substituting equation 13-13.2 into equation 13-12.13 gives the following expression for the transformed Love function:

$$\bar{Z}^0 = \frac{P}{2\pi \xi^3}[2\nu + \xi z]e^{-\xi z}. \tag{13-13.3}$$

Taking the inverse transform 13-12.6 and 13-12.7 yields

$$\sigma_{zz} = \int_0^\infty \frac{P\xi}{2\pi}\{1 + \xi z\}e^{-\xi z} J_0(\xi r)\, d\xi, \tag{13-13.4}$$

$$\sigma_{rz} = \int_0^\infty \frac{P\xi}{2\pi} \xi z e^{-\xi z} J_1(\xi r)\, d\xi. \tag{13-13.5}$$

The two following integrals are useful for evaluation of these Hankel inverses:

$$\begin{aligned} \int_0^\infty x^n e^{-\lambda x} J_m(\alpha x)\, dx &= \frac{(n - m)!}{(\lambda^2 + \alpha^2)^{1/2(n+1)}} P_n^m\left(\frac{\lambda}{(\lambda^2 + \alpha^2)^{1/2}}\right), \qquad m \leq n, \\ &= \frac{(n + m)!}{(\lambda^2 + \alpha^2)^{1/2(n+1)}} P_n^{-m}\left(\frac{\lambda}{(\lambda^2 + \alpha^2)^{1/2}}\right), \qquad m > n, \end{aligned} \tag{13-13.6}$$

where P_n^m is the associated Legendre function.

Consider, for example, the case when $m = 0, n = 1, \lambda = z$, and $\alpha = r$:

$$\int_0^\infty x e^{-zx} J_0(xr)\, dx = \frac{1}{r^2 + z^2} P_1^0\left(\frac{z}{(r^2 + z^2)^{1/2}}\right).$$

$P_1(z) = z$; therefore,

$$P_1^0\left(\frac{z}{(r^2 + z^2)^{1/2}}\right) = \frac{z}{(r^2 + z^2)^{1/2}},$$

and the integral becomes

$$\int_0^\infty xe^{-zx}J_0(xr)\,dx = \frac{z}{(r^2+z^2)^{3/2}}. \qquad (13\text{-}13.7)$$

If we differentiate both sides of 13-13.7 with respect to z, we obtain

$$\int_0^\infty x^2e^{-zx}J_0(xr)\,dx = \frac{2z^2-r^2}{(z^2+r^2)^{5/2}}. \qquad (13\text{-}13.8)$$

Differentiating both sides of 13-13.7 with respect to r yields

$$\int_0^\infty x^2e^{-zx}J_1(xr)\,dx = \frac{3rz}{(r^2+z^2)^{5/2}}. \qquad (13\text{-}13.9)$$

Using 13-13.7 and 13-13.8 in 13-13.4 yields

$$\sigma_{zz} = \left[\frac{z}{(r^2+z^2)^{3/2}} + \frac{2z^3-r^2z}{(r^2+z^2)^{5/2}}\right] = \frac{3P}{2\pi}\frac{z^3}{R^5}, \qquad (13\text{-}13.10)$$

where $R = (r^2+z^2)^{1/2}$. In a similar manner, we obtain

$$\sigma_{rz} = \frac{3}{2\pi}\frac{Prz^2}{R^5}, \qquad (13\text{-}13.11)$$

$$\sigma_{rr} = \frac{P}{2\pi}\left\{\frac{3r^2z}{R^5} - \frac{(1-2\nu)}{R^2+Rz}\right\}, \qquad (13\text{-}13.12)$$

$$\sigma_{\theta\theta} = \frac{(1-2\nu)P}{2\pi}\left\{-\frac{z}{R^3} + \frac{1}{R^2+Rz}\right\}, \qquad (13\text{-}13.13)$$

$$u_r = \frac{P}{4\pi G}\left\{-\frac{rz}{R^3} + \frac{(1-2\nu)r}{R^2+Rz}\right\}, \qquad (13\text{-}13.14)$$

$$u_z = -\frac{P}{4\pi G}\left\{\frac{2(1-\nu)}{R} + \frac{z^2}{R^3}\right\}. \qquad (13\text{-}13.15)$$

The last expressions requires integrations of equation 13-13.7 with respect to z. Lamb* and Terazawa constructed the solution in this form.

This solution can be superimposed to give other loadings on the half-space by replacing r with $(x^2+y^2)^{1/2}$ and shifting the origin and integrating.

14 Contact problems

One of the most practical problems requiring solutions in the half-space is when the deformation u_z is specified over some part of the boundary. One encounters this types of problem when punches of some shape are brought into contact with the half-space; this is therefore referred to as a

* H. Lamb, *Proc. London Math. Soc.*, **34**, 276, 1902; and K. Terazawa, *J. Coll. Sci. Univ. Tokyo*, **37**, 1916.

contact problem. This problem involves a mixed specification of boundary condition; that is, we shall specify u_z over part of the boundary and σ_{zz} over the rest. When the punch is considered frictionless, the boundary conditions become

$$u_z = f\left(\frac{r}{a}\right) + b, \qquad 0 \leq r \leq a, \tag{13-14.1}$$

$$\sigma_{zz} = 0, \qquad r > a, \tag{13-14.2}$$

$$\sigma_{rz} = 0, \qquad 0 \leq r < \infty. \tag{13-14.3}$$

b is a constant to be specified later.

The transformed solution is again taken in the form of 13-12.3:

$$\bar{Z}^0 = (A + B\xi z)e^{-\xi z}. \tag{13-14.4}$$

Equation 13-14.3 yields the result

$$\xi^3(1 - \nu)\bar{Z}^0 + \nu\xi \left.\frac{d^2\bar{Z}^0}{dz^2}\right|_{z=0} = 0. \tag{13-14.5}$$

This leads to the condition

$$A = 2\nu B. \tag{13-14.6}$$

Substituting 13-14.6 into 13-14.4 yields

$$\bar{Z}^0 = B(2\nu + \xi z)e^{-\xi z}. \tag{13-14.7}$$

Boundary conditions 13-14.1 and 13-14.2 are written in the following form by taking the inverse transforms of equations 13-12.4 and 13-12.6:

$$\int_0^\infty \xi\left[-\xi^2 2(1 - \nu)\bar{Z}^0 + (1 - 2\nu)\frac{d^2\bar{Z}^0}{dz^2}\right]_{z=0} J_0(\xi r)\, d\xi = 2G\left[f\left(\frac{r}{a}\right) + b\right], \quad r \leq a, \tag{13-14.8}$$

$$\int_0^\infty \xi\left[-\xi^2(2 - \nu)\frac{d\bar{Z}^0}{dz} + (1 - \nu)\frac{d^3\bar{Z}^0}{dz^3}\right]_{z=0} J_0(\xi r)\, d\xi = 0, \ \pi > a.$$

We must resort to a new technique, because u_z and σ_{zz} are not known a priori everywhere on the surface, and we are unable to take the usual transform of the boundary conditions. It is necessary to express these boundary conditions in terms of the transformed Love function, and equation 13-14.8 accomplishes this. Substituting 13-14.7 into 13-14.8 yields

$$\int_0^\infty -2\xi^3(1 - \nu)BJ_0(\xi r)\, d\xi = 2G\left[f\left(\frac{r}{a}\right) + b\right], \qquad r \leq a,$$

$$\int_0^\infty \xi^4 BJ_0(\xi r)\, d\xi = 0, \qquad r > a. \tag{13-14.9}$$

We can write these equations in a different form by introducing the new variables $\beta = a\xi$ and $\rho = r/a$:

$$C(\beta) = \beta^3 B, \qquad g(\rho) = -\frac{a^4 f(r/a)G}{(1-\nu)}, \qquad b' = -\frac{bGa^4}{(1-\nu)},$$

$$\begin{aligned} &\int_0^\infty C(\beta)J_0(\beta\rho)\,d\beta = g(\rho) + b', \qquad 0 \le \rho \le 1, \\ &\frac{1}{a^5}\int_0^\infty \beta C(\beta)J_0(\beta\rho)\,d\beta = 0, \qquad \rho > 1. \end{aligned} \tag{13-14.10}$$

Equations 13-14.10 are called *dual integral equations*. Extensive work has been done by Sneddon on solutions of equations of this type. In particular, equations 13-14.10 are solved by use of some relations between the Hankel and Fourier transforms. We shall digress to present these relations and then return to show how these can be used to solve equations 13-14.10. Consider the transform

$$\begin{aligned} \mathcal{H}_0[\xi^{-1}\mathcal{F}_c\{g(t);\xi\};r] &= \int_0^\infty J_0(\xi r)\,d\xi\sqrt{\frac{2}{\pi}}\int_0^\infty g(t)\cos\xi t\,dt \\ &= \sqrt{\frac{2}{\pi}}\int_0^\infty g(t)\,dt\int_0^\infty J_0(\xi r)\cos\xi t\,d\xi. \end{aligned} \tag{13-14.11}$$

The last integral can be evaluated in closed form, yielding

$$\int_0^\infty J_0(\xi r)\cos\xi t\,d\xi = \frac{H(r-t)}{(r^2-t^2)^{1/2}},$$

where $H(r-t)$ is the Heavyside step function. This function is equal to one for a positive argument and equal to zero for a negative argument. We therefore obtain the identity

$$\begin{aligned} \sqrt{\frac{2}{\pi}}\int_0^\infty J_0(\xi r)\,d\xi\int_0^\infty g(t)\cos(\xi t)\,dt &= \sqrt{\frac{2}{\pi}}\int_0^\infty \frac{g(t)H(r-t)}{(r^2-t^2)^{1/2}}\,dt \\ &= \sqrt{\frac{2}{\pi}}\int_0^r \frac{g(t)\,dt}{(r^2-t^2)^{1/2}}. \end{aligned} \tag{13-14.12}$$

In a similar manner, we obtain

$$\begin{aligned} \mathcal{H}_0[\xi^{-1}\mathcal{F}_s\{g(t);\xi\};r] &= \int_0^\infty J_0(\xi r)\,d\xi\sqrt{\frac{2}{\pi}}\int_0^\infty g(t)\sin\xi t\,dt \\ &= \sqrt{\frac{2}{\pi}}\int_0^\infty g(t)\,dt\int_0^\infty \sin\xi t J_0(\xi r)\,d\xi \\ &= \sqrt{\frac{2}{\pi}}\int_r^\infty \frac{g(t)}{(t^2-r^2)^{1/2}}\,dt. \end{aligned} \tag{13-14.13}$$

The integral relation,

$$\int_0^\infty J_0(\xi r) \sin \xi t \, d\xi = \frac{H(t - r)}{(t^2 - r^2)^{1/2}},$$

has been used to obtain 13-14.13.

From the relationship of the sine and cosine transforms of the derivative of $g(t)$, we obtain

$$\mathfrak{F}_s\{g'(t); \xi\} = -\xi \mathfrak{F}_c\{g(t); \xi\}, \qquad (13\text{-}14.14)$$

$$\mathfrak{F}_c\{g'(t); \xi\} = \xi \mathfrak{F}_s\{g(t); \xi\}, \qquad \text{if } g(0) = 0. \qquad (13\text{-}14.15)$$

Therefore, we may obtain the relations

$$\begin{aligned} \mathcal{H}_0[\mathfrak{F}_c\{g(t); \xi\}; r] &= -\mathcal{H}_0[\xi^{-1}\mathfrak{F}_s\{g'(t); \xi\}; r] \\ &= -\sqrt{\frac{2}{\pi}} \int_r^\infty \frac{g'(t)}{(t^2 - r^2)^{1/2}} \, dt, \end{aligned} \qquad (13\text{-}14.16)$$

$$\begin{aligned} \mathcal{H}_0[\mathfrak{F}_s\{g(t); \xi\}; r] &= \mathcal{H}_0[\xi^{-1}\mathfrak{F}_c\{g'(t); \xi\}; r] \\ &= \sqrt{\frac{2}{\pi}} \int_0^r \frac{g'(t)}{(r^2 - t^2)^{1/2}} \, dt, \qquad \text{if } g(0) = 0. \end{aligned} \qquad (13\text{-}14.17)$$

Now consider the special case when $g(t) = \psi(t)H(1 - t)$ in equations 13-14.12 and 13-14.16 ,which yields

$$\mathcal{H}_0\left[\xi^{-1} \int_0^1 \psi(t) \cos \xi t \, dt; r\right] = \int_0^r \frac{\psi(t) \, dt}{(r^2 - t^2)^{1/2}}, \qquad 0 \le r \le 1, \qquad (13\text{-}14.18)$$

$$\mathcal{H}_0\left[\int_0^1 \psi(t) \cos \xi t \, dt; r\right] = 0, \qquad r > 1. \qquad (13\text{-}14.19)$$

Similar relations may be obtained from 13-14.13 and 13-14.17.

Returning to the dual integral equations 13-14.10, we note that the second is satisfied if

$$C(\beta) = \int_0^1 \psi(t) \cos \beta t \, dt, \qquad (13\text{-}14.20)$$

and the first becomes

$$\int_0^\rho \frac{\psi(t)}{(\rho^2 - t^2)^{1/2}} \, dt = b' + g(\rho), \qquad 0 \le \rho \le 1. \qquad (13\text{-}14.21)$$

This is an Abel integral equation, which has the solution

$$\begin{aligned} \psi(t) &= \frac{2}{\pi} \frac{d}{dt} \int_0^t \frac{b'\rho + \rho g(\rho)}{(t^2 - \rho^2)^{1/2}} \, d\rho \\ &= \frac{2b'}{\pi} + \frac{2}{\pi} \frac{d}{dt} \int_0^t \frac{\rho g(\rho)}{(t^2 - \rho^2)^{1/2}} \, d\rho. \end{aligned} \qquad (13\text{-}14.22)$$

Historically, Abel's integral equation was one of the first integral equations studied. It arises in the physical problem of finding the time of descent of a particle starting at rest and sliding down a smooth curve under the infleunce of gravity. Consider the curve shown in Figure 13-6. If the particle starts at x, its velocity at any point ξ is given by

$$v = \sqrt{2g(x - \xi)} = \frac{ds}{dt}.$$

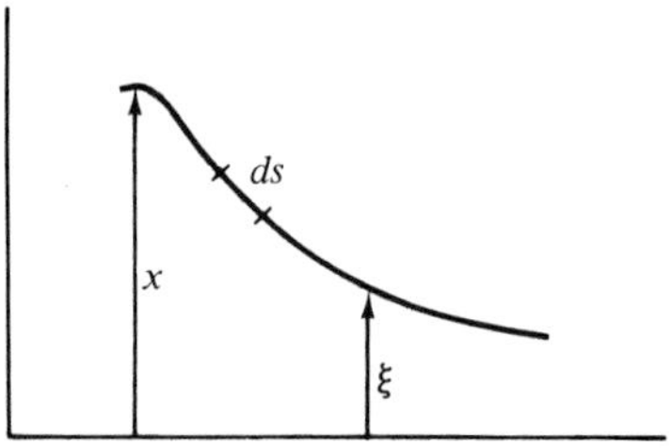

Figure 13-6

The time to reach that point is

$$\int dt = \int \frac{ds}{[2g(x - \xi)]^{1/2}}.$$

The time to descend to 0 is

$$t = -\int_0^x \frac{ds}{[2g(x - \xi)]^{1/2}} = \int_0^x \frac{\psi(\xi)}{(x - \xi)^{1/2}}\, d\xi, \qquad \text{where } \psi(\xi) = \frac{1}{\sqrt{2g}} \frac{ds}{d\xi}.$$

If the path is known, the time of descent can be computed. However, if the time is given and the path corresponding to this time is unknown, we obtain Able's integral equation:

$$T(x) = \int_0^x \frac{\psi(\xi)}{(x - \xi)^{1/2}}\, d\xi.$$

The general form of this equation is

$$f(x) = \int_0^x \frac{\psi(\xi)}{(x - \xi)^{1/2}}\, d\xi.$$

If we multiply both sides by $1/(t - x)^{1/2}\, dx$ and integrate from 0 to t, we obtain

$$\int_0^t \frac{f(x)\, dx}{(t - x)^{1/2}} = \int_0^t \frac{1}{(t - x)^{1/2}} \left[\int_0^x \frac{\psi(\xi)}{(x - \xi)^{1/2}}\, d\xi \right] dx.$$

We can reverse the order of integration by noting that

$$\int_0^t \int_0^x g(x, \xi)\, d\xi\, dx = \int_0^t \left[\int_\xi^t g(x, \xi)\, dx \right] d\xi.$$

(See Figure 13-7)

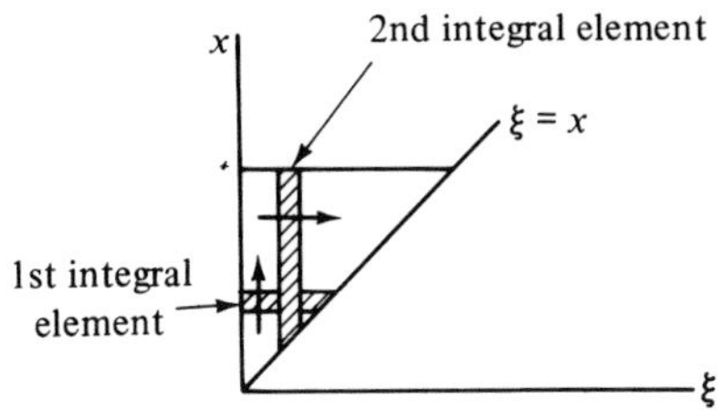

Figure 13-7

The integral becomes

$$\int_0^t \frac{f(x)\,dx}{(t-x)^{1/2}} = \int_0^t \psi(\xi)\left[\int_\xi^t \frac{dx}{[(x-\xi)(t-x)]^{1/2}}\right] d\xi.$$

A change of variables places the inner integral in the form

$$\int_\xi^t \frac{dx}{[(x-\xi)(t-x)]^{1/2}} = \int_0^1 \frac{dy}{[y(y-\xi)]^{1/2}} = \pi, \qquad \text{where } y = \frac{x-\xi}{t-\xi}.$$

Abel's integral equation is solved now by differentiation of both sides:

$$\psi(t) = \frac{1}{\pi}\frac{d}{dt}\int_0^t \frac{f(x)\,dx}{(t-x)^{1/2}}.$$

The general form of this equation can be written

$$f(x) = \int_0^x \frac{\psi(\xi)}{(x-\xi)^{1/2}}\,d\xi \Longrightarrow \psi(x) = \frac{1}{\pi}\frac{d}{dx}\int_0^x \frac{f(\xi)\,d\xi}{(x-\xi)^{1/2}}.$$

This form can be used in the solution of the punch problem. The constant b in equation 13-14.1 is physically interpreted as the depth of penetration at the origin:

$$u_z(0, 0) = b + f(0).$$

If the profile of the punch is "smooth," the stress σ_{zz} must be finite under the contact surface. This condition is used to specify b and therefore b'. This requirement is equivalent to the condition that

$$\text{for } r < a,\ \rho < 1, \quad \beta\int_0^\infty \xi^4 B J_0(\xi r)\,d\xi = \frac{1}{a^5}\int_0^\infty C(\beta)J_0(\beta\rho)\,d\beta.$$

These integrals must remain finite. I. N. Sneddon has shown in his technical report, "The Use of Transform Methods in Elasticity,"* that for smooth punches this requirement can be met by imposing the condition

$$\psi(1) = 0. \qquad (13\text{-}14.23)$$

* *AFORS Tech. Rept. 64–1789*, North Carolina State College, Department of Mathematics, Applied Mathematics Research Group, Raleigh, N.C., 1964.

Integrating by parts and using Liebnitz's rule, equation 13-14.22 is written

$$\psi(t) = \frac{2}{\pi}\left\{b' + g(0) + t\int_0^t \frac{g'(\rho)\,d\rho}{(t^2 - \rho^2)^{1/2}}\right\}. \tag{13-14.24}$$

For smooth punches, b' becomes, by 13-14.23,

$$b' = -g(0) - \int_0^1 \frac{g'(\rho)\,d\rho}{(1 - \rho^2)^{1/2}}. \tag{13-14.25}$$

Instead of seeking a complete solution of this problem at this time, we shall show that some very useful information can be obtained in a simple manner. The total penetration at the tip of the punch is one such parameter:

$$b + f(0) = -\frac{(1 - \nu)}{Ga^4}[b' + g(0)] = \frac{(1 - \nu)}{Ga^4}\int_0^1 \frac{g'(\rho)\,d\rho}{(1 - \rho^2)^{1/2}}. \tag{13-14.26}$$

$$D = \text{depth of penetration} = -\int_0^1 \frac{f'(\rho)\,d\rho}{(1 - \rho^2)^{1/2}}.$$

Another useful parameter is the total force P that must be applied to the punch to give this penetration:

$$P = -2\pi\int_0^a r\sigma_{zz}(r, 0)\,dr, \tag{13-14.27}$$

$$\begin{aligned} \sigma_{zz}(r, 0) &= \int_0^\infty \xi^4 B(\xi) J_0(\xi r)\,d\xi \\ &= \frac{1}{a^5}\int_0^\infty \beta C(\beta) J_0\left(\beta\frac{r}{a}\right)d\beta, \end{aligned} \tag{13-14.28}$$

$$\begin{aligned} P &= -\frac{2\pi}{a^5}\int_0^\infty \beta C(\beta)\,d\beta \int_0^a rJ_0\left(\frac{\beta}{a}r\right)dr \\ &= -\frac{2\pi}{a^3}\int_0^\infty C(\beta) J_1(\beta)\,d\beta. \end{aligned} \tag{13-14.29}$$

From 13-14.20, 13-14.29 becomes

$$P = -\frac{2\pi}{a^3}\int_0^\infty \int_0^1 \psi(t)\cos\beta t\,dt\,J_1(\beta)\,d\beta. \tag{13-14.30}$$

Noting that

$$\int_0^\infty J_1(\beta)\cos\beta t\,d\beta = 1, \qquad \text{for } 0 \le t \le 1,$$

equation 13-14.30 becomes

$$P = -\frac{2\pi}{a^3}\int_0^1 \psi(t)\,dt. \tag{13-14.31}$$

Substituting 13-14.22 into 13-14.31, we obtain

$$P = -\frac{4}{a^3}\left[b' + \int_0^1 \frac{\rho g(\rho)\,d\rho}{(1-\rho^2)^{1/2}}\right]$$
$$= \frac{4aG}{(1-\nu)}\left[b + \int_0^1 \frac{\rho f(\rho)\,d\rho}{(1-\rho^2)^{1/2}}\right]. \qquad (13\text{-}14.32)$$

The stress $\sigma_{zz}(r, 0)$ is obtained from equation 13-14.28:

$$\sigma_{zz} = +\frac{1}{a^5}\int_0^\infty \beta J_0(\beta\rho)\,d\beta \int_0^1 \psi(t)\cos\beta t\,dt. \qquad (13\text{-}14.33)$$

Using the relation between Hankel transforms that

$$\frac{1}{\rho}\frac{d}{d\rho}\{\rho\mathcal{H}_1[\beta^{-1}f(\beta);\rho]\} = \mathcal{H}_0\{f(\beta);\rho\}, \qquad (13\text{-}14.34)$$

equation 13-14.33 becomes

$$\sigma_{zz} = \frac{1}{a^5}\left\{\frac{\psi(1)}{(1-\rho^2)^{1/2}} - \int_\rho^1 \frac{\psi'(t)\,dt}{(t^2-\rho^2)^{1/2}}\right\}. \qquad (13\text{-}14.35)$$

Equations 13-14.26, 13-14.32, and 13-14.35 are used to obtain most of the information needed in practical applications. This development was done first by Sneddon and was applied to punches of many different shapes.

As an example, consider a punch of a rigid cylindrical shape pushed into the half-space a distance Δ (Figure 13-8). The punch is not smooth and

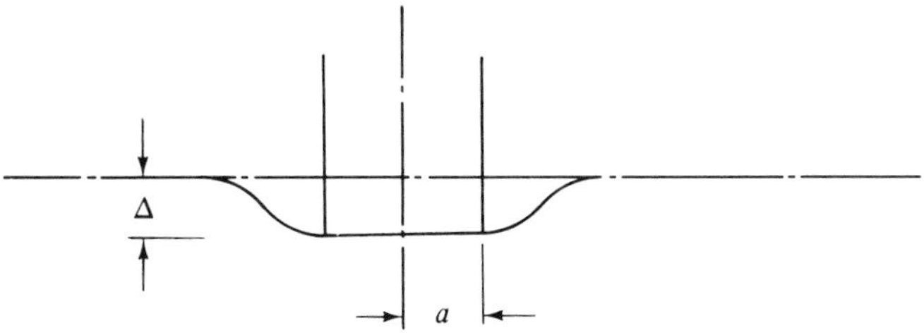

Figure 13-8

the condition 13-14.23 is not necessary. The displacement condition at the boundary is written as

$$b = \Delta, \qquad f\left(\frac{r}{a}\right) = 0,$$
$$b' = -\frac{\Delta G a^4}{(1-\nu)}, \qquad g(\rho) = 0.$$

Equation 13-14.22 yields

$$\psi(t) = -\frac{2\Delta G a^4}{(1-\nu)\pi}.$$

Therefore, from 13-14-20 we obtain

$$C(\beta) = -\frac{2\Delta G a^4}{\pi(1-\nu)} \frac{\sin\beta}{\beta}.$$

The stresses are obtained from 13-14.35 after noting that $\beta = r/a$:

$$\sigma_{zz} = -\frac{2G\Delta}{\pi(1-\nu)} \frac{1}{(a^2 - r^2)^{1/2}},$$

$$P = \frac{4aG}{(1-\nu)}\Delta.$$

15 Short cylinders—multiple Fourier–Bessel series analysis

We have considered in section 7 the case of long solid cylinders when the boundary conditions at $z =$ constant were approximated in the Saint-Venant sense. In section 9 thick plates were considered where the boundary conditions at $r =$ constant were approximated in the same way. We wish now to consider the more difficult problem in which the ratio of length to diameter, l/R, is close to 1. This means that the Saint-Venant approximation cannot be used on any boundary. This type of problem is solved by a combination or superposition of the two solutions considered in sections 7 and 9. This is not, however, just the addition of these two solutions, as they couple, as was the case in two-dimensional problems using multiple Fourier series.*

Consider the solid cylinder of length $2l$ and radius R shown in Figure 13-9. If the loading is torsionless and axisymmetric, we again use the Love strain function to solve the problem. We can separate this function into a part that is even in z and a part that is odd in z. Anticipating the need to satisfy all boundary conditions, we take the solution for Z, odd in z, to be (body forces are zero)

$$Z^{\text{odd}} = \sum \sin\alpha z\{AI_0(\alpha r) + B\alpha r I_1(\alpha r)\} + \sum J_0(\gamma r)\{C \operatorname{sh}\gamma z + D\gamma z \operatorname{ch}\gamma z\}. \tag{13-15.1}$$

The even solution is

$$Z^{\text{even}} = \sum \cos\alpha z\{AI_0(\alpha r) + B\alpha r I_1(\alpha r)\} + \sum J_0(\gamma r)\{C \operatorname{ch}\gamma z + D\gamma z \operatorname{sh}\gamma z\}. \tag{13-15.2}$$

* This method of soluton was proposed by G. Pickett, *J. Appl. Mech.*, **11**, 1944. This method has been reviewed by many authors and is shown in the infinite case to be equivalent to the use of eigenfunctions developed by Papkovich and Prokopov. See discussion of this in J. L. Klemm and R. W. Little, "Saint-Venant Principle," *Tech. Rept. 10.1*, Division of Engineering Research, Michigan State University, 1971. Also see Moghe and Neff, "Elastic deformations of constrained cylinders," *J. Appl. Mech.*, **E 38**, June 1971.

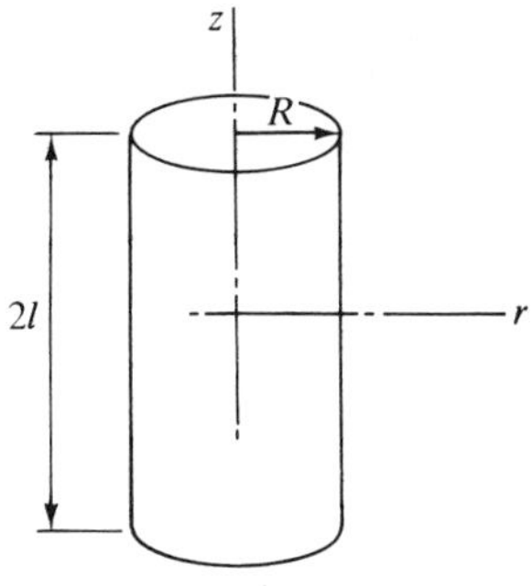

Figure 13-9

To illustrate the use of these solutions, consider the following boundary conditions for Z^{odd}:

$$
\begin{aligned}
z &= \pm l, \qquad \sigma_{zz} = f(r), \qquad \sigma_{rz} = 0, \\
r &= R, \qquad \sigma_{rr} = \sigma_{rz} = 0.
\end{aligned} \tag{13-15.3}
$$

By use of 13-15.1, the stresses become

$$
\begin{aligned}
\sigma_{rz} = {} & \sum \alpha^3\{AI_1(\alpha r) + B[\alpha r I_0(\alpha r) + 2(1 - \nu)I_1(\alpha r)]\} \sin \alpha z \\
& + \sum \gamma^3\{D[2\nu \operatorname{sh} \gamma z + \gamma z \operatorname{ch} \gamma z] + C \operatorname{sh} \gamma z\} J_1(\gamma r),
\end{aligned} \tag{13-15.4}
$$

$$
\begin{aligned}
\sigma_{zz} = {} & \sum \alpha^3\{AI_0(\alpha r) + B[\alpha r I_1(\alpha r) + 2(2 - \nu)I_0(\alpha r)]\} \cos \alpha z \\
& + \sum \gamma^3\{D[1 - 2\nu) \operatorname{ch} \gamma z - \gamma z \operatorname{sh} \gamma z] - C \operatorname{ch} \gamma z\} J_0(\gamma r),
\end{aligned} \tag{13-15.5}
$$

$$
\begin{aligned}
\sigma_{rr} = {} & -\sum \alpha^3 \left\{ A\left[I_0(\alpha r) - \frac{I_1(\alpha r)}{\alpha r} \right] + B[(1 - 2\nu)I_0(\alpha r) + \alpha r I_1(\alpha r)] \right\} \cos \alpha z \\
& + \sum \gamma^3 \left\{ C \operatorname{ch} \gamma z \left[J_0(\gamma r) - \frac{J_1(\gamma r)}{\gamma r} \right] + D\left[J_0(\gamma r)\{(1 + 2\nu) \operatorname{ch} \gamma z \right.\right. \\
& \left.\left. + \gamma z \operatorname{sh} \gamma z\} - \frac{J_1(\gamma r)}{\gamma r}\{\operatorname{ch} \gamma z + \gamma z \operatorname{sh} \gamma z\} \right] \right\}.
\end{aligned} \tag{13-15.6}
$$

We can satisfy the condition that σ_{rz} is zero at all boundaries by choosing

$$
\sin \alpha l = 0, \quad \Longrightarrow \quad \alpha_n = n\pi/l, \tag{13-15.7}
$$

$$
J_1(\gamma R) = 0, \quad \Longrightarrow \quad \gamma_k R \text{ are the roots of } J_1. \tag{13-15.8}
$$

(As was done in section 9, a zero term from equation 13-6.25 must be added to equation 13-15.1 corresponding to the zero eigenvalve.)

$$
A = -B\left\{ \frac{\alpha R I_0(\alpha R)}{I_1(\alpha R)} + 2(1 - \nu) \right\}, \tag{13-15.9}
$$

$$
C = -D\left\{ 2\nu + \gamma l \frac{\operatorname{ch} \gamma l}{\operatorname{sh} \gamma l} \right\}. \tag{13-15.10}
$$

Substituting equations 13-15.9 and 13-15.10 into the equations for stresses yields

$$\sigma_{rz} = \sum \frac{\alpha_n^3 B_n}{I_1(\alpha_n R)}\{\alpha_n r I_0(\alpha_n r)I_1(\alpha_n R) - \alpha_n R I_0(\alpha_n R)I_1(\alpha_n r)\} \sin \alpha_n z$$

$$+ \sum \frac{\gamma_k^3 D_k}{\operatorname{sh} \gamma_k l}\{\gamma_k z \operatorname{ch} \gamma_k z \operatorname{sh} \gamma_k l - \gamma_k l \operatorname{ch} \gamma_k l \operatorname{sh} \gamma_k z\}J_1(\gamma_k r), \qquad (13\text{-}15.11)$$

$$\sigma_{zz} = \sum \frac{\alpha_n^3 B_n}{I_1(\alpha_n R)}\{\alpha_n r I_1(\alpha_n R)I_1(\alpha_n r) + [2I_1(\alpha_n R) - \alpha_n R I_0(\alpha_n R)]I_0(\alpha_n r)\} \cos \alpha_n z$$

$$+ \sum \frac{\gamma_k^3 D_k}{\operatorname{sh} \gamma_k l}\{(\gamma_k l \operatorname{ch} \gamma_k l + \operatorname{sh} \gamma_k l) \operatorname{ch} \gamma_k z - \gamma_k z \operatorname{sh} \gamma_k l \operatorname{sh} \gamma_k z\}J_0(\gamma_k r)$$

$$- 6C_0 + (14 - 10\nu)D_0, \qquad (13\text{-}15.12)$$

$$\sigma_{rr} = -\sum \frac{\alpha_n^3 B_n}{I_1(\alpha_n R)}\Big\{-[\alpha_n R I_0(\alpha_n R) - I_1(\alpha_n R)]I_0(\alpha_n r) + [(\alpha_n r)^2 I_1(\alpha_n R)$$

$$+ 2(1 - \nu)I_1(\alpha_n R) + \alpha_n R I_0(\alpha_n R)]\frac{I_1(\alpha_n r)}{\alpha_n r}\Big\} \cos \alpha_n z$$

$$+ \sum \frac{\gamma_k^3 D_k}{\operatorname{sh} \gamma_k l}\Big\{[\operatorname{ch} \gamma_k z \operatorname{sh} \gamma_k l - \gamma_k l \operatorname{ch} \gamma_k l \operatorname{ch} \gamma_k z$$

$$+ \gamma_k z \operatorname{sh} \gamma_k z \operatorname{sh} \gamma_k l]J_0(\gamma_k r) - [(1 + 2\nu) \operatorname{sh} \gamma_k l \operatorname{ch} \gamma_k z + \gamma_k l \operatorname{ch} \gamma_k l \operatorname{ch} \gamma_k z$$

$$+ \gamma_k z \operatorname{sh} \gamma_k z \operatorname{sh} \gamma_k l]\frac{J_1(\gamma_k r)}{\gamma_k r}\Big\} + 3C_0 - (10\nu - 2)D_0. \qquad (13\text{-}15.13)$$

The two remaining boundary conditions become

$\sigma_{rr} = 0 \qquad$ at $r = R$,

$$\sum_{n=1} \frac{\alpha_n^3 B_n}{I_1(\alpha_n R)}\Big\{\alpha_n R[I_1^2(\alpha_n R) - I_0^2(\alpha_n R)] + 2I_0(\alpha_n R)I_1(\alpha_n R)$$

$$+ \frac{2(1 - \nu)I_1^2(\alpha_n R)}{\alpha_n R}\Big\} \cos \alpha_n z + 3C_0 - (10\nu - 2)D_0 \qquad (13\text{-}15.14)$$

$$= \sum_k \gamma_k^3 D_k\{(1 - \gamma_k l \coth \gamma_k l) \operatorname{ch} \gamma_k z + \gamma_k z \operatorname{sh} \gamma_k z\} J_0(\gamma_k R),$$

$\sigma_{zz} = f(r) \qquad$ at $z = l$,

$$\sum_{k=1} \gamma_k^3 D_k\left[\frac{\gamma_k l + \operatorname{sh} \gamma_k l \operatorname{ch} \gamma_k l}{\operatorname{sh} \gamma_k l}\right]J_0(\gamma_k r) - 6C_0 + (14 - 10\nu)D_0$$

$$= f(r) - \sum \alpha_n^3 B_n\left\{\alpha_n r I_1(\alpha_n r) + \left(2 - \frac{\alpha_n R I_0(\alpha_n R)}{I_1(\alpha_n R)}\right) I_0(\alpha_n r)\right\} \cos \alpha_n l. \qquad (13\text{-}15.15)$$

The zero eigenvalue constants become

$$3C_0 = (10\nu - 2)D_0,$$

$$-6C_0 + (14 - 10\nu)D_0 = 2\int_0^R r f(r)\, dr.$$

Taking the finite Fourier cosine transform of equation 13-15.14 yields

$$B_n = \frac{1}{K_n} \sum_k \frac{2\gamma_k^4 \alpha_n^2(-1)^n}{(\gamma_k^2 + \alpha_n^2)^2} \operatorname{sh} \gamma_k l D_k, \qquad (13\text{-}15.16)$$

where

$$K_n = \frac{\alpha_n^3 l}{I_1(\alpha_n R)}\Big\{\alpha_n R[I_1^2(\alpha_n R) - I_0^2(\alpha_n R)] + 2I_0(\alpha_n R)I_1(\alpha_n R) + \frac{2(1-\nu)I_1^2(\alpha_n R)}{\alpha_n R}\Big\}.$$

The integrals in Appendix B are used to evaluate these transforms.

Taking the finite Hankel transform of order zero of equation 13-15.15 yields

$$D_k = \frac{1}{G_k}\Big\{\int_0^R f(r) r J_0(\gamma_k r)\,dr - \sum_{n=1} \frac{2\alpha_n^4 \gamma_k^2}{(\alpha_n^2 + \gamma_k^2)^2} I_1(\alpha_n R) J_0(\gamma_k R) B_n\Big\},$$

where

$$G_k = \frac{2\,\mathrm{sh}\,\gamma_k l}{\gamma_k^3(\gamma_k l + \mathrm{sh}\,\gamma_k l\,\mathrm{ch}\,\gamma_k l)R^2 J_0(\gamma_k R)}.$$

Equations 13-15.16 and 13-15.17 form a double infinity set of equations for the double infinity set of unknowns B_n and D_k. This system can be truncated and solved for a finite number of coefficients. Substituting these coefficients into the equations for the stresses and for the Love strain function yields the desired solution. Examining the case when $l/R \ll 1$, we observe from 13-15.16 that B_n approaches zero and the solution approaches that obtained when the Saint-Venant approximation was used.

The method of multiple Fourier–Bessel series is used to solve any problem of this type. One should use caution when specifying displacements on one side and stresses at the adjacent boundary, because singularities may exist at these corners.

16 End-loading on a semi-infinite cylinder

The importance of the Saint-Venant approximation has been apparent throughout this chapter. In the previous section, the interaction of this boundary with other loadings was indicated by coupling of the Fourier series. The end problem of a cylinder loaded with self-equilibrated stresses has been studied by many authors* to determine the width of the Saint-Venant boundary region and the rate of the decay within that region.

Consider a semi-infinite cylinder occupying the region, $0 \le r \le 1$,

* F. H. Murray, *U.S. At. Energy Comm. Rept. AECD-2966*, 1945; Horvay, Giaever, and Mirabel, *Ingen. Archiv.*, **27**, 1959, *J. Appl. Mech.*, **25**, 1958; Mendelson and Roberts, 8th Midwestern Mech. Conf., Dev. in Mech., **2**, 1963; R. W. Little and S. B. Childs, *Quart. Appl. Math*, 1967; J.L. Klemm and R. W. Little, *SIAM. J. Appl. Math.*, 1970, and others.

$0 \le z < \infty$. The boundary conditions on the semi-infinite surface are

$$\sigma_{rr} = \sigma_{rz} = 0, \qquad r = 1. \tag{13-16.1}$$

At the edge $z = 0$, we can specify any of the following conditions:

$$\text{given} \quad \begin{matrix} \sigma_{zz}, u_r \\ \sigma_{rz}, u_z \\ \sigma_{zz}, \sigma_{rz} \\ u_r, u_z. \end{matrix} \tag{13-16.2}$$

The particular problem of most importance is when self-equilibrated end loads are applied. This leads to a measurement of the Saint-Venant approximation. We assume a decaying solution, such that all stresses and displacements approach zero as z approaches infinity. The solution takes the following form:

$$Z = \sum A_n M_n(r) e^{-\alpha_n z}. \tag{13-16.3}$$

The function $M_n(r)$ is obtained by solving the biharmonic equation and is

$$M_n = J_0(\alpha_n r) + B_n \alpha_n r J_1(\alpha_n r). \tag{13-16.4}$$

The boundary condition, 13-16.1, is satisfied by the following function:

$$M_n(r) = -\left[2(1-\nu)\frac{J_1(\alpha_n)}{\alpha_n} + J_0(\alpha_n)\right] J_0(\alpha_n r) - J_1(\alpha_n) r J_1(\alpha_n r), \tag{13-16.5}$$

where α_n are the roots of the transcendental equation,

$$J_0^2(\alpha) + \left\{1 - \frac{2(1-\nu)}{\alpha^2}\right\} J_1^2(\alpha) = 0. \tag{13-16.6}$$

For $\nu = 0.3$, the lowest eigenvalues are

$$2.722 + 1.36i \quad \text{and} \quad 6.06 + 1.63i.$$

The asymptotic formula for large eigenvalues is

$$\alpha_n = n\pi - \frac{\ln(4n\pi)}{4n\pi} + \frac{1}{2}\ln(4n\pi)i. \tag{13-16.7}$$

We can identify a vector $\mathbf{f}$ in terms of the boundary stress and displacements as follows:

$$\mathbf{f} = \begin{bmatrix} \sigma_{zz} \\ \sigma_{rz} \\ \dfrac{Eu_r}{(1+\nu)} \\ \dfrac{Eu_z}{(1+\nu)} \end{bmatrix} = \sum_n A_n \boldsymbol{\varphi}_n(r) e^{-\alpha_n z}. \tag{13-16.8}$$

The components of the vector $\boldsymbol{\varphi}_n$ are derivatives of M_n and become

$$
\begin{aligned}
\varphi_n^1 &= [2\alpha_n^2 J_1(\alpha_n) - \alpha_n^3 J_0(\alpha_n)]J_0(\alpha_n r) - \alpha_n^3 J_1(\alpha_n) r J_1(\alpha_n r),\\
\varphi_n^2 &= \alpha_n^3 J_1(\alpha_n) r J_0(\alpha_n r) - \alpha_n^3 J_0(\alpha_n) J_1(\alpha_n r),\\
\varphi_n^3 &= -\alpha_n^2 J_1(\alpha_n) r J_0(\alpha_n r) + [2(1-\nu)\alpha_n J_1(\alpha_n) + \alpha_n^2 J_0(\alpha_n)]J_1(\alpha_n r),\\
\varphi_n^4 &= [\alpha_n^2 J_0(\alpha_n) - 2(1-\nu)\alpha_n J_1(\alpha_n)]J_0(\alpha_n r) + \alpha_n^2 r J_1(\alpha_n) J_1(\alpha_n r).
\end{aligned}
\tag{13-16.9}
$$

At the end, $z = 0$, the boundary functions are given in terms of $\boldsymbol{\varphi}_n$ as

$$
\mathbf{f}^b = \sum_n A_n \boldsymbol{\varphi}_n(r). \tag{13-16.10}
$$

The functions $\boldsymbol{\varphi}_n$ are not orthogonal, but both Little and Childs and Klemm and Little have developed biorthogonal functions $\mathbf{w}_n(r)$, which satisfy the condition

$$
\int_0^1 \mathbf{w}_k(r)\boldsymbol{\varphi}_j(r) r\, dr = \begin{cases} 0, & j \neq k \\ N_j, & j = k \end{cases}, \tag{13-16.11}
$$

where

$$
N_j = (1-\nu)[-4\alpha_j J_0^2(\alpha_j) - 2\alpha_j J_1^2(\alpha_j) + 4(1-\nu)J_0(\alpha_j)J_1(\alpha_j)]. \tag{13-16.12}
$$

The biorthogonal vectors can be obtained by development of the adjoint problem (see Appendix A and section 7-9) or by use of a multiple Fourier analysis involving a Fourier–Bessel series and a Fourier sine or cosine integral. Details of these developments are available in the cited references. The biorthogonal vector has the following components:

$$
\begin{aligned}
w_n^1 &= \frac{1}{2\alpha_n J_1^2(\alpha_n)}\{[\alpha_n J_0(\alpha_n) - 2(1-\nu)J_1(\alpha_n)]J_0(\alpha_n r)\\
&\qquad + \alpha_n J_1(\alpha_n) r J_1(\alpha_n r)\},\\
w_n^2 &= \frac{1}{2\alpha_n J_1^2(\alpha_n)}\{\alpha_n r J_1(\alpha_n) J_0(\alpha_n r) - [2(1-\nu)J_1(\alpha_n)\\
&\qquad + \alpha_n J_0(\alpha_n)]J_1(\alpha_n r)\},\\
w_n^3 &= \frac{1}{2\alpha_n J_1^2(\alpha_n)}\{-\alpha_n^2 r J_1(\alpha_n) J_0(\alpha_n r) + \alpha_n^2 J_0(\alpha_n) J_1(\alpha_n r)\},\\
w_n^4 &= \frac{1}{2\alpha_n J_1^2(\alpha_n)}\{[2\alpha_n J_1(\alpha_n) - \alpha_n^2 J_0(\alpha_n)]J_0(\alpha_n r)\\
&\qquad - \alpha_n J_1(\alpha_n)\alpha_n r J_1(\alpha_n r)\}.
\end{aligned}
\tag{13-16.13}
$$

The coefficients of the series, equation 13-16.10, are formally obtained by use of the biorthogonality condition

$$
A_j = \frac{1}{N_j}\int_0^1 \mathbf{w}_j \cdot \mathbf{f}^b r\, dr. \tag{13-16.14}
$$

As illustrated in section 7-9, the particular boundary vector $\mathbf{f}^b$ will contain two components in terms of specified functions and the remaining two in terms of series. For example, when the stresses are given at $z = 0$, the boundary condition 13-16.10 becomes

$$\begin{bmatrix} \sigma_{zz} \\ \sigma_{rz} \\ \sum A_j \varphi_j^3 \\ \sum A_j \varphi_j^4 \end{bmatrix} = \sum_n A_n \boldsymbol{\varphi}_n(r). \qquad (13\text{-}16.15)$$

Equation 13-16.15 yields

$$\begin{aligned} A_j = {} & \frac{1}{N_j}\int_0^1 [w_j^1 \sigma_{zz}(r, 0) + w_j^2 \sigma_{rz}(r, 0)] r\, dr \\ & + \frac{1}{N_j}\sum_k A_k \int_0^1 [w_j^3 \varphi_k^3 + w_j^4 \varphi_k^4] r\, dr. \end{aligned} \qquad (13\text{-}16.16)$$

This equation can be written as a set of infinite equations in infinite unknowns, which is truncated and solved:

$$A_j = F_j + \sum_k S_{jk} A_k, \qquad (13\text{-}16.17)$$

where

$$F_j = \frac{1}{N_j}\int_0^1 [w_j^1 \sigma_{zz} + w_j^2 \sigma_{rz}] r\, dr,$$

$$S_{jk} = \frac{1}{N_j}\int_0^1 [w_j^3 \varphi_k^3 + w_j^4 \varphi_k^4] r\, dr.$$

17 Axisymmetric torsion

Another component of the Galerkin vector deserves comment when discussing axisymmetric problems. The Love strain function was independent of θ and did not contribute to $\sigma_{\theta z}$, $\sigma_{\theta r}$, or u_θ. We refer to this as axisymmetric and torsionless. We can assume that the θ component of the Galerkin vector G_θ is independent of θ and produce a case of axisymmetric torsion.

Consider the case when the Galerkin vector is written

$$\bar{G} = G_\theta \hat{i}_{(\theta)}, \qquad G_\theta = \psi(r, z). \qquad (13\text{-}17.1)$$

For problems where the body forces are zero, ψ is governed by

$$\left(\nabla^2 - \frac{1}{r^2}\right)^2 \psi = 0. \qquad (13\text{-}17.2)$$

The displacements are calculated by use of the equation 13-3.10:

$$\bar{u} = \frac{1}{2G}\{2(1 - \nu)\nabla^2 \bar{G} - \nabla\nabla \cdot \bar{G}\}. \qquad (13\text{-}17.3)$$

Noting that

$$\nabla^2\bar{G} = \left[\left(\nabla^2 - \frac{1}{r^2}\right)\psi\right]\hat{i}_{(\theta)},$$
$$\bar{\nabla}\cdot\bar{G} = 0,$$

the only nonzero displacement is

$$u_\theta = \frac{1-\nu}{G}\left(\nabla^2 - \frac{1}{r^2}\right)\psi. \tag{13-17.4}$$

The normal stresses σ_{rr}, $\sigma_{\theta\theta}$, and σ_{zz} and the shear stress σ_{rz} are zero. The nonzero stresses are

$$\sigma_{r\theta} = Gr\frac{\partial}{\partial r}\frac{u_\theta}{r} = (1-\nu)r\frac{\partial}{\partial r}\left[\frac{1}{r}\left(\nabla^2 - \frac{1}{r^2}\right)\psi\right], \tag{13-17.5}$$

$$\sigma_{z\theta} = G\frac{\partial}{\partial z}u_\theta = (1-\nu)\frac{\partial}{\partial z}\left[\left(\nabla^2 - \frac{1}{r^2}\right)\psi\right]. \tag{13-17.6}$$

Note that the axisymmetric ψ function contributes only to those displacements and stresses which are zero for the axisymmetric Love function, so that the two axisymmetric cases decouple. The nonzero stresses and displacements do not depend directly upon ψ but upon $[\nabla^2 - (1/r^2)]\psi$.

If we write the axisymmetric Laplacian operator as

$$\nabla^2 = \mathfrak{L} + \frac{\partial^2}{\partial z^2},$$

where

$$\mathfrak{L} = \frac{\partial^2}{\partial r^2} + \frac{1}{r}\frac{\partial}{\partial r}, \tag{13-17.7}$$

we obtain solutions to equation 13-17.2 in terms of harmonic functions by examining the following operator identity:

$$\left(\mathfrak{L} - \frac{1}{r^2}\right)\frac{\partial}{\partial r} = \frac{\partial}{\partial r}\mathfrak{L}. \tag{13-17.8}$$

The following axisymmetric identity holds:

$$\left(\nabla^2 - \frac{1}{r^2}\right)\frac{\partial\varphi}{\partial r} = \frac{\partial}{\partial r}\nabla^2\varphi. \tag{13-17.9}$$

If φ is harmonic, then $\partial\varphi/\partial r$ satisfies the equation

$$\left(\nabla^2 - \frac{1}{r^2}\right)\frac{\partial\varphi}{\partial r} = 0. \tag{13-17.10}$$

Using the properties of biharmonic functions, a solution for ψ^{odd} can now be written:

$$\psi = \sum J_1(\alpha r)[A \operatorname{sh}\alpha z + B\alpha z \operatorname{ch}\alpha z] + \sum C\alpha r J_0(\alpha r)\operatorname{sh}\alpha z + \sum \sin\gamma z[DI_1(\gamma r) + E\gamma r I_0(\gamma r)]. \tag{13-17.11}$$

We consider axisymmetric torsion problems in which the stresses producing the torque are given on the boundary $r =$ constant or $z =$ constant. As an example, consider the problem shown in Figure 13-10. The boundary condition is

$$\sigma_{r\theta} = \tau z, \qquad -b \leq z \leq b, \qquad r = R.$$

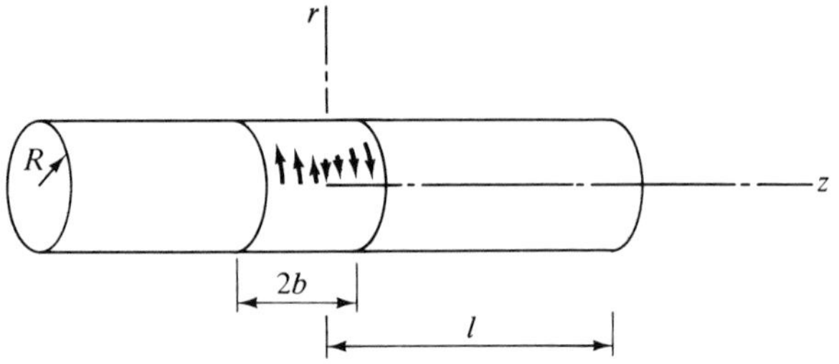

Figure 13-10

If $l/R \gg 1$, we can use a Saint-Venant approximation on the ends $z = \pm l$. The boundary conditions indicate that $\sigma_{r\theta}$ is odd in z and $\sigma_{z\theta}$ is even in z. This leads us to choose ψ as an odd function of z:

$$\psi = \sum \sin \gamma z[DI_1(\gamma r) + E\gamma r I_0(\gamma r)]. \qquad (13\text{-}17.12)$$

The first term does not contribute to the the displacement or stresses and need not be considered. The function ψ becomes

$$\psi = \sum \sin \gamma z[E\gamma r I_0(\gamma r)]. \qquad (13\text{-}17.13)$$

The stresses and displacement become

$$u_\theta = \frac{1-\nu}{G}\{\sum 2\gamma^2 E I_1(\gamma r) \sin \gamma z\}, \qquad (13\text{-}17.14)$$

$$\sigma_{r\theta} = 2(1-\nu) \sum \gamma^3 E\left[I_0(\gamma r) - \frac{2I_1(\gamma r)}{\gamma r}\right] \sin \gamma z, \qquad (13\text{-}17.15)$$

$$\sigma_{z\theta} = 2(1-\nu) \sum \gamma^3 E I_1(\gamma r) \cos \gamma z. \qquad (13\text{-}17.16)$$

We choose γ such that $\sigma_{z\theta}$ is zero at $z = \pm l$:

$$\cos \gamma_m l = 0 \implies \gamma_m = \frac{2m+1}{2}\frac{\pi}{l}. \qquad (13\text{-}17.17)$$

The boundary conditions at $r = R$ are satisfied by Fourier analysis techniques:

$$2(1-\nu)\gamma_m^3 E_m\left[I_0(\gamma_m R) - \frac{2I_1(\gamma_m R)}{\gamma_m R}\right] = \frac{2}{l}\int_0^l \sigma_{r\theta}|_{r=R} \sin \gamma_m z \, dz. \qquad (13\text{-}17.18)$$

The particular specified stress gives

$$2(1-\nu)\gamma_m^3 E_m\left[I_0(\gamma_m R) - \frac{2I_1(\gamma_m R)}{\gamma_m R}\right] = \frac{2\tau}{l}\left[\frac{1}{\gamma_m^2}\sin\gamma_m b - \frac{b}{\gamma_m}\cos\gamma_m b\right]. \tag{13-17.19}$$

This yields the required expression for E_m, which is used to calculate the stress and displacement.

It is clear that the Love function Z and the axisymmetric torsion ψ function can be combined to satisfy the boundary conditions on any axisymmetric problem. The methods of multiple Fourier–Bessel series analysis and integral transforms may easily be applied to the axisymmetric torsion case in a manner similar to that used for the Love function. The details will not be presented here.

18 Asymmetric loadings

R. Muki* recommended the addition of a harmonic function to the Love strain function to consider asymmetric problems. In section 6-4 the Galerkin approach was motivated by use of the Helmholtz theorem. If we retain an additional irrotational element, we obtain Muki's solution:

$$\bar{u} = \frac{1}{2G}\{2(1-\nu)\nabla^2\bar{G} - \vec{\nabla}(\vec{\nabla}\cdot\bar{G}) + \vec{\nabla}\times\bar{A}\}, \tag{13-18.1}$$

where $\bar{G}$ is a biharmonic vector and $\bar{A}$ is a harmonic vector. Muki proposed single z components for both $\bar{G}$ and $\bar{A}$:

$$\begin{aligned}\bar{G} &= Z(r,\theta,z)\hat{i}_{(z)},\\ \bar{A} &= 2\psi(r,\theta,z)\hat{i}_{(z)}.\end{aligned} \tag{13-18.2}$$

For this case, the displacements become

$$\begin{aligned}2Gu_r &= -\frac{\partial^2 Z}{\partial r\,\partial z} + \frac{2}{r}\frac{\partial\psi}{\partial\theta},\\ 2Gu_\theta &= -\frac{1}{r}\frac{\partial^2 Z}{\partial\theta\,\partial z} - 2\frac{\partial\psi}{\partial r},\\ 2Gu_z &= \left\{2(1-\nu)\nabla^2 - \frac{\partial^2}{\partial z^2}\right\}Z.\end{aligned} \tag{13-18.3}$$

The stresses become

* In I. N. Sneddon and R. Hill, eds., *Progress in Solid Mechanics*, North-Holland Publishing Company, Amsterdam, 1960, pp. 401–439.

$$
\begin{aligned}
\sigma_{rr} &= \frac{\partial}{\partial z}\left(\nu\nabla^2 - \frac{\partial^2}{\partial r^2}\right)Z + \frac{\partial}{\partial\theta}\left(\frac{2}{r}\frac{\partial}{\partial r} - \frac{2}{r^2}\right)\psi, \\
\sigma_{\theta\theta} &= \frac{\partial}{\partial z}\left(\nu\nabla^2 - \frac{1}{r}\frac{\partial}{\partial r} - \frac{1}{r^2}\frac{\partial^2}{\partial\theta^2}\right)Z - \frac{\partial}{\partial\theta}\left(\frac{2}{r}\frac{\partial}{\partial r} - \frac{\partial}{r^2}\right)\psi, \\
\sigma_{zz} &= \frac{\partial}{\partial z}\left([2 - \nu]\nabla^2 - \frac{\partial^2}{\partial z^2}\right)Z, \\
\sigma_{\theta z} &= \frac{1}{r}\frac{\partial}{\partial\theta}\left([1 - \nu]\nabla^2 - \frac{\partial^2}{\partial z^2}\right)Z - \frac{\partial^2\psi}{\partial r\,\partial z}, \\
\sigma_{zr} &= \frac{\partial}{\partial r}\left([1 - \nu]\nabla^2 - \frac{\partial^2}{\partial z^2}\right)Z + \frac{1}{r}\frac{\partial^2\psi}{\partial\theta\,\partial z}, \\
\sigma_{r\theta} &= \frac{1}{r}\frac{\partial^2}{\partial\theta\,\partial z}\left(\frac{1}{r} - \frac{\partial}{\partial r}\right)Z - \left(2\frac{\partial^2}{\partial r^2} - \frac{\partial^2}{\partial z^2}\right)\psi.
\end{aligned}
\qquad (13\text{-}18.4)
$$

Z and ψ are defined by the biharmonic and harmonic equations, respectively, in the absence of body forces as

$$
\begin{aligned}
\nabla^2\nabla^2 Z &= 0, \\
\nabla^2\psi &= 0.
\end{aligned}
\qquad (13\text{-}18.5)
$$

The solution of the harmonic and biharmonic equations can be written

$$
\begin{aligned}
Z(r, \theta, z) &= \sum_{m=0}^{\infty} Z_{cm}(r, z)\cos m\theta + Z_{sm}(r, z)\sin m\theta, \\
\psi(r, \theta, z) &= \sum_{m=0}^{\infty} \psi_{cm}(r, z)\cos m\theta + \psi_{sm}(r, z)\sin m\theta.
\end{aligned}
\qquad (13\text{-}18.6)
$$

The functions of r and z satisfy the following partial differential equations:

$$
\begin{aligned}
\nabla_m^2\nabla_m^2 Z_m &= \left(\frac{\partial^2}{\partial r^2} + \frac{1}{r}\frac{\partial}{\partial r} - \frac{m^2}{r^2} + \frac{\partial^2}{\partial z^2}\right)^2 Z_m = 0, \\
\nabla_m^2\psi &= \left(\frac{\partial^2}{\partial r^2} + \frac{1}{r}\frac{\partial}{\partial r} - \frac{m^2}{r^2} + \frac{\partial^2}{\partial z^2}\right)\psi_m = 0.
\end{aligned}
\qquad (13\text{-}18.7)
$$

Examining the expressions for the stresses and displacements leads to grouping Z_{cm} and ψ_{sm} together for an even dependency in normal stresses in θ, and Z_{sm} and ψ_{cm} together for the odd dependency. We can work with either of these to indicate the form of the solution. The other is obtained by interchange of sines and cosines with care being taken to examine sign changes.

The solution of 13-18.7 is taken such that we obtain orthogonal functions in either the r or z variable. A general form, similar to equation 13-6.6, is written as follows:

$$
\begin{aligned}
Z_m = \{&AJ_m(\alpha r) + BY_m(\alpha r) + C\alpha rJ_{m+1}(\alpha r) + D\alpha rY_{m+1}(\alpha r) \\
&+ E\alpha zJ_m(\alpha z) + F\alpha zY_m(\alpha z)\}e^{\pm\alpha z},
\end{aligned}
\qquad (13\text{-}18.8)
$$

$$
\psi_m = \{GJ_m(\alpha r) + HY_m(\alpha r)\}e^{\pm\alpha z}. \qquad (13\text{-}18.9)
$$

The eigenvalue α is chosen to be imaginary if orthogonal functions are desired in the z direction. The constants, E and F, are set equal to zero for this case. This form of solution is used for long cylinders under asymmetric loading conditions and uses a Saint-Venant approximation on the ends, z equal to a constant. If α is real, the constants, C and D, are set equal to zero and problems of thick plates may be considered.

The method of solution of particular problems is similar to those used for axisymmetric problems. The theta dependency is expanded in a Fourier series as indicated by equation 13-18.6, and the individual terms are handled by the methods outlined previously.

As an example of solutions of this type, consider a loading tangential to the surface of the half-plane, as shown in Figure 13-11. The boundary

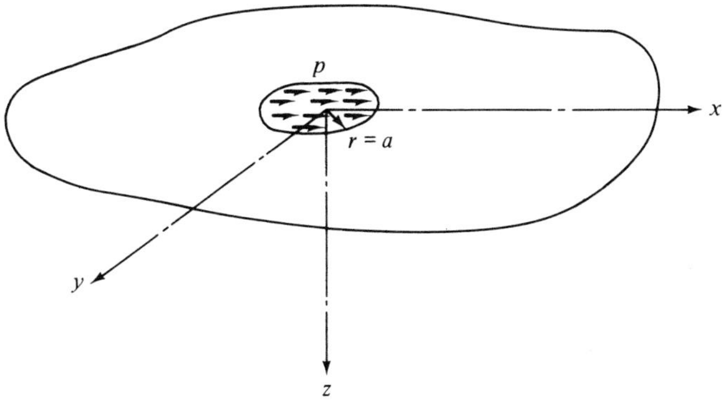

Figure 13-11

conditions at the plane, $z = 0$, are

$$\sigma_{zz} = 0,$$

$$\sigma_{rz} = \frac{P}{\pi a^2} \cos \theta, \qquad (13\text{-}18.10)$$

$$\sigma_{\theta z} = -\frac{P}{\pi a^2} \sin \theta.$$

P is a force in the x direction and approaches a concentrated load as $a \longrightarrow 0$. This problem is known as Cerruti's problem and has been solved by potential functions by Love and Luré* and by Westergaard† by use of the *twinned-gradient* method.

We assume a solution in the form of Z_{c1} and ψ_{s1}, and the stresses and

* See Luré, *Three-Dimensional Problems of Elasticity.*

† See Westergaard, *Theory of Elasticity and Plasticity.*

displacements become

$$
\begin{aligned}
2Gu_r &= \left[-\frac{\partial^2 Z_{c1}}{\partial r\,\partial z} + \frac{2}{r}\psi_{s1}\right]\cos\theta,\\
2Gu_\theta &= \left[\frac{1}{r}\frac{\partial Z_{c1}}{\partial z} - 2\frac{\partial \psi_{s1}}{\partial r}\right]\sin\theta,\\
2Gu_z &= \left[2(1-\nu)\nabla_1^2 - \frac{\partial^2}{\partial z^2}\right]Z_{c1}\cos\theta,\\
\sigma_{rr} &= \left\{\frac{\partial}{\partial z}\left[\nu\nabla_1^2 - \frac{\partial^2}{\partial r^2}\right]Z_{c1} + \left[\frac{2}{r}\frac{\partial}{\partial r} - \frac{2}{r^2}\right]\psi_{s1}\right\}\cos\theta,\\
\sigma_{\theta\theta} &= \left\{\frac{\partial}{\partial z}\left[\nu\nabla_1^2 - \frac{1}{r}\frac{\partial}{\partial r} + \frac{1}{r^2}\right]Z_{c1} - \left[\frac{2}{r}\frac{\partial}{\partial r} - \frac{2}{r^2}\right]\psi_{s1}\right\}\cos\theta,\\
\sigma_{zz} &= \frac{\partial}{\partial z}\left\{(2-\nu)\nabla_1^2 - \frac{\partial^2}{\partial z^2}\right\}Z_{c1}\cos\theta,\\
\sigma_{\theta z} &= \left[-\frac{1}{r}\left\{(1-\nu)\nabla_1^2 - \frac{\partial^2}{\partial z^2}\right\}Z_{c1} - \frac{\partial^2\psi_{s1}}{\partial r\,\partial z}\right]\sin\theta,\\
\sigma_{zr} &= \left[\frac{\partial}{\partial r}\left\{(1-\nu)\nabla_1^2 - \frac{\partial^2}{\partial z^2}\right\}Z_{c1} + \frac{1}{r}\frac{\partial\psi_{s1}}{\partial z}\right]\cos\theta,\\
\sigma_{r\theta} &= \left[\frac{\partial^2}{\partial r\,\partial z}\left(\frac{Z_{c1}}{r}\right) - 2\frac{\partial^2\psi_{s1}}{\partial r^2} - \frac{\partial^2\psi_{s1}}{\partial z^2}\right]\sin\theta,
\end{aligned}
\qquad (13\text{-}18.11)
$$

where
$$
\nabla_1^2 = \frac{\partial^2}{\partial r^2} + \frac{1}{r}\frac{\partial}{\partial r} - \frac{1}{r^2} + \frac{\partial^2}{\partial z^2}.
$$

The defining equations for Z_{c1} and ψ_{s1} are

$$
\begin{aligned}
\nabla_1^2\nabla_1^2 Z_{c1} &= 0,\\
\nabla_1^2\psi_{s1} &= 0.
\end{aligned}
\qquad (13\text{-}18.12)
$$

The boundary conditions at the surface become

$$
\begin{aligned}
\frac{\partial}{\partial z}\left\{(2-\nu)\nabla_1^2 - \frac{\partial^2}{\partial z^2}\right\}Z_{c1}\bigg|_{z=0} &= 0,\\
\frac{\partial}{\partial r}\left\{(1-\nu)\nabla_1^2 - \frac{\partial^2}{\partial z^2}\right\}Z_{c1} + \frac{1}{r}\frac{\partial\psi_{s1}}{\partial z}\bigg|_{z=0} &= \frac{P}{\pi a^2},\\
\frac{1}{r}\left\{(1-\nu)\nabla_1^2 - \frac{\partial^2}{\partial z^2}\right\}Z_{c1} + \frac{\partial^2\psi_{s1}}{\partial r\,\partial z}\bigg|_{z=0} &= \frac{P}{\pi a^2}.
\end{aligned}
\qquad (13\text{-}18.13)
$$

We can use the operational methods outlined in section 11 to solve this problem. The operator ∇_m^2 is transformed by use of the mth-order Hankel transform:

$$
\begin{aligned}
\mathcal{H}_m[\nabla_m^2\psi(r, z); \xi] &= \left(\frac{d^2}{dz^2} - \xi^2\right)\bar{\psi}^m(\xi, z),\\
\mathcal{H}_m[\nabla_m^2\nabla_m^2 Z(r, z); \xi] &= \left(\frac{d^2}{dz^2} - \xi^2\right)^2\bar{Z}^m(\xi, z).
\end{aligned}
\qquad (13\text{-}18.14)
$$

Letting Z designate Z_{c1} and ψ designate ψ_{s1}, equation 13-18.12 becomes

$$\left(\frac{d^2}{dz^2} - \xi^2\right)\bar{\psi}^1 = 0,$$
$$\left(\frac{d^2}{dz^2} - \xi^2\right)^2 \bar{Z}^1 = 0, \qquad (13\text{-}18.15)$$

This first boundary condition, equation 13-18.13, becomes

$$\bar{\sigma}^1_{zz}|_{z=0} = \frac{d}{dz}\left\{(2-\nu)\left(\frac{d^2}{dz^2} - \xi^2\right) - \frac{d^2}{dz^2}\right\}\bar{Z}^1|_{z=0} = 0. \qquad (13\text{-}18.16)$$

The other two boundary conditions do not lend themselves to direct transformation, but their sum and difference can be transformed. The sum of the last two boundary conditions is

$$\frac{1}{r}\frac{\partial}{\partial r}\left\{r\left[(1-\nu)\nabla_1^2 Z - \frac{\partial^2 Z}{\partial z^2} + \frac{\partial \psi}{\partial z}\right]\right\}\Bigg|_{\substack{z=0\\ r<a}} = \frac{2P}{\pi a^2}. \qquad (13\text{-}18.17)$$

Using equation 13-11.13, when the order of the transform is zero, yields

$$\xi\left\{(1-\nu)\left(\frac{d^2}{dz^2} - \xi^2\right)\bar{Z}^1 - \frac{d^2}{dz^2}\bar{Z}^1 + \frac{d\bar{\psi}^1}{dz}\right\}\Bigg|_{z=0} = \int_0^a \frac{2P}{\pi a^2} r J_0(\xi r)\, dr. \qquad (13\text{-}18.18)$$

The difference of the last two bounadry conditions is

$$r\frac{\partial}{\partial r}\left\{\frac{1}{r}\left[(1-\nu)\nabla_1^2 Z - \frac{\partial^2 Z}{\partial z^2} + \frac{\partial \psi}{\partial z}\right]\right\}\Bigg|_{z=0} = 0. \qquad (13\text{-}18.19)$$

Using equation 13-11.13, $\nu = 2$, yields

$$-\xi\left\{(1-\nu)\left(\frac{d^2}{dz^2} - \xi^2\right)\bar{Z}^1 - \frac{d^2}{dz^2}\bar{Z}^1 - \frac{d\bar{\psi}^1}{dz}\right\}\Bigg|_{z=0} = 0. \qquad (13\text{-}18.20)$$

The solutions of equation 13-18.15 are written

$$\bar{\psi}^1 = Ae^{-\xi z},$$
$$\bar{Z}^1 = (B + C_{\xi z})e^{-\xi z}. \qquad (13\text{-}18.21)$$

Equation 13-18.16 yields

$$\xi^3[B + (1-2\nu)C] = 0. \qquad (13\text{-}18.22)$$

The constant B can be expressed in trems of C to satisfy this condition:

$$B = -(1-2\nu)C. \qquad (13\text{-}18.23)$$

Equations 13-18.18 and 13-18.20 yield

$$\xi^2[(2C\nu - B)\xi - A] = \int_0^a \frac{2P}{\pi a^2} r J_0(\xi r)\, dr,$$
$$\xi^2[(2C\nu - B)\xi + A] = 0. \qquad (13\text{-}18.24)$$

Equations 13-18.23 and 13-18.24 give

$$A = -\frac{1}{2\xi^2}\int_0^a \frac{2P}{\pi a^2} r J_0(\xi r)\,dr,$$

$$B = -\frac{(1-2\nu)}{2\xi^3}\int_0^a \frac{2P}{\pi a^2} r J_0(\xi r)\,dr, \tag{13-18.25}$$

$$C = \frac{1}{2\xi^3}\int_0^a \frac{2P}{\pi a^2} r J_0(\xi r)\,dr.$$

For a concentrated load, equation 13-18.25 can be written in the following form after integration and taking the limit as a approaches zero:

$$A = -\frac{P}{2\pi\xi^2},$$

$$B = -\frac{(1-2\nu)}{2\pi\xi^3}P, \tag{13-18.26}$$

$$C = \frac{P}{2\pi\xi^3}.$$

The stresses σ_{zz}, $\sigma_{\theta z}$, and σ_{rz} are calculated from their inverse expressions, using equations 13-18.11:

$$\bar{\sigma}^1_{zz}\cos\theta = \frac{d}{dz}\left\{(2-\nu)\left(\frac{d^2}{dz^2}-\xi^2\right)-\frac{d^2}{dz^2}\right\}\bar{Z}^1\cos\theta, \tag{13-18.27}$$

$$\frac{\bar{\sigma}^2_{zr}}{\cos\theta}+\frac{\bar{\sigma}^2_{\theta z}}{\sin\theta} = \xi\left\{(1-\nu)\left(\frac{d^2}{dz^2}-\xi^2\right)\bar{Z}^1-\frac{d^2}{dz^2}\bar{Z}^1+\frac{d\bar{\psi}^1}{dz}\right\}, \tag{13-18.28}$$

$$\frac{\bar{\sigma}^0_{zr}}{\cos\theta}-\frac{\bar{\sigma}^0_{\theta z}}{\sin\theta} = -\xi\left\{(1-\nu)\left(\frac{d^2}{dz^2}-\xi^2\right)\bar{Z}^1-\frac{d^2}{dz^2}\bar{Z}^1-\frac{d\bar{\psi}^1}{dz}\right\}. \tag{13-18.29}$$

Equation 13-18.27 yields

$$\sigma_{zz} = \int_0^\infty \frac{P\xi^2 z}{2\pi}e^{-\xi z}\cos\theta J_1(\xi r)\,d\xi = \frac{3P}{2\pi}\frac{rz^2\cos\theta}{(r^2+z^2)^{5/2}}. \tag{13-18.30}$$

Equation 13-18.28 gives

$$\begin{aligned}\frac{\sigma_{zr}}{\cos\theta}+\frac{\sigma_{\theta z}}{\sin\theta} &= -\int_0^\infty \frac{P}{2\pi}\xi^2 z e^{-\xi z}J_2(\xi r)\,d\xi \\ &= -\frac{P}{2\pi}\frac{3r^2 z}{(z^2+r^2)^{5/2}}.\end{aligned} \tag{13-18.31}$$

$$\begin{aligned}\frac{\sigma_{zr}}{\cos\theta}-\frac{\sigma_{\theta z}}{\sin\theta} &= \frac{P}{2\pi}\left\{\int_0^\infty 2\xi e^{-\xi z}J_0(\xi r)\,d\xi - \int_0^\infty \xi^2 z e^{-\xi z}J_0(\xi r)\,d\xi\right\} \\ &= \frac{P}{2\pi}\left\{\frac{2z}{(z^2+r^2)^{3/2}}-\frac{2z^3-r^2 z}{(z^2+r^2)^{5/2}}\right\} \\ &= \frac{P}{2\pi}\frac{3r^2 z}{(z^2+r^2)^{5/2}}.\end{aligned} \tag{13-18.32}$$

Solving these two equations yields

$$\sigma_{rz} = \frac{P}{2\pi}\frac{3r^2 z}{(z^2 + r^2)^{5/2}}\cos\theta, \tag{13-18.33}$$

$$\sigma_{\theta z} = 0. \tag{13-18.34}$$

These integrals were evaluated by use of equation 13-13.6.

The remaining stresses and displacements are obtained from the following expressions:

$$\begin{aligned}
\frac{\bar{u}_r^2}{\cos\theta} + \frac{\bar{u}_\theta^2}{\sin\theta} &= \frac{\xi}{2G}\left\{\frac{d\bar{Z}^1}{dz} + 2\bar{\psi}^1\right\},\\
\frac{\bar{u}_r^0}{\cos\theta} - \frac{\bar{u}_\theta^0}{\sin\theta} &= -\frac{\xi}{2G}\left\{\frac{d\bar{Z}^1}{dz} - 2\bar{\psi}^1\right\},\\
\frac{\bar{\sigma}_{rr}^1}{\cos\theta} + \frac{\bar{\sigma}_{\theta\theta}^1}{\cos\theta} &= \left\{2\nu\frac{d^3}{dz^3} + (1-2\nu)\xi^2\frac{d}{dz}\right\}\bar{Z}^1,\\
\frac{\bar{\sigma}_{rr}^1}{\cos\theta} + \frac{2G\bar{u}_r^1}{r\cos\theta} + \frac{2G\bar{u}_\theta^1}{r\sin\theta} &= \left\{\nu\frac{d^3}{dz^3} + (1-\nu)\xi^2\frac{d}{dz}\right\}\bar{Z}^1,\\
\frac{\bar{\sigma}_{r\theta}^1}{\sin\theta} + \frac{2G\bar{u}_r^1}{r\cos\theta} + \frac{2G\bar{u}_\theta^1}{r\sin\theta} &= \xi^2\bar{\psi}^1,\\
\frac{2G\bar{u}_z^1}{\cos\theta} &= \left\{(1-2\nu)\frac{d^2}{dz^2} - 2(1-\nu)\xi^2\right\}\bar{Z}^1.
\end{aligned} \tag{13-18.35}$$

These six transformed expressions can be inverted and the remaining six unknowns obtained.

The Cartesian components of the stresses and displacements for Cerruti's problem become

$$\begin{aligned}
u_x &= \frac{P}{4\pi GR}\left\{1 + \frac{x^2}{R^2} + (1-2\nu)\left[\frac{P}{R+z} - \frac{x^2}{(R+z)^2}\right]\right\},\\
u_y &= \frac{P}{4\pi GR}\left\{\frac{xy}{R^2} - \frac{(1-2\nu)xy}{(R+z)^2}\right\},\\
u_z &= \frac{P}{4\pi GR}\left\{\frac{xy}{R^2} + \frac{(1-2\nu)x}{R+z}\right\}.\\
\sigma_{xx} &= \frac{Px}{2\pi R^3}\left\{-\frac{3x^2}{R^2} + \frac{(1-2\nu)}{(R+z)^2}\left(R^2 - y^2 - \frac{2Ry^2}{R+z}\right)\right\},\\
\sigma_{yy} &= \frac{Px}{2\pi R^3}\left\{-\frac{3y^2}{R^2} + \frac{(1-2\nu)}{(R+z)^2}\left(3R^2 - x^2 - \frac{2Rx^2}{R+z}\right)\right\},\\
\sigma_{zz} &= -\frac{3}{2\pi}\frac{Pxz^2}{R^5},\\
\sigma_{xy} &= \frac{Py}{2\pi R^3}\left\{-\frac{3x^2}{R^2} + \frac{(1-2\nu)}{(R+z)^2}\left(-R^2 + x^2 + \frac{2Rx^2}{R+z}\right)\right\},\\
\sigma_{yz} &= \frac{3Pxyz}{2\pi R^5},\\
\sigma_{xz} &= -\frac{3Px^2 z}{2\pi R^5},
\end{aligned} \tag{13-18.36}$$

where $$R = (x^2 + y^2 + z^2)^{1/2}.$$

PROBLEMS

13.1. Verify equation 13-2.26.

13.2. Determine the stress distribution for a cylinder of length $2l$ and radius R, which is loaded by a uniform radial load of width $2b$ at its center.

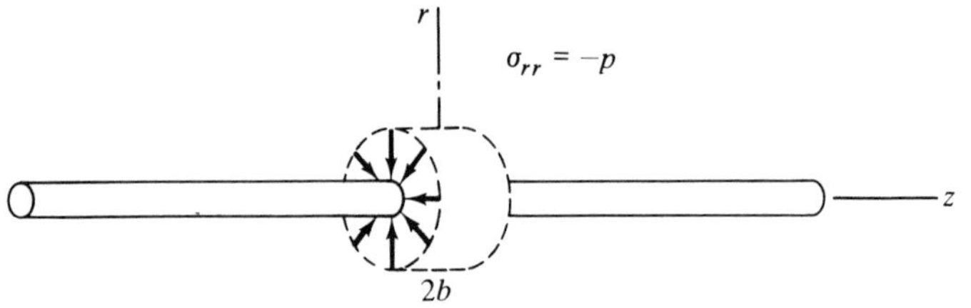

Problem 13-2

13.3. Determine the general expressions for an infinite cylinder loaded with a prescribed shear on the outer boundary and zero radial stress.

13.4. Solve the even problem corresponding to section 9.

13.5. Solve the even problem corresponding to section 10.

13.6. Determine the force P required to push the punch in and the distance b (smooth punch at $r/a = 1$) and the stress distribution under the punch.

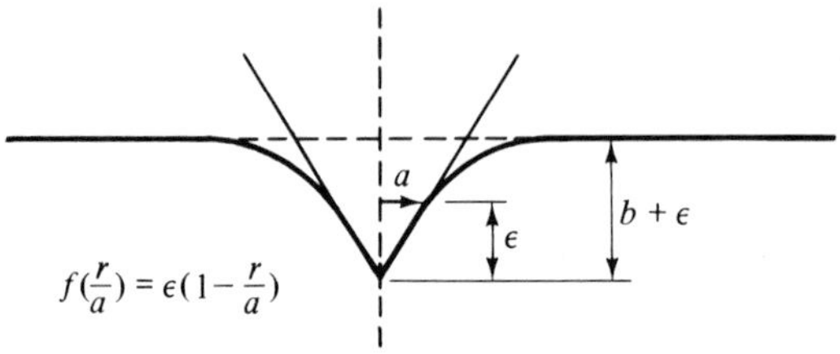

Problem 13-6

13.7. Construct the solution of the distribution of stress in a solid with a penny-shaped crack under uniform tension p. Note this is equivalent to the sum of the solutions of the problem of a solid without the crack under uniform tension (solution: $\sigma_{zz} = p$, all other stresses equal to zero) and the problem shown below:

$$z = 0,$$

$$\sigma_{zz} = -p, \qquad 0 \leq r < 1,$$

$$u_z = 0, \qquad r > 1,$$

$$\sigma_{rz} = 0, \qquad 0 \leq r < \infty.$$

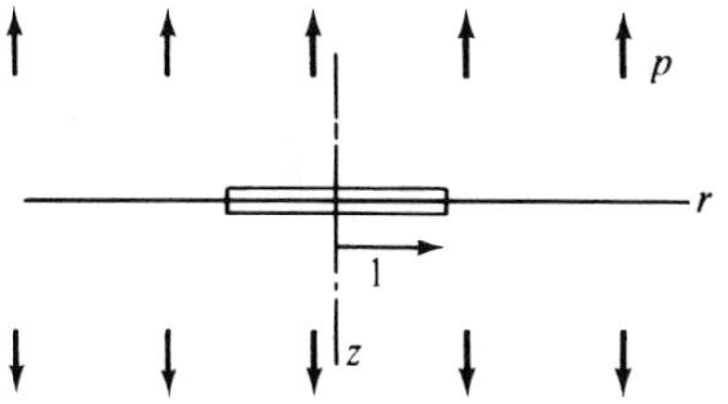

Problem 13-7.1

Solve this second problem using a method similar to the one used to solve the dual integral equations in the punch problems. Determine the shape of the crack (u_z for r less than 1).

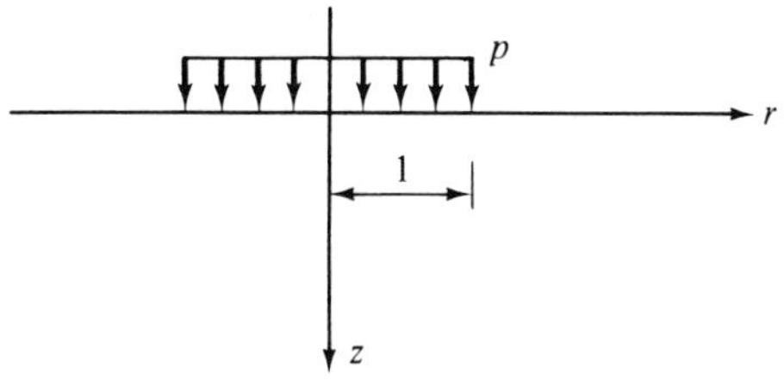

Problem 13-7.2

13.8. Set up the solution for a long cylinder ($-\infty < z < \infty$) under the loading

$$\sigma_{rr} = -p \begin{cases} -\beta \leq \theta \leq \beta, & r = 1 \\ \pi - \beta \leq \theta \leq \pi + \beta, & -c \leq z \leq +c \end{cases}$$

$$\sigma_{rz} = 0, \quad r = 1, \quad -\infty < z < +\infty, \qquad 0 \leq \theta < 2\pi.$$

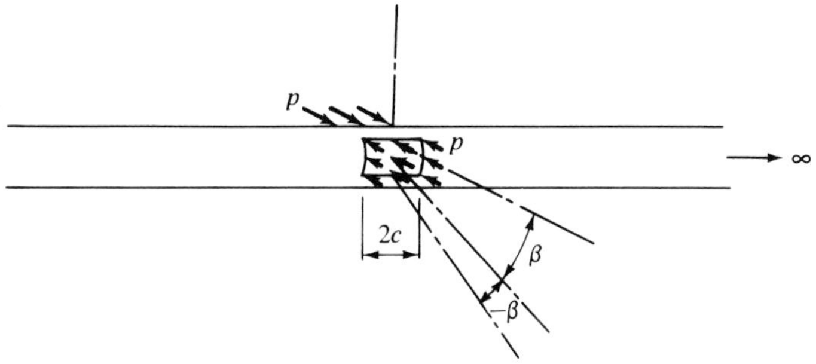

Problem 13-8

CHAPTER 14

PAPKOVICH-NEUBER SOLUTION

1 Introduction

Papkovich in 1932* and Neuber in 1934† independently developed solutions of the Navier equations of elastostatics in terms of a harmonic vector and a harmonic function. The Navier equation can be written as

$$\nabla^2\bar{u} + \frac{1}{(1-2\nu)}\vec{\nabla}\vec{\nabla}\cdot\bar{u} + \frac{\bar{F}}{G} = 0. \tag{14-1.1}$$

By use of the Helmholz theorem, any vector field that vanishes at infinity may be represented as the sum of a solenoidal and irrotational vector:

$$\bar{u} = \bar{u}_S + \bar{u}_{IR} + \bar{u}_0, \tag{14-1.2}$$

* P. F. Papkovich, "An expression for a general integral of the equations of the theory of elasticity in terms of harmonic functions," *Izvest. Akad. Nauk SSSR, Ser. Matem. k. estestv. neuk*, No. 10, 1932.

† H. Neuber, "Ein neuer Censatz zur Lösung räumlicher Probleme der Elastezetatstheorie," *Z. Angew. Math. Mech.* **14**, 1934.

where

$$\vec{\nabla}\cdot\bar{u}_S = 0,$$
$$\vec{\nabla} \times u_{IR} = 0.$$

The vector $\bar{u}_0$ is the particular solution satisfying the equation

$$\nabla^2\bar{u}_0 + \frac{1}{(1-2\nu)}\vec{\nabla}\vec{\nabla}\cdot\bar{u}_0 = -\frac{\bar{F}}{G} = 0. \tag{14-1.3}$$

Equations 14-1.1, 14-1.2, and 14-1.3 yield

$$\nabla^2(\bar{u}_{IR} + \bar{u}_S) + \frac{1}{(1-2\nu)}\vec{\nabla}(\vec{\nabla}\cdot\bar{u}_{IR}). \tag{14-1.4}$$

Noting the vector identity

$$\nabla^2\bar{A} = \vec{\nabla}\vec{\nabla}\cdot\bar{A} - \vec{\nabla}\times(\vec{\nabla}\times\bar{A}), \tag{14-1.5}$$

equation 14-1.4 can be separated into two equations:

$$\begin{aligned} 2(1-\nu)\bar{\nabla}\bar{\nabla}\cdot u_{IR} &= 0, \\ \bar{\nabla}\times(\bar{\nabla}\times\bar{u}_S) &= 0. \end{aligned} \tag{14-1.6}$$

Using equation 14-1.5 we can note that both $\bar{u}_{IR}$ and $\bar{u}_S$ are harmonic vectors. The vector $\bar{u}_{IR}$ is irrotational and can be derived from a potential function:

$$\bar{u}_{IR} = \bar{\nabla}\chi. \tag{14-1.7}$$

We are thus motivated to choose $\bar{u}$ in the form

$$\bar{u} = \bar{u}_0 + \bar{B} + \bar{\nabla}\chi, \tag{14-1.8}$$

where $\bar{u}_0$ satisfies equation 14-1.3 and $\bar{B}$ is a harmonic vector

$$\nabla^2\bar{B} = 0. \tag{14-1.9}$$

Substitution of equation 14-1.8 into 14-1.1 yields the following interrelation for $\bar{B}$ and χ:

$$2(1-\nu)\bar{\nabla}\nabla^2\chi + \bar{\nabla}\bar{\nabla}\cdot\bar{B} = 0, \tag{14-1.10}$$

$$\nabla^2\chi = -\frac{1}{2(1-\nu)}\bar{\nabla}\cdot\bar{B}. \tag{14-1.11}$$

We take the solution of 14-1.11 to be

$$\chi = -\frac{1}{4(1-\nu)}(\bar{R}\cdot\bar{B} + B_0), \tag{14-1.12}$$

where $\bar{R}$ is the position vector and B_0 is an arbitrary harmonic function.

We can verify that 14-1.12 is the solution to 14-1.11 by examining these equations in Cartesian component notation:

$$\chi = -\frac{1}{4(1-\nu)}(x_iB_i + B_0),$$

$$\frac{\partial^2\chi}{\partial x_k\,\partial x_k} = -\frac{1}{4(1-\nu)}\frac{\partial}{\partial x_k}\left[\delta_{ik}B_i + x_i\frac{\partial B_i}{\partial x_k} + \frac{\partial B_0}{\partial x_k}\right]$$
$$= -\frac{1}{2(1-\nu)}\frac{\partial B_k}{\partial x_k}.$$

Substitution of equation 14-1.12 in 14-1.8 yields

$$\bar{u} = \bar{u}_0 + \bar{B} - \frac{1}{4(1-\nu)}\vec{\nabla}(\bar{R}\cdot\bar{B} + B_0). \qquad (14\text{-}1.13)$$

The strain tensor is obtained from equation 14-1.13:

$$\bar{\bar{\epsilon}} = \tfrac{1}{2}[\vec{\nabla}\bar{u} + \bar{u}\overleftarrow{\nabla}], \qquad (14\text{-}1.14)$$

$$\bar{\bar{\epsilon}} = \bar{\bar{\epsilon}}_0 + \frac{1}{2}[\vec{\nabla}\bar{B} + \bar{B}\overleftarrow{\nabla}] - \frac{1}{4(1-\nu)}\vec{\nabla}\vec{\nabla}(\bar{R}\cdot\bar{B} + B_0), \qquad (14\text{-}1.15)$$

where $\bar{\bar{\epsilon}}_0$ is the strain associated with body forces. The first invariant of this tensor, E, is the trace of equation 14-1.15:

$$E = E_0 + \frac{(1-2\nu)}{2(1-\nu)}\bar{\nabla}\cdot\bar{B}. \qquad (14\text{-}1.16)$$

The stresses are calculated using Hooke's law:

$$\bar{\bar{\sigma}} = 2G\left\{\bar{\bar{\epsilon}} + \frac{\nu}{1-2\nu}E\bar{\bar{\mathrm{I}}}\right\}. \qquad (14\text{-}1.17)$$

One needs to retain only three of the four functions for a particular problem. A discussion of this point is found in Sokolnikoff's* *Mathematical Theory of Elasticity*, where he indicates that one should not arbitrarily set any one component equal to zero. The choice of a zero function depends upon the topology of the domain. In Cartesian coordinates each of the four functions B_0, B_x, B_y, and B_z are harmonic. In curvilinear coordinate systems, in general only B_0 will be harmonic, because the components of the harmonic vector need not be harmonic. We shall retain all four functions for the present to give us certain flexibility in choosing the form of the solution.

2 Concentrated force in the infinite solid

Isolation of the type of singularity associated with a concentrated force within the infinite elastic space is an important problem. We use the solution of this problem to generate the solution of body force distributions and other types of concentrated loadings.

Consider a concentrated force $\bar{P}$ applied at the origin within an elastic medium. This is called the Kelvin problem† (Figure 14-1). Let $\bar{R}$ designate a position vector with unit vector $\hat{i}_{(R)}$. Consider an arbitrary volume enclosing the origin and let $\bar{t}^n$ be a surface traction acting on an element of surface of this volume. By equilibrium conditions,

* See Sokolnikoff, *op. cit.*

† W. Thomson (Lord Kelvin), "Note on the integration of the equations of equilibrium of an elastic solid," *Mathematical and Physical Papers*, Vol. 1, Cambridge, 1882.

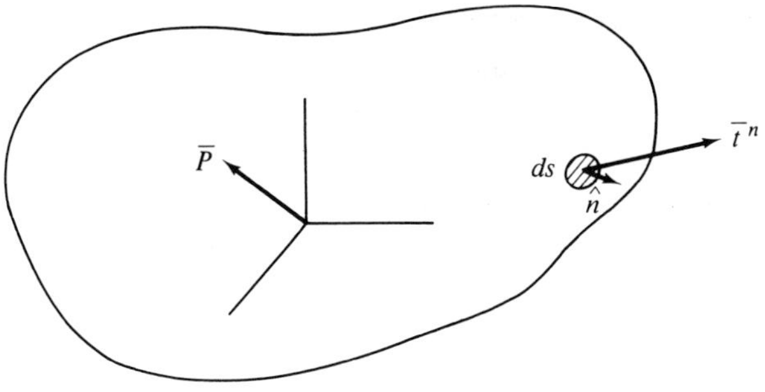

Figure 14-1

$$\bar{P} = -\int_S \bar{t}^n\, dS, \tag{14-2.1}$$

$$\bar{M} = 0 = \int_S \bar{R} \times \bar{t}^n\, dS. \tag{14-2.2}$$

If the enclosing surface is a sphere of radius R, then the element of surface area becomes

$$dS = R^2 \sin \varphi\, d\varphi\, d\theta, \tag{14-2.3}$$

and the normal vector n becomes $\hat{i}_{(R)}$. The stress traction vector is related to the stress tensor as follows:

$$\bar{t}^n = \hat{n} \cdot \bar{\bar{\sigma}}. \tag{14-2.4}$$

Equation 14-2.1 becomes

$$\bar{P} = -\iint_S \hat{i}_{(R)} \cdot \bar{\bar{\sigma}} R^2 \sin \varphi\, d\varphi\, d\theta. \tag{14-2.5}$$

If the integral is to be independent of the radius of the sphere, the components of the stress tensor involved must vary inversely proportional to the square of the radius R. The displacement vector therefore varies inversely as the radius R.

Starting with the Papkovich–Neuber formulation in the form

$$\bar{u} = \bar{B} - \frac{1}{4(1-\nu)} \vec{\nabla}(\bar{R} \cdot \bar{B} + B_0), \tag{14-2.6}$$

we seek a solution of the form

$$\bar{B} = \bar{P}\varphi, \qquad B_0 = 0, \tag{14-2.7}$$

where φ is a harmonic function and $\bar{P}$ is the constant force vector.

Substitution of equation 14-2.7 into equation 14-2.6 after examination of the gradient of the dot product of the position vector with $\bar{B}$ yields

$$\begin{aligned} \vec{\nabla}(\bar{R} \cdot \bar{B}) &= \vec{\nabla}(\bar{R} \cdot \bar{P}\varphi) \\ &= \bar{\bar{I}} \cdot \bar{P}\varphi + (\bar{R} \cdot \bar{P})\bar{\nabla}\varphi \\ &= \bar{P}\varphi + (\bar{R} \cdot \bar{P})\bar{\nabla}\varphi. \end{aligned} \tag{14-2.8}$$

The displacement vector becomes

$$\bar{u} = \frac{3 - 4\nu}{4(1 - \nu)}\bar{P}\varphi - \frac{1}{4(1 - \nu)}(\bar{R}\cdot\bar{P})\bar{\nabla}\varphi. \qquad (14\text{-}2.9)$$

As we have seen previously, we expect φ to be a harmonic function that is singular at the origin and approaches infinity as $1/R$. These conditions indicate that we should assume a solution in the form

$$\varphi = \frac{A}{R}. \qquad (14\text{-}2.10)$$

The function φ is harmonic, and in spherical coordinates Laplace's equation becomes

$$\nabla^2\varphi = \frac{1}{R^2}\frac{\partial}{\partial R}\left(R^2\frac{\partial\varphi}{\partial R}\right) = 0.$$

To determine the constant A, we calculate the stress traction on a surface with unit normal $\hat{n}$:

$$\bar{t}^n = \hat{n}\cdot\bar{\bar{\sigma}} = \hat{n}\cdot\left[2G\left\{\bar{\bar{\epsilon}} + \frac{\nu}{1 - 2\nu}E\bar{\bar{I}}\right\}\right]. \qquad (14\text{-}2.11)$$

The first term is

$$\hat{n}\cdot\bar{\bar{\epsilon}} = \hat{n}\cdot\vec{\nabla}\bar{u} + \tfrac{1}{2}\hat{n}\cdot(\bar{u}\overleftarrow{\nabla} - \vec{\nabla}\bar{u}).$$

The last term can be written in terms of the curl of $\bar{u}$ by use of the dual vector of the skew-symmetric tensor. This expression becomes

$$\hat{n}\cdot\bar{\bar{\epsilon}} = \hat{n}\cdot\vec{\nabla}\bar{u} + \tfrac{1}{2}\hat{n}\times(\vec{\nabla}\times\bar{u}).$$

$$\begin{aligned}\hat{n}\times(\vec{\nabla}\times\bar{u}) &\Longrightarrow e_{ijk}e_{klm}n_j\nabla_l u_m \\ &= (\delta_{il}\delta_{jm} - \delta_{im}\partial_{jl})n_j\nabla_l u_m \\ &= n_j\nabla_i u_j - n_j\nabla_j u_i \Longrightarrow \hat{n}\cdot(\bar{u}\overleftarrow{\nabla} - \vec{\nabla}\bar{u}).\end{aligned}$$

Equation 14-2.11 can now be written

$$\bar{t}^n = 2G\left\{\hat{n}\cdot\bar{\nabla}\bar{u} + \tfrac{1}{2}\hat{n}\times(\bar{\nabla}\times\bar{u}) + \frac{\nu}{(1 - 2\nu)}\hat{n}\bar{\nabla}\cdot\bar{u}\right\}. \qquad (14\text{-}2.12)$$

Substitution of equation 14-2.10 into 14-2.9 yields the following expression for $\bar{u}$:

$$\bar{u} = \frac{A}{4(1 - \nu)}\left\{(3 - 4\nu)\frac{\bar{P}}{R} + \frac{(\bar{R}\cdot\bar{P})\bar{R}}{R^3}\right\}. \qquad (14\text{-}2.13)$$

Substitution of 14-2.13 into 14-2.12 involves considerable mathematical manipulations, but yields the stress traction vector as follows:

$$\begin{aligned}\bar{t}^n = -\frac{GA(1 - 2\nu)}{2(1 - \nu)R^3}\Big\{&-(\bar{P}\cdot\bar{R})\hat{n} + (\hat{n}\cdot\bar{P})\bar{R} + \bar{P}(\hat{n}\cdot\bar{R}) \\ &+ \frac{3}{1 - 2\nu}\frac{1}{R^2}(\bar{P}\cdot\bar{R})(\hat{n}\cdot\bar{R})\bar{R}\Big\}.\end{aligned} \qquad (14\text{-}2.14)$$

When $\hat{n} = \hat{\imath}_R$ for a spherical surface, equation 14-2.14 becomes

$$\bar{t}^R = -\frac{GA(1-2\nu)}{2(1-\nu)R^2}\left\{\bar{P} + \frac{3}{1-2\nu}(\bar{P}\cdot\hat{\imath}_{(R)})\hat{\imath}_{(R)}\right\}. \qquad (14\text{-}2.15)$$

For a spherical surface of radius R, substitution of 14-2.15 into 14-2.1 yields

$$\bar{P} = \frac{GA(1-2\nu)}{2(1-\nu)R^2}\iint_S \left[\bar{P} + \frac{3}{1-2\nu}(\bar{P}\cdot\hat{\imath}_{(R)})\hat{\imath}_{(R)}\right]R^2\, dS^*, \qquad (14\text{-}2.16)$$

where $dS^* = \sin\varphi\, d\varphi\, d\theta$ (the surface element for a unit sphere).

The first term under the integral sign yields

$$\iint_S \bar{P}R^2\, dS^* = \bar{P}R^2 \iint_S dS^* = 4\pi R^2 \bar{P}.$$

The second term can be evaluated by use of the divergence theorem if, in the surface integral, $\hat{\imath}_{(R)}$ is replaced to its value on the boundary of a unit sphere, that is, $\bar{R}$:

$$\frac{3R^2}{(1-2\nu)}\iint_S \hat{\imath}_{(R)}(\bar{P}\cdot\hat{\imath}_{(R)})\, dS^* = \frac{3R^2}{(1-2\nu)}\iint_S \hat{\imath}_{(R)}(\bar{P}\cdot\bar{R})\, dS^*$$
$$= \frac{3R^2}{1-2\nu}\iiint_V \bar{\nabla}(\bar{P}\cdot\bar{R})\, dV^* = \frac{3R^2}{1-2\nu}\frac{4\pi}{3}\bar{P}.$$

Equation 14-2.16 becomes

$$\bar{P} = \frac{GA(1-2\nu)}{2(1-\nu)}\left[4\pi + \frac{4\pi}{1-2\nu}\right]\bar{P} = 4\pi GA\bar{P}.$$

The constant A becomes

$$A = \frac{1}{4\pi G}. \qquad (14\text{-}2.17)$$

The displacement vector becomes

$$\bar{u} = \frac{1}{16\pi G(1-\nu)}\frac{1}{R}\{(3-4\nu)\bar{P} + (\hat{\imath}_{(R)}\cdot\bar{P})\hat{\imath}_{(R)}\}. \qquad (14\text{-}2.18)$$

The stresses are calculated from the stress traction vector by noting that

$$\sigma_{nm} = \bar{t}^n\cdot\hat{m}. \qquad (14\text{-}2.19)$$

Substituting 14-2.17 into 14-2.14 yields

$$\bar{t}^n = -\frac{(1-2\nu)}{8\pi(1-\nu)R^2}\left\{-(\bar{P}\cdot\hat{\imath}_{(R)})\hat{n} + (\hat{n}\cdot\bar{P})\hat{\imath}_{(R)} + \bar{P}(\hat{n}\cdot\hat{\imath}_{(R)}) + \frac{3}{1-2\nu}(\bar{P}\cdot\hat{\imath}_{(R)})(\hat{n}\cdot\hat{\imath}_{(R)})\hat{\imath}_{(R)}\right\}. \qquad (14\text{-}2.20)$$

From 14-2.19, we obtain

$$\sigma_{nm} = -\frac{(1-2\nu)}{8\pi(1-\nu)R^2}\left\{-(\bar{P}\cdot\hat{\imath}_{(R)})(\hat{n}\cdot\hat{m}) + (\hat{n}\cdot\bar{P})(\hat{\imath}_{(R)}\cdot\hat{m}) + (\hat{m}\cdot\bar{P})(\hat{\imath}_{(R)}\cdot\hat{n}) + \frac{3}{1-2\nu}(\bar{P}\cdot\hat{\imath}_{(R)})(\hat{n}\cdot\hat{\imath}_{(R)})(\hat{m}\cdot\hat{\imath}_{(R)})\right\}, \qquad (14\text{-}2.21)$$

where both $\hat{m}$ and $\hat{n}$ are unit base vectors of an orthogonal coordinate system. As an example, consider the following case:

$$\bar{P} = -P\hat{k}, \qquad \hat{n} = \hat{k},$$

$$\sigma_{zz} = \frac{1-2\nu}{8\pi(1-\nu)}\frac{P}{R^2}\left\{\frac{z}{R} + \frac{3}{1-2\nu}\frac{z^3}{R^3}\right\},$$

$$\sigma_{zx} = \frac{1-2\nu}{8\pi(1-\nu)}\frac{P}{R^2}\left\{\frac{x}{R} + \frac{3}{1-2\nu}\frac{xz^2}{R^3}\right\}, \qquad (14\text{-}2.22)$$

$$\sigma_{zy} = \frac{1-2\nu}{8\pi(1-\nu)}\frac{P}{R^2}\left\{\frac{y}{R} + \frac{3}{1-2\nu}\frac{yz^2}{R^3}\right\}.$$

When $\nu = \frac{1}{2}$, the Kelvin solution reduces to the Boussinesq solution.

3 Concentrated load not at the origin

Note that $\bar{R}' = \bar{R} - \bar{\rho}$ and that $R' = [(x_1 - \rho_1)^2 + (x_2 - \rho_2)^2 + (x_3 - \rho_3)^2]^{1/2}$. The displacement vector can be written from equation 14-2.18 in the form

$$\bar{u} = \frac{1}{16\pi G(1-\nu)}\left\{\frac{(3-4\nu)\bar{P}}{R'} + \frac{(\bar{R}'\cdot\bar{P})\bar{R}'}{(R')^3}\right\}. \qquad (14\text{-}3.1)$$

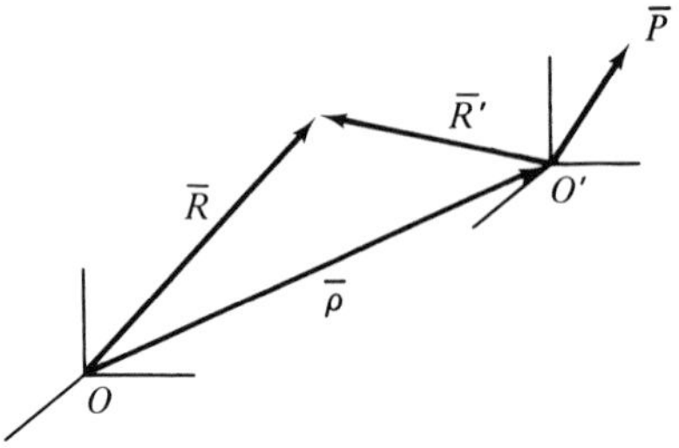

Figure 14-2

If the concentrated force is distributed or smeared across some region, we obtain the solution by integrating the solution given above. The displacement vector becomes

$$\bar{u} = \frac{1}{16\pi G(1-\nu)}\iiint_V \left\{(3-4\nu)\frac{\bar{P}(\xi)}{R'} + \frac{[\bar{R}'\cdot\bar{P}(\xi)]\bar{R}'}{(R')^3}\right\} dV, \qquad (14\text{-}3.2)$$

where

$$R' = [(x_1 - \xi_1)^2 + (x_2 - \xi_2)^2 + (x_3 - \xi_3)^2]^{1/2}$$

and

$$dV = d\xi_1\, d\xi_2\, d\xi_3.$$

This is a particular integral of the Navier equation obtained by Lord

Kelvin in 1848. A different approach to this type of singularity is presented by Luré (*Three Dimensional Problems of the Theory of Elasticity*).* In this approach he considers also the concentrated moment and center of dilatation. The center of dilatation is considered in the next section and is a useful building block for the construction of other solutions.

4 Other special singular solutions

Two particular solutions to the Navier equations represent singularities associated with centers of rotation and dilatation. These become important when these centers are not part of the elastic domain. In particular, we consider the cases when these centers are continuously distributed along semi-infinite lines. We shall consider the case of a center of dilatation at the origin first. For a center of dilatation the displacements are irrotational and and singular at the origin. Because the displacement vector is irrotational,

$$\bar{\nabla} \times \bar{u} = 0. \tag{14-4.1}$$

A potential function is defined

$$\bar{u} = \vec{\nabla}\varphi. \tag{14-4.2}$$

This formulation was initially considered by G. Lamé in 1852 and led to consideration of the problem of internal pressure in a hollow cylinder, which bears his name. We have seen in section 1 that the irrotational part and the solenoidal parts of the displacement vector are harmonic. Therefore, we write

$$\vec{\nabla}\nabla^2\varphi = 0, \tag{14-4.3}$$

$$\nabla^2\varphi = \text{constant}. \tag{14-4.4}$$

If we seek some solution and not a general solution, we shall make the particular choice of this constant as zero. φ is a harmonic function and is taken to be

$$\varphi = \frac{A}{R}, \tag{14-4.5}$$

$$\bar{u} = A\vec{\nabla}\left(\frac{1}{R}\right). \tag{14-4.6}$$

If this center is located at some point $0'$ removed from the origin, and $\bar{R}'$ designates a vector measured from this point, the displacement vector becomes

$$\bar{u} = A\vec{\nabla}\left(\frac{1}{R'}\right). \tag{14-4.7}$$

Now consider the case when the center of dilatation is continuously distributed along a line running from the origin to infinity. We designate by

* See Luré, *Three-Dimensional Problems of Elasticity*.

λ a length along this line and the direction of the line by the unit vector $\hat{i}$ (Figure 14-3).

$$\bar{R}' = \bar{R} - \lambda\hat{i}, \qquad R' = [R^2 + \lambda^2 - 2\lambda R\cos\gamma]^{1/2}, \qquad (14\text{-}4.8)$$

where γ is the angle between $\bar{R}$ and $\hat{i}$.

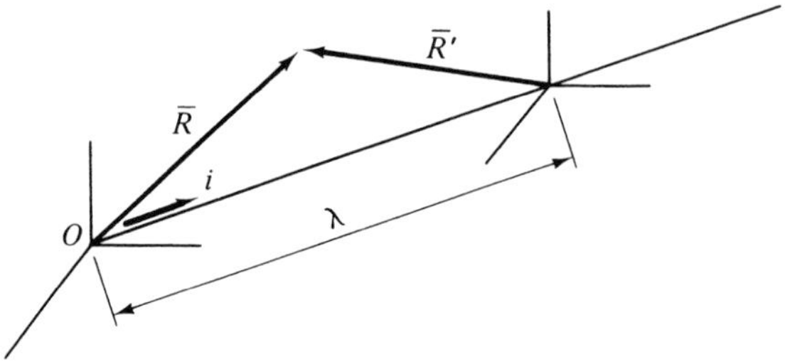

Figure 14-3

To obtain the solution of the centers of dilatation along this line, we consider the integral

$$\int_0^\infty d\lambda\, \vec{\nabla}\left(\frac{1}{R'}\right) = \vec{\nabla}\int_0^\infty \frac{d\lambda}{R'}. \qquad (14\text{-}4.9)$$

This integral can not be evaluated directly and is evaluated by a limiting process allowing the upper limit to go to infinity and noting the unbounded term cancels, due to the gradient. The integral becomes

$$\int_0^\infty d\lambda\, \vec{\nabla}\left(\frac{1}{R'}\right) = -\vec{\nabla}\log(R - R\cos\gamma) = -\vec{\nabla}\log(R - \bar{R}\cdot\hat{i}). \qquad (14\text{-}4.10)$$

The solution takes the form

$$\bar{u} = -A\vec{\nabla}\log(R - \bar{R}\cdot\hat{i}). \qquad (14\text{-}4.11)$$

This solution is called the Boussinesq solution of the second type.

The solution corresponding to a continuously distributed line of centers of rotation is obtained in a similar manner and is a solenoidal vector:

$$u = \bar{C} \times \vec{\nabla}\log(R - \bar{R}\cdot\hat{i}), \qquad (14\text{-}4.12)$$

where $\bar{C}$ is a constant vector.

Equations 14-4.11 and 14-4.12 correspond to solutions of the equations of elasticity outside the region of a cone of an arbitrary small opening angle whose apex is at the origin and whose axis coincides with the line $\lambda\hat{i}$. This excludes the region where $R = \bar{R}\cdot\hat{i}$.

The stresses in each of these cases are calculated by considering the stress tractions, as shown in section 13-2. The stress traction can be obtained by use of equation 14-2.12. For the case of the center of dilatations, equation

14-4.11 yields

$$\bar{t}^n = 2GA(\hat{n}\cdot\vec{\nabla})\vec{\nabla}\log(R - \bar{R}\cdot\hat{\imath}). \qquad (14\text{-}4.13)$$

If one considers the case when the center of dilatations lies along the negative z axis, $\hat{\imath} = -\hat{k}$, the stress traction on the z plane becomes

$$\bar{t}^z = -2GA\left(\frac{\bar{R}}{R^3}\right), \qquad (14\text{-}4.14)$$

and the stresses related to this vector become

$$\sigma_{xz} = -2GA\frac{x}{R^3}, \qquad \sigma_{yz} = -2GA\frac{y}{R^3},$$
$$\sigma_{zz} = -2GA\frac{z}{R^3}. \qquad (14\text{-}4.15)$$

The Boussinesq solution may be constructed by use of the two singular solutions of a concentrated load in the elastic space and the line of center of dilatations lying along the negative z axis. The constant A is chosen so that the boundary stresses at $z = 0$ are zero, except at the origin. From equation 14-2.22 and equation 14-4.15, we obtain

$$A = -\frac{P(1 - 2\nu)}{16\pi G(1 - \nu)}. \qquad (14\text{-}4.16)$$

If the load P is taken to be $4(1 - \nu)F$, we obtain the Boussinesq solution, as shown in section 13-13. We could have obtained this solution by use of a Papkovich–Neuber solution of the form

$$\bar{u} = \frac{F}{\pi G}\left\{(1 - \nu)\frac{\hat{k}}{R} - \frac{1}{4}\vec{\nabla}\left(\frac{z}{R} + (1 - 2\nu)\log(R + z)\right)\right\}. \qquad (14\text{-}4.17)$$

These displacements are

$$u_x = \frac{Fx}{4\pi G}\left[\frac{z}{R^3} - (1 - 2\nu)\frac{1}{R(R + z)}\right],$$
$$u_y = \frac{Fy}{4\pi G}\left[\frac{z}{R^3} - (1 - 2\nu)\frac{1}{R(R + z)}\right], \qquad (14\text{-}4.18)$$
$$u_z = \frac{F}{4\pi G}\left[\frac{z^2}{R^3} + 2(1 - \nu)\frac{1}{R}\right].$$

As indicated in section 13-13, this solution can be generalized by considering the load to be applied at some point other than the origin. For example, if the coordinates of the loading point were $(\xi, \zeta, 0)$, the solution becomes

$$u_x = \frac{F(x - \xi)}{\pi G}\left[\frac{z}{(R')^3} - (1 - 2\nu)\frac{1}{R'(R' + z)}\right],$$
$$u_y = \frac{F(y - \zeta)}{\pi G}\left[\frac{z}{(R')^3} - (1 - 2\nu)\frac{1}{R'(R' + z)}\right], \qquad (14\text{-}4.19)$$
$$u_z = \frac{F}{4\pi G}\left[\frac{z^2}{(R')^3} + 2(1 - \nu)\frac{1}{R'}\right],$$

where

$$R' = [(x - \xi)^2 + (y - \zeta)^2 + z^2]^{1/2}.$$

This solution can be smeared to give any normal loading on the half-plane by considering F to be a function of ξ and ζ and integrating. Note that we are now free from any axisymmetric constraint on loading normal to the half-plane.

Westergaard observed that the Kelvin solution satisfied the required boundary conditions for the Boussinesq problem when $\nu = \frac{1}{2}$. He developed a perturbation method on ν to obtain the Kelvin solution. When $\nu = \frac{1}{2}$ the material is incompressible, and it is not too surprising that the Kelvin and Boussinesq problems are related through the solution of center of dilatations.

5 Concentrated load tangential to the surface of the half-space—Cerruti's problem

Westergaard noted that the Kelvin solution (equation 14-2.21) reduces to the Cerruti solution if the material is incompressible ($\nu = \frac{1}{2}$). We can show this by setting $\bar{P} = P\hat{\imath}$ and observing the stresses on a plane $z =$ constant. From equation 14-2.21, we obtain

$$\begin{aligned} \sigma_{zx} &= -\frac{(1-2\nu)P}{8\pi(1-\nu)R^2}\left\{\frac{z}{R} + \frac{3}{(1-2\nu)}\frac{x^2 z}{R^3}\right\}, \\ \sigma_{zy} &= -\frac{(1-2\nu)P}{8\pi(1-\nu)R^2}\left\{\frac{3}{(1-2\nu)}\frac{xyz}{R^3}\right\}, \\ \sigma_{zz} &= -\frac{(1-2\nu)P}{8\pi(1-\nu)R^2}\left\{-\frac{x}{R} + \frac{3}{(1-2\nu)}\frac{xz^2}{R^3}\right\}. \end{aligned} \tag{14-5.1}$$

If $\nu = \frac{1}{2}$, equation 14-5.1 becomes

$$\begin{aligned} \sigma_{zx} &= -\frac{3P}{4\pi R^2}\frac{x^2 z}{R^3}, \\ \sigma_{zy} &= -\frac{3P}{4\pi R^2}\frac{xyz}{R^3}, \\ \sigma_{zz} &= -\frac{3P}{4\pi R^2}\frac{xz^2}{R^3}. \end{aligned} \tag{14-5.2}$$

The boundary condition that the surface $z = 0$ is stress free except at the origin is satisfied by equation 14-5.2. It is not difficult to imagine that the Cerruti solution can be considered to be a combination of the Kelvin solution and other singular solutions. Originally, Lord Kelvin, Boussinesq, and

Cerruti obtained their solutions by use of singularities from potential theory. These solutions were obtained in 1848, 1878, and 1882, respectively.

Luré shows that this solution can be obtained by the sum of three singularities: the Kelvin solution, a center of rotations along the negative z axis, and a different singularity referred to as *double line of centers of dilatation.* The development of the solution by that approach will not be shown here. The displacements corresponding to a concentrated load in the x direction were obtained in section 13-18 and will be repeated here for reference only:

$$u_x = \frac{P}{4\pi G}\left\{\frac{1}{R} + \frac{x^2}{R^3} + (1-2\nu)\left[\frac{1}{(R+z)} - \frac{x^2}{R(R+z)^2}\right]\right\},$$
$$u_y = \frac{P}{4\pi G}\left\{\frac{xy}{R^3} - (1-2\nu)\frac{xy}{R(R+z)^2}\right\}, \qquad (14\text{-}5.3)$$
$$u_z = \frac{P}{4\pi G}\left\{\frac{xz}{R^3} + (1-2\nu)\frac{x}{R(R+z)}\right\}.$$

This is one of the important solutions in three-dimensional elasticity because it can be smeared by integration across the half-space, generating many more solutions.

These solutions can be generalized by combining the Boussinesq solution and the Cerruti solution in both the x and y directions. The displacements for this combination are

$$\begin{aligned}4\pi G u_x = {} & P_x\left[\frac{1}{R} + \frac{x^2}{R^3} + (1-2\nu)\left(\frac{1}{R+z} - \frac{x^2}{R(R+z)^2}\right)\right] \\ & + P_y\left[\frac{xy}{R^3} - (1-2\nu)\frac{xy}{R(R+z)^2}\right] \\ & + P_z\left[\frac{xz}{R^3} - (1-2\nu)\frac{x}{R(R+z)}\right],\end{aligned}$$

$$\begin{aligned}4\pi G u_y = {} & P_x\left[\frac{xy}{R^3} - (1-2\nu)\frac{xy}{R(R+z)^2}\right] \\ & + P_y\left[\frac{1}{R} + \frac{y^2}{R^3} + (1-2\nu)\left(\frac{1}{R+z} - \frac{y^2}{R(R+z)^2}\right)\right] \qquad (14\text{-}5.4) \\ & + P_z\left[\frac{yz}{R^3} - (1-2\nu)\frac{y}{R(R+z)}\right],\end{aligned}$$

$$\begin{aligned}4\pi G u_z = {} & P_x\left[\frac{xz}{R^3} + (1-2\nu)\frac{x}{R(R+z)}\right] \\ & + P_y\left[\frac{yz}{R^3} + (1-2\nu)\frac{y}{R(R+z)}\right] \\ & + P_z\left[2(1-\nu)\frac{1}{R} + \frac{z^2}{R^3}\right].\end{aligned}$$

These relations can be written in vector form as

$$4\pi G\bar{u} = (3 - 4\nu)\frac{\bar{P}\cdot}{R} + \frac{\bar{R}(\bar{R}\cdot\bar{P})}{R^3} + 2(1 - 2\nu)(\bar{P} \times \hat{k}) \times \vec{\nabla} \log (R + z) - (1 - 2\nu)(\bar{P}\cdot\vec{\nabla})\vec{\nabla}[z \log (R + z) - R]. \qquad (14\text{-}5.5)$$

The first two terms correspond to the Kelvin solution, the third term to the centers of rotation along the negative z axis, and the last term to a double line of centers of dilatation which also lies along the negative z axis. Equation 14-5.5 can be reduced to the following expression:

$$4\pi G\bar{u} = \frac{\bar{P}}{R} + \frac{\bar{R}(\bar{R}\cdot\bar{P})}{R^3} + (1 - 2\nu)\bar{P}\cdot\vec{\nabla}\left[\frac{x\hat{i} + y\hat{j}}{R + z} + \hat{k} \log (R + z)\right]. \qquad (14\text{-}5.6)$$

Finally, this solution can be spread over the half-space by considering $\bar{P}$ to be a function of ξ and ζ and replacing x by $(x - \xi)$, y by $(y - \zeta)$, and R by R', where $R' = [(x - \xi)^2 + (y - \zeta)^2 + z^2]^{1/2}$. This solution can be used to solve problems having any distribution of loading on the half-space.

6 Concentrated loads at the tip of an elastic cone

We now consider the problem of an elastic cone with a concentrated load at its apex (Figure 14-4). The problem is formulated in spherical coordinates to simplify the boundary conditions. If the cone angle is designated by α, the boundary conditions become

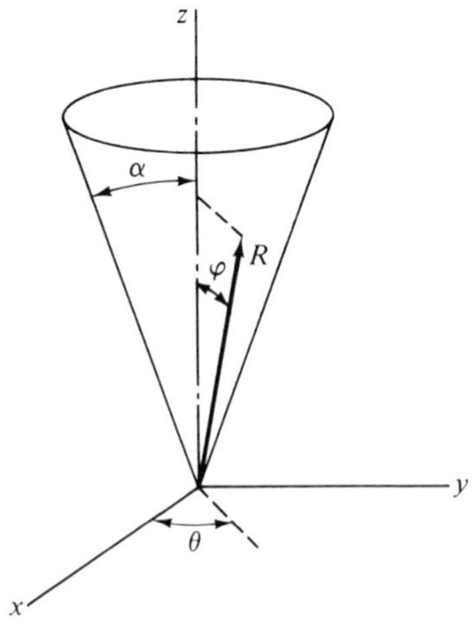

Figure 14-4

$$\varphi = \alpha, \qquad \sigma_{\varphi\varphi} = \sigma_{R\varphi} = \sigma_{\theta\varphi} = 0. \tag{14-6.1}$$

If we consider a load $\bar{F}$ applied at the apex of the cone along the cone axis, we note that the Boussinesq solution corresponds to this special case when $\alpha = \pi/2$. We can seek a solution in the form of a sum of the Kelvin solution and center of dilatations lying along the negative z direction. To facilitate calculation of the boundary conditions, the stresses for the Kelvin problem (equation 14-2.21) and centers of dilatation (equation 14-4.13) are written in spherical coordinates. The Kelvin solution, for a load $\bar{P} = -P\hat{k}$, in spherical coordinates is

$$\begin{aligned}
\sigma_{RR} &= \frac{2(2-\nu)P}{8\pi(1-\nu)R^2}\cos\varphi,\\
\sigma_{\varphi\varphi} &= \frac{(1-2\nu)P}{8\pi(1-\nu)R^2}\cos\varphi,\\
\sigma_{\theta\theta} &= \frac{(1-2\nu)P}{8\pi(1-\nu)R^2}\cos\varphi,\\
\sigma_{R\varphi} &= \frac{(1-2\nu)P}{8\pi(1-\nu)R^2}\sin\varphi,\\
\sigma_{R\theta} &= \sigma_{\theta\varphi} = 0.
\end{aligned} \tag{14-6.2}$$

Equation 14-4.13 for a line of center of dilatations is

$$\begin{aligned}
\bar{t}^n = 2GA\Big\{\hat{i}\frac{\bar{R}\cdot\hat{n} - \bar{R}(\hat{n}\cdot\hat{i})}{R(R-\bar{R}\cdot\hat{i})^2} + \frac{\hat{n}}{R(R-\bar{R}\cdot\hat{i})}\\
+ \bar{R}\frac{R^2(\hat{n}\cdot\hat{i}) - 2R(\bar{R}\cdot\hat{n}) + (\bar{R}\cdot\hat{n})(\bar{R}\cdot\hat{i})}{R^3(R-\bar{R}\cdot\hat{i})^2}\Big\},
\end{aligned} \tag{14-6.3}$$

$$\sigma_{nm} = \bar{t}^n\cdot\hat{m}. \tag{14-6.4}$$

Letting $\hat{i} = -\hat{k}$, the stress components in spherical coordinates become

$$\begin{aligned}
\sigma_{RR} &= -\frac{2GA}{R^2}, \qquad \sigma_{R\varphi} = \frac{2GA}{R^2}\frac{\sin\varphi}{1+\cos\varphi}, \qquad \sigma_{R\theta} = 0,\\
\sigma_{\varphi\varphi} &= \frac{2GA}{R^2}\frac{\cos\varphi}{1+\cos\varphi}, \qquad \sigma_{\theta\varphi} = 0, \qquad \sigma_{\theta\theta} = \frac{2GA}{R^2}\frac{1}{1+\cos\varphi}.
\end{aligned} \tag{14-6.5}$$

To satisfy the boundary conditions of equation 14-6.1, the following relation must hold:

$$\frac{2GA}{1+\cos\alpha} + \frac{P}{8\pi}\frac{(1-2\nu)}{(1-\nu)} = 0, \tag{14-6.6}$$

$$2GA = -\frac{P(1-2\nu)}{8\pi(1-\nu)}(1+\cos\alpha). \tag{14-6.7}$$

Note that, when $\alpha = \pi/2$, this is in agreement with equation 14-4.16 for the Boussinesq problem.

The stresses for the combined solution using equations 14-6.2, 14-6.5, and 14-6.7 become

$$\begin{aligned}
\sigma_{RR} &= \frac{P(1-2\nu)}{8\pi(1-\nu)R^2}\left\{1+\cos\alpha - \frac{2(2-\nu)}{1-2\nu}\cos\varphi\right\},\\
\sigma_{\varphi\varphi} &= \frac{P(1-2\nu)}{8\pi(1-\nu)R^2}\frac{\cos\varphi(\cos\varphi-\cos\alpha)}{1+\cos\varphi},\\
\sigma_{\theta\theta} &= \frac{P(1-2\nu)}{8\pi(1-\nu)R^2}\left\{\frac{\cos\varphi-\cos\alpha}{1+\cos\alpha} - 1 + \cos\varphi\right\},\\
\sigma_{R\varphi} &= \frac{P(1-2\nu)}{8\pi(1-\nu)R^2}\frac{\sin\varphi(\cos\varphi-\cos\alpha)}{1+\cos\varphi},\\
\sigma_{\theta\varphi} &= \sigma_{R\theta} = 0.
\end{aligned} \qquad (14\text{-}6.8)$$

P can now be related to the applied force F by the equilibrium condition

$$F + \int_0^{2\pi} d\theta \int_0^{\alpha} R^2 \sin\varphi(\sigma_{RR}\cos\varphi - \sigma_{R\varphi}\sin\varphi)\, d\varphi = 0. \qquad (14\text{-}6.9)$$

Equation 14-6.9 is the summation of forces in the z direction. Using 14-6.8, equation 14-6.9 yields

$$P = \frac{4F(1-\nu)}{(1-\cos^3\alpha) - (1-2\nu)\cos\alpha(1-\cos\alpha)}. \qquad (14\text{-}6.10)$$

This is in agreement with the relationship between P and F given after equation 14-4.16.

The solution for a cone with the load applied normal to the cone axis can be obtained through use of the three singular solutions used previously for the Cerruti problem. If a force F is applied in the x direction, the solution becomes

$$\begin{aligned}
\sigma_{RR} &= \frac{A\sin\varphi\cos\theta}{R^2}\left[-\frac{2(2-\nu)}{1-2\nu}\frac{(1-\cos\alpha)}{(1+\cos\alpha)\sin\alpha} + \frac{2(1-\cos\alpha)}{\sin\alpha(1+\cos\varphi)}\right],\\
\sigma_{\varphi\varphi} &= \frac{A\sin\varphi\cos\theta}{R^2}\left[\frac{1-\cos\alpha}{(1+\cos\alpha)\sin\alpha} - \frac{2(1-\cos\alpha)}{\sin\alpha(1+\cos\varphi)}\right.\\
&\quad \left. + \frac{\sin\alpha(1-\cos\varphi)}{\sin^2\varphi(1+\cos\varphi)}\right],\\
\sigma_{\theta\theta} &= \frac{A\sin\varphi\cos\theta}{R^2}\left[\frac{1-\cos\alpha}{(1+\cos\alpha)\sin\alpha} - \frac{\sin\alpha(1-\cos\varphi)}{\sin^2\varphi(1+\cos\varphi)}\right],\\
\sigma_{R\varphi} &= \frac{A\cos\theta}{R^2}\left[-\frac{(1-\cos\alpha)\cos\varphi}{(1+\cos\alpha)\sin\alpha}\right.\\
&\quad \left. + \frac{(1-\cos\alpha)}{\sin\alpha}\frac{(2\cos\varphi+1)}{(1+\cos\varphi)} - \frac{\sin\alpha}{1+\cos\varphi}\right],\\
\sigma_{R\theta} &= \frac{A\sin\theta}{R^2}\left[\frac{1-\cos\alpha}{(1+\cos\alpha)\sin\alpha}\right.\\
&\quad \left. - \frac{(1-\cos\alpha)}{\sin\alpha}\frac{(2\cos\varphi+1)}{(1+\cos\varphi)} + \frac{\sin\alpha}{1+\cos\varphi}\right],
\end{aligned} \qquad (14\text{-}6.11)$$

$$\sigma_{\theta\varphi} = \frac{A \sin\theta}{R^2(1+\cos\varphi)}\left[\frac{1-\cos\alpha}{\sin\alpha}\sin\varphi - \frac{1-\cos\varphi}{\sin\varphi}\sin\alpha\right],$$

where
$$A = \frac{F(1-2\nu)(1+\cos\alpha)\sin\alpha}{2\pi(1-\cos\alpha)^3(1+\nu\cos\alpha)}.$$

7 Symmetrically loaded spheres

Problems of symmetrically loaded spheres can be solved by use of the Papkovich–Neuber formulation. A spherical coordinate system (R, φ, θ) is used for these problems (Figure 14-5).

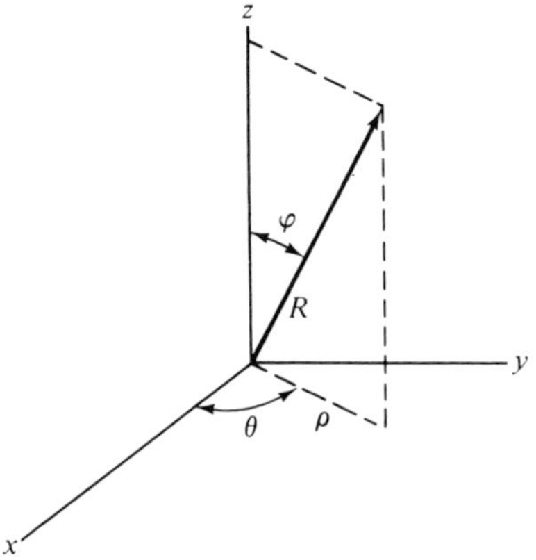

Figure 14-5

The symmetrical loading condition allows us to consider the problem to be independent of θ. The displacement vector exclusive of body forces can be written

$$\bar{u} = 4(1-\nu)\bar{B} - \vec{\nabla}(\bar{R}\cdot\bar{B} + B_0). \qquad (14\text{-}7.1)$$

This is the same form of solution as given in equation 14-1.13 with an adjustment of constants. $\bar{B}$ is a harmonic vector and B_0 is a harmonic function. We have noted previously that only in Cartesian coordinates are all the components of $\bar{B}$ harmonic functions. These components can be related to those in other coordinate systems, thereby relating the nonharmonic components to harmonic ones. The Cartesian vector components are related to those in cylindrical coordinates (ρ, θ, z) as follows:

$$B_x = B_\rho \cos\theta - B_\theta \sin\theta, \qquad (14\text{-}7.2)$$

$$B_y = B_\rho \sin\theta + B_\theta \cos\theta. \qquad (14\text{-}7.3)$$

In complex notation this can be written

$$B_x + iB_y = (B_\rho + iB_\theta)e^{i\theta}. \tag{14-7.4}$$

Both B_x and B_y are harmonic and all the vector components are independent of θ:

$$\nabla^2(B_\rho e^{i\theta}) = 0, \qquad \nabla^2(B_\theta e^{i\theta}) = 0. \tag{14-7.5}$$

These expressions are obtained directly from the vector equations if the components are independent of θ:

$$\begin{aligned}\nabla^2(B_\rho \hat{i}_{(\rho)}) &= (\nabla^2 B_\rho)\hat{i}_{(\rho)} + 2(\vec{\nabla} B_\rho)\cdot(\vec{\nabla}\hat{i}_{(\rho)}) + B_\rho(\nabla^2 \hat{i}_{(\rho)}),\\ &= (\nabla^2 B_\rho)\hat{i}_{(\rho)} + \frac{2}{\rho^2}\frac{\partial B_\rho}{\partial\theta}\hat{i}_{(\theta)} - \frac{B_\rho}{\rho^2}\hat{i}_{(\rho)},\\ &= \left[\left(\nabla^2 - \frac{1}{\rho^2}\right)B_\rho\right]\hat{i}_{(\rho)}.\end{aligned} \tag{14-7.6}$$

$$\left[\left(\nabla^2 - \frac{1}{\rho^2}\right)B_\rho\right]e^{i\theta} = \nabla^2(B_\rho e^{i\theta}) \qquad \text{if } B_\rho = B_\rho(\rho, z). \tag{14-7.7}$$

The dot product of the position vector and $\bar{B}$ becomes

$$\vec{R}\cdot\bar{B} = \rho B_\rho + zB_z. \tag{14-7.8}$$

The displacements become

$$\begin{aligned}u_\rho &= 4(1-\nu)B_\rho - \frac{\partial}{\partial\rho}(\rho B_\rho + zB_z + B_0),\\ u_\theta &= 4(1-\nu)B_\theta,\\ u_z &= 4(1-\nu)B_z - \frac{\partial}{\partial z}(\rho B_\rho + zB_z + B_0).\end{aligned} \tag{14-7.9}$$

As observed earlier when the Galerkin vector was used, the axisymmetric torsion problem decouples from the axisymmetric torsionless problem. We shall consider torsionless problems ($B_\theta = 0$).

The Papkovich–Neuber solution could have been used instead of the Love strain function for axisymmetric cylindrical bodies if $B_\theta = 0$. The equation for B_ρ is

$$\left(\nabla^2 - \frac{1}{\rho^2}\right)B_\rho = 0. \tag{14-7.10}$$

This equation is amenable to a separation of variables technique of solution and B_ρ becomes

$$B_\rho = [AJ_1(\alpha\rho) + BY_1(\alpha\rho)]e^{\pm\alpha z}. \tag{14-7.11}$$

B_0 and B_z are both harmonic terms and their solution is taken as

$$\begin{aligned}B_0 &= [CJ_0(\alpha\rho) + DY_0(\alpha\rho)]e^{\pm\alpha z},\\ B_z &= [EJ_0(\alpha\rho) + FY_0(\alpha\rho)]e^{\pm\alpha z}.\end{aligned} \tag{14-7.12}$$

The displacements are computed by use of equation 14-7.9 for this case, and they compare after adjustment of constants to those obtained by substitution of equation 13-6.6 into equation 13-6.3 for the Love function.

In spherical coordinates, the displacements become

$$u_R = 4(1-\nu)B_R - \frac{\partial}{\partial R}(RB_R + B_0),$$

$$u_\theta = 0, \qquad \text{axisymmetric torsionless,} \tag{14-7.13}$$

$$u_\varphi = 4(1-\nu)B_\varphi - \frac{1}{R}\frac{\partial}{\partial\varphi}(RB_R + B_0).$$

B_R and B_φ are related to the cylindrical components as follows:

$$B_R = B_\rho \sin\varphi + B_z \cos\varphi, \qquad B_\varphi = B_\rho \cos\varphi - B_z \sin\varphi. \tag{14-7.14}$$

We have now related the components in spherical coordinates to a harmonic function B_z and the function B_ρ. The nonzero strains are

$$\epsilon_{RR} = \frac{\partial u_R}{\partial R}, \qquad \epsilon_{\varphi\varphi} = \frac{1}{R}\frac{\partial u_\varphi}{\partial R} + \frac{u_R}{R},$$

$$\epsilon_{\theta\theta} = \frac{1}{R}(u_\varphi \cot\varphi + u_R), \tag{14-7.15}$$

$$\epsilon_{R\varphi} = \frac{1}{2}\left[\frac{1}{R}\frac{\partial u_R}{\partial\varphi} + R\frac{\partial}{\partial R}\left(\frac{u_\varphi}{R}\right)\right].$$

8 Spherical harmonics

In spherical coordinates the Laplace equation becomes

$$\nabla^2\psi = \frac{1}{R^2}\frac{\partial}{\partial R}\left(R^2\frac{\partial\psi}{\partial R}\right) + \frac{1}{R^2\sin\varphi}\frac{\partial}{\partial\varphi}\left(\sin\varphi\frac{\partial\psi}{\partial\varphi}\right) + \frac{1}{R^2\sin^2\varphi}\frac{\partial^2\psi}{\partial\theta^2} = 0. \tag{14-8.1}$$

When ψ is independent of θ, the solution for integer values of n can be written by separation of variables as

$$\psi = AR^nP_n(\zeta) + B\frac{1}{R^{n+1}}P_n(\zeta), \tag{14-8.2}$$

where $\zeta = \cos\varphi$ and $P_n(\zeta)$ is the Legendre polynomial of degree n of the first kind and satisfies the equation

$$(\zeta^2 - 1)P_n''(\zeta) + 2\zeta P_n'(\zeta) - n(n+1)P_n(\zeta) = 0.$$

By Rodrigues's formula, $P_n(\zeta)$ can be written as

$$P_n(\zeta) = \frac{1}{2^n n!}\frac{d^n(\zeta^2-1)^n}{d\zeta^n}. \tag{14-8.3}$$

The following recurrence relations are useful:

$$
\begin{aligned}
(n+1)P_{n+1}(\zeta) &= (2n+1)\zeta P_n(\zeta) - nP_{n-1}(\zeta),\\
(\zeta^2 - 1)P_n'(\zeta) &= n\zeta P_n(\zeta) - nP_{n-1}(\zeta),\\
\zeta P_n'(\zeta) &= nP_n(\zeta) + P_{n-1}'(\zeta),
\end{aligned}
\tag{14-8.4}
$$

where

$$P_n' = \frac{dP_n}{d\zeta}.$$

From equation 14-8.3, the first three polynomials become

$$
\begin{aligned}
P_0(\zeta) &= 1,\\
P_1(\zeta) &= \zeta,\\
P_2(\zeta) &= \tfrac{1}{2}(3\zeta^2 - 1).
\end{aligned}
\tag{14-8.5}
$$

The most important feature of these polynomials is that they form an orthogonal set across the region $0 < \varphi < \pi$ or $-1 < \zeta < 1$. The following orthogonality condition holds:

$$\int_{-1}^{+1} P_m(\zeta)P_n(\zeta)\,d\zeta = \frac{\delta_{mn}}{[n + (1/2)]}. \tag{14-8.6}$$

It can easily be shown that if $R^nP_n(\zeta)$ satisfies the θ independent Laplacian equation, the following expression is also harmonic:

$$R^n \frac{d}{d\varphi}[P_n(\zeta)]e^{i\theta} = -R^nP_n'(\zeta)\sin\varphi e^{i\theta},$$

where (14-8.7)

$$P_n'(\zeta) = \frac{d}{d\zeta}[P_n(\zeta)].$$

The following is also harmonic:

$$\frac{1}{R^{n+1}}\frac{d}{d\varphi}[P_n(\zeta)]e^{i\theta} = -\frac{1}{R^{n+1}}P_n'(\zeta)\sin\varphi e^{i\theta}. \tag{14-8.8}$$

We take the following as spherical components of a harmonic vector by examination of equations 14-7.5 and 14-7.14:

$$
\begin{aligned}
B_R &= \sum_n R^n\left\{A_n\frac{d}{d\varphi}[P_n(\zeta)\sin\varphi + A_n'[P_n(\zeta)]\cos\varphi\right\},\\
&= \sum_n R^n\{A_n(\zeta^2 - 1)P_n'(\zeta) + A_n'\zeta P_n(\zeta)\}.
\end{aligned}
\tag{14-8.9}
$$

$$
\begin{aligned}
B_\varphi &= \sum_n R^n\left\{A_n\frac{d}{d\varphi}[P_n(\zeta)]\cos\varphi - A_n'P_n(\zeta)\sin\varphi\right\},\\
&= -\sum_n R^n\sin\varphi\{A_n\zeta P_n'(\zeta) + A_n'P_n(\zeta)\}.
\end{aligned}
\tag{14-8.10}
$$

Using the recurrence relations, we write the spherical components as

$$
\begin{aligned}
B_R &= \sum_n R^n[-A_nnP_{n-1}(\zeta) + \zeta P_n(\zeta)\{A_nn + A_n'\}],\\
B_\varphi &= -\sum_n R^n[A_nP_{n-1}'(\zeta) + P_n(\zeta)\{A_nn + A_n'\}]\sin\varphi.
\end{aligned}
\tag{14-8.11}
$$

We seek a harmonic function B_0 that will be compatible with the boundary conditions at a constant radius. We take B_0 in the form

$$B_0 = -\sum B_n R^{n-1} P_{n-1}(\zeta). \tag{14-8.12}$$

We can write these in a slightly different form by replacing n by $(n+1)$, thereby yielding the following (for convience A_n is the former A_{n+1}):

$$B_R = \sum_n R^{n+1}\{-A_n(n+1)P_n(\zeta) + \zeta P_{n+1}(\zeta)[A_n(n+1) + A_n']\}, \tag{14-8.13}$$

$$B_\varphi = -\sum_n R^{n+1}\{A_n P_n'(\zeta) + P_{n+1}(\zeta)[A_n(n+1) + A_n']\} \sin\varphi, \tag{14-8.14}$$

$$B_0 = -\sum_n B_n R^n P_n(\zeta). \tag{14-8.15}$$

The displacements are now computed by substituting equations 14-8.13 to 14-8.15 into equation 14-7.13:

$$\begin{aligned} u_R = \sum_n \{&4(1-\nu)R^{n+1}[-A_n(n+1)P_n(\zeta) + \zeta\{A_n(n+1) + A_n'\}P_{n+1}(\zeta)] \\ &- R^{n+1}[-A_n(n+2)(n+1)P_n(\zeta) \\ &+ \zeta(n+2)\{A_n(n+1) + A_n'\}P_{n+1}(\zeta)] + R^{n-1}nB_nP_n(\zeta)\}, \end{aligned}$$

$$\begin{aligned} u_R = \sum_n \{&[A_n(n+1)(n-2+4\nu)R^{n+1} + B_n n R^{n-1}]P_n(\zeta) \\ &+ [A_n(n+1) + A_n'](2-n-4\nu)\zeta R^{n+1}P_{n+1}(\zeta)\}, \end{aligned} \tag{14-8.16}$$

$$\begin{aligned} u_\varphi = \sum_n \{&-4(1-\nu)R^{n+1}[A_nP_n'(\zeta) + \{A_n(n+1) + A_n'\}P_{n+1}(\zeta)] \\ &+ R^{n+1}[-A_n(n+1)P_n'(\zeta) + \{A_n(n+1) + A_n'\}\{P_{n+1}(\zeta) + \zeta P_{n+1}'(\zeta)\}] \\ &- R^{n-1}B_nP_n'(\zeta)\}\sin\varphi. \end{aligned}$$

Using the fact that

$$\zeta P_{n+1}'(\zeta) = (n+1)P_{n+1}(\zeta) + P_n'(\zeta), \tag{14-8.17}$$

this can be written

$$\begin{aligned} u_\varphi = \sum_n \Big\{&[A_n(n+1)(n+5-4\nu)R^{n+1} + B_nR^{n-1} \\ &- \{A_n(n+1) + A_n'\}R^{n+1}]\frac{dP_n(\zeta)}{d\varphi} \\ &+ \{A_n(n+1) + A_n'\}(n-2+4\nu)R^{n+1}P_{n+1}(\zeta)\sin\varphi\Big\}. \end{aligned} \tag{14-8.18}$$

Note that the constant A_n' always appears in combination with $A_n(n+1)$; thus this combination can be replaced by a new constant C_n and the displacements finally become

$$\begin{aligned} u_R = \sum_n \{&[A_n(n+1)(n-2+4\nu)R^{n+1} + B_n n R^{n-1}]P_n(\zeta) \\ &+ C_n(2-n-4\nu)\zeta R^{n+1}P_{n+1}(\zeta)\}, \end{aligned} \tag{14-8.19}$$

$$\begin{aligned} u_\varphi = \sum_n \Big\{&[A_n(n+1)(n+5-4\nu)R^{n+1} + B_nR^{n-1} - C_nR^{n+1}]\frac{d}{d\varphi}P_n(\zeta) \\ &+ C_n(n-2+4\nu)R^{n+1}P_{n+1}(\zeta)\sin\varphi\Big\}. \end{aligned} \tag{14-8.20}$$

It is now apparent that we can set one of the constants to zero, because we need to specify only two functions (stresses or displacements) on the boundary of the sphere. We specify these on a boundary $R =$ constant and choose C_n to be zero to have orthogonal functions on the boundary. The displacements now become

$$u_R = \sum_n [A_n(n+1)(n-2+4\nu)R^{n+1} + B_n nR^{n-1}]P_n(\zeta), \tag{14-8.21}$$

$$u_\varphi = \sum_n [A_n(n+1)(n+5-4\nu)R^{n+1} + B_n R^{n-1}]\frac{d}{d\varphi}P_n(\zeta). \tag{14-8.22}$$

If we were working with bodies with a conical geometry, we would have been concerned with boundaries, $\varphi =$ constant, and would have made a different selection at this point. We shall consider this type of problem later and may also, at that time, wish to examine the harmonic function B_0 again. It is now clear that our choice of B_0 for this case was influenced by the desire to have orthogonal functions on the boundary.

Equations 14-8.21 and 14-8.22 represent the most general solution of the "interior problem" or an axisymmetrically loaded solid sphere. We have not considered the solution corresponding to the second part of equations 14-8.2 and 14-8.8, as these solutions become unbounded at the origin. These solutions correspond to the "exterior problem" or axisymmetric loading on a spherical hole in an infinite solid. Both solutions are required to examine axisymmetrically loaded hollow spherical bodies.

The external solution can be obtained from the internal solution by replacing n by $-(n+1)$ and noting that

$$P_{-(n+1)} = P_n.$$

This can be obtained directly from the defining differential equation for the Legendre polynomial. The external solution becomes

$$u_R = \sum_n \left[\frac{C_n}{R^n}n(n+3-4\nu) - \frac{D_n(n+1)}{R^{n+2}}\right]P_n(\zeta), \tag{14-8.23}$$

$$u_\varphi = \sum_n \left[\frac{C_n}{R^n}(-n+4-4\nu) + \frac{D_n}{R^{n+2}}\right]\frac{d}{d\varphi}P_n(\zeta). \tag{14-8.24}$$

The stresses become for the internal problem

$$\begin{aligned}
\frac{1}{2G}\sigma_{RR} &= \sum_n [A_n(n+1)(n^2-n-2-2\nu)R^n + B_n n(n-1)R^{n-2}]P_n(\zeta),\\
\frac{1}{2G}\sigma_{R\varphi} &= \sum_n [A_n(n^2+2n-1+2\nu)R^n + B_n(n-1)R^{n-2}]\frac{d}{d\varphi}P_n(\zeta),\\
\frac{1}{2G}\sigma_{\varphi\varphi} &= -\sum_n \Big[\{A_n(n^2+4n+2+2\nu)(n+1)R^n + B_n n^2 R^{n-2}\}P_n(\zeta)\\
&\qquad + \{A_n(n+5-4\nu)R^n + B_n R^{n-2}\}\cot\varphi\frac{d}{d\varphi}P_n(\zeta)\Big],\\
\frac{1}{2G}\sigma_{\theta\theta} &= \sum_n \Big[\{A_n(n+1)(n-2-2\nu-4\nu n)R^n + B_n nR^{n-2}\}P_n(\zeta)\\
&\qquad + \{A_n(n+5-4\nu)R^n + B_n R^{n-2}\}\cot\varphi\frac{d}{d\varphi}P_n(\zeta)\Big].
\end{aligned} \tag{14-8.25}$$

The stresses for the external problem become

$$\frac{1}{2G}\sigma_{RR} = \sum_n \left[-\frac{C_n n}{R^{n+1}}(n^2 + 3n - 2\nu) + \frac{D_n(n+1)(n+2)}{R^{n+3}} \right] P_n(\zeta),$$

$$\frac{1}{2G}\sigma_{R\varphi} = \sum_n \left[\frac{C_n}{R^{n+1}}(n^2 - 2 + 2\nu) - \frac{D_n(n+2)}{R^{n+3}} \right] \frac{d}{d\varphi} P_n(\zeta),$$

$$\frac{1}{2G}\sigma_{\varphi\varphi} = \sum_n \left\{ \left[\frac{C_n n}{R^{n+1}}(n^2 - 2n - 1 + 2\nu) - \frac{D_n(n+1)^2}{R^{n+3}} \right] P_n(\zeta) - \left[\frac{C_n}{R^{n+1}}(-n + 4 - 4\nu) + \frac{D_n}{R^{n+3}} \right] \cot\varphi \frac{d}{d\varphi} P_n(\zeta) \right\}, \tag{14-8.26}$$

$$\frac{1}{2G}\sigma_{\theta\theta} = \sum_n \left\{ \left[\frac{C_n n}{R^{n+1}}(n + 3 - 4n\nu - 2\nu) - \frac{D_n(n+1)}{R^{n+3}} \right] P_n(\zeta) + \left[\frac{C_n}{R^{n+1}}(-n + 4 - 4\nu) + \frac{D_n}{R^{n+3}} \right] \cot\varphi \frac{d}{d\varphi} P_n(\zeta) \right\}.$$

9 The internal problem

Equations 14-8.21 and 14-8.22 give the displacements for the internal problem and equation 14-8.25 gives the corresponding stresses. These expansions are used to expand any stress or displacement distribution that corresponds to an axisymmetric loading system in equilibrium. We can check the equilibrium conditions by computing the resultant of the stress tractions in the x, y, and z directions. In the x direction this resultant is

$$R_x = \int_0^{2\pi} \left[\int_0^{\pi} (\sigma_{RR} \sin\varphi \cos\theta + \sigma_{R\varphi} \cos\varphi \cos\theta) R^2 \sin\varphi \, d\varphi \right] d\theta. \tag{14-9.1}$$

The stresses are independent of θ and this resultant is zero. In a similar manner, we show that the resultant force in the y direction is zero.

The resultant force in the z direction is

$$R_z = \int_0^{2\pi} d\theta \int_0^{\pi} (\sigma_{RR} \cos\varphi - \sigma_{R\varphi} \sin\varphi) R^2 \sin\varphi \, d\varphi = 2\pi R^2 \int_{-1}^{+1} (\sigma_{RR}\zeta - \sigma_{R\varphi} \sin\varphi) \, d\zeta. \tag{14-9.2}$$

The integrals take the form

$$\int_{-1}^{+1} \zeta P_n(\zeta) \, d\zeta = \int_{-1}^{+1} P_1(\zeta) P_n(\zeta) \, d\zeta, \tag{14-9.3}$$

and

$$\int_{-1}^{+1} \frac{dP_n(\zeta)}{d\varphi} \sin\varphi \, d\zeta = \int_{-1}^{+1} n(\zeta P_n - P_{n-1}) \, d\zeta = n \int_{-1}^{+1} P_1 P_n \, d\zeta - n \int_{-1}^{+1} P_0 P_{n-1} \, d\zeta. \tag{14-9.4}$$

By use of the orthogonality condition, both of these integrals are zero for

$n \geq 2$ and for $n = 0$. The only value of n that needs to be considered in detail is when $n = 1$. From 14-8.25 the stresses become

$$\begin{aligned} \frac{1}{2G}\sigma_{RR} &= -4(1-\nu)A_1 R P_1(\zeta), \\ \frac{1}{2G}\sigma_{R\varphi} &= 2(1+\nu)A_1 R\frac{d}{d\varphi}P_1(\zeta), \\ \sigma_{RR} &= -8GA_1(1+\nu)R\cos\varphi, \\ \sigma_{R\varphi} &= -4GA_1(1+\nu)R\sin\varphi. \end{aligned} \tag{14-9.5}$$

Substituting these stresses into the expression for R_z yields

$$\begin{aligned} R_z &= -8\pi R^3 GA_1(1+\nu)\int_{-1}^{+1}[2\zeta^2-(1-\zeta^2)]\,d\zeta, \\ &= -8\pi R^3 GA_1(1+\nu)\int_{-1}^{+1}[3\zeta^2-1]\,d\zeta = 0. \end{aligned} \tag{14-9.6}$$

The expansion does correspond to a system that is in equilibrium, as suspected.

We consider problems in which either the stress traction, the displacement vector, or a combination of these is prescribed on the surface of a sphere of radius R_0. When the stresses are given, the boundary conditions take the form

$$\sigma_{RR} = f(\varphi), \qquad \sigma_{R\varphi} = g(\varphi), \qquad \begin{matrix} R = R_0 \\ 0 < \varphi < \pi \end{matrix}. \tag{14-9.7}$$

Equation 14-8.25 yields

$$\begin{aligned} f(\varphi) &= 2G\sum_n [A_n(n+1)(n^2-n-2-2\nu)R_0^n + B_n n(n-1)R_0^{n-2}]P_n(\zeta) \\ &= \sum_n F_n P_n(\zeta), \end{aligned} \tag{14-9.8}$$

$$\begin{aligned} g(\varphi) &= 2G\sum_n [A_n(n^2+2n-1+2\nu)R_0^n + B_n(n-1)R_0^{n-2}]\frac{d}{d\varphi}P_n(\zeta) \\ &= \sum G_n\frac{d}{d\varphi}P_n(\zeta) = -\sum_n G_n\frac{dP_n}{d\zeta}\sin\varphi. \end{aligned} \tag{14-9.9}$$

Equations 14-9.8 and 14-9.9 represent expansions of $f(\varphi)$ and $g(\varphi)$ in spherical harmonics. To examine how these expansions are used, we first consider the more general spherical harmonics, the associated Legendre functions of the first kind. These functions satisfy the following differential equation:

$$(1-\zeta^2)\frac{d^2}{d\zeta^2}P_\nu^\mu(\zeta) - 2\zeta\frac{d}{d\zeta}P_\nu^\mu(\zeta) + \left[\nu(1+\nu) - \frac{\mu^2}{1-\zeta^2}\right]P_\nu^\mu(\zeta) = 0. \tag{14-9.10}$$

$P_\nu^\mu(\zeta)$ is the associated Legendre function of the first kind of degree ν and order μ. When these functions are of integer order ($\mu = m = 0, 1, 2, \ldots$), they are related to the Legendre functions of zero order as follows:

$$P_\nu^m(\zeta) = (1-\zeta^2)^{(1/2)m}\frac{d^m}{d\zeta^m}P_\nu(\zeta). \tag{14-9.11}$$

A factor of $(-1)^m$ is sometimes inserted into this relation by convention. This will not affect our use of these functions, but care should be taken to check definitions before using tables of these functions or integrals of them. The associated functions satisfy similar recurrence relations to those of the Legendre polynomials:

$$\begin{aligned}(\nu-\mu+1)P^{\mu}_{\nu+1}(\zeta) &= (2\nu+1)\zeta P^{\mu}_{\nu}(\zeta)-(\nu+\mu)P^{\mu}_{\nu-1}(\zeta),\\ (\zeta^2-1)\frac{d}{d\zeta}P^{\mu}_{\nu}(\zeta) &= \nu\zeta P^{\mu}_{\nu}(\zeta)-(\nu+\mu)P^{\mu}_{\nu-1}(\zeta).\end{aligned} \qquad (14\text{-}9.12)$$

The most important property of these functions is that an arbitrary function $f(\varphi, \theta)$ given on the surface of a sphere can be expanded in a series expansion of spherical harmonics as follows:

$$f(\varphi,\theta)=\sum_{n=0}^{\infty}\sum_{m=0}^{\infty}(A_{nm}\cos m\theta+B_{nm}\sin m\theta)P^m_n(\zeta), \qquad (14\text{-}9.13)$$

where $\zeta = \cos\varphi$.

The associated Legendre functions have the following orthogonality property:

$$\int_{-1}^{+1}P^m_i(\zeta)P^m_j(\zeta)\,d\zeta=\frac{\delta_{ij}}{[i+(1/2)]}\frac{(i+m)!}{(i-m)!}. \qquad (14\text{-}9.14)$$

Using this orthogonality property and that of the trigonometric functions, the coefficients A_{mn} and B_{nm} are evaluated by Fourier analysis:

$$\begin{aligned}A_{nm} &= \frac{(n-m)!}{(n+m)!}\frac{2n+1}{2\pi\lambda_m}\int_0^{2\pi}d\theta\int_0^{\pi}f(\theta,\varphi)\cos m\theta P^m_n(\cos\varphi)\sin\varphi\,d\varphi,\\ B_{nm} &= \frac{(n-m)!}{(n+m)!}\frac{2n+1}{2\pi}\int_0^{2\pi}d\theta\int_0^{\pi}f(\theta,\varphi)\sin m\theta P^m_n(\cos\varphi)\sin\varphi\,d\varphi,\end{aligned} \qquad (14\text{-}9.15)$$

where
$$\lambda_m=\begin{cases}1, & m\neq 0\\ 2, & m=0\end{cases}.$$

We evaluate the coefficients in equations 14-9.8 and 14-9.9, without resorting to the general expression 14-9.15, by equation 14-9.11:

$$P^1_n(\zeta)=\sin\varphi\frac{d}{d\zeta}P_n(\zeta)=-\frac{d}{d\varphi}P_n(\zeta). \qquad (14\text{-}9.16)$$

Both P^1_n and P_n form sets of orthogonal functions across the sphere. Applying equation 14-9.14 to 14-9.8 and 14-9.9 yields

$$F_n=\frac{2n+1}{2}\int_0^{\pi}f(\varphi)P_n(\cos\varphi)\sin\varphi\,d\varphi, \qquad (14\text{-}9.17)$$

$$G_n=\frac{2n+1}{2}\int_0^{\pi}g(\varphi)\frac{dP_n}{d\varphi}\sin\varphi\,d\varphi. \qquad (14\text{-}9.18)$$

$f(\varphi)$ and $g(\varphi)$ must be boundary values of σ_{RR} and $\sigma_{R\varphi}$ which correspond to a self-equilibrating set of tractions on the sphere's surface. This condition becomes obvious when one attempts to solve for A_n and B_n for $n = 1$. For $n = 0$, $G_0 = 0$, F_0 is related to A_0 as follows:

$$F_0 = -4GA_0(1 + \nu) \implies A_0 = -\frac{F_0}{4G(1 + \nu)}. \qquad (14\text{-}9.19)$$

If $n = 1$, we obtain two expressions for A_1:

$$A_1 = \frac{F_1}{8GR_0(1 + \nu)} = \frac{G_1}{4GR_0(1 + \nu)} \implies F_1 + 2G_1 = 0. \qquad (14\text{-}9.20)$$

This condition corresponds to the condition that the resultant force in the z direction be zero. The remaining coefficients are obtained from

$$\begin{aligned} 2G[A_n(n + 1)(n^2 - n - 2 - 2\nu)R_0^n + B_n n(n - 1)R_0^{n-2}] &= F_n, \\ 2G[A_n(n^2 - 2n - 1 + 2\nu)R_0^n + B_n(n - 1)R_0^{n-2}] &= G_n. \end{aligned} \qquad (14\text{-}9.21)$$

A final word needs to be said about the two coefficients B_0 and B_1, which are not obtained from 14-9.19 to 14-9.21. Examination of the displacement expansions shows that B_0 does not enter into the problem and may be ignored. The displacements corresponding to B_1 are

$$\begin{aligned} u_R &= B_1 \cos \varphi, \\ u_\varphi &= -B_1 \sin \varphi. \end{aligned} \qquad (14\text{-}9.22)$$

These correspond to a rigid-body translation in the z direction, and the coefficient B_1 can be taken as zero without altering the state of stress within the body. When displacements are given on the boundary, this coefficient will take care of the rigid-body motions.

The evaluation of the coefficients A_n and B_n gives a formal series solution for the displacements and stresses for the internal problem. For "smooth" loadings these solutions are easily obtained and the series converges. If concentrated loads or loadings with finite discontinuities are specified, the series will converge slowly or may even diverge (concentrated polar loads). By use of the Boussinesq solution these singularities can be isolated and the balance of the solution expanded in series form. This approach is similar to the one used in two-dimensional theory, but unfortunately does not yield simple closed solutions.

10 Interior problem with body forces

We wish now to consider a solid sphere which, in addition to being subjected to surface tractions, may be under the influence of axisymmetric body forces. A general form of solution will be developed and applied to two special cases. This solution is combined with the homogeneous solution discussed in section 9 to match the boundary conditions. In particular we consider the case where the body forces may be derived from a potential function. A more general case can be obtained by smearing the Kelvin solution throughout the body as body forces. For the case of conservative body forces, the Navier equation can be written as

$$\nabla^2\bar{u} + \frac{1}{1-2\nu}\vec{\nabla}\vec{\nabla}\cdot\bar{u} = \frac{1}{G}\vec{\nabla}\Gamma, \tag{14-10.1}$$

where Γ and the body force $\bar{F}$ are related as follows:

$$\bar{F} = -\vec{\nabla}\Gamma. \tag{14-10.2}$$

A particular solution for equation 14-10.1 is obtained by noting that this solution is irrotational:

$$\nabla^2\bar{u}_{\text{IR}} = \frac{1-2\nu}{2(1-\nu)G}\vec{\nabla}\Gamma. \tag{14-10.3}$$

If $\bar{u}$ is irrotational, we can relate it to a potential function χ as follows:

$$\bar{u}_{\text{IR}} = \vec{\nabla}\chi. \tag{14-10.4}$$

Equation 14-10.3 can be written as

$$\nabla^2\chi = \frac{1-2\nu}{2(1-\nu)G}\Gamma. \tag{14-10.5}$$

If the potential Γ depends only upon R, equation 14-10.5 can be integrated to obtain χ.

The Poisson equation 14-10.5 can also be easily solved when Γ is a harmonic function. The axisymmetric function Γ, in this case, is

$$\Gamma = C_n R^n P_n(\zeta). \tag{14-10.6}$$

χ is a biharmonic function and can be written as R^2 times a harmonic function (see Section 13-5):

$$\chi = D_n R^2[R^n P_n(\zeta)]. \tag{14-10.7}$$

The Laplacian operating on equation 14-10.7 yields, using equation 13-5.12,

$$\begin{aligned}\nabla^2\chi &= D_n[6R^nP_n(\zeta) + 4\bar{R}\cdot\vec{\nabla}(R^nP_n(\zeta))] \\ &= D_n(6+4n)R^nP_n(\zeta).\end{aligned} \tag{14-10.8}$$

Substituting equations 14-10.6 and 14-10.8 into 14-10.5 yields

$$2(3+2n)D_n = \frac{1-2\nu}{2(1-\nu)G}C_n. \tag{14-10.9}$$

Substituting D_n from equation 14-10.9 into equation 14-10.7 yields the particular solution for this case:

$$\chi = \frac{(1-2\nu)}{4(2n+3)(1-\nu)G}C_nR^{n+2}P_n(\zeta). \tag{14-10.10}$$

The displacements are

$$\bar{u} = \bar{\nabla}\chi,$$

$$u_R = \frac{(1-2\nu)(n+2)}{4(1-\nu)(2n+3)G} C_n R^{n+1} P_n(\zeta),$$

$$u_\varphi = \frac{(1-2\nu)}{4(1-\nu)(2n+3)G} C_n R^{n+1} \frac{dP_n(\zeta)}{d\varphi}. \tag{14-10.11}$$

The stresses are

$$\sigma_{RR} = \frac{C_n}{2(1-\nu)(2n+3)}[(4n+6)\nu + (n+2)(n+1)(1-2\nu)]R^n P_n(\zeta),$$

$$\sigma_{R\varphi} = \frac{(1-2\nu)(n+1)}{2(1-\nu)(2n+3)} C_n R^n \frac{dP_n}{d\varphi},$$

$$\sigma_{\varphi\varphi} = \frac{C_n}{2(1-\nu)(2n+3)}[(4n+6)\nu + (2-n^2)(1-2\nu)]R^n P_n(\zeta) - \frac{(1-2\nu)C_n}{2(1-\nu)(2n+3)} R^n \frac{dP_n}{d\varphi} \cot\varphi, \tag{14-10.12}$$

$$\sigma_{\theta\theta} = \frac{C_n}{2(1-\nu)(2n+3)}[(4n+6)\nu + (n+2)(1-2\nu)]R^n P_n(\zeta) + \frac{(1-2\nu)C_n}{2(1-\nu)(2n+3)} R^n \frac{dP_n}{d\varphi} \cot\varphi.$$

11 Gravitating sphere

An example of a body force independent of θ and obtained from a potential function is seen in the geophysical problem of a gravitating sphere. If the surface of the sphere is free of stress, all the internal stresses are induced by the body forces. We shall solve this problem in two parts, first obtaining the particular solution for the body forces and then considering the homogeneous solution of Navier's equation to remove the unwanted surface tractions that arise because of body force solution.

The body force is

$$\bar{F} = -C\bar{R}. \tag{14-11.1}$$

The potential function becomes

$$\Gamma = \frac{CR^2}{2}. \tag{14-11.2}$$

If R_0 is the radius of the sphere and γ is the weight per unit volume of the sphere on the surface, equation 14-11.2 can be written

$$\Gamma = \frac{\gamma}{R_0}\frac{R^2}{2}. \tag{14-11.3}$$

Equation 14-10.5 yields

$$\frac{d^2}{dR^2}\chi + \frac{2\cdot d\chi}{R\,dR} = \frac{1}{R^2}\frac{d}{dR}\left[R^2\frac{d\chi}{dR}\right] = \frac{(1-2\nu)}{2G(1-\nu)}\frac{\gamma}{R_0}\frac{R^2}{2}. \tag{14-11.4}$$

Integrating equation 14-11.4 yields

$$\chi = \frac{(1-2\nu)}{80G(1-\nu)}\frac{\gamma R^4}{R_0}. \tag{14-11.5}$$

Equation 14-10.4 gives the only nonzero displacement, u_R, as follows:

$$u_R = \frac{(1-2\nu)}{20G(1-\nu)}\frac{\gamma R^3}{R_0}. \tag{14-11.6}$$

This is the particular solution for the body force and we add solutions obtained in section 9 to remove the unwanted surface tractions. The body force solution gives the following strains and surface stresses:

$$\epsilon_{RR} = \frac{3\gamma(1-2\nu)R^2}{20G(1-\nu)R_0}, \qquad \epsilon_{\varphi\varphi} = \epsilon_{\theta\theta} = \frac{\gamma(1-2\nu)R^2}{20G(1-\nu)R_0}, \qquad \epsilon_{R\varphi} = 0,$$

$$\sigma_{RR} = \frac{\gamma(1-2\nu)R^2}{10G(1-\nu)R_0}\left(3 + \frac{5\nu}{1-2\nu}\right) = \frac{\gamma(3-\nu)R^2}{10(1-\nu)R_0}.$$

If the sphere is stress free on the surface, the boundary conditions on the homogeneous solution are

$$\sigma_{RR}|_{R=R_0} = f(\varphi) = -\frac{\gamma(3-\nu)R_0}{10(1-\nu)}, \qquad \sigma_{R\varphi}|_{R=R_0} = g(\varphi) = 0. \tag{14-11.7}$$

Substituting equation 14-11.7 into equations 14-9.17 and 14-9.18 yields

$$F_0 = -\frac{\gamma(3-\nu)R_0}{10(1-\nu)},$$
$$F_n = 0, \qquad n \neq 0, \tag{14-11.8}$$
$$G_n = 0.$$

The coefficients A_n and B_n for the stress and displacement equations 14-8.12, 14-8.22, and 14-8.25 are

$$A_0 = \frac{R_0\gamma(3-\nu)}{40G(1+\nu)(1-\nu)}, \qquad A_n = 0, \qquad n \neq 0, \tag{14-11.9}$$
$$B_n = 0.$$

Substituting equation 14-11.9 into equation 14-8.21 yields the desired solution:

$$u_R = -\frac{(1-2\nu)(3-\nu)RR_0}{40G(1-\nu)(1+\nu)}. \tag{14-11.10}$$

Adding equations 14-11.6 and 14-11.10 yields the complete solution:

$$u_R = -\frac{\gamma R_0 R(1-2\nu)}{20G(1-\nu)}\left[\frac{3-\nu}{1+\nu} - \frac{R^2}{R_0^2}\right], \tag{14-11.11}$$

$$u_\varphi = 0. \tag{14-11.12}$$

The strains and stresses can now be calculated without difficulty. If the sphere is the earth, the magnitude of the stresses involved may cause one to question the validity of the use of linear elasticity theory. However, the core is liquid and the question is purely academic except near the earth's surface.

12 Rotating sphere

If a solid sphere is rotated at constant angular velocity about its z axis, axisymmetric body forces will be present. The body forces are proportional to the mass of an element of the sphere, the square of the angular velocity, and the distance from the axis of rotation. The body forces are

$$F = \rho\omega^2 R \sin\varphi[\hat{i}_{(R)} \sin\varphi + \hat{i}_{(\varphi)} \cos\varphi]. \qquad (14\text{-}12.1)$$

This force field is conservative and is related to a potential function as follows:

$$\Gamma = -\frac{\rho\omega^2 R^2}{2} \sin^2\varphi. \qquad (14\text{-}12.2)$$

We can rewrite this potential function as a function of R plus a harmonic function:

$$\Gamma = -\frac{\rho\omega^2 R^2}{3} + \frac{\rho\omega^2 R^2}{3} P_2(\zeta). \qquad (14\text{-}12.3)$$

The particular solution in this case is obtained by considering the sum of the two special cases discussed in section 13-10. We shall use the notation that $\bar{u}^1$ corresponds to the particular solution of the first term of the potential function, $\bar{u}^2$ the second term, and $\bar{u}^3$ the homogeneous solution satisfying the boundary conditions. By analogy of equation 14-11.3 to 14-12.3, $\bar{u}^1$ becomes

$$\bar{u}^1 = -\frac{(1-2\nu)}{60G(1-\nu)} \rho\omega^2 R^3 \hat{i}_{(R)}. \qquad (14\text{-}12.4)$$

From equation 14-10.11, $\bar{u}^2$ becomes

$$\bar{u}^2 = \frac{(1-2\nu)}{28(1-\nu)G} \frac{\rho\omega^2}{3} R^3 \left\{ 4P_n(\zeta)\hat{i}_{(R)} + \frac{dP_n(\zeta)}{d\varphi} \hat{i}_{(\varphi)} \right\}. \qquad (14\text{-}12.5)$$

The stresses on the boundary from $\bar{u}^1$ and $\bar{u}^2$ are

$$\sigma_{RR} = \frac{\rho\omega^2 R_0^2}{6(1-\nu)} \left\{ -\frac{(3-\nu)}{5} + \frac{12-10\nu}{7} P_2(\zeta) \right\}, \qquad (14\text{-}12.6)$$

$$\sigma_{R\varphi} = \frac{\rho\omega^2 R_0^2}{(1-\nu)} \frac{(1-2\nu)}{14} \frac{dP_2(\zeta)}{d\varphi}. \qquad (14\text{-}12.7)$$

The boundary conditions for $\bar{u}^3$ are

$$f(\varphi) = -\frac{\rho\omega^2 R_0^2}{6(1-\nu)}\left\{-\frac{(3-\nu)}{5} + \frac{12-10\nu}{7}P_2(\zeta)\right\},$$
$$g(\varphi) = -\frac{\rho\omega^2 R_0^2}{14(1-\nu)}(1-2\nu)\frac{dP_2(\zeta)}{d\varphi}. \tag{14-12.8}$$

From equations 14-9.8 and 14-9.9 we obtain

$$-4GA_0(1+\nu) = \frac{\rho\omega^2 R_0^2(3-\nu)}{30(1-\nu)},$$
$$2G[-A_2 6\nu R_0^2 + 2B_2] = -\frac{\rho\omega^2 R_0^2(12-10\nu)}{42(1-\nu)}, \tag{14-12.9}$$
$$2G[A_2(7+2\nu)R_0^2 + B_2] = -\frac{\rho\omega^2 R_0^2(1-2\nu)}{14(1-\nu)}.$$

Solving these equations yields

$$A_0 = -\frac{\rho\omega^2 R_0^2(3-\nu)}{120G(1-\nu)(1+\nu)},$$
$$A_2 = \frac{\rho\omega^2(3+\nu)}{84G(1-\nu)(7+5\nu)}, \qquad B_2 = -\frac{(3+2\nu)\rho\omega^2 R_0^2}{6(7+5\nu)G}.$$

The constants 14-12.10 are substituted into equations 14-8.21 and 14-8.22, yielding the desired solution for $\bar{u}^3$:

$$\bar{u}_R^3 = \frac{\rho\omega^2}{G}\left\{\frac{(3-\nu)(1-2\nu)RR_0^2}{60(1-\nu)(1+\nu)} + \left[\frac{(3+\nu)\nu R^3}{7(1-\nu)(7+5\nu)} - \frac{3+2\nu RR_0^2}{3(7+5\nu)}\right]\right\},$$
$$\bar{u}_\varphi^3 = \frac{\rho\omega^2}{G}\left\{\frac{(3+\nu)(7-4\nu)R^3}{28(1-\nu)(7+5\nu)} - \frac{(3+2\nu)RR_0^2}{6(7+5\nu)}\right\}\frac{dP_2(\zeta)}{d\varphi}. \tag{14-12.11}$$

The final solution for the rotating sphere is the sum of equations 14-12.4, 14-12.5, and 14-12.11.

13 The external problem

Equations 14-8.23, 14-8.24, and 14-8.26 give the displacements and stresses for the external problem. These equations can be used for the region $R_0 \leq R < \infty$ if the body is under axisymmetric loading. All the resultant forces applied to the spherical boundary $R = R_0$ need not be zero in this case as was the condition for the internal problem. The resultant force in the x and y directions will be zero as the stresses are independent of θ, and we would be led to similar equations as those in section 9. The resultant force in the z direction, however, needs further consideration. The resultant force on the boundary is

$$R_z = -\int_0^{2\pi} d\theta \int_0^{\pi} (\sigma_{RR} \cos\varphi - \sigma_{R\varphi} \sin\varphi) R_0^2 \sin\varphi \, d\varphi,$$
$$= -2\pi R_0^2 \int_{-1}^{+1} (\sigma_{RR}\zeta - \sigma_{R\varphi} \sin\varphi) \, d\zeta. \tag{14-13.1}$$

Note that the sign of this resultant is different from that of the internal problem as the signs on the stresses are reversed. Due to the orthogonality of the Legendre functions, only the case $n = 1$ needs to be considered. Substituting σ_{RR} and $\sigma_{R\varphi}$ from equation 14-8.26 into equation 14-13.1 yields

$$R_z = \tfrac{4}{3}\pi R_0^2(F_1 + 2G_1),$$

where

$$F_1 = -\frac{2C_1}{R_0^2}(2 - \nu) + \frac{6D_1}{R_0^4}, \tag{14-13.2}$$
$$G_1 = -\frac{C_1(1 - 2\nu)}{R_0^2} - \frac{3D_1}{R_0^4}.$$

In the internal problem we obtained (equation 14-9.20) two equations for the same constant A_1, necessitating a relation between F_1 and G_1. This relation represented the constraint on the boundary tractions that they be self-equilibrating over the surface of the sphere. Here we obtain two equations for C_1 and D_1 that do not require any such constraint on the boundary tractions.

The formal solution for the external problem for the case when stresses are specified on the boundary $R = R_0$ is

$$\sigma_{RR} = f(\varphi), \qquad \sigma_{R\varphi} = g(\varphi), \qquad \begin{array}{l} R = R_0, \\ 0 < \varphi < \pi. \end{array} \tag{14-13.3}$$

Equation 14-8.26 yields

$$f(\varphi) = 2G \sum \left\{ -\frac{C_n n}{R_0^{n+1}}(n^2 + 3n - 2\nu) + \frac{D_n(n+1)(n+2)}{R_0^{n+3}} \right\} P_n(\zeta),$$
$$g(\varphi) = 2G \sum \left\{ \frac{C_n}{R_0^{n+1}}(n^2 - 2 + 2\nu) - \frac{D_n(n+2)}{R_0^{n+3}} \right\} \frac{dP_n}{d\varphi}. \tag{14-13.4}$$

As before, we write this in terms of two new constants F_n and G_n as

$$f(\varphi) = \sum F_n P_n(\zeta),$$
$$g(\varphi) = \sum G_n \frac{dP_n(\zeta)}{d\varphi}. \tag{14-13.5}$$

Due to the orthogonality characteristics of the associated Legendre functions, the constants F_n and G_n are obtained by use of equations 14-9.17 and 14-9.18. The formal solution of the external problem is given by these equations for any smooth loading of the surface of a spherical cavity in the infinite space.

14 Stress concentration due to a spherical cavity

As an example of the use of the external formulation, let us consider an infinite body having a spherical cavity and subjected to a uniaxial tension in the z direction. The state in the absence of the cavity is

$$\sigma_{zz} = T, \qquad \text{all other stresses zero.} \tag{14-14.1}$$

In spherical coordinates, this corresponds to the following stresses:

$$\begin{aligned} \sigma_{RR} &= T\cos^2\varphi, \\ \sigma_{\varphi\varphi} &= T\sin^2\varphi, \\ \sigma_{R\varphi} &= -T\sin\varphi\cos\varphi. \end{aligned} \tag{14-14.2}$$

Now consider the external problem where the negatives of these stresses are specified on the boundary of the spherical cavity. The desired solution of the problem is obtained by taking a linear combination of the two problems. We consider the following external problem, $R_0 \le R < \infty$:

$$R = R_0: \quad \begin{aligned} \sigma_{RR} &= -T\cos^2\varphi = -\frac{T}{3}[1 + 2P_2(\zeta)], \\ \sigma_{R\varphi} &= T\sin\varphi\cos\varphi = -\frac{T}{3}\frac{dP_2}{d\varphi}. \end{aligned} \tag{14-14.3}$$

Equating coefficients of the Legendre functions in equation 14-8.26 yields

$$\begin{aligned} \frac{4GD_0}{R_0^3} &= -\frac{T}{3}, \\ 2G\left[-\frac{2C_2}{R_0^3}(10 - 2\nu) + \frac{12D_2}{R_0^5}\right] &= -\frac{2}{3}T, \\ 2G\left[\frac{C_2}{R_0^3}(2 + 2\nu) - \frac{4D_2}{R_0^5}\right] &= -\frac{T}{3}. \end{aligned} \tag{14-14.4}$$

Solving equation 14-14.4 for the constants yields

$$D_0 = -\frac{TR_0^3}{12G}, \qquad D_2 = \frac{T}{2G}\frac{R_0^5}{(7 - 5\nu)}, \qquad C_2 = \frac{5T}{12G}\frac{R_0^3}{(7 - 5\nu)}. \tag{14-14.5}$$

The evaluation of these constants completes the formal solution of the problem, and we can substitute these into the displacement and stress solutions for the external problem. To obtain the stress concentration factors, one needs to examine the stresses $\sigma_{\varphi\varphi}$ and $\sigma_{\theta\theta}$ on the surface of the cavity. The values of these stresses are

$$\begin{aligned} \sigma_{\varphi\varphi}\big|_{R=R_0} &= \frac{T(27 - 15\nu)}{2(7 - 5\nu)} - \frac{15T}{(7 - 5\nu)}\cos^2\varphi, \\ \sigma_{\theta\theta}\big|_{R=R_0} &= \frac{T(15\nu - 3)}{2(7 - 5\nu)} - \frac{15\nu T}{(7 - 5\nu)}\cos^2\varphi. \end{aligned} \tag{14-14.6}$$

At the point $\varphi = \pi/2$, the stresses produce the maximum effect:

$$\sigma_{\varphi\varphi}\left(R_0, \frac{\pi}{2}\right) = \frac{T}{2}\frac{27 - 15\nu}{7 - 5\nu}, \qquad \sigma_{\theta\theta}\left(R_0, \frac{\pi}{2}\right) = \frac{T}{2}\frac{(15\nu - 3)}{(7 - 5\nu)}. \qquad (14\text{-}14.7)$$

For Poisson's ratio of 0.3, these values would be

$$\sigma_{\varphi\varphi}\left(R_0, \frac{\pi}{2}\right) = 2.05T, \qquad \sigma_{\theta\theta}\left(R_0, \frac{\pi}{2}\right) = 0.14T. \qquad (14\text{-}14.8)$$

Without the cavity $\sigma_{\varphi\varphi}$ would have been equal to T and $\sigma_{\theta\theta}$ to zero. At $\varphi = 0$, the spherical cavity causes a compressive stress of the form

$$\sigma_{\varphi\varphi}(R_0, 0) = \sigma_{\theta\theta}(R_0, 0) = -\frac{3 + 15\nu}{7 - 5\nu}\frac{T}{2}. \qquad (14\text{-}14.9)$$

For $\nu = 0.3$,

$$\sigma_{\varphi\varphi}(R_0, 0) = \sigma_{\theta\theta}(R_0, 0) = -0.68T. \qquad (14\text{-}14.10)$$

15 Hollow sphere

The general solution of a hollow sphere under axisymmetric loadings does not, in general, produce any difficulties other than algebraic complexities. The solution is the combination of both previously discussed solutions for the internal and external problems. Adding equations 14-8.25 and 14-8.26, one obtains orthogonal functions for the stresses on the boundaries $R = R_1$ and $R = R_2$ with four sets of constants to be determined. The boundary tractions can be expanded in Fourier–Legendre expansions and the coefficients A_n, B_n, C_n, and D_n obtained. The solution is straightforward and the details will not be shown here.

16 End effects in a truncated semi-infinite cone

The solution of end loading of the truncated semi-infinite cone, shown in Figure 14-6, gives some insight into the Saint-Venant approximation for this geometry.*

The displacement field is related to the Papkovich–Neuber function by equation 14-1.13. The solution can be developed in a manner similar to that used for spherical bodies, but the functions are not orthogonal on the curved boundaries. It is desirable to group terms in the solution that contribute to the same power of R, so that boundary conditions on the stress-free surface of the cone may be met.

* For details see T. R. Thompson and R. W. Little, "End effects in a truncated semi-infinite cone," *Quart. Meth. Appl. Math.*, **23**, pt. 2, 1970.

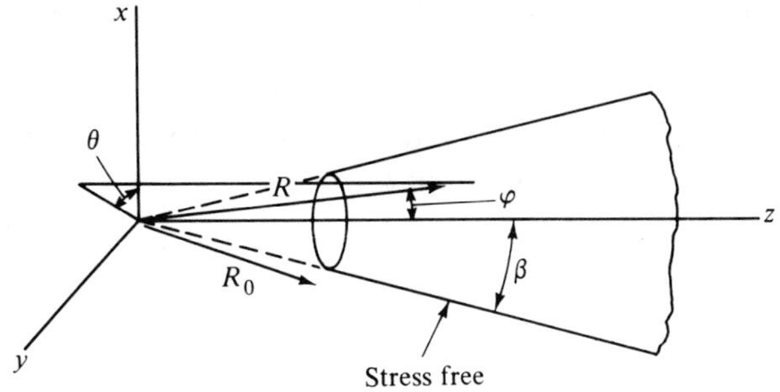

Figure 14-6

The solution for the displacements are

$$u_R = \sum_n R^{-\alpha_n-1}[C_n(1 + k\alpha_n)\zeta P_{\alpha_n}(\zeta) - \alpha_n D_n P_{\alpha_n-1}(\zeta)], \qquad (14\text{-}16.1)$$

$$u_\varphi = \sum_n R^{-\alpha_n-1}\Big[(D_n - kC_n)\frac{d}{d\varphi}P_{\alpha_n-1}(\zeta) - C_n\{1 - k(\alpha_n + 1)\}P_{\alpha_n}(\zeta)\sin\varphi\Big], \qquad (14\text{-}16.2)$$

where $k = \frac{1}{4}(1-\nu)^{-1}$. The power R and the degree of the Legendre function are not integers in this example and the eigenvalues α_n are not yet determined. If α_n is zero, a solution corresponding to this zero value of the separation constant must be included. The stresses are obtained from equations 14-16.1 and 14-16.2, and the zero term is included:

$$\sigma_{RR} = \sum_n \left(\frac{R}{R_0}\right)^{-\alpha_n-1}\{[-(\alpha_n^2 + 5\alpha_n + 2(2-\nu))\zeta P_{\alpha_n}(\zeta) + 2\nu\alpha_n P_{\alpha_n-1}(\zeta)]kC_n + \alpha_n(\alpha_n+1)P_{\alpha_n-1}(\zeta)D_n\} + C_0\left(\frac{R}{R_0}\right)^{-2}\left\{1 + \cos\beta + \frac{2(2-\nu)}{1-2\nu}\cos\varphi\right\},$$

$$\sigma_{\theta\theta} = \sum_n \left(\frac{R}{R_0}\right)^{-\alpha_n-2}\{[(1-2\nu)(1+2\alpha_n)\zeta P_{\alpha_n}(\zeta) + 2\nu\alpha_n P_{\alpha_n-1}(\zeta) + P'_{\alpha_n-1}(\zeta)]kC_n - [\alpha_n P_{\alpha_n-1}(\zeta) + \zeta P'_{\alpha_n-1}(\zeta)]D_n\} + C_0\left(\frac{R}{R_0}\right)^{-2}\left\{\frac{\cos\varphi - \cos\beta}{1+\cos\varphi} + \cos\varphi - 1\right\},$$

$$\sigma_{\varphi\varphi} = \sum_n \left(\frac{R}{R_0}\right)^{-\alpha_n-2}\{[(\alpha_n^2 - \alpha_n + (1-2\nu))\zeta P_{\alpha_n}(\zeta) + 2(1-\nu)\alpha_n P_{\alpha_n-1}(\zeta) - \zeta P'_{\alpha_n-1}(\zeta)]kC_n - [\alpha_n^2 P_{\alpha_n-1}(\zeta) - \zeta P'_{\alpha_n-1}(\zeta)]D_n\} + C_0\left(\frac{R}{R_0}\right)^{-2}\left\{\frac{\cos\varphi - \cos\beta}{1+\cos\varphi}\right\}, \qquad (14\text{-}16.3)$$

$$\sigma_{R\varphi} = \sum_n \left(\frac{R}{R_0}\right)^{-\alpha_n-2}\sin\varphi\{-[\{(\alpha_n+1)^2 - 2(1-\nu)\}P_{\alpha_n}(\zeta) + \{(\alpha_n+1) + 2(1-\nu)\}P'_{\alpha_n-1}(\zeta)]kC_n + (\alpha_n+1)P'_{\alpha_n-1}(\zeta)D_n\} + C_0\left(\frac{R}{R_0}\right)^{-2}\sin\varphi\left\{\frac{\cos\varphi - \cos\beta}{1+\cos\varphi}\right\}.$$

The boundary conditions on the semi-infinite surface $\varphi = \beta$ are

$$\sigma_{\varphi\varphi}\big|_{\varphi=\beta} = 0, \qquad \sigma_{R\varphi}\big|_{\varphi=\beta} = 0. \tag{14-16.4}$$

Substitution of equations 14-16.3 into these boundary conditions yields the following transcendental equation for α_n:

$$\begin{aligned}\alpha_n^2[&\cos\beta\{2k(\cos^2\beta - 1)\alpha_n^2 + 2k(\cos^2\beta - 1)\alpha_n + \cos^2\beta\}P_{\alpha_n}^2(\cos\beta) \\ &+ \cos\beta\{2k(\cos^2\beta - 1)\alpha_n^2 + 2k(\cos^2\beta - 1)\alpha_n + 1\}P_{\alpha_n-1}^2(\cos\beta) \\ &- 2k\{2\cos^2\beta(\cos^2\beta - 1)\alpha_n^2 + (\cos^2\beta - 1)(3\cos^2\beta - 1)\alpha_n \\ &+ [\cos^4\beta + 2(1 - 2\nu)\cos^2\beta + 1]\}P_{\alpha_n}(\cos\beta)P_{\alpha_n-1}(\cos\beta)] = 0.\end{aligned} \tag{14-16.5}$$

This equation has roots at zero and paired complex roots α_n and $\bar{\alpha}_n$. The lower values of the nonzero roots for different cone angles are given in Table 14-1. As the cone angle becomes larger the decay length (measured by the lowest nonzero eigenvalue) becomes longer.

Boundary conditions on the boundary $R = R_0$ may be specified by a least-squares procedure.

Table 14-1. Roots of Equation 14-16.5

	$\beta = 15°$		$\beta = 30°$
n	α_n	n	α_n
1	$9.9170 + i5.0850$	1	$4.7409 + i2.3589$
2	$22.6618 + i6.1604$	2	$11.1025 + i2.9327$
3	$34.9063 + i6.8925$	3	$17.2181 + i3.3041$
	$\beta = 45°$		$\beta = 60°$
n	α_n	n	α_n
1	$3.0371 + i1.3520$	1	$2.2189 + i0.7425$
2	$7.2610 + i1.7771$	2	$5.3518 + i1.1135$
3	$11.3298 + i2.0309$	3	$8.3933 + i1.3111$
	$\beta = 75°$		
n	α_n		
1	$1.7530 + i0.0460$		
2	$4.2211 + i0.5707$		
3	$6.6413 + i0.7429$		

17 General solution of nonaxisymmetric problems by use of spherical harmonics

The application of the method of series to isotropic elastic bodies was first investigated by Lamé and E. Clapeyion (1833). This has been the method used mostly in this book and can now be employed to solve more general problems. Lamé (1854) considered spherical bodies on which surface tractions are prescribed. He attempted to use this solution to investigate geophysical problems concerning the earth's rigidity. Lord Kelvin continued these investigations in 1892 by considering solutions in which the dilatation and the displacements are related to the spherical harmonics. Papkovich continued this form of solution, employing the Papkovich–Neuber form of the solution of the equations of elasticity. Luré formalized the notation and presented solutions for both the solid sphere and a spherical cavity in the infinite space in his book on elasticity.

We shall review Kelvin's development and present a similar one as that used by Luré and show its applications to bodies having spherical and nonspherical boundaries. We know the displacement components can be expressed in terms of four harmonic functions, one of which is frequently chosen to be zero. We shall consider these harmonic functions to be related to the spherical harmonics. The fourth harmonic function can be related to the others to simplify the solution when displacements or surface tractions are specified on the surface of a sphere.

A three-dimensional harmonic function $\varphi(x, y, z)$ can be represented as the sum of the homogeneous harmonic polynomials $\varphi_n(x, y, z)$:

$$\varphi = \sum_{n=0}^{\infty} \varphi_n, \tag{14-17.1}$$

where φ_n is

$$\varphi_n(x, y, z) = R^n Y_n(\varphi, \theta). \tag{14-17.2}$$

Y_n are the spherical harmonics:

$$Y_n(\varphi, \theta) = \sum_{m=0}^{n} P_n^m(\zeta)[A_{mn} \cos m\theta + B_{nm} \sin m\theta], \qquad \zeta = \cos \varphi. \tag{14-17.3}$$

$P_n^m(\zeta)$ are the associated Legendre functions.

Equations 14-17.1 to 14-17.3 are used to represent harmonic functions inside the spherical domain $R \leq R_0$. φ_n are homogeneous polynomials of degree n, and $\partial\varphi_n/\partial x$ is a homogeneous polynomial of degree $(n - 1)$:

$$\frac{\partial \varphi_n}{\partial x} = R^{n-1} Y_{n-1}(\varphi, \theta). \tag{14-17.4}$$

Other useful properties of the homogeneous harmonic polynomals are

$$\bar{R} \cdot \vec{\nabla} \varphi_n = n\varphi_n, \tag{14-17.5}$$

$$\nabla^2(R^m \varphi_n) = m(m + 2n + 1)R^{m-2}\varphi_n. \tag{14-17.6}$$

Lord Kelvin obtained a solution in terms of harmonics directly from the Navier equations by noting that $\bar{\nabla} \cdot \bar{u}$ is a harmonic function and $\nabla^2 \bar{u}$ is a harmonic vector in the absence of body forces. The Navier equation can be written

$$G\left[\nabla^2 \bar{u} + \frac{1}{1 - 2\nu} \bar{\nabla}\bar{\nabla} \cdot \bar{u}\right] = 0. \tag{14-17.7}$$

$\bar{\nabla}\cdot\bar{u}$ can now be expressed in terms of the homogeneous harmonic polynomials:

$$\bar{\nabla}\cdot\bar{u} = \sum \varphi_n. \qquad (14\text{-}17.8)$$

Noting that $\partial\varphi_n/\partial x$ is a polynomial of degree $n - 1$, φ_{n-1}, equation 14-17.6 yields

$$\nabla^2\left(R^m\frac{\partial\varphi_n}{\partial x}\right) = m(m + 2n - 1)R^{m-2}\frac{\partial\varphi_n}{\partial x}. \qquad (14\text{-}17.9)$$

We write

$$\nabla^2(R^m\vec{\nabla}\varphi_n) = m(m + 2n - 1)R^{m-2}\vec{\nabla}\varphi_n. \qquad (14\text{-}17.10)$$

Substituting equation 14-17.8 into 14-17.7 and using equation 14-17.10, when $m = 2$, yields

$$\nabla^2\bar{u} + \frac{1}{1 - 2\nu}\sum_n \frac{1}{2(2n + 1)}\nabla^2(R^2\vec{\nabla}\varphi_n). \qquad (14\text{-}17.11)$$

Equation 14-17.11 is integrated as follows:

$$\bar{u} = -\frac{1}{1 - 2\nu}\sum_n \frac{1}{2(2n + 1)}R^2\vec{\nabla}\varphi_n + \bar{\chi}, \qquad (14\text{-}17.12)$$

where $\bar{\chi}$ is a harmonic vector. Noting that

$$\vec{\nabla}\cdot(R^2\vec{\nabla}\varphi_n) = 2\vec{R}\cdot\vec{\nabla}\varphi_n + R^2\nabla^2\varphi_n = 2n\varphi_n,$$

the divergence of equation 14-17.12 yields

$$\vec{\nabla}\cdot\bar{u} = -\frac{1}{1 - 2\nu}\sum_n \frac{n}{2n + 1}\varphi_n + \vec{\nabla}\cdot\bar{\chi}. \qquad (14\text{-}17.13)$$

Using equation 14-17.8 we relate φ_n to $\vec{\nabla}\cdot\bar{\chi}$ as follows:

$$\sum_n \frac{3n + 1 - 2\nu(2n + 1)}{(1 - 2\nu)(2n + 1)}\varphi_n = \vec{\nabla}\cdot\bar{\chi}. \qquad (14\text{-}17.14)$$

Finally, noting that the components of $\bar{\chi}$ are harmonic in Cartesian coordinates, we write

$$\bar{\chi} = \sum \bar{\chi}_n,$$

where $\bar{\chi}_n$ are homogeneous harmonic polynomial vectors of degree n. This means that each component can be written in the form of equations 14-17.2 and 14-17.3 with different constants A_{nm} and B_{nm} in each component. Equation 14-17.14 yields

$$\varphi_n = \frac{(1 - 2\nu)(2n + 1)}{3n + 1 - 2\nu(2n + 1)}\vec{\nabla}\cdot\bar{\chi}_n, \qquad (14\text{-}17.15)$$

and the displacements are written as

$$\bar{u} = \sum_n \bar{\chi}_n - \frac{R^2}{2[3n + 1 - 2\nu(2n + 1)]}\bar{\nabla}(\bar{\nabla}\cdot\bar{\chi}_n). \qquad (14\text{-}17.16)$$

We write each component of the displacement vector in terms of the six constants ($3A$'s and $3B$'s) that can be solved by some technique. Note that the theta dependency splits into even and odd functions corresponding to A and B, respectively. A. E. H. Love shows in his text, *A Treatise on the Mathematical Theory of Elasticity* (Dover Publications, Inc, New York, 1944), a method of determining these constants in closed form when displacements are specified on the surface of a sphere. As indicated earlier, we shall present simplier Luré development here.

Let the harmonic vector in the Papkovich–Neuber development be written

$$\bar{B} = \sum_{n=0}^{\infty} \bar{B}_n, \tag{14-17.17}$$

where $\bar{B}_n$ are homogeneous harmonic vectors of degree n. In a similar manner, the harmonic function B^0 is chosen:

$$B^0 = \sum_{n=0}^{\infty} B^0_{n-1}. \tag{14-17.18}$$

The displacement vector becomes

$$\bar{u} = \sum_{n=0}^{\infty} \bar{u}_n = \sum_{n=0}^{\infty} [4(1-\nu)\bar{B}_n - \nabla(\bar{R}\cdot\bar{B}_n) - \vec{\nabla}(B^0_{n-1})]. \tag{14-17.19}$$

Luré shows after lengthy calculations that

$$\begin{aligned}\vec{\nabla}(\bar{R}\cdot\bar{B}_n) = \frac{1}{2n+1}\Bigg\{&\vec{\nabla}\left[R^{2n+3}\vec{\nabla}\cdot\left(\frac{\bar{B}_n}{R^{2n+1}}\right)\right] \\ &+ \frac{2}{2n-1}R^{2n+1}\vec{\nabla}\left[\frac{1}{R^{2n-1}}\nabla\cdot\bar{B}\right]\Bigg\} \\ &+ \frac{1}{2n-1}R^2\vec{\nabla}(\vec{\nabla}\cdot\bar{B}_n).\end{aligned} \tag{14-17.20}$$

The first two terms in equation 14-17.20 are harmonic homogeneous vectors of degree n, which suggests that equation 14-17.19 can be written

$$\bar{u}_n = \bar{U}_n - \frac{1}{2n+1}R^2\vec{\nabla}(\vec{\nabla}\cdot B_n) - \vec{\nabla}(B^0_{n-1}), \tag{14-17.21}$$

where

$$\bar{U}_n = 4(1-\nu)\bar{B}_n + \frac{1}{2n+1}\left\{\vec{\nabla}\left[R^{2n+3}\vec{\nabla}\cdot\left(\frac{\bar{B}_n}{R^{2n+1}}\right)\right] + \frac{2R^{2n+1}}{2n-1}\bar{\nabla}\cdot\left(\frac{\vec{\nabla}\cdot\bar{B}_n}{R^{2n-1}}\right)\right\}. \tag{14-17.21a}$$

$\bar{U}_n$ is a homogeneous harmonic vector of degree n.

We seek B^0_{n-1} such that the last two terms in equation 14-17.21 vanish on the boundary of a sphere ($R = R_0$):

$$\left[\frac{1}{2n-1}R^2\vec{\nabla}(\vec{\nabla}\cdot B_n) + \vec{\nabla}(B^0_{n-1})\right]_{R=R_0} = 0. \tag{14-17.22}$$

This can be obtained by letting the following relation hold:

$$\vec{\nabla}B^0_{n-1} = -\frac{R_0^2}{2n-1}\vec{\nabla}\vec{\nabla}\cdot\bar{B}_n. \tag{14-17.23}$$

This relation is possible as $\vec{\nabla}\cdot\bar{B}_n$ is a polynomical of degree $(n-1)$. $\vec{\nabla}(\vec{\nabla}\cdot\bar{B}_n)$ vanishes if n is either 0 or 1. Equation 14-17.21 becomes

$$\bar{u} = \bar{U}_n + \frac{1}{2n-1}(R_0^2 - R^2)\vec{\nabla}\vec{\nabla}\cdot\bar{B}_n. \qquad (14\text{-}17.24)$$

$\bar{U}_n$ and $\bar{B}_n$ are related by equation 14-17.21a, and an explicit form of this relation is obtained by taking the divergence of the equation 14-17.21a:

$$\vec{\nabla}\cdot\bar{U}_n = \left[4(1-\nu) - \frac{2n(2n+1)}{4n^2-1}\right]\vec{\nabla}\cdot\bar{B}_n, \qquad (14\text{-}17.25)$$

where the following relations have been used:

$$\bar{\nabla}\cdot\bar{\nabla}\left(R^{2n+3}\bar{\nabla}\cdot\frac{\bar{B}_n}{R^{2n+1}}\right) = 0,$$

$$\bar{\nabla}\cdot\left[R^{2n+1}\vec{\nabla}\left(\frac{1}{R^{2n-1}}\bar{\nabla}\cdot\bar{B}_n\right)\right] = -(2n+1)n\vec{\nabla}\cdot\bar{B}_n.$$

Equation 14-17.25 can now be written

$$\bar{\nabla}\cdot\bar{B}_n = \frac{2n-1}{2[2(1-\nu)(2n-1)-n]}\bar{\nabla}\cdot\bar{U}_n, \qquad (14\text{-}17.26)$$

and the displacement vector becomes

$$\bar{u} = \sum_{n=0}^{\infty}\bar{U}_n + \frac{1}{2}(R_0^2 - R^2)\sum_{n=2}^{\infty}\frac{1}{(3-4\nu)n-2+2\nu}\vec{\nabla}(\bar{\nabla}\cdot\bar{U}_n). \qquad (14\text{-}17.27)$$

Note that the Luré solution 14-17.27 and the Kelvin solution 14-17.16 are similar, but the Luré solution has the advantage that on the boundary, $R = R_0$, the boundary displacements become expressible in terms of the spherical harmonics directly. $\bar{U}_n$ may be nondimensionalized as follows:

$$\bar{U}_n = \left(\frac{R}{R_0}\right)^n \bar{Y}_n(\varphi, \theta). \qquad (14\text{-}17.28)$$

For the exterior problem $(R \geq R_0)$ these equations become, by replacing n by $-(n+1)$,

$$\bar{u} = \sum_{n=0}^{\infty}\bar{U}_{-(n+1)} - \frac{R_0^2 - R^2}{2[(3-4\nu)(n+1)+2-2\nu]}\bar{\nabla}\bar{\nabla}\cdot\bar{U}_{-(n+1)}, \qquad (14\text{-}17.29)$$

where

$$\bar{U}_{-(n+1)} = \left(\frac{R_0}{R}\right)^{n+1}\bar{Y}_n(\varphi, \theta). \qquad (14\text{-}17.30)$$

The importance of these solutions lies not only in solution of non-axisymmetric spherical bodies, but as general solutions for problems of other geometries using a point-matching or least-squares approximation on the boundaries.

PROBLEMS

14.1. Set up the form of solution for circular cylinders under axisymmetric loadings by use of the Papkovich–Neuber approach.

14.2. Set up the solution of a sphere under a pressure band about its center: $R = 1$, $(\pi/2) - \beta \leq \varphi \leq (\pi/2) + \beta$, $\sigma_{RR} = -p$ and free of stress elsewhere on its surface.

14.3. Set up the solution for a hollow sphere under loading, similar to that given in problem 14.2.

14.4. Find the general asymmetric form solution for circular cylinders similar to section 13-18 by use of the Papkovich–Neuber approach.

APPENDIX A

EIGENFUNCTION ANALYSIS

Many problems presented in this text use eigenfunction expansions as the basic method of solution. It is, therefore, appropriate to present a brief note on the theory of eigenfunctions.

For any problem the choice of coordinate systems is made such that the boundaries of the elastic body correspond to one or more of the coordinate surfaces. In most of the coordinate systems considered, the partial differential equations can be solved by separation methods involving one or more separation constants. The role that these separation constants play in satisfying the boundary conditions of the partial differential equations indicates the usefulness of eigenfunction analysis. The choice of coordinates has been such that on the boundary one coordinate variable is a constant. For example, in two-dimensional Cartesian coordinates, the coordinate surface $y =$ constant involves only the x variables. The solution on this surface involves not only the variable x but the separation constant. This form of solution occurs in Section 7-4, where $\varphi = C \sin \beta x$ on the boundaries $y =$ constant. In this case β is the separation constant. This function on the boundaries must also satisfy simple boundary conditions at the ends of the boundary, $x = a$ and $x = b$. This means that only for particular values of the separation constant will these end conditions be satisfied. For this reason, these constants are called eigenvalues or characteristic values, and the boundary functions containing them are called eigenfunctions. It is assumed that a series of these eigenfunctions, for all possible eigenvalues, should be able to represent any specified boundary function. We have shown with particular functions that we can indeed represent such boundary functions. For orthogonal functions, the constants multiplying each eigenfunction can be easily obtained (see Section 7-5.)

Perhaps a few more comments are needed on the simple boundary conditions that lead to the definition of the eigenvalues. In general, these will be simple homogeneous conditions specifying that the function or its derivative is zero at the end points. They may, however, be the more general homogeneous condition specifying that a linear combination of the function and its derivative is zero at the ends.

Sturm–Liouville problem

Separation of the partial differential equation

$$\nabla^2\varphi + k^2\varphi = 0$$

leads to the following ordinary differential equation:

$$\frac{d}{dx}\left[p(x)\frac{d\varphi}{dx}\right] + [q(x) + \lambda r(x)]\varphi = 0.$$

This equation is called the Liouville equation and, as many of our solutions involve the harmonic or biharmonic equations, we encounter the Liouville equation frequently. The functions p, q, and r are characteristics of the coordinate system chosen and λ is the separation constant. Points where p is zero are singular points and can never occur within the region considered; therefore, p may always be taken as positive.

The Liouville equation is a self-adjoint equation. Consider the general second-order differential equation

$$a(x)\frac{d^2}{dx^2}\psi + b(x)\frac{d}{dx}\psi + [c(x) + \lambda d(x)]\psi = 0.$$

The adjoint equation is written as

$$\frac{d^2}{dx^2}(a\psi) - \frac{d}{dx}(b\psi) + [c + \lambda d]\psi = 0.$$

Examination of the Liouville equation shows that it equals its adjoint equation and is, therefore, called self-adjoint. We may use this property of the Liouville equation to examine some of the characteristics of eigenvalues and eigenfunctions arising from it. Consider the eigenvalue problem when the boundary conditions at $x = a$ and $x = b$ are of the homogeneous type. Let us compare two solutions corresponding to two different eigenvalues $\lambda_1 \neq \lambda_2$. If we multiply the equation for φ_1 by φ_2 and the equation for φ_2 by φ_1 and subtract, we obtain

$$\varphi_2 \frac{d}{dx}\left[p\frac{d\varphi_1}{dx}\right] + \varphi_2[q + r\lambda_1]\varphi_1 - \varphi_1 \frac{d}{dx}\left[p\frac{d\varphi_2}{dx}\right] - \varphi_1[q + r\lambda_2]\varphi_2 = 0.$$

This equation may be written

$$\frac{d}{dx}\left[\varphi_2 p\frac{d\varphi_1}{dx} - \varphi_1 p\frac{d\varphi_2}{dx}\right] = (\lambda_2 - \lambda_1)r\varphi_1\varphi_2.$$

If we integrate over the entire region $a \leq x \leq b$, we obtain

$$\left[\varphi_2 p\frac{d\varphi_1}{dx} - \varphi_1 p\frac{d\varphi_2}{dx}\right]_a^b = (\lambda_2 - \lambda_1)\int_a^b r\varphi_1\varphi_2\,dx.$$

For homogeneous boundary conditions, the left side of this equation is zero and the eigenfunctions form an orthogonal set. For a nonself-adjoint differential equation, this procedure would not work, but we may define adjoint functions and boundary conditions such that these adjoint functions are biorthogonal to the original eigenfunctions (see section 7-9 for an example).

APPENDIX B

USEFUL MATHEMATICAL FORMULAS

1 Trigonometric functions

$$\int \sinh au \sin bu \, du = \frac{-b \sinh au \cos bu + a \cosh au \sin bu}{a^2 + b^2}$$

$$\int \cosh au \sin bu \, du = \frac{-b \cosh au \cos bu + a \sinh au \sin bu}{a^2 + b^2}$$

$$\int \cosh au \cos bu \, du = \frac{b \cosh au \sin bu + a \sinh au \cos bu}{a^2 + b^2}$$

$$\int \sinh au \cos bu \, du = \frac{b \sinh au \sin bu + a \cosh au \cos bu}{a^2 + b^2}$$

$$\int u \sinh au \sin bu \, du = \frac{-bu \sinh au \cos bu + au \cosh au \sin bu}{a^2 + b^2} - \frac{-2\,ab \cosh au \cos bu + (a^2 - b^2) \sinh au \sin bu}{(a^2 + b^2)^2}$$

$$\int u \cosh au \sin bu\, du = \frac{-bu \cosh au \cos bu + au \sinh au \sin bu}{a^2 + b^2} - \frac{-2\,ab \sinh au \cos bu + (a^2 - b^2) \cosh au \sin bu}{(a^2 + b^2)^2}$$

$$\int u \cosh au \cos bu\, du = \frac{bu \cosh au \sin bu + au \sinh au \cos bu}{a^2 + b^2} - \frac{2\,ab \sinh au \sin bu + (a^2 - b^2) \cosh au \cos bu}{(a^2 + b^2)^2}$$

$$\int u \sinh au \cos bu\, du = \frac{bu \sinh au \sin bu + au \cosh au \cos bu}{a^2 + b^2} - \frac{2\,ab \cosh au \sin bu + (a^2 - b^2) \sinh au \cos bu}{(a^2 + b^2)^2}$$

$$\int_{-\pi}^{+\pi} \sin mx \sin nx\, dx = \begin{cases} 0, & m \neq n \\ \pi, & m = n \end{cases}$$

$$\int_{-\pi}^{+\pi} \cos mx \cos nx\, dx = \begin{cases} 0, & m \neq n \\ \pi, & m = n \neq 0 \\ 2\pi, & m = n = 0 \end{cases}$$

2 Fourier transform

$$\mathfrak{F}[f(x); \beta] = \frac{1}{\sqrt{2\pi}} \int_{-\infty}^{+\infty} f(x) e^{i\beta x}\, dx$$

$$f(x) = \frac{1}{\sqrt{2\pi}} \int_{-\infty}^{+\infty} \mathfrak{F}(\beta) e^{-i\beta x}\, d\beta$$

3 Fourier sine transform

$$F_s(\beta) = \sqrt{\frac{2}{\pi}} \int_0^{\infty} f(x) \sin \beta x\, dx$$

$$f(x) = \sqrt{\frac{2}{\pi}} \int_0^{\infty} F_s(\beta) \sin \beta x\, d\beta$$

4 Fourier cosine transform

$$F_c(\beta) = \sqrt{\frac{2}{\pi}} \int_0^{\infty} f(x) \cos \beta x \, dx$$

$$f(x) = \sqrt{\frac{2}{\pi}} \int_0^{\infty} F_c(\beta) \cos \beta x \, d\beta$$

5 Inverse transform

$$\int_{-\infty}^{+\infty} F(\beta)G(\beta)e^{-i\beta x} \, d\beta = \int_{-\infty}^{+\infty} f(u)g(x-u) \, du$$

$$\mathfrak{F}^{-1}[(1 - |\beta| y)e^{-|\beta| y}] = 2^{3/2}\pi^{-1/2}x^2 y(x^2 + y^2)^{-2}$$

$$\mathfrak{F}^{-1}[(1 + |\beta| y)e^{-|\beta| y}] = 2^{3/2}\pi^{-1/2}y^3(x^2 + y^2)^{-2}$$

$$\mathfrak{F}^{-1}[i\beta y e^{-|\beta| y}] = 2^{3/2}\pi^{-1/2}xy^2(x^2 + y^2)^{-2}$$

$$\mathfrak{F}^{-1}[i(\text{sign}\beta)(2 - |\beta| y)e^{-|\beta| y}] = 2^{3/2}\pi^{-1/2}x^3(x^2 + y^2)^{-2}$$

6 Complex variable formulas

Cauchy–Riemann equations:

$$\frac{\partial u}{\partial y} = -\frac{\partial v}{\partial x}, \quad \frac{\partial u}{\partial x} = \frac{\partial v}{\partial y}$$

If $f(z)$ is holomorphic on and inside C

$$\oint_C f(z) \, dz = 0, \qquad \text{Cauchy's integral theorem.}$$

For C including, but not passing through, a:

$$\oint_C (z-a)^n dz = 0, \qquad n \neq -1,$$

$$\oint_C \frac{dz}{z-a} = 2\pi i.$$

$F(z)$ is analytic inside C:

$$f(z) = \frac{1}{2\pi i} \oint_C \frac{f(\alpha)}{\alpha - z} d\alpha, \qquad z \text{ inside } C,$$

$$0 = \frac{1}{2\pi i} \oint_C \frac{f(\alpha)}{\alpha - z} d\alpha, \qquad z \text{ outside } C.$$

7 Bessel functions

$$C_n = J_n(x) + \beta Y_n(x)$$

Recurrence relations:

$$C_{n-1}(x) + C_{n+1}(x) = \frac{2n}{x} C_n(x)$$

$$C_{n-1}(x) - C_{n+1}(x) = 2C_n'(x)$$

$$C_n'(x) = C_{n-1} - \frac{n}{x} C_n(x)$$

$$C_n'(x) = -C_{n+1}(x) + \frac{n}{x} C_n(x)$$

Orthogonality condition:

$$\int_0^1 rJ_n(\alpha_k r)J_n(\alpha_j r)\,dr = 0, \qquad k \neq j,$$

$$= \tfrac{1}{2}[J_n'(\alpha_j)]^2, \qquad k = j$$

if α_j are the roots of $J_n(x) = 0$,

$$= \frac{1}{2\alpha_j}\left[\left(\frac{a}{b}\right)^2 + \alpha_j^2 - n^2\right][J_n(\alpha_j)]^2, \qquad k = j$$

if α_j are the roots of $[aJ_n(x) + bxJ_n'(x)] = 0$.

Modified Bessel functions:

$$I_n(x) = e^{-1/2 n\pi i} J_n(ix)$$

Recurrence relations:

$$I_{n-1}(x) - I_{n+1}(x) = \frac{2n}{x} I_n(x)$$

$$I_n'(x) = I_{n-1}(x) - \frac{n}{x} I_n(x)$$

$$I_{n-1}(x) + I_{n+1}(x) = 2I_n'(x)$$

$$I_n'(x) = I_{n+1}(x) + \frac{n}{x} I_n(x)$$

Integrals:

$$\int_0^a J_0(\alpha r)I_0(\beta r)r\,dr = \frac{1}{\alpha^2 + \beta^2}[\alpha a I_0(\beta a)J_1(\alpha a) + \beta a I_1(\beta a)J_0(\alpha a)]$$

$$\int_0^a J_0(\alpha r)\beta r I_1(\beta r)r\,dr = \frac{a^2}{\alpha^2 + \beta^2}[\alpha\beta I_1(\beta a)J_1(\alpha a) + \beta^2 I_0(\beta a)J_0(\alpha a)] - \frac{2\beta^2}{(\alpha^2 + \beta^2)^2}[\beta a I_1(\beta a)J_0(\alpha a) + \alpha a I_0(\beta a)J_1(\alpha a)]$$

8 Hankel transform

$$\bar{f}^{\nu}(\xi) = \mathcal{H}_{\nu}[f(x); \xi] = \int_0^{\infty} x f(x) J_{\nu}(\xi x)\, dx$$

$$f(x) = \int_0^{\infty} \bar{f}^{\nu}(\xi) \xi J_{\nu}(\xi x)\, d\xi$$

Inverse transforms:

$$\int_0^{\infty} x^n e^{-\lambda x} J_m(\alpha x)\, dx = \frac{(n-m)!}{(\lambda^2+\alpha^2)^{1/2(n+1)}} P_n^m\left[\frac{\lambda}{(\lambda^2+\alpha^2)^{1/2}}\right], \qquad m \leq n,$$

$$= \frac{(n+m)!}{(\lambda^2+\alpha^2)^{1/2(n+1)}} P_n^{-m}\left[\frac{\lambda}{(\lambda^2+\alpha^2)^{1/2}}\right], \qquad m > n.$$

$$\int_0^{\infty} J_0(\xi r) \cos \xi t\, d\xi = \frac{H(r-t)}{(r^2-t^2)^{1/2}}$$

$$\int_0^{\infty} J_0(\xi r) \sin \xi t\, d\xi = \frac{H(t-r)}{(t^2-r^2)^{1/2}}$$

9 Legendre functions

$$P_n(\zeta) = \frac{1}{2^n n!} \frac{d^n}{d\zeta^n} (\zeta^2 - 1)$$

Recurrence relations:

$$(n+1)P_{n+1}(\zeta) = (2n+1)\zeta P_n(\zeta) - nP_{n-1}(\zeta)$$

$$(\zeta^2 - 1)P_n'(\zeta) = n\zeta P_n(\zeta) - nP_{n-1}(\zeta)$$

$$\zeta P_n'(\zeta) = nP_n(\zeta) + P_{n-1}'(\zeta)$$

Orthogonality:

$$\int_{-1}^{+1} P_m(\zeta) P_n(\zeta)\, d\zeta = \frac{\delta_{mn}}{n + \frac{1}{2}}$$

Associated Legendre functions:

$$P_{\nu}^m(\zeta) = (1-\zeta^2)^{1/2m} \frac{d^m}{d\zeta^m} P_{\nu}(\zeta)$$

Recurrence relations:

$$(\nu - \mu + 1)P_{\nu+1}^{\mu}(\zeta) = (2\nu+1)\zeta P_{\nu}^{\mu}(\zeta) - (\nu+\mu)P_{\nu-1}^{\mu}(\zeta)$$

$$(\zeta^2 - 1)\frac{d}{d\zeta} P_{\nu}^{\mu}(\zeta) = \nu\zeta P_{\nu}^{\mu}(\xi) - (\nu+\mu)P_{\nu-1}^{\mu}(\zeta)$$

Orthogonality:

$$\int_{-1}^{+1} P_i^m(\xi) P_j^m(\zeta)\, d\zeta = \frac{\delta_{ij}(i+m)!}{[i+(1/2)](i-m)!}$$

SUBJECT INDEX

AUTHOR INDEX